TECHNICAL COMMUNICATION

TECHNICAL COMMUNICATION

SECOND CANADIAN EDITION

JOHN M. LANNON
University of Massachusetts

DON KLEPP
Okanagan University College

Longman

Toronto

National Library of Canada Cataloguing in Publication Data

Lannon, John M.
    Technical communication / John M. Lannon, Don Klepp. — 2nd Canadian ed.

Includes bibliographical references and index.
ISBN 0-201-78971-X

    1. Technical writing. 2. Communication of technical information.
I. Klepp, Don, 1944–   . II. Title.

T11.L36 2003   808'.0666   C2002-902592-3

ISBN 0-201-78971-X

Vice-President, Editorial Director: Michael J. Young
Marketing Manager: Toivo Pajo
Developmental Editor: Dawn du Quesnay
Senior Production Editor: Joe Zingrone
Copy Editor: Judy Phillips
Proofreader: Claudia Forgas
Senior Production Coordinator: Peggy Brown
Page Layout: Christine Velakis
Art Director: Mary Opper
Cover Design: Julia Hall
Cover Art: Pat Durr, "Not Quite What It Seems" (1995)

2 3 4 5   06 05 04 03

Printed and bound in the United States of America.

Part IV   Descriptive Writing   311

Detailed Contents

Brief Contents

http://www.ouc.bc.ca/libr
http://www.mtroyal.ab.ca/programs/academserv/lib/Indexlibweb.htm

CHAPTER 7

These sites provide wide-ranging advice on how to critically evaluate Web-based material.

http://biome.ac.uk/guidelines/eval/howto.html
http://library.usm.maine.edu/guides/webeval.html
http://servercc.oakton.edu/~wittman/find/eval.htm
http://discoveryschool.com/schrockguide/ppoint.html

CHAPTER 8

A tour of the following sites reveals that documentation "experts" have not yet agreed about formats for citing electronic information sources. I recommend the Guffey site, the first listed.

http://www.westwords.com./guffey/documentation.html
http://www.english.uiuc.edu/cws/wworkshop/bibliography.htm
http://www.columbia.edu/cu/cup/cgos/idx_basic.html
http://www.bedfordstmartins.com/online/citeappxa.html

CHAPTER 9

The next two sites discuss summary writing.

http://www.chss.iup.edu/wc/resources/summary.html
http://web.uvic.ca/wguide/Pages/MasterToc.html#Summaries

CHAPTER 10

The University of Toronto's Engineering Communication Centre has useful advice on using outlines and other aspects of effective writing. (Click on the appropriate sub-topic.)

http://www.ecf.utoronto.ca/~writing/handbook.html

CHAPTER 11

Many websites provide advice about writing clear prose and exercises to improve one's editing techniques. Here is a representative sampling:

http://owl.english.purdue.edu/handouts/index.html
http://quintcareers.com/writing
http://andromeda.rutgers.edu/~jlynch/Writing/
http://www.io.com/~hcexres/tcm1603/acchtml/lists.html
http://europa.eu.int/comm/translation/en/ftfog/index.htm

Note: *probably the most useful of the above sites is the last one, a site sponsored by the European Commission's Translation Service.*

CHAPTER 12

The following websites are less commercial than most which deal with graphics.

http://www.arcm.com/illustra.html
http://members.aol.com/macbloom/
http://www.nidus-corp.com/

Web Connect

The following Web connections will take you to sites that provide useful advice and examples. Please note that, at any time, a site might change its URL or the site might disappear.

CHAPTER 1

The following links discuss the nature of technical writing, technical writing careers and job listings for technical writers, collaborative writing techniques, and hints for effective technical documents.

http://staff.bmcc.cc.or.us/~gberlie/wr227/intro.htm
http://www.chss.iup.edu/wc/resources/
http://www.stcloudstate.edu/~scogdill/collaboration/biblio.html
http://www.uwec.edu/jerzdg/orr/handouts/TW/proj/index.html
This site provides "Internet resources for technical communicators":
http://www.soltys.ca/techcomm.html

CHAPTER 2

The following site usefully discusses audience analysis.

http://www.chss.iup.edu/wc/resources/audience.html

CHAPTER 3

Try the Clarkson University Writing Center's tips for writing efficiently and effectively; look for "General Writing Skills," half way down the Center's home page.

http://www.clarkson.edu/~wcenter/index.html

CHAPTER 4

These two sites focus on essay writing, but their advice can be applied to business writing, also.

http://www.tc.cc.va.us/writcent/handouts/writing/argument.htm
http://www.wuacc.edu/services/zzcwwctr/you-attitude.txt

CHAPTER 5

This website covers a wide range of ethics issues for technical people trying to cope with difficult situations. In particular, see the "Famous Cases" and "Recent Engineering Cases" secondary sites.

http://onlineethics.org

CHAPTER 6

The first of the following sites offers a comprehensive guide to Web search engines. Next, the Megaresearch site defines focus groups. The Okanagan University College and Mount Royal College sites advise how to use a variety of traditional and electronic research tools.

http://www.monash.com/spidap.html
http://www.megresearch.com/focus.html

http://www.nas.nasa.gov/About/Media/graphics.html
http://publishing.grc.nasa.gov/graphics/graphics.htm

CHAPTER 14

The first of the following site discusses operational definitions. The last two sites illustrate operational and categorical definitions of terms used in physics and in medicine.

http://www.kcmetro.cc.mo.us/longview/ctac/psychexer1.htm
http://www.lhup.edu/~dsimanek/glossary.htm
http://www.io.com/~hcexres/tcm1603/acchtml/def_ex.html

CHAPTER 15

The first of the following sites presents an excellent example of extended mechanism description.
Then the next two sites discuss how to write effective descriptions of mechanisms.

http://www.cisco.com/warp/public/cc/so/neso/voso/voip/ptelb_rg.htm
http://english.ttu.edu/grad/AHanson/2309/description.html
http://www.uwec.edu/Academic/Curric/jerzdg/English305/formats/mechanism.htm

CHAPTER 16

The Cisco site contains process description as well as mechanism description. The next two sites show process descriptions from medicine (heart transplant evaluations) and computer software operation (true type fonts). The fourth site illustrates a set of instructions. The fifth site deals with a variety of descriptive writing, including processes and instructions.

http://www.cisco.com/warp/public/cc/so/neso/voso/voip/ptelb_rg.htm
http://www.cpmc.org/advanced/heart/patients/topics/evaluation.html
http://www.microsoft.com/opentype/otspec/ttch01.htm
http://www.fishingnorthwest.com/flypterns.htm
http://www.dsu.edu/departments/liberal/english/techwrit/

CHAPTER 17

The first of the following sites explains why it's important to write readable, usable manuals. Next, Peter Ring's website on user-friendly manuals is mostly commercial in nature, but the site does contain a very useful list of free information sources. Then, Jakob Nielsen's site discusses "the usability life cycle" and the Society for Technical Communicators presents its views on usability.

http://www.westworld.com/~ayale/TechWrtg.html
http://www.prc.dk/user-friendly-manuals/
http://www-106.ibm.com/developerworks/library/it-nielsen3/
http://www.stcsig.org/usability/

CHAPTER 18

First, you'll find U.S. and Canadian government sites, which provide RFPs for suppliers and contractors. Then, sites sponsored by Dakota State and the University of Wisconsin discuss proposal writing, among other things. Also, check out the U of Toronto's advice.

http://www.uspto.gov/web/offices/ac/comp/proc/acquisitions/itpa/itpa-rfp.htm
http://contractscanada.gc.ca

http://www.dsu.edu/departments/liberal/english/techwrit/
http://tc.engr.wisc.edu/zwickel/397/
http://www.ecf.utoronto.ca/~writing/handbook.html

Chapter 19

The first site, a Netscape listing of technical reports, is very interesting. Then, the next three sites will take you to examples of technical reports, most of which have an analytical purpose. The final site, from Kent State, gives report writing advice.

http://search.netscape.com/nscp_results.adp?source=NSCPIndex&query=technical+reports
http://www.cics.uvic.ca/climate/change/cimpact.htm
http://www.ec.gc.ca
http://www.epc-pcc.gc.ca/research/scie_tech/
http://jaring.nmhu.edu/Chaut98/rwojtecki/report.html

Chapter 20

None

Chapter 21

The listed sites discuss, and in some cases illustrate, a variety of job-related reports.

http://www.io.com/~hcexres/tcm1603/acchtml/progrepx2a_non.html
http://www.ecf.toronto.edu/~writing/handbook-shrtrept.html
http://www.rpi.edu/web/writingcenter/handouts.html
http://www.io.com/~hcexres/tcm1603/acchtml/progrep.html
http://www.io.com/~hcexres/tcm1603/acchtml/feas.html
http://www.io.com/~hcexres/tcm1603/acchtml/props.html

Chapter 22

Letters And Memos

http://www.ecf.utoronto.ca/~writing/handbook-memo.html
http://www.wuacc.edu:80/services/zzcwwctr/you-attitude.txt

E-mail usage

http://www.pimall.com/nais/n.pgp.faq.html
http://www.webfoot.com
http://www.uwec.edu/jerzdg/orr/handouts/TW/e-mail.htm

Chapter 23

Résumés: Hundreds of Internet sites dispense advice on résumé writing; Susan Ireland's site is particularly helpful regarding electronic résumés.

http://susanireland.com
http://www.contractengineering.com/data1/ResCivil.htm
http://www.headhunter.net/jobseeker/index.htm?siteid=cmhome
http://www.provenresumes.com
http://www.stantongp.com/
http://www.uwec.edu/jerzdg/orr/handouts/TW/resume.htm

Job market research
http://www.monster.com and http://www.monster.ca
http://www.kenevacorp.mb.ca/
http://www.headhunter.net
http://www.jobtrak.com
http://www.workinfonet.ca
http://www.careerexchange.com
http://www.careeredge.org/
http://www.rileyguide.com/ (the Riley Report)
http://www.cacee.com
http://www.canadajobsearch.ca

Cover letters
http://jobsmart.org/tools/resume/cletters.htm#Good
http://www.uwec.edu/jerzdg/writing/resume/covers.html

CHAPTER 24

Preparing Speeches
http://www.speeches.com/index.shtml
http://www.eeicom.com/eye/shyness.html
http://www.uwec.edu/jerzdg/orr/handouts/TW/oral.htm

Technical talks
http://tc.engr.wisc.edu/zwickel/397/HW05.html

Preparing and delivering conference papers
http://www.dfrc.nasa.gov/organizations/techinfo/manual/confDetails.html

Planning meetings and speaking effectively at meetings
http://www.effectivemeetings.com/
http://www.triangle.org/howto/meetings/elements.html
http://www.meetingwizard.org/meetings/index.cfm
http://www.powerfulpresentations.net/article1017.html

CHAPTER 25
Designing Web pages is fully discussed in the many linked sites listed at the following sites.

http://trace.wisc.edu/world/web/
http://www.world-ready.com/readlist.htm
http://www.uwec.edu/jerzdg/orr/handouts/TW/web/index.html
http://useit.com/alertbox/9710a.html
http://www-106.ibm.com/developerworks/library/it-nielsen3/
http://www.uwec.edu/jerzdg/English305/web/index.htm

Preface

This Second Canadian Edition of *Technical Communication* continues to provide a comprehensive, flexible introduction to technical and professional communication. Although this book focuses on writing, it also discusses interpersonal and group collaborative processes, oral communications, intercultural communication, and information gathering methods.

Originally commissioned as a Canadian version of John Lannon's U.S. edition of *Technical Writing*, this edition has increasingly assumed its own character. Continuing to meet the requirements of Canadian students and workplace writers, this new edition presents current examples and a Canadian perspective throughout.

ORGANIZATION

The text has five main sections:

Part I: Writing for Readers in the Workplace (Chapters 1–5) considers the bases of successful technical and workplace communication: the nature of technical writing and other forms of workplace communication; factors in successful collaboration; rhetorical analysis (of the sender's purpose and the receiver's needs and priorities); and a process for writing effective documents quickly. Factors in persuasive communication and the ethics of persuading others round out Part I.

Part II: Information Gathering, Analysis, and Manipulation (Chapters 6–9) illustrates various stages of successful research. Students learn to formulate useful research questions; to explore primary and secondary sources (including electronic sources); to record, evaluate, interpret, and document their findings; and to summarize for economy, accuracy, and emphasis.

Part III: Structural, Style, and Format Elements (Chapters 10–13) demonstrates strategies for organizing and expressing messages that readers can follow and understand. Students learn to control their material and to develop a style that connects with readers. Also, students learn to enhance a document's readability and appeal by using page design, graphics, and visuals.

Part IV: Descriptive Writing (Chapters 14–17) applies principles of audience and purpose analysis, style, and format to descriptive writing: definitions, descriptions, instructions, and manuals. Students will learn to clearly distinguish among varieties of descriptive writing in order to choose effective structures and phrasing; they will also learn how to test the usability of their documents.

Part V: Applications (Chapters 18–25) applies earlier concepts and strategies to the preparation of proposals, reports, and workplace correspondence, all of which are illustrated by examples taken from workplace and student communication. The concluding chapters on job-search communications, oral communications, and Internet documents demonstrate strategies for "real-world" success.

THE FOUNDATIONS OF TECHNICAL COMMUNICATION

This book has been conceived and designed with the following principles in mind.

- Workplace communication is complex. Each report, proposal, correspondence, meeting, or talk requires the successful communicator to be aware of specific interpersonal, ethical, legal, and cultural demands. Writers with rhetorical awareness (of audience, purpose, and situation) make decisions crucial for effective writing. Without rhetorical awareness, "real-world" writers seldom succeed.

◆ Today's workplace professionals are more than communicators. Increasingly, they consume information, so they need to be skilled in the methods of enquiry, retrieval, evaluation, and interpretation that make up the research process.

◆ Successful workplace writing requires an organized, sensible, efficient approach. Only rarely does the workplace writer use the discovery process frequently used by the academic essay writer. People at work can't afford the time to write, and rewrite, and rewrite again!

◆ A technical writing classroom typically contains an assortment of students with varied backgrounds. This textbook, then, offers thorough explanations, broadly intelligible examples and models, and practical methods for dealing with workplace demands. Moreover, the book means to be flexible enough to allow for various course plans.

UNIQUE TO BOTH CANADIAN EDITIONS

◆ Concepts, structures, and techniques are illustrated by current Canadian examples.

◆ Chapter 2 fully explains the factors to consider in analyzing the audiences for one's written and spoken messages.

◆ Chapter 3 shows writers how to produce documents efficiently.

◆ New information about Internet research techniques is accompanied by advice on how to evaluate Internet sources.

◆ Chapter 11 includes objective indexes designed to help a writer evaluate and improve the readability of his or her writing.

◆ Chapter 13's advice about headings format is designed to make the sample headings adaptable to a wide range of applications, and to reflect current usage.

◆ Chapter 16 clearly distinguishes among process analyses, instructions, and procedures.

◆ Chapter 17 discusses the content and structure of a variety of manuals and describes a process for writing manuals.

◆ Chapter 18 shows how to plan and produce proposals.

◆ Chapter 19 illustrates a practical process for producing formal analytical reports.

◆ Chapter 21 presents a practical method of writing the short reports that represent the bulk of writing done by technical professionals. The chapter features full coverage of progress reports and other commonly written reports. All of these reports use a variation of the adaptable "Action Structure."

◆ The chapter on workplace correspondence, Chapter 22, discusses letters, memos, e-mail, and faxes from a current Canadian perspective. Canadian format standards, sensible advice, and an "Action Structure" for workplace messages highlight this chapter.

◆ In Chapter 23, the step-by-step process for gaining a position will help the successful candidate find job satisfaction, not just a job with a salary. The chapter includes self-help material and current examples of job-search materials.

◆ Chapter 24 follows a positive path in preparing for oral presentations. The chapter adds advice about impromptu speaking and a self-evaluation exercise.

◆ Chapter 25 advises how to write, design, and evaluate Web-based documents.

NEW TO THIS EDITION

◆ First and foremost, the brilliant new full-colour design is both fresh and functional, making reading *Technical Communication,* Second Canadian Edition, a more user-friendly experience.

◆ Each of the five Part Openers features an interview with a Canadian technical professional who researches and writes as part (or all) of his or her daily work. Each of the five employs a practical approach suited to the users of this book.

◆ Dozens of new, updated Canadian examples and 20 new document samples appear throughout the book. Also, all Web Connect references have been updated.

- The latest requirements for MLA, APA, and CBE documentation, especially of electronic sources, appear in this edition.
- Increased coverage of successful collaboration is found in Chapter 1.
- Chapter 10 contains more comprehensive and more logically organized coverage of paragraphing techniques.
- Chapter 17 now contains detailed, practical advice on testing the usability of manuals and other documents.
- Chapter 21 incorporates advice on compliance reporting.
- Chapter 22 features the results of my Canada-wide survey of current e-mail practices.
- Chapter 24 contains new material on using presentation software such as PowerPoint. Also, the chapter now discusses meetings—how to plan and prepare for them, and how to speak effectively at meetings.

SUPPLEMENTS TO THE TEXT:

An **Instructor's Manual** containing chapter overviews, additional exercises, supplementary lecture material, transparency masters, and quizzes accompanies this edition of *Technical Communication*.

A website for both students and instructors may be found at **www.pearsoned.ca/lannon**.

ACKNOWLEDGMENTS

John Lannon's seventh edition of *Technical Writing* formed the original basis of the Canadian editions.

Garth Homer, Sue-Tina Kong, Marilyn Riley-Nault, Brian Schnitzler, and Judith White-head have significantly contributed to this edition by sharing their hard-won wisdom.

Comments by the following perceptive reviewers have significantly contributed to the Canadian editions: David Wiens, Kwantlen College; Judy Cox, Algonquin College; Marnie Squire, Algonquin College; Mary Balser, New Brunswick Community College; Jennifer MacLennan, University of Saskatchewan; Sandra Dorely, George Brown College; Hilde Clovechok, Southern Alberta Institute of Technology; Trish Campbell, Red Deer College; Ann Gasior, Okanagan University College; Anne Mackenzie-Rivers, George Brown College; Rod Macpherson, Lance Moen, and Patricia Puderak, Saskatchewan Institute of Applied Science and Technology; and Mary Silas, Concordia University.

Developmental editor Dawn du Quesnay has wisely guided this edition's revision process. I have appreciated her perceptive comments and good humour. I would also like to thank acquisitions editor Sophia Fortier, senior production editor Joe Zingrone, senior production coordinator Peggy Brown, formatter Christine Velakis, and copy editor Judy Phillips for their hard work on this project.

Student contributors Dean Ashby, Sandy Hagel, Chris Hendsbee, Matt Nakazawa, Grant Perkins, Thanh Pham, Brandin Slonski, Todd Trann, and Curtis Willis have generously allowed their work to be included in this book. Students in the Civil, Electronic and Mechanical Engineering Technology programs at Okanagan University College have made many useful comments about the first edition.

Ross Tyner and Garth Homer, Okanagan University College librarians, have been very helpful in identifying the new array of research tools available to student and workplace researchers.

My wife, Betty Chan Klepp, has provided an excellent sounding board for ideas and approaches.

Finally, I acknowledge the enduring influence of University of Saskatchewan professors Dr. Ron Marken and Dr. Rob Scott, whose love of teaching and respect for the English language are reflected in this book's philosophy and practice.

Don Klepp

PART
I

Writing for Readers in the Workplace

Marilyn Riley Nault, Montreal

In Marilyn Riley Nault's rich, varied career as educator, communications professional, editor, and writer, two themes emerge: her need to grow and to learn, and her desire to share what she's learned. That sharing gets expressed in her writing.

Ms. Nault is conscious of the writing process she uses: "I usually start with a written goal, based on my understanding of the reader's needs. I never assume that readers know what I'm 'talking about,' so I research what the reader already knows." Ms. Nault uses "an evolving detailed outline, a draft Table of Contents, and a draft Introduction" to start writing longer documents. She writes first drafts in longhand and edits as she keys in those drafts.

Not all her writing is solitary—she's collaborated with others on complex documents. She says that "collaborative writing means having to accommodate different perspectives and needs ... That's why the group must agree on common goals and an overall framework. Then the members must follow style guidelines, to produce a consistent document."

Most writing has a persuasive element, she says. "The key to persuasive writing is to be sincere and honest ... and to *connect* on some level with your reader ... you must provide something the reader wants."

CHAPTER 1

Introduction to Technical Writing

As a technical writer, you communicate and interpret specialized information for your readers' use. Readers may need your information to perform a task, answer a question, solve a problem, or make a decision. Whether you write a memo, letter, report, or manual, the document must advance the goals of your readers and of the company or organization you represent.

TECHNICAL WRITING SERVES PRACTICAL NEEDS

Unlike poetry and fiction, which appeal mainly to our *imagination,* technical documents appeal to our *understanding.* Technical writing therefore rarely seeks to entertain, create suspense, or invite differing interpretations. Those of you who have written any type of lab or research report already know that technical writing has little room for ambiguity.

To serve practical needs in the workplace, technical documents must be reader oriented and efficient.

TECHNICAL DOCUMENTS MEET READER NEEDS

Instead of focusing on the writer's desire for self-expression, a technical document addresses the reader's desire for information. This doesn't mean your writing should sound like something produced by a robot, without any personality (or *voice*) at all. Your document may in fact reveal a lot about you (your competence, knowledge, integrity, and so on), but it rarely focuses on you personally. Readers are interested in *what you have done,* in *what you recommend,* or in *how you speak for your company;* they have only a professional interest in *who you are* (your feelings, hopes, dreams, visions). A personal essay, then, would not be technical writing. Consider this essay fragment:

Focuses on the feelings

> Computers are not a particularly forgiving breed. The wrong key struck or the wrong command entered is almost sure to avenge itself on the inattentive user by banishing the document to some electronic trash can.

This personal view communicates a good deal about the writer's resentment and anxiety, but very little about computers themselves.

The following example can be called technical writing because it focuses (see italics) on the subject, on what the writer has done, and on what the reader should do:

Focuses on the subject, actions taken, and actions required

> On MK 950 terminals, *the BREAK key* is adjacent to keys used for text editing and special functions. Too often, users inadvertently strike the BREAK key, causing the program to quit prematurely. To prevent the problem, *we have modified all database management terminals:* to quit a program, *you must now strike BREAK twice successively.*

This next example also can be called technical writing because it focuses on what the writer recommends:

Focuses on the recommendation

> I recommend that our Linux server be upgraded. This expansion will (1) increase the number of simultaneous users from 60 to 80, (2) increase the system's responsiveness, and (3) provide sorely needed disk storage for word processing and company databases.

As the above example illustrates, documents should not make the writer "disappear," but they should focus on that which is most important to the readers.

TECHNICAL DOCUMENTS STRIVE FOR EFFICIENCY

Educators read to *test* our knowledge; colleagues, customers, and supervisors read to *use* our knowledge. Workplace readers hate waste and demand efficiency; instead of reading a document from beginning to end, they are more likely to use it for reference and want only as much as they need: "When it comes to memos, letters, proposals, and reports, there's no extra credit for extra words. And no praise for elegant prose. Bosses want employees to get to the point—quickly, clearly, and concisely" (Spruell 32). Efficient documents save time, energy, and money in the workplace.

No reader should have to spend 10 minutes deciphering a message worth only 5 minutes. Consider, for example, this wordy message:

An inefficient message

> At this point in time, we are presently awaiting an on-site inspection by vendor representatives relative to electrical utilization adaptations necessary for the new computer installation. Meanwhile, all staff are asked to respect the off-limits designation of said location, as requested, due to liability insurance provisions requiring the on-line status of the computer.

Inefficient documents drain a reader's energy; they are too easily misinterpreted; they waste time and money. Notice how hard you had to work with the previous message to extract information that could be expressed this efficiently:

A more efficient message

> Hardware consultants soon will inspect our new computer room to recommend appropriate wiring. Because our insurance covers only an *operational* computer, this room must remain off limits until the computer is fully installed.

When readers sense they are working too hard, they tune out the message—or they stop reading altogether.

Inefficient documents have varied origins. Even when the information is accurate, errors like the following make readers work too hard:

Causes of inefficient documents

- more (or less) information than readers need
- irrelevant or uninterpreted information
- confusing organization
- jargon or vague technical expressions readers cannot understand
- more words than readers need
- uninviting appearance or confusing layout
- no visual aids when readers need or expect them

An efficient document sorts, organizes, and interprets its information to suit the audience's needs, abilities, and interests.

Instead of merely happening, an efficient document is carefully designed to include these elements:

Elements of efficient documents

- *content* that makes the document worth reading
- *organization* that guides readers and emphasizes important material

- *style* that is economical and easy to read
- *visuals* (graphs, diagrams, pictures) that clarify concepts and relationships, and that substitute for words whenever possible
- *format* (layout, typeface) that is accessible and appealing
- *supplements* (abstracts, appendices) that enable readers with different needs to read only those sections required for their work

Reader orientation and efficiency are more than abstract rules: Writers are accountable for their documents. In questions of liability, faulty writing is no different from any other faulty product. If your inaccurate, or unclear, or incomplete information leads to injury, or damage, or loss, *you* can be held legally responsible.

WRITING IS PART OF MOST CAREERS

Although you might not anticipate a career as a "writer," your writing skills will be tested routinely in situations like these:

Ways in which your career may test your writing skills

- proposing various projects to management or to clients
- writing progress reports to supervisors, managers, and executives
- contributing articles to employee newsletters
- describing a product to employees or customers
- writing procedures and instructions for employees or customers
- justifying to management a request for funding or personnel
- editing and reviewing documents written by colleagues
- designing material that will be read on a computer screen or transformed into sound and pictures

So, whatever your career plans, you can expect to write as part of your job.

Your value to any organization will depend on how clearly and persuasively you communicate. Many working professionals spend at least 40 percent of their time writing or dealing with someone else's writing. The higher their position, the more they write (Barnum and Fisher 9–11). Here is a corporate executive's description of some audiences you can expect:

Writing is an indicator of job performance

> *The technical graduate entering industry today will, in all probability, spend a portion of his or her career explaining technology to lawyers—some friendly and some not— to consumers, to legislators or judges, to bureaucrats, to environmentalists and to representatives of the press.* (Florman 23)

These audiences, among countless others, expect to read efficient documents.

Here is what two top managers for an automaker say about the effect a document can have on the organization *and* on the writer:

> *A written report is often the only record which is made of results that have come out of years of thought and effort. It is used to judge the value of the person's work and serves as the foundation for all future action on the project. If it is written clearly and precisely, it is accepted as the result of sound reasoning and careful observation. If it is poorly written, the results presented in it are placed in a bad light and are often dismissed as the work of a careless or incompetent worker.* (Richards and Richards 6)

Good writing gives you and your ideas *visibility* and *authority* within your organization. Bad writing, on the other hand, is not only useless to readers and politically damaging to the writer, but also expensive: Written communication[1] in North American business and industry costs billions of dollars yearly. More than 60 percent of the writing is inefficient; it is unclear, misleading, irrelevant, deceptive, or otherwise wasteful of time and money (Max 5–6).

In your career, you'll have to write well. Employers first judge your writing by your application letter and résumé. In a large organization, your future may be decided by executives you've never met. One concrete measure of your job performance will be your letters, memos, and reports. As you advance, communication skill becomes more important than technical background. The higher your goals, the better you need to communicate.

THE INFORMATION AGE REQUIRES EXCELLENT WRITING SKILLS

Desktop publishing (DTP) systems have eliminated clerical support positions in countless organizations. Gone are the days in which writers could count on secretarial help in "fixing up" a document. In today's electronic office, each writer is responsible for a document's creation, proofreading, editing, page layout, and distribution.

More importantly, information has become our prized commodity:

Information is the ultimate product

> *The new source of wealth is not material; it is information, knowledge applied to work to create value. The pursuit of wealth is now largely the pursuit of information, and the application of information to the means of production.* (Wriston 8)

In education, industry, business, and government, people create, gather, analyze, and distribute information electronically around the world. But whether the information itself finally appears on a printed page or a computer screen, it usually needs to be *written*. A computer can transmit information, but it cannot give *meaning* to the information—only the writer and reader can do so.

Computer networks have increased the speed and volume of communication, but excessive information actually can *impede* or *prevent* communication. Despite our advances in communication technology, information still needs to be processed by the human brain. Unfortunately, it is easy for people to process information erroneously (to wrongly interpret the *meaning*).

Accidents that make headlines often result from human error in processing information—situations in which the complexity of information overwhelms the person receiving it (Wickens 2). The 1984 chemical explosion in Bhopal, India; the 1986 meltdown at the Chernobyl nuclear plant; the 1989 runway collision at Los Angeles airport—these disasters occurred because vital information had been misunderstood. Similar but less publicized disasters occur routinely in the workplace. Today, more than ever, writers need to sort, organize, and interpret their material so readers can understand it.

---

1. "Written communication" includes the full range of activities involved in preparing, producing, processing, storing, and retrieving documents for re-use.

Today's readers often lack technical expertise, but they do *use* the technology. Some non-technical workers who rely on technology include the manager using a teleconferencing network, the data-entry clerk using a computer terminal, or the bank teller using a cheque-verification system. Writers tell laypeople how to use software, or how to operate hardware, medical equipment, telephone systems, precision instruments, and all types of automated products, from computer games to microwave ovens. With so much information required, and so much available, no writer can afford to "let the facts speak for themselves."

In Brief WRITING REACHES A GLOBAL AUDIENCE

Electronically linked, our global community shares social, political, and financial interests that demand cooperation as well as competition. Multi-national corporations often use parts that are manufactured in one country and shipped to another for assembly into a product that will be marketed elsewhere. For example, cars may be assembled in North America for a German automaker, or farm equipment may be manufactured in East Asia for a Canadian company. Medical, environmental, and other research crosses national boundaries, and professionals in all fields transact with colleagues from other cultures.

What kinds of documents might address global audiences? Here is a sample (Weymouth 143):

◆ scientific reports and articles on AIDS and other diseases
◆ studies of global pollution and industrial emissions
◆ specifications for hydroelectric dams and other engineering projects
◆ operating instructions for appliances and electronic equipment
◆ catalogues, promotional literature, and repair manuals
◆ contracts and business agreements

To communicate effectively across cultural and national boundaries, any document must respect not only language differences but also *cultural* differences. One writer offers this helpful definition of *culture:*

Our accumulated knowledge and experiences, beliefs and values, attitudes and roles—in other words, our cultures—shape us as individuals and differentiate us as a people. Our cultures, inbred through family life, religious training, and educational and work experiences . . . manifest themselves . . . in our thoughts and feelings, our actions and reactions, and our views of the world.

Most important for communicators, our cultures manifest themselves in our information needs and our styles of communication. In other words, our cultures define our expectations as to how information should be organized, what should be included in its content, and how it should be expressed. (Hein 125)

Cultures differ over which behaviours seem appropriate or inappropriate for social interaction, business relationships, contract negotiation, and communication practices. A communication style considered perfectly acceptable in one culture may be offensive elsewhere.

Effective communicators recognize these differences but withhold judgment or evaluation, focusing instead on *similarities*. For example, needs assessment and needs satisfaction know no international boundaries; technical solutions are technical solutions without regard to nationality, creed, or language; courtesy and goodwill are universal values. In the diverse global context, the writer's challenge is to establish trust and enhance human relationships.

TECHNICAL WRITERS FACE INTERRELATED CHALLENGES

No matter how sophisticated our communication technology, computers cannot *think* for us. More specifically, computers cannot solve the challenges faced by all people who write in the technical professions. These challenges include:

- *the information challenge:* different readers in different situations have different information needs
- *the persuasion challenge:* people often disagree about what the information means and about what should be done
- *the ethics challenge:* the interests of your company may conflict with the interests of your readers
- *the global context challenge:* diverse people work together on information for a diverse audience

An earlier section of this chapter explains that information has to have meaning for its audience. But people differ in their interpretations of facts, and so they may need persuading that one viewpoint is preferable to another. Persuasion, however, is a powerful and sometimes unethical strategy. Even the most useful and efficient document could deceive or harm. Therefore, solving the persuasion challenge doesn't mean manipulating readers by using "whatever works," but rather building a case from honest and reasonable interpretation of the facts. Figure 1.1 offers one way of visualizing how these challenges relate.

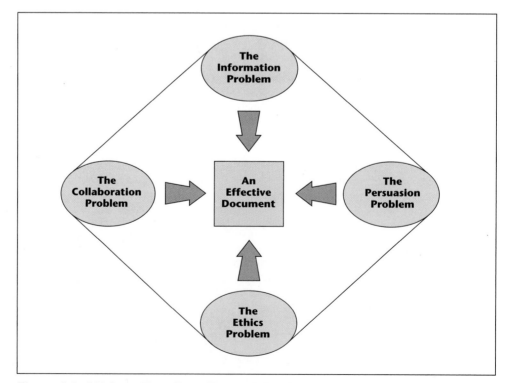

Figure 1.1   Writers Face Four Related Challenges

The scenarios that follow illustrate how a typical professional confronts the challenges of communicating in the workplace.

The Information Analysis Challenge

Sarah Habib was hired two months ago as a chemical engineer for Millisun, a leading maker of cameras, multi-purpose film, and photographic equipment. Sarah's first major assignment is to evaluate the plant's incoming and outgoing water. (Waterborne contaminants can taint film during production, and the production process itself can pollute outgoing water.) Management wants an answer to this question: How often should we change water filters? The filters are very expensive and difficult to change, halting production for up to a day at a time. The company wants as much usage as possible from these filters, without incurring government fines or tainting its film production.

"Can I provide accurate and useful information?"

Sarah will study endless printouts of chemical analysis, review current research, do some testing of her own, and consult with her colleagues. When she finally decides on what all the data mean, Sarah will prepare a recommendation report for her manager.

Later she will collaborate with the company training manager and the maintenance supervisor to prepare a manual, instructing employees how to check and change the filters. Trying to cut administrative assistant and printing costs, the company has asked Sarah to design this manual using its desktop publishing system. ▲

Sarah's report, above all, needs to be accurate; otherwise, the company gets fined or lowers production. Once she has processed all of the information, she has to solve the challenge of giving readers what they need. She has to answer questions such as: *How much explaining should I do? How will I organize the information? Do I need visuals?* And so on.

In other situations, Sarah will face a persuasion challenge as well: when decisions must be made or actions taken on the basis of incomplete or inconclusive facts or conflicting interpretations (Hauser 72). In these instances, Sarah will seek reader acceptance for *her* view. Her writing will have to be persuasive as well as informative.

The Persuasion Challenge

Millisun and other electronics producers are located on the shores of a small harbour, the port for a major fishing fleet. For 20 years, these companies discharged directly into the harbour effluents containing metal compounds, PCBs, and other toxins. Sarah is on a multi-company team, assigned to work with the Canadian Environmental Assessment Agency to clean up the harbour. Much of the team's collaboration occurs via e-mail.

"Can I influence readers to see things my way?"

Enraged local citizens and environmental groups are demanding immediate action, and the companies themselves are anxious to end this public-relations nightmare. But the team's analysis reveals that any type of clean-up would stir up harbour sediment, possibly dispersing the solution into surrounding waters and the atmosphere. (Many of the contaminants are airborne.) Premature action actually might *increase* danger, but team members disagree on the degree of risk and on how to proceed.

Sarah's communication here takes on a persuasive dimension: She and her team members first have to resolve their own conflicts and produce an environmental-impact report that reflects the team's consensus. If the report recommends further study, Sarah will have to justify the delays to her manager and the public relations office. She will have to make readers understand the dangers as well as she understands them. ▲

In the above situation, the facts are neither complete nor conclusive, and views differ about what these facts mean. Sarah will have to balance the various political pressures and make a case for *her* interpretation. Moreover, as company *spokesperson*, Sarah will be expected to take a position that protects her company's interests. Some elements of Sarah's persuasion challenge are: *Are other interpretations possible? Is there a better way? Can I expect political fallout?*

Whenever she writes, Sarah will have to reckon with the ethical implications of her writing, with the question of "doing the right thing." Some situations will present hard choices. For instance, Sarah might feel pressured to overlook, sugarcoat, or suppress facts that would be costly or embarrassing to her company. Sometimes the best technical solution to a challenge might be a poor solution in human terms (as when a heavy industry decreases local pollution by building a smokestack to disperse the emissions over hundreds of miles).

The Ethics Challenge

"Can I be honest and still keep my job?"

To ensure compliance with OHS[2], WCB[3], and WHMIS[4] regulations for worker safety, Sarah is assigned to test the air purification system in Millisun's chemical division. After finding the filters hopelessly clogged, she decides to test the air quality and discovers dangerous levels of benzene (a potent carcinogen). She reports these findings in a memo to the production manager, with an urgent recommendation that all employees be tested for benzene poisoning. The manager phones and tells Sarah to "have the filters replaced, and forget about it." Now Sarah has to decide what to do next: bury the memo in some file cabinet or defy her manager and send copies to other readers who might take action. ▲

Situations that jeopardize truth and fairness present the hardest choices of all: remain silent and look the other way or speak out and risk being dismissed. Some elements of Sarah's ethics challenge are: *Is this fair? Who might benefit or suffer? What other consequences could this have?*

In addition to meeting these various challenges, Sarah has to reckon with the implications of communication technology: Much of her writing, done at the computer, will be produced in collaboration with others (editors, managers, graphic artists), and her audience will extend beyond her own culture.

The Global Context Challenge

"Can I connect with all these different audiences?"

Recent mergers have transformed Millisun into a multi-national corporation with branches in 11 countries, all connected by computer network. Sarah can expect to collaborate with co-workers from diverse cultures on research and development; and with government agencies of the host countries on safety issues, patents and licensing rights, product liability laws, and environmental concerns.

---

2. Occupational Health and Safety
3. Workers' Compensation Board
4. Workplace Hazardous Materials Information System

> In order to standardize the sensitive management of the toxic, volatile, and even ex-
> plosive chemicals used in film production, Millisun is developing automated procedures for
> quality control, troubleshooting, and emergency response to chemical leakage. Sarah has
> been assigned to work with a group of colleagues to prepare Web-based instructional
> packages for all personnel involved in Millisun's chemical management worldwide. ▲

Sarah will have to develop working relationships with people she has never met, people she knows only via an electronic medium.

MANY TECHNICAL WRITERS COLLABORATE

Electronic communication allows more and more writing to be done collaboratively. Workplace documents (especially long reports, proposals, or manuals) are produced by teams who share information, expertise, ideas, and responsibilities. Production of a software manual, for instance, relies on writers, programmers, software engineers, graphic artists, editors, reviewers, marketing personnel, and lawyers (Debs, "Recent Research" 477).

Successful collaboration brings together the best that each team member has to offer. It enhances critical thinking by providing feedback, new perspectives, group support, and the chance to test one's ideas in group discussion.

Not all members of a collaborative team do the actual writing; some might research, edit, proofread, or test the document's *usability*. But even a document with a single author can be a collaborative product, as depicted in Figure 1.2. The more important the document, the more it will be reviewed.

As in any group effort, collaborative writers can experience the types of conflict depicted in Figure 1.3. Members might fail to get along because of differences in personality, working style, commitment, standards, or ability to take criticism. Some might disagree about exactly what or how much the group should accomplish, who should do what, or who should have the final say. Some might feel intimidated or hesitant to speak out.[5] These interpersonal problems actually can worsen when the group communicates exclusively via e-mail.

In any group, members have to find ways of expressing their views persuasively, of accepting constructive criticism, or getting along and reaching agreement with others who hold different views. These skills are essential for overcoming personal differences so that the group can achieve its goal.

GUIDELINES for Writing Collaboratively	The following guidelines[6] focus on projects in which people meet face to face, but can apply as well to electronically mediated collaboration. 1. *Appoint a project or group manager.* This person assigns tasks, enforces deadlines, conducts meetings, consults with supervisors, and generally runs the show.

5. Adapted from Bogert and Butt 51; Burnett 533–34; Hill-Duin 45–46; Debs, "Collaborative Writing" 38; Nelson and Smith 61.
6. Adapted from Debs, "Collaborative Writing" 38–41; Hill-Duin 45–46; Hulbert, "Developing" 53–54; McGuire 467–68; Morgan 540–41.

2. *Define a clear and definite goal.* Compose a purpose statement that spells out the project's goal and the group's plan for achieving it. Be sure each member understands the goal.

3. *Decide how the group will be organized.* Here are two possibilities:

   a. The group researches and plans together, but each person writes a different part of the document.

   b. Some members plan and research; one person writes a complete draft; others review, edit, revise, and produce the final version.

   Whatever the arrangement, the final revision should display a consistent style throughout—as if written by one person only.

4. *Divide the task.* Who will be responsible for which parts of the document or which phases of the project? Should one person alone do the final revision? Which jobs are the most difficult? Who is best at doing what (writing, editing, layout, graphics, oral presentation)? Who will make final decisions?

5. *Establish a timetable.* Specific completion dates for each phase will keep everyone focused on what is due, and when.

6. *Decide on a meeting schedule and format.* How often will the group meet, and for how long? In or out of office? Who will take notes (or minutes)? Will the supervisor attend or participate?

7. *Establish a procedure for responding to the work of other members.* Will reviewing and editing be done in writing, face to face, as a group, one on one, or via computer? Will this process be supervised by the project manager?

8. *Establish procedures for dealing with group problems.* How will gripes and disagreements be aired (to the manager, the whole group, the offending individual)? How will disputes be resolved (by vote, the manager)? How will irrelevant discussion be avoided or curtailed? Expect some conflict, but try to use it positively, and try to identify a natural peacemaker in the group.

9. *Decide how to evaluate each member's contribution.* Will the manager assess each member's performance and, in turn, be evaluated by each member? Will members evaluate each other? What are the criteria?

   Figure 1.4 depicts one possible form for a manager's evaluation of team members. Equivalent criteria for evaluating the manager include open mindedness, ability to organize the team, fairness in assigning tasks, ability to resolve conflicts, or ability to motivate. (Members might keep a journal of personal observations for overall evaluation of the project.)

10. *Prepare a project management plan.* Figure 1.5 depicts a sample plan sheet. Distribute completed copies to members.

11. *Submit progress reports regularly.* Progress reports enable everyone to track activities, problems, and rate of progress.

    Beyond these guidelines, respect for other people's views and willingness to listen are essential ingredients for successful collaboration.

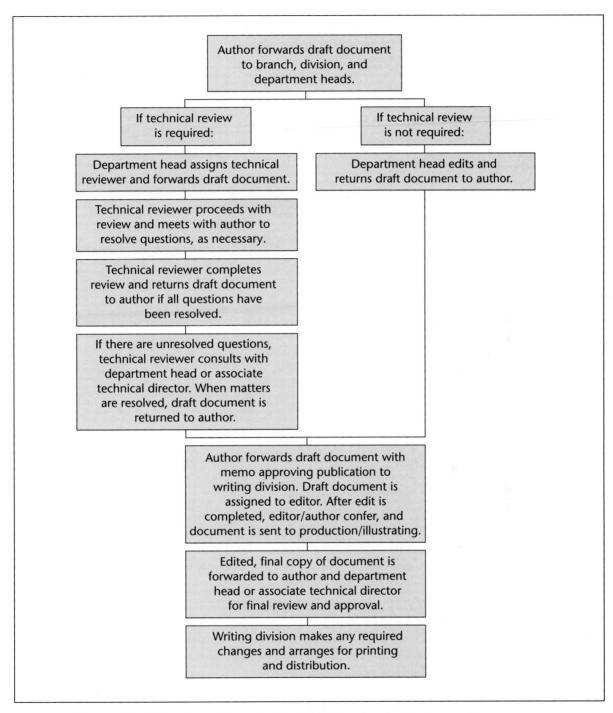

Figure 1.2   How One Organization Collaborates to Produce a Report

Source: Adapted from *NUSC Technical Publication Guidelines.* Naval Underwater Systems Center, Technical Information Dept.

Figure 1.3 How Pressures of a Group Effort Can Add Up

EFFECTIVE ROLES IN GROUPS[7]

Groups function best when certain key roles are established and followed. Two main types of roles have been identified: *task roles* and *maintenance roles*. Often, such roles are established informally.

Task Roles.   In order to accomplish its set task, a group has to have its members successfully complete a number of task roles. Some people play one or two of these roles almost exclusively, but most people slide easily in and out of most of the roles.

- *Initiators* propose and define tasks; they also suggest solutions to problems or ways of solving problems.
- *Information seekers* notice where facts are needed; they push the group to find those facts. Information givers have the information at hand or they know how to find the needed information.
- *Opinion seekers* actively canvas, group members for their ideas and opinions concerning a problem. *Opinion givers* volunteer input, respond readily when asked, and help set the criteria for solving problems.

7. Barker, Larry J. et al. *Groups in Process*. 3rd ed. Englewood Cliffs, New Jersey: Prentice-Hall, 1987.

Performance appraisal for _____

(After each item, place an X in the column that applies.)

	Superior	Acceptable	Unacceptable
Dependability			
Cooperation			
Effort			
Quality of work			
Ability to meet deadlines			

_____
Project manager's signature

Figure 1.4 Sample Form for Evaluating Team Members

- *Summarizers* draw together various ideas, facts, and opinions into a coherent whole; they review solutions, decisions, or problem-solving criteria. Often, the summarizing role is assumed by the group's elected chair or appointed manager.

Group Maintenance Roles.   All successful groups need a supportive group climate in order for group members to give their best efforts. A positive group climate does not usually happen by accident; members have to consciously perform some of the following roles in order to develop and sustain a good working relationship.

- *Encouragers* help other group members feel accepted and valued. In particular, they go out of their way to reward those group members who are shy about contributing to the group's tasks.
- *Feeling expressers* try to get all group members to state their feelings about the group and those members' roles in the group.
- *Harmonizers* help deal with unproductive conflict by removing the personal, emotional aspects of disputes and concentrating on the objective issues. They recognize the good points made by the respective disputants and they help the "warring members" recognize the merits of each other's positions.
- *Gatekeepers* try to draw quiet group members into the discussion so that the louder, more aggressive group members do not dominate. When everyone contributes, two main advantages result: (1) the group has a better chance of getting the ideas and information it needs and (2) group morale improves.

Management Plan Sheet

Project title:
Audience:
Project manager:
Team members:
Purpose statement:

Specific Assignments   **Due Dates**

   Research: Research due:
   Planning: Plan and outline due:
   Drafting: First draft due:
   Revising: Reviews due:
   Preparing final document: Revisions due:
   Presenting oral briefing: Final document due:
                                            Progress report(s) due:

Work Schedule

Group meetings:	Date	Place	Time	Note-taker
#1				
#2				
#3				
etc.				

Mtgs. w/manager
   #1
   #2
   etc.

Miscellaneous

   How will disputes and grievances be resolved?
   How will performances be evaluated?
   Other matters (Internet searches, e-mail routing, computer conferences, etc.)?

Figure 1.5   Sample Plan Sheet for Managing a Collaborative Project

In Brief GENDER AND CULTURAL DIFFERENCES IN COLLABORATIVE GROUPS

Any collaborative effort stands the best chance of succeeding when each group member feels included. A big mistake is to ignore personal differences, to assume everyone shares one viewpoint, one communication style, one approach to problem solving.

Collaboration often involves working with peers—those considered of equal status, rank, and expertise. In addition to the personality differences that lead to group conflict, gender and cultural differences in collaborative groups can create perceptions of inequality.

Gender Differences

Research on ways men and women communicate in meetings indicates a definite gender gap. Communication specialist Kathleen Kelley-Reardon offers this assessment of gender differences in workplace communication:

> Women and men operate according to communication rules for their gender, what experts call "gender codes." They learn, for example, to show gratitude, ask for help, take control, and express emotion, deference, and commitment in different ways. (88–89)

Professor Kelley-Reardon describes specific elements of a female gender code: Women are more likely than men to take as much time as needed to explore an issue, build consensus and relationship among members, use tact in expressing views, use care in choosing their words, consider the listener's feelings, speak softly, allow interruptions, make requests instead of giving commands ("Could I have the report by Friday?" versus "Have this ready by Friday."), preface assertions in ways that avoid offending ("I don't want to seem disagreeable here, but . . .").

One study of mixed-gender interaction among peers indicates that women tend to be more agreeable, solicit and admit the merits of other opinions, ask questions, and express uncertainty (e.g., with qualifiers such as *maybe, probably, it seems as if*) more often than men (Wojahn 747).

None of these traits, of course, is gender specific. Some people—regardless of gender—are more soft-spoken, contemplative, and

reflective. But such traits most often are attributed to the "feminine" stereotype. Moreover, any woman who breaches the gender code, for instance, by being assertive, may be seen by peers as "too controlling" (Kelley-Reardon 6). In fact, studies suggest that women have less freedom than male peers to alter their communication strategies: Less-assertive males often are still considered persuasive, whereas more-assertive females often are not (Perloff 273).

Cultural Differences

International business expert David A. Victor describes cultural codes that influence interaction in collaborative groups: Some cultures value silence more than speech, intuition and ambiguity more than hard evidence or data, politeness and personal relationships more than business relationships.

Cultures differ in their perceptions of time. Some want to get the job done immediately; others take as long as needed to weigh all the issues, engage in small talk and digressions, enquire about family, health, and other personal matters.

Cultures may differ in their willingness to express disagreement, question or be questioned, leave things unstated, touch, shake hands, kiss, hug, or backslap.

Direct eye contact is not always a good indicator of listening. In some cultures it is offensive. Other eye movements, such as squinting, closing the eyes, staring away, or staring at legs or other body parts are acceptable in some cultures but insulting in others.

Influence of On-line Communication

Some observers argue that on-line communication eliminates many such problems encountered in face-to-face meetings, because every participant, in effect, has as much time as needed to contribute to the electronically mediated discussion. Also, "status cues" such as age, gender, appearance, or ethnicity virtually disappear online. Other observers disagree, arguing that the interpersonal chemistry of communication transcends specific media (Wojahn 747–48).

MANAGING GROUP CONFLICT

No team can afford to assume that all members share one viewpoint, one communication style, one approach to problem solving. Before any group can reach final agreement, conflicts must be expressed and addressed openly. Pointing out that "conflict can be good for an organization—as long as it's resolved quickly," management expert David House offers these strategies for overcoming personal differences (Warshaw 48):

How to manage group conflict

- ◆ Give everyone a chance to be heard.
- ◆ Take everyone's feelings and opinions seriously.
- ◆ Don't be afraid to disagree.
- ◆ Offer and accept constructive criticism.
- ◆ Find points of agreement with others who hold different views.
- ◆ When the group does make a decision, support it fully.

Central to all these strategies is the ability to *listen* to what others have to say.

GUIDELINES

for Active Listening*

1. *Don't dictate.* If you are leading the group, don't express your view until everyone else has had a chance.

2. *Be receptive.* Instead of resisting different views, develop a "learner's" mindset: take it all in first and evaluate it later.

3. *Keep an open mind.* Judgment stops thought (Hayakawa 42). Reserve judgment until everyone has had his or her say.

4. *Be courteous.* Don't smirk, roll your eyes, whisper, or wisecrack.

5. *Show genuine interest.* Eye contact is vital, and so is body language (nodding, smiling, leaning toward the speaker). Make it a point to remember everyone's name.

6. *Hear the speaker out.* Instead of "tuning out" a message you find disagreeable, allow the speaker to continue without interruption (except to ask for clarification). Delay your own questions, comments, and rebuttals until the speaker has finished.

7. *Focus on the message.* Instead of thinking about what you want to say next, try to get a clear understanding of the speaker's position.

8. *Ask for clarification.* If anything is unclear, say so: "Can you run that by me again?" To ensure accuracy, paraphrase the message: "So what you're saying is. . . . Is that right?"

9. *Be agreeable.* Don't turn the conversation into a contest, and don't insist on having the last word.

10. *Observe the 90/10 rule.* You rarely go wrong spending 90 percent of your time listening and 10 percent speaking. Former U.S. President Calvin Coolidge claimed that "Nobody ever listened himself out of a job." Some historians would argue that "Silent Cal" listened himself right into the White House.

---

*Adapted from Armstrong 248; Bashein and Markus 37; Cooper 78–84; Dumont and Lannon 648–51; Pearce, Johnson, and Barker 28–32.

REVIEWING AND EDITING OTHERS' WORK

Documents produced collaboratively are reviewed and edited extensively. Reviewing means evaluating how well a document connects with its intended audience and meets its intended purpose. Reviewers typically examine a document for these specific qualities:

What reviewers look for

- accurate, appropriate, useful, and legal content
- material organized for the reader's understanding
- clear, easy-to-read, and engaging style
- effective visuals and page design
- a document that is safe, dependable, and easy to use

In reviewing, you explain to the writer how you respond as a reader. This commentary helps writers think about ways of revising. Criteria for reviewing various documents appear in checklists throughout this book. (See also Chapter 17, on Usability).

Editing means actually "fixing" copy by making it more precise and readable. Editors typically suggest improvements like these:

Ways in which editors "fix" copy

- rephrasing or reorganizing sentences
- clarifying a topic sentence
- choosing a better word or phrase
- correcting spelling, usage, or punctuation, and so on

Criteria for editing appear in Chapter 11, Appendix A, and inside the back cover.

GUIDELINES

for Peer Reviewing and Editing

1. *Read the entire piece at least twice before you comment.* Develop a clear sense of the document's purpose and its intended audience. Try to visualize the document as a whole before you evaluate specific parts or features.

2. *Remember that correctness does not guarantee effectiveness.* Poor usage, punctuation, or mechanics do distract readers and harm the writer's credibility. However, a "correct" piece of writing might still contain faulty rhetorical elements (inappropriate content, confusing organization, wordy style, or the like).

3. *Understand the acceptable limits of editing.* In the workplace, editing can range from cleaning up and fine tuning to an in-depth rewrite (in which case editors are cited prominently as consulting editors or co-authors). In school, however, rewriting someone else's piece to the extent that it ceases to belong to that writer may constitute plagiarism (see page 20).

4. *Be honest but diplomatic.* Most of us are sensitive to criticism—even when it is constructive—and we all respond more favourably to encouragement. Begin with something positive before moving to material needing improvement. Be supportive instead of judgmental.

GUIDELINES

for Peer
Reviewing and
Editing

(continued)

5. *Always explain why something doesn't work.* Instead of "this paragraph is confusing," say "because this paragraph lacks a clear topic sentence, I had trouble discovering the main idea" (see page 215 for sample criteria). Help the writer identify the cause of the problem.

6. *Make specific recommendations for improvements.* Write out suggestions in enough detail for the writer to know what to do.

7. *Be aware that not all feedback has equal value.* Even professional reviewers and editors can disagree. Keep in mind that your job as a reviewer or editor is to help clarify and enhance a document—without altering its original meaning.

In Brief   ETHICAL ISSUES IN WORKPLACE COLLABORATION

Our *lean* and *downsized* corporate world spells competition among workers—along with this dilemma:

> Many companies send mixed signals . . . saying they value teamwork while still rewarding individual stars, so that nobody has any real incentive to share the glory. (Fisher, "My Team Leader" 291)

The resulting mistrust interferes with fair and open teamwork and promotes unethical behaviour.

Intimidating Co-workers

A dominant personality may intimidate peers into silence or agreement, or the group leader may allow no other viewpoints (Matson, "The Seven Sins" 30). Intimidated employees resort to "mimicking"—merely repeating what the boss says (Haskin, "Meetings without Walls" 55).

Claiming Credit for Others' Work

Workplace plagiarism occurs when the team or project leader claims all the credit. Even with good intentions, "the person who speaks for a team often gets the credit, not the people who had the ideas or did the work" (Nakache 287–88).

Team expert James Stern offers this solution:

> Some companies list "core" and "contributing" team members, to distinguish those who did most of the heavy lifting from those who were less involved. (quoted in Fisher, "My Team Leader" 291)

Stern advises groups to decide beforehand—and in writing—what credit will be given for which contributions.

Hoarding Information

Surveys reveal that the biggest obstacle to workplace collaboration is employees' tendency "to hoard their own know-how" (Cole-Gomolski 6). Examples:

◆ Whom do we contact for what?

◆ Where do we get the best price, the quickest repair, the most dependable service?

◆ What's the best way to do X?

People might withhold exclusive information when they think it provides job security or in an attempt to sabotage peers.

EXERCISES

1. Locate a brief example of a technical document (or a section of one). Make a photocopy, bring it to class, and explain why your selection can be called technical writing.

2. Research the kinds of writing you will do in your future career. (Begin with the Dictionary of Occupational Titles in your library.) You might interview a member of your chosen profession. Why will you write on the job? For whom will you write? Explain in a memo to your instructor. (See Chapter 22 for memo elements and format.)

3. In a memo to your instructor, describe the skills you seek to develop in your technical writing course. How exactly will you apply these skills to your career?

4. Assume a friend in your major thinks that writing skills are obsolete, and that administrative assistants or word processors can fix any writing. Write your friend a letter, explaining why you think these assumptions are mistaken. Use examples to support your position. (See Chapter 22 for letter elements and format.)

COLLABORATIVE PROJECT

Divide into small groups of mixed genders. Read "In Brief" (page 17) and complete the following tasks to test the hypothesis that women and men communicate differently in the workplace.

Each group member prepares the following brief messages—without consulting with other members:

- A thank-you note to a co-worker who has done you a favour.
- A note asking a co-worker for help with a problem or project.
- A note asking a collaborative peer to be more co-operative or stop interrupting or complaining.
- A note expressing impatience, frustration, confusion, or satisfaction to members of your group.
- A recommendation for a friend who is applying for a position with your company.
- A note offering support to a good friend and co-worker.
- A note to a new colleague, welcoming this person to the company.
- A request for a raise, based on your hard work.
- The meeting is out of hand, so you decide to take control. Write what you would say.
- Some members of your group are dragging their feet on a project. Write what you would say.

As a group, compare messages, draw conclusions about the original hypothesis, and appoint one member to present the findings to the class.

CHAPTER 2

Preparing to Write: Audience/Purpose Analysis

USE A COMMUNICATIONS MODEL

ASSESS READERS' INFORMATION NEEDS

IDENTIFY LEVELS OF TECHNICALITY

DEVELOP AN AUDIENCE/PURPOSE PROFILE

As a technical writer, you'll need to know your readers' needs, in order to meet those needs. At the same time, you'll need to consider your own role needs. Perhaps you want to persuade readers to support a proposal; perhaps you are simply responding to your supervisor's request for information about your progress on a project. Whatever you write, you must first think very carefully about your audience and your purpose for writing.

Actually, you should analyze your audience and purpose for *any* kind of technical or business communication, written or spoken. For example, when you prepare for a job interview, you need to consider more than the points you hope to make about your qualifications and skills; you also need to consider what the interviewer hopes to accomplish by interviewing you. By analyzing the interviewer's role-dominated needs, you'll be able to anticipate some of the tough questions that the interviewer necessarily asks.

Can you know for certain how a listener will react to your statements in a job interview? Can you know how a reader will interpret your report's statements of fact and conclusions? The answer to both questions is that you can't know *for certain*. However, you *can* make some shrewd guesses. The following communications model can help you anticipate receiver reactions so that you can successfully adapt the content and presentation of your messages.

USE A COMMUNICATIONS MODEL

Whether you have a quiet talk with a friend at a doughnut shop or whether you write a high-powered report for an important client, certain key factors affect the nature and outcome of that communication. We will now place those factors in a communications model, which will be developed in stages, so that you will be able to fully understand how the various factors interact and contribute to the success of the communications exchange.

Let's analyze a conversation between a consulting mechanical engineer, Daphne McCrae, and McCrae's client, Max Lauder, the maintenance supervisor for the Trendmark Ski Resort. Lauder has heard about a neighbouring ski hill's problem with the massive cast metal gripping mechanisms that clamp the chairs of a high-speed chairlift onto the cable of that lift. Lauder wants McCrae's advice about whether to replace the grips on Trendmark's high-speed quad lift. Thus, McCrae has tested several grips and is giving a preliminary oral report of her findings.

Identify the primary message

First, in order to develop a communications model, let's identify the conversation's primary *message* as Daphne McCrae's semi-technical answer to Lauder's question. McCrae is the *sender* of that message; Lauder is the *receiver*. Their conversation, which occurs in the ski resort's maintenance building, uses an oral verbal *channel* and several non-verbal *channels*.

The conversation between McCrae and Lauder seems to depend on the verbal channel of speech. After all, McCrae uses words to convey her knowledge about the factors that might contribute to cast-alloy failure. However, several non-verbal channels such as facial expressions, vocal inflections, posture, and gestures send accompanying messages about the sender's confidence in her own knowledge, her degree of certainty, and her level of concern for her client's current problem. How Max Lauder interprets these non-verbal messages affects how he judges the accuracy and value of McCrae's words.

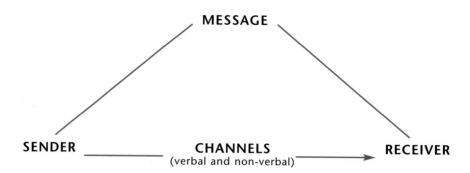

Has Daphne McCrae chosen an appropriate channel to convey her initial reaction to her client's question? The answer is "yes," especially if she wants to alleviate her receiver's immediate concerns. However, if McCrae were to communicate a full technical analysis of the grips she has tested, she should use a formal written report to present her findings. McCrae's choice of communications channels illustrates a basic communications principle: Senders need to choose their primary communication channels carefully in order to reach their receivers.

Here are some other questions to help assess which channels should be used in a given situation:

1. *Has the sender chosen a channel used by the receiver?* For example, Daphne McCrae won't reach Max Lauder via e-mail on days when Lauder is working on the lifts, away from his office computer.
2. *Has the sender taken advantage of the chosen channel's particular strengths?* The main advantage of presenting McCrae's initial analysis *orally* is that it allows her to accompany her semi-technical comments with reassuring vocal tones and other non-verbal messages.
3. *Have both the sender and receiver blocked out external "noise"?* The maintenance building could be very noisy, as workers repair and maintain equipment, so perhaps the conversation should be held in Lauder's office. In this case, the two participants need to examine the equipment as they talk, so they meet in the maintenance shed, where they need to block out the noise in order to focus on each other's messages.
4. *Has the sender considered both the verbal and non-verbal aspects of the transmission?* McCrae's subsequent written analysis will need to use a formal report format, to correspond to the subject's serious nature and to signal the writer's credibility.

Choosing the right channel is only the beginning. The sender also has to encode the message so that the receiver has at least a hope of decoding it as the sender intended. Proper encoding means choosing the right words, sentence patterns, and message structures to suit the sender's purpose and the reader's interests and needs. This collection of choices requires careful thought; that's why this text emphasizes audience/purpose analysis for all writing and speaking assignments.

Decode messages carefully

Does Max Lauder understand Daphne McCrae's explanations? Perhaps. If McCrae uses a great deal of technical jargon, Lauder may be perplexed. And if McCrae oversimplifies her message, Lauder won't get the full picture and thus will not understand McCrae's advice about the grips.

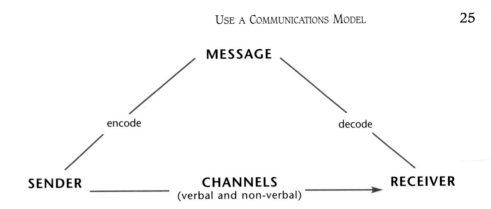

Indeed, decoding messages is often harder than encoding them. To start with, many receivers do not know how to listen or read effectively and so they miss much of the intended message. Even the alert, skilled receiver may have difficulty following the sender's train of thought if the encoded message has been poorly organized, or if the sender has chosen an inappropriate level of phrasing or detail, or if the sender's verbal and non-verbal messages contradict each other.

Even when the receiver is confident that she or he has understood the message, communication can break down. That's because senders and receivers often have different meanings for the same words. (Your English professor, for example, might intend the word "decode" to interpret written and spoken words, but you might think of deciphering Morse code and other coded messages if you have a military background or a passion for spy novels.)

Now let's introduce more psychology to our communications model:

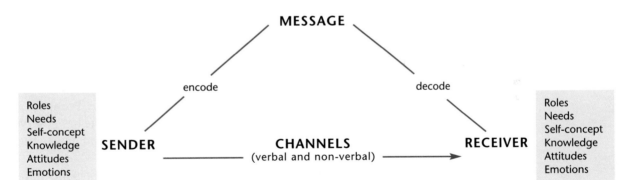

The terms beside *sender* and *receiver* in the model refer to factors that may affect each person at the time the communication occurs. These terms will now be discussed. The discussion will analyze the McCrae/Lauder conversation, but you might also like to stop at various points to think of a recent conversation you've had, or a letter you've sent, in order to see how each of the following factors may have affected the communication's outcome in each of those cases.

Consider the factors that affect the encoding and decoding processes

1. *Role.* Daphne McCrae's role is the expert analyst, which compels her to think and speak carefully and rationally. In that role, she must not jump to conclusions or make rash statements. Why? First, her job role requires her to make sense of the available technical data. Second, she has to maintain credibility in order to convince her listener to consider her expert advice.

Max Lauder's role requires him to understand and absorb McCrae's analysis and advice. Thus, Max listens very attentively.

Now, think of how your role influences your behaviour at this moment. Probably, you're reading this page because your student role requires you to do so. So, your receiving behaviour is role-dominated. But what if you find yourself getting really interested in this subject? What if you are now starting to think of how *role* affects your communications with your friends and your colleagues at work or school? If so, your *personal needs* are starting to influence your receiving behaviour.

2. *Needs.* It's obvious that we all have strong reasons for communicating with others: *practical needs* associated with making a living, basic *physical needs,* and *social needs* for acceptance, affection, and control. Also, *identity needs* seem always to influence our behaviour, even when we're taking care of our basic physical needs or our practical business needs. Much of our communications has us trying to determine who we are and then trying to assert that identity to others. If we allow our identity needs to dominate, however, problems may result.

For example, if Daphne McCrae were to have a strong identity need to appear forceful and infallible, she might be driven to make definite conclusions even if there's insufficient data to support such conclusions. If her identity need overruns her role requirement, she might provide disastrous advice to her client.

3. *Self-concept.* Our self-beliefs affect our sending and receiving behaviours. If, for example, Daphne McCrae sees herself as analytical and intelligent, her word choices and speaking pace will reflect those self-perceptions. And if Max Lauder sees himself as very practical but lacking education, he may defer to some of McCrae's conclusions even if Max's instincts tell him that McCrae is wrong.

4. *Knowledge.* A sender like Daphne McCrae has to know her subject well before she can successfully explain, for example, resistance to stress in cast metal chairlift grips. Similarly, Max Lauder has to have a rudimentary knowledge in order to understand any of the complexities in McCrae's analysis. From a practical viewpoint, Max probably understands these stresses well: he has observed the chairlift in action. However, he probably does not understand how to measure shear forces or how metal breaks down.

5. *Attitudes.* In the conversation we've been analyzing, both participants have a serious attitude toward the topic being discussed. They concentrate on the technical and practical aspects of a potential problem.

Another factor lies in their degree of respect for each other. As it turns out, each of them is in his/her late 30s with over 15 years' experience in their field. Each is aware of the other's expertise, so each chooses words carefully.

Actually, our attitudes are usually shown non-verbally. In the McCrae/Lauder conversation, the most likely channels revealing their attitudes toward the situation and each other would be posture, facial expressions, and tone of voice.

6. *Emotions.* You have experienced many situations where your emotions have affected how you spoke or how you listened. Indeed, one's emotions can totally block effective communications. In business, we should not allow that to happen, and it's not likely to happen in the conversation between Max and Daphne.

Communication can
break down

As this chapter has hinted, communication can break down for many reasons:

- poor choice of channels
- receiver inattentiveness
- poor sender decoding
- lack of knowledge by sender or receiver
- conflicting roles or contrasting personal needs

However, most misunderstandings can be prevented by timely and useful *feedback* that is so important for two-way channels such as face-to-face conversations, telephone calls, and e-mail. Let's introduce that feedback loop next.

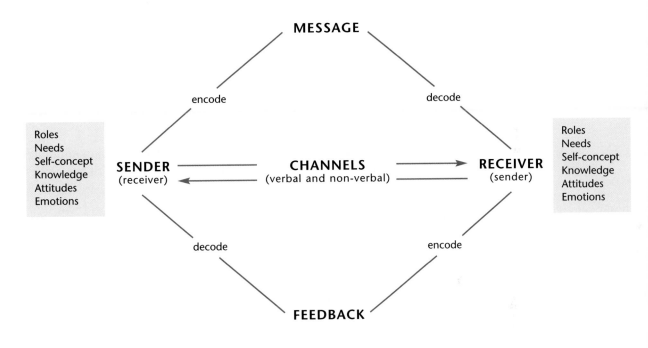

Feedback is useful

Feedback could be particularly useful in the McCrae/Lauder meeting—Max could immediately signal that he didn't understand some aspect of the message he heard and Daphne could use different terminology or a different order of explanation to convey her message.

While two-way channels provide timely, direct feedback, written channels (letters, memos, reports) do not. That's why you should consider the reader's role, needs, and knowledge levels when you choose the content, structure, and style of your written messages.

Let's look at assessing your readers' information needs now.

ASSESS READERS' INFORMATION NEEDS

Good writing connects with readers by recognizing their different backgrounds, needs, and preferences. A single message may appear in several versions for several audiences. For instance, an article describing a new cancer treatment might appear in a medical journal read by doctors and nurses. A less technical version might appear in a medical textbook read by medical and nursing students. An even simpler version might appear in *Reader's Digest*. All three versions treat the same topic, but each meets the needs of a different audience.

Technical writing is intended to be *used*. You become the teacher and the reader becomes the student. Because your readers may know less than you, they may have questions.

TYPICAL READER QUESTIONS ABOUT WORKPLACE DOCUMENTS

- What is the purpose of this document?
- Who should read the document?
- What is being described or explained?
- What does it look like?
- How do I do it?

- How did you do it?
- Why did it happen?
- When will it happen?
- Why should we do it?
- How much will it cost?
- What are the risks?

You always write to enable a specific audience to grasp the information and follow the discussion. In order to be useful, the writing must connect with the reader's level of understanding.

IDENTIFY LEVELS OF TECHNICALITY

When you write for a close acquaintance (friend, computer crony, co-worker, professor, or supervisor), you know a good deal about your reader's background. You deliberately adapt your document to that reader's knowledge, interests, and needs. But sometimes you write for less defined audiences, particularly when the audience

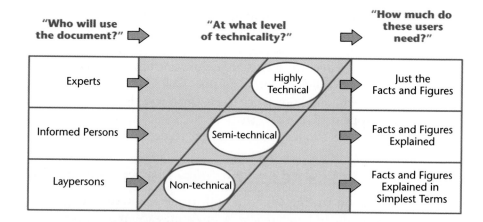

Figure 2.1   Deciding on a Document's Level of Technicality

is large (when you are writing a journal article, a computer manual, a set of first-aid procedures, or a report of an accident). When you have only a general notion about your audience's background, you must decide whether your document should be *highly technical, semi-technical,* or *non-technical,* as depicted in Figure 2.1.

THE HIGHLY TECHNICAL DOCUMENT

Readers at the specialized level expect the technical facts and figures they need, without long explanations. The following report of treatment given to a heart attack victim is highly technical. The writer, an emergency room physician, is reporting to the patient's doctor. This reader needs an exact record of the patient's symptoms and treatment.

A Highly Technical Version

Expert readers need merely the facts and figures, which they can interpret for themselves

The patient was brought to the emergency room by ambulance at 0100 hours, September 27, 2004. The patient complained of severe chest pains, dyspnea, and vertigo. Auscultation and EKG revealed a massive cardiac infarction and pulmonary edema marked by pronounced cyanosis. Vital signs: blood pressure, 80/40; pulse. 140/min; respiration, 35/min. Lab: WBC, 20 000; elevated serum transaminase; urea nitrogen, 60 mg%. Urinalysis showed 4+ protein and 4+ granular casts/field, indicating acute renal failure secondary to the hypotension.

The patient received 10 mg of morphine stat, subcutaneously, followed by nasal oxygen and D_5W intravenously. At 0125 the cardiac monitor recorded an irregular sinus rhythm, indicating left ventricular fibrillation. The patient was defibrillated stat and given a 50 mg bolus of Xylocaine intravenously. A Xylocaine drip was started, and sodium bicarbonate administered until a normal heartbeat was established. By 0300, the oscilloscope was recording a normal sinus rhythm.

As the heartbeat stabilized and cyanosis diminished, the patient received 5 cc of heparin intravenously, to be repeated every six hours. By 0500 the BUN had fallen to 20 mg% and vital signs had stabilized: blood pressure, 110/60; pulse, 105/min; respiration, 22/min. The patient was now conscious and responsive. ▲

This highly technical report is clear only to the medical expert. Because her reader has extensive background, this writer defines no technical terms (pulmonary edema, sinus rhythm). Nor does she interpret lab findings (4+ protein, elevated serum transaminase). She uses abbreviations her reader understands (WBC, BUN, D_5W). Because her reader knows the reasons for specific treatments and medications (defibrillation, Xylocaine drip), she includes no theoretical background. Her report answers concisely the main questions she can anticipate from her reader: What happened? What treatment was given? What were the results?

THE SEMI-TECHNICAL DOCUMENT

One broad class of readers may have some technical background, but less than the experts. For instance, first-year medical students have specialized knowledge, but not as much as second-, third-, and fourth-year students. Yet students in all four groups could be considered semi-technical readers. When you write for a semi-technical audience, identify the *lowest* level of understanding in the group, and write to that level. Too much explanation is better than too little.

Here is a partial version of the earlier medical report. Written at a semi-technical level, it might appear in a textbook for first-year medical or nursing students, in a report for a medical social worker, in a patient's history for the medical technology department, or in a monthly report for the hospital administration.

A Semi-technical Version

Informed but non-expert readers need enough explanations to understand what the facts mean

Examination by stethoscope and electrocardiogram revealed a massive failure of the heart muscle along with fluid buildup in the lungs, which produced a cyanotic **discolouration of the lips and fingertips from lack of oxygen.**

The patient's blood pressure at 80 mm Hg (systolic)/40 mm Hg (diastolic) was **dangerously below its normal measure of 130/70.** A pulse rate of 140/minute was **almost twice the normal rate of 60–80.** Respiration at 35/minute was more than **twice the normal rate of 12–16.**

Laboratory blood tests yielded a white blood cell count of 20 000/cu mm (normal value: 5,000–10,000), **indicating a severe inflammatory response by the heart muscle.** The elevated serum transaminase enzymes **(produced in quantity only when the heart muscle fails)** confirmed the earlier diagnosis. A blood urea nitrogen level of 60 mg% (normal value: 12–16 mg%) indicated **that the kidneys had ceased to filter out metabolic waste products.** The 4+ protein and casts reported from the urinalysis (normal value: 0) **revealed that the kidney tubules were degenerating as a result of the lowered blood pressure.**

The patient immediately received morphine **to ease the chest pain,** followed by oxygen **to relieve strain on the cardiopulmonary system,** and an intravenous solution of dextrose and water **to prevent shock.** ▲

The version explains (in boldface) the raw data. Exact dosages are not mentioned because the readers are not treating the patient. Normal values of lab tests and vital signs, however, make interpretation easier. (Expert readers would know these values.) Knowing what medications the patient received would be especially important to the lab technician, because some medications affect test results. For a non-technical audience, however, the message needs further translation.

THE NON-TECHNICAL DOCUMENT

Readers with no specialized training expect technical data to be translated into terms they understand. Non-technical readers are impatient with abstract theories but want enough background to help them make the right decision or take the right action. They are bored by long explanations but frustrated by bare facts not explained or interpreted. They expect a report that is clear on first reading, not one that requires review or study.

The following is a non-technical version of our medical report. The physician might write this version for the patient's spouse who is overseas on business, or as part of a script for a documentary film about emergency room treatment.

A Non-technical Version

Heart sounds and electrical impulses both were abnormal, **indicating a massive heart attack caused by failure of a large part of the heart muscle.** The lungs

General readers
need everything
translated into
terms they
understand

were swollen with fluid and the lips and fingertips showed a bluish discolouration from **lack of oxygen**.

Blood pressure was dangerously low, **creating the risk of shock**. Pulse and respiration were **almost twice the normal rate, indicating that the heart and lungs were being overworked** in keeping oxygenated blood circulating freely.

Blood tests confirmed the heart attack diagnosis and **indicated that waste products usually filtered out by the kidneys were building up in the bloodstream. Urine tests showed that the kidneys were failing as a result of the lowered blood pressure**.

The patient was given **medication to ease the chest pain, oxygen to ease the strain on the heart and lungs**, and **intravenous solution to prevent the blood vessels from collapsing and causing irreversible shock**. ▲

This non-technical version explains (in boldface) the situation using everyday language. It omits any mention of medications, lab tests, or normal values because these have no meaning for the reader. The writer merely summarizes events and explains the causes of the crisis and the reasons for the particular treatment.

In some other situation, however (say, in a jury trial for malpractice), the non-technical audience might need information about specific medication and treatment. Such a report would, of course, be much longer—a short course in emergency coronary treatment.

Each version of the medical report is useful *only* to readers at a specific level. Doctors and nurses have no need for the explanations in the two latter versions, but they do need the specialized data in the first. Beginning medical students and paramedics might be confused by the first version and bored by the third. Non-technical readers would find both the first and second versions meaningless.

PRIMARY AND SECONDARY READERS

Whenever you prepare a single document for multiple readers, classify your readers as *primary* or *secondary*. Primary readers usually are those who requested the document and who will use it as a basis for decisions or actions. Secondary readers are those who will carry out the project, who will advise the primary readers about their decision, or who will somehow be affected by this decision. They will read your document (or perhaps only part of it) for information that will help them get the job done, for educated advice, or to keep up with new developments.

Often these two audiences differ in technical background. Primary readers may require highly technical messages, and secondary readers may need semi-technical or non-technical messages—or vice versa. When you must write for audiences at different levels, follow these guidelines:

How to tailor a
single document to
multiple readers

1. If the document is short (a letter, memo, or anything less than two pages), rewrite it at various levels for various readers.
2. If the document exceeds two pages, address the primary readers. Then provide appendices for secondary readers (technical appendices when secondary readers are technical, or vice versa). Letters of transmittal, informative abstracts, and glossaries are other supplements that help non-specialized audiences understand a highly technical report. (See Chapter 20 for how to use and prepare appendices and other supplements.)

The next scenario shows how some documents must be tailored to both primary and secondary readers.

Tailoring a Document to Different Readers

When Daphne McCrae writes the results of her tests on potentially damaged ski-lift grips, her primary reader will be the client, Trendmark Ski Resort. That audience will include Max Lauder, the maintenance supervisor who has some technical knowledge of metal and metal fatigue. The audience will also include Trendmark's management board and its lawyer, all of whom have little or no technical knowledge.

Daphne's report may eventually have legal implications, so it must be presented in meticulous detail. But her non-specialist readers will need explanations of her testing methods. These readers will also need photos of the test equipment to fully understand the processes involved. The report will have to define specialized terms such as "fractographs" (microscopic photographs of fractured surfaces) and "HSLA" (high-strength, low-alloy) steel such as ASTM grade A 242 (which has 0.4% copper alloyed to the steel to provide greater weathering resistance).

The report's secondary readers will include Daphne McCrae's supervisor and outside consulting engineers who may evaluate Daphne's test procedures and assess the validity of her findings. Consultants will focus on various parts of the report to verify that Daphne's procedure has been exact and faultless. For these readers, she will have to include appendices spelling out the technical details of her analysis: *how* light-microscopic fractographs revealed the presence and direction of fractures, and *how* the pattern of these fractures indicates a casting flaw in the original grip, not torsional fatigue. Finally, Daphne will need to present the technical details of her finding that only one grip has the casting flaw and that the other grips are safe for operation. ▲

In Daphne McCrae's situation, primary readers need to know *what her findings mean,* whereas secondary readers need to know *how she arrived at her conclusions.* Unless she serves the needs of each group independently, her information will be worthless.

DEVELOP AN AUDIENCE/PURPOSE PROFILE

When you write for a particular reader or a small group of readers, you can focus sharply on your audience by asking the questions listed below. To answer these questions consider the suggestions that follow and use a version of the Audience/Purpose Profile Sheet (Figure 2.2 on page 34) for all of your writing.

READER CHARACTERISTICS

Identify the primary readers by name, job title, and specialty (e.g., Martha Jones, Director of Quality Control, B.S. and M.S. in mechanical engineering). Are they superiors, colleagues, or subordinates? Are they inside or outside your organization? What is their attitude toward this topic likely to be? Are they apt to accept or reject your conclusions and recommendations? Will your report be good or bad news? For readers from other cultures, how might cultural differences affect their expectations and interpretations?

Identify also those secondary readers who might be interested in or affected by your document, or who will affect the primary readers' perceptions or use of your document.

QUESTIONS ABOUT A DOCUMENT'S INTENDED AUDIENCE AND PURPOSE

- Who wants the document? Who else will read it?
- Why do they want the document? How will they use it?
- What is the purpose of my document? What do I want to achieve?
- What is the technical background of the primary audience? Of the secondary audience?
- How might cultural differences shape readers' expectations and interpretations?
- How much does the audience already know about the subject? What material will have informative value?
- What exactly does the audience need to know, and in what format? How much is enough?
- When is the document due?

PURPOSE OF THE DOCUMENT

Learn why readers want the document and how they will use it. Do they merely want a record of activities or progress? Do they expect only raw data, or conclusions and recommendations as well? Will readers act immediately on the information? Do they need step-by-step instructions? Will the document be read and discarded, filed, published, distributed electronically? In your audience's view, *what* is most important? What purpose should this document achieve?

READERS' TECHNICAL BACKGROUND

Colleagues who speak your technical language will understand raw data. Supervisors responsible for several technical areas may want interpretations and recommendations. Managers who have limited technical knowledge expect definitions and explanations. Clients with no technical background expect versions that spell out what the facts mean to *them* (to their health, pocketbook, business prospects). However, none of these generalizations might apply to *your* situation. When in doubt, aim for low technicality.

READERS' CULTURAL BACKGROUND

Some information needs can be culturally determined. For example, readers in certain cultures might value thoroughness and complexity above all: lists of data, with every relevant detail included and explained. Readers in other cultures might prefer an overview of the material, with multiple perspectives and liberal use of graphics (Hein 125–26).

North American business culture generally values plain talk that spells out the meaning directly, but some cultures prefer indirect and somewhat ambiguous messages, which leave explanations and interpretation for readers to decipher (Leki 151; Martin and Chaney 276–77). To avoid seeming impolite, some readers might hesitate to request clarification or additional information. Even disagreement or refusal might be expressed as "We will do our best" or "This is very difficult," instead of "No"—to avoid offending and to preserve harmony (Rowland 47).

AUDIENCE/PURPOSE PROFILE

Audience Identity and Needs

Primary reader(s): _____ *(name, title)*

Secondary reader(s): _____

Relationship: _____ *(client, employer, other)*

Intended use of document: _____ *(perform a task, solve a problem, other)*

Prior knowledge about this topic: _____ *(knows nothing, a few details, other)*

Additional information needed: _____ *(background, only bare facts, other)*

Probable questions: _____

_____

_____

_____

_____

_____

Audience's Probable Attitude and Personality

Attitude toward topic: _____ *(indifferent, skeptical, other)*

Probable objections: _____ *(cost, time, none, other)*

Probable attitude toward this writer: _____ *(intimidated, hostile, receptive, other)*

Persons most affected by this document: _____

Temperament: _____ *(cautious, impatient, other)*

Probable reaction to document: _____ *(resistance, approval, anger, guilt, other)*

Risk of alienating anyone: _____

_____

Audience Expectations About the Document

Reason document originated: _____ *(audience request, my idea, other)*

Acceptable length: _____ *(comprehensive, concise, other)*

Material important to this audience: _____ *(interpretations, costs,*

_____ *conclusions, other)*

Most useful arrangement: _____ *(problem-causes-solutions, other)*

Tone: _____ *(businesslike, apologetic, enthusiastic, other)*

Intended effect on this audience: _____ *(win support, change behaviour, other)*

Due date: _____

Figure 2.2   Audience/Purpose Profile Sheet (see www.pearsoned.ca/lannon for an on-line version of this form)

Correspondence practices vary from culture to culture. In British business letters, for example, the salutation is followed by a comma (Dear Ms. Morrison,); in North America, it is followed by a colon (Dear Ms. Morrison:). Also, European data formats vary from North American practices, as Table 2.1 illustrates.

Table 2.1   Typical Data Formats

	Canada/U.S.	United Kingdom	France	Germany	Portugal
Date	May 15, 2004 5/15/04	15th May, 2004 15/5/04	15 Mai 2004 15.05.04	15. Mai 2004 15.5.04	15/5/04
Time	10:32 p.m.	10:32 p.m.	22.32 22 h 32	22.32 Uhr Uhr 22.32	22H32m
Currency	$123.45 C$123.45 Can $123.45 123,45 $(Quebec) US$123.45	£123.45 GB£123.45	123F45 123,45 F	DM 123,45 123,45 DM	123$45 ESC 123.45
Large Number	1,234,567.89	1,234,567.89	1. 234 .567,89	1.234.567,89	1.234.567,89
Phone Number	(905) 555-1234	(081) 987 1234 0255 876543	(15) 61-87-34-02 (15) 61.87.34.02	(089) 2 61 39 12	056-244 33 056 45 45 45

Source: Adapted from Guffey, M.E. et al. *Business Communication.* 2nd Canadian Edition. Toronto: Nelson, 1999, p. 59.

The following general advice may help with your intercultural communication:

1. Where feasible, hire a translator to convert your proposal or report to the *buyer's* language.
2. Where that's not feasible, or where your reader prefers to receive your document in English, keep your sentences and paragraphs short. Use direct, simple, precise words. Use relative pronouns such as "that," "which," and "who" to introduce clauses. Avoid technical jargon and North American idioms ("hit the ceiling," "at the drop of a hat," "been there, done that").
3. Pay special attention to the openings of letters and memos, and to the introductions of reports—North Americans like to get straight to the point; Asians and Latin Americans like to build relationships first.
4. When writing recommendations, consider whether your client's culture favours careful, deliberate team consultation or quick individual action.
5. Look at the physical format of reports produced in the reader's culture, and incorporate some aspects of that format in your report.
6. Learn whether it's best to deliver reports in person, by messenger, by courier, or by mail.

So far, we've discussed readers outside North America. But we also need to remember that our own culture is not homogeneous. Canadians and Americans have different attitudes and political systems. Within Canada, each region has distinctive characteristics that could affect how a reader interprets a given document.

READERS' KNOWLEDGE OF THE SUBJECT

Do not waste time rehashing information readers already have. Readers expect something *new* and *significant*. Writing has informative value[1] when it (1) conveys knowledge that will be new *and* worthwhile to the intended audience; (2) reminds the audience about something they know but ignore; or (3) offers fresh insight about something familiar.

The informative value of any document is measured by its relevancy to the writer's purpose and the audience's needs. As a member of this book's audience, for instance, you expect to learn about technical writing, and our purpose is to help you do so. In this situation, which of these statements would you find useful?

1. Technical writing is hard work.
2. Technical writing is a process of making deliberate decisions in response to a specific situation. In this process, you discover important meanings in your topic and give your readers the information they need to understand your meanings.[2]

Statement 1 offers no news to anyone who has ever picked up a pencil, and so it has no informative value for you. But statement 2 offers a new perspective on something familiar. No matter how much you might have struggled through decisions about punctuation, organization, and grammar, chances are you haven't viewed writing as entailing the critical thinking discussed in this book. Because statement 2 provides new insight, you can say it has informative value.

The more non-essential information readers receive, the more they are likely to overlook or misinterpret the important material. Take the time to determine what your readers need, and try to give them just that.

APPROPRIATE DETAILS AND FORMAT

The amount of detail in your document *(How much is enough?)* will depend on what you have learned about your readers' needs. Were you asked to "keep it short" or to "be comprehensive"? Can you summarize, or does everything need spelling out? What length will they tolerate? Are the primary readers most interested in conclusions and recommendations, or do they want all of the details? Have they requested a letter, a memo, a short report, or a long, formal report with supplements (title page, table of contents, appendices, and so on)? What kinds of visuals (charts, graphs, drawings, photographs) make this material more accessible? What level of technicality will connect with primary readers?

---

1. Adapted from James L. Kinneavy's assertion that discourse ought to be unpredictable, in *A Theory of Discourse* (Englewood Cliffs: Prentice, 1971).
2. Our thanks to Robert M. Hogge, Weber State University, for this definition.

For example, which level of technicality in the following example means more to you? Which level would be more appropriate for an automotive sales brochure?

High Technicality　The diesel engine generates 10 BTUs per gallon of fuel, as opposed to the conventional gas engine's 8 BTUs.

Low Technicality　The diesel engine yields 25 percent better fuel mileage than its gas-burning counterpart.

DUE DATE

Does your document have a deadline? Workplace documents almost always do. Allow plenty of time to collect data, to write, and to revise. Whenever possible, ask primary readers to review an early draft and to suggest improvements.

EXERCISES

1. Locate a short article from your field (or part of a long article or a selection from your textbook for an advanced course). Choose a piece written at the highest level of technicality you understand and then translate the piece for a layperson, as in the example on page 30. Exchange translations with a colleague from a different major. Read your colleague's translation and write a paragraph evaluating its level of technicality. Submit to your instructor a copy of the original, your translated version, and your evaluation of your colleague's translation.

2. Assume that a new employee is taking over your job because you have been promoted. Identify a specific problem in the job that could cause difficulty for the new employee. Write for the employee instructions for avoiding or dealing with the problem. Before writing, create an audience/purpose profile by answering (on paper) the questions on page 34. Then brainstorm for details. Submit to your instructor your audience/purpose analysis, brainstorming list, and instructions.

COLLABORATIVE PROJECT

Form teams according to major (electrical engineering, biology, etc.) and respond to the following situation.

Assume your team has received the following assignment from your major department's chairperson: An increasing number of first-year students are dropping out of the major because of low grades, stress, or inability to keep up with the work. Your task is to prepare a "Survival Guide," for distribution to incoming students. This one- or two-page memo should focus on the challenges and the pitfalls and should include a brief motivational section, along with whatever else team members think readers need.

ANALYZE YOUR AUDIENCE

Use Figure 2.2 as a guide for developing your audience/purpose profile.

Writing Efficiently

At work, writers need to produce effective documents quickly. Most employers will not tolerate inefficient work habits, including writing habits. Here are three scenarios that illustrate that it's just as important to write efficiently as it is to create effective documents.

> *Bill,* a Halifax civil engineering technologist, returns to his office from a site inspection. As Bill sits down at his desk, his office manager tells him that he must write a proposal that afternoon for a soil-testing contract. Bill checks his watch; he has two hours to gather relevant data from company files and produce a two-page proposal before he has to catch a plane for a company meeting in Toronto.
>
> Bill spends 20 minutes gathering, choosing, and arranging the data and supporting arguments. Then, working from a standard proposal structure for soil-testing contracts, he composes a 550-word proposal on his personal computer in 35 minutes. He spends another 15 minutes polishing the document and sends it to two colleagues for proofreading. They find four mechanical errors and two minor errors in logic. Bill corrects the errors and prints three copies of the proposal; he takes two copies to the office manager. In total, Bill produces the proposal in 90 minutes.
>
> *George* works for a national research organization in Saskatoon. He has a master's degree in chemistry and a doctorate in biology. His company bills his services for $85 an hour, a rate that his company's clients are glad to pay because George's research methods, data, and analyses are thorough and accurate. His reports feature clear structures and phrasing.
>
> However, George is required to take a three-day technical writing course because typically it takes him a full work week (plus his own time in the evenings) to produce a project completion report that other researchers could produce in half the time.
>
> *Rita* has a background in electronics and computers; she has a 10-month contract to write software documentation manuals for an Ottawa-based software development company. The company is very pleased with the quality of Rita's work, but its director of development, Martin Lefebvre, has told Rita that her contract will be renewed only if she can decrease the time it takes her to produce a manual. Martin suggests a minimum improvement of 25 percent in writing efficiency. ▲

Bill is an efficient writer. George and Rita are not, partly because both have perfectionist tendencies, but more because each has learned bad writing and time-wasting habits. Rita, for example, writes through a discovery method that requires several complete rewrites of each document. George's main problem is that he frequently reorganizes his reports as he composes drafts and thus wastes time re-writing whole sections in order to make the material read smoothly.

Why does Bill write more efficiently than Rita and George? Bill has learned to:

- ◆ identify several related but separate writing tasks
- ◆ focus on one task at a time and perform each task well
- ◆ identify the best sequence for completing the various writing tasks
- ◆ reduce writing time by starting quickly and by writing a first draft that requires relatively little revision

Efficient writers such as Bill have learned, primarily through trial and error, to use a writing process that is broken down into the following stages.

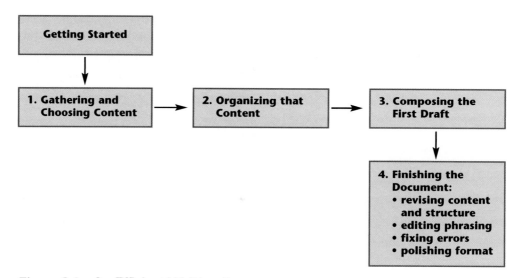

Figure 3.1   An Efficient Writing Process

GETTING STARTED

Figure 3.1 shows the first section, "Getting Started," as preliminary to the other four stages. Technical and business writers benefit from immediately asking: *What does my reader need and expect?* and *What purpose am I trying to fulfill?* (The audience/purpose profile described in Chapter 2 provides an excellent starting point for people who write business correspondence and technical documents.)

Technical writing differs from essay writing. Essayists often have to discover their subject as they progress. This process of writing until the writer discovers what she or he really wants to say is called *free-writing*; it may be necessary, but it results in many crumpled pages and in lengthy writing sessions.

Rewriting draft after draft is simply not necessary for most business and technical writing projects. Nor do you have to wait for that first "golden phrase" to start the river of words flowing. As a technical/business writer, you can be productive within 30 seconds of sitting down to write. Here's how:

- *Use an audience/purpose profile* to determine the types of information and analysis to include. List the types of questions your reader would ask (or which your reader has already asked).
- *Choose an appropriate, proven structure* and then "fill in the blanks." Several of this book's chapters suggest structures for frequently written documents; progress reports, proposals, feasibility reports, and application letters are among those described. Then, use elements of an audience/purpose profile to modify the suggested structure to meet your readers' needs and preferences.

The foregoing two methods will help you start writing almost any job-related document. However, occasionally you'll tackle a subject that is not clearly defined, or perhaps your readers' needs and priorities will be difficult to pin down. For those "open-ended" situations, here are two methods for getting your mind in gear:

1. *Brainstorm a list of ideas and topics.* A random listing of possible topics and ideas works precisely because it takes advantage of the natural chaos that exists in our minds. Often, we are most creative when we allow free thought association to generate a series of loosely related points and topics. It's important to simply record these points as they come, and not to edit them. Later, when the creative frenzy has abated, you can discard the points that don't seem relevant. Then, you can organize the material that remains.

   Now let's look at a list of topics and ideas that might be generated by the writer of an in-house product description of a new camera-style scanner. (The writer is part of a design team at GlobeTech that has recently developed the device in response to a request from GlobeTech's general manager to produce a new consumer product. GlobeTech is seeking to diversify.)

   The writer, Thanh Pham, has not previously written such a product description. Also, the product is radically different from the industrial controls manufactured by his company, so he doesn't have a model to emulate. Here's Thanh's list:

   - limitation of previous scanners—why inadequate for scanning images
   - solution:
     – appearance and size
     – weight
     – features and controls
   - how the new device works (just an overview): (1) the optical process and (2) storage of images—floppy disk; on-board storage
   - downloading images to a PC
   - performance specs: resolution; size of images that can be scanned; storage capacity

   Notice that Thanh's thoughts contain some order and "connections," even though he was just letting the ideas "flow." The first items in the list reflect a problem-solution sequence, and the last three items a chronological pattern.

2. *Brainstorm ideas in a cluster diagram.* Cluster diagramming suits people who think visually. It also suits those who are used to following hypertext links through the Internet. Here's how Thanh could use clustering to generate ideas for his product description. He could:

   - circle the main topic ("Cam Scan"), which he writes in the centre of a clean page
   - record any ideas that pop into his mind and circle those ideas
   - avoid censoring ideas, but simply record them
   - join related ideas with lines, but not make these organizational connections his main priority. Instead, he should keep on recording ideas until the flow stops

The resulting diagram might look like this:

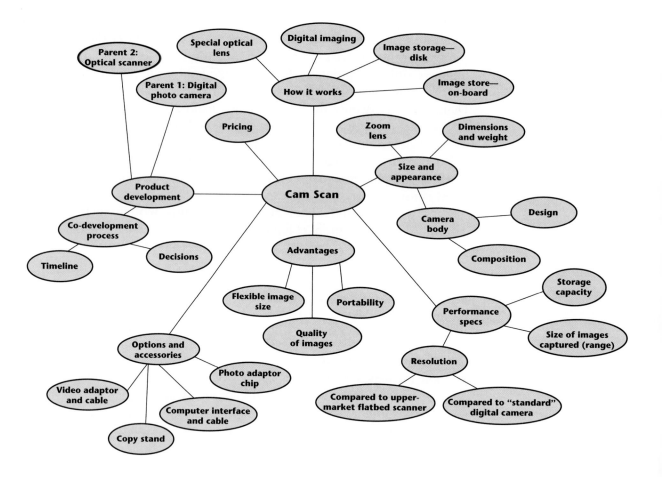

Figure 3.2   Sample Cluster Diagram

HOW TO SAVE WRITING TIME

Whether you have generated ideas by brainstorming or by one of the more structured approaches described earlier, you are now well underway and you can complete these four writing tasks in turn:

1. *Choose the content*, based on—

   ◆ technical or research notes
   ◆ personal observations
   ◆ arguments and evidence
   ◆ deductions and conclusions
   ◆ available illustrations

Then, check to see if anything has been omitted, or if any material should be deleted.

> **Revising content at this point takes less time than revising material later in the process.**

2. *Organize the blocks of material.* If you're writing a letter, decide what goes in each paragraph. If you're writing a longer document, start with larger blocks of material and work your way to the paragraph level. Next, organize the ideas and information *within* each block.

   Again, determine whether any material should be added, rearranged, or omitted; **such changes take much more time after a draft's been phrased**.

3. *Write the first draft.* Since you know what to include, and where to place it, you will be able to concentrate on the best way to phrase each sentence, and you will understand where to inject transitional statements. If you have done steps 1 and 2 properly, this draft will be close to a finished product. The very act of phrasing sentences can sometimes change your perception of your message, so you might have to modify your writing plan as you go.

   By working from an outline, you will save composition time. After all, you only have to think of *how* to phrase things; you've already chosen the content and arranged it. Your mind is free to concentrate on phrasing; you're not burdened by three writing tasks at once.

4. *Evaluate, revise, edit, proofread, and correct.* Wait as long as possible before polishing the writing. You may gain a new perspective on the best way to structure and express parts of the document. Also, you'll proofread more effectively if you distance yourself from the material.

   Use objective indexes such as the Fog Index (Chapter 11, pages 249–50) to help evaluate your writing and to make it more readable. Finally, proofread the material at least three times to locate errors. Ideally, you should use at least two other proofreaders in addition to yourself.

   Compared to a process of writing and rewriting (and more rewriting), this four-stage writing process will save you a great deal of revision and editing. **Investing a little time early in the process pays large dividends later in the process!**

   Writing efficiently in the manner just described also will help you produce more *effective* writing; concentrating on one task at a time allows you to better perform each of those tasks.

COMPOSING WITH A WORD PROCESSOR

If you type as fast as you handwrite, and if you're accustomed to reading a computer screen, consider using a word processor to plan and compose the first draft of a document. Composing with a computer has several advantages:

- *Brainstorming* lists of ideas suits word processing, especially if you type quickly.
- Choosing content is easy to do—you can add, delete, or move points with little effort.
- Today's sophisticated word-processing software makes it easy to arrange the *chosen content into an outline*. Most software includes an outliner function that helps you divide topics into main headings and levels of subheadings. The

computer tracks levels in the outline so that you can easily add, subtract, or rearrange parts of the outline. The outliner functions are especially useful for long, complicated reports.

◆ *Key phrases* can be placed within the outline to represent paragraphs. Later, when you compose each paragraph, each key phrase gets expanded into a topic sentence for its corresponding paragraph. Essentially, then, the computer eliminates the drudgery of retyping headings and key phrases. You simply fill in the paragraphs under each heading or key phrase. (Some writers type several key notes for each intended paragraph and then expand the notes into a series of closely linked sentences. They find that a premeditated list of points for each paragraph helps them write tight, clear paragraphs that require little subsequent revision.)

◆ Word processors help *identify and correct errors in spelling, punctuation, and grammar.* However, automated checkers have many limitations: a synonym

---

GUIDELINES

for Using
Word
Processors and
Printers

1. *Decide whether to create the draft by hand or computer.* Experiment to learn which works best for you.

2. *Beware of computer junk.* The ease of cranking out words on a computer can result in long, windy pieces that say nothing. Edit final drafts to eliminate anything that fails to advance your meaning.

3. *Never confuse style with substance.* With laser printers, desktop publishing, and graphics software, documents can be made highly attractive, but not even the most attractive design can redeem a document whose content is worthless or inaccessible.

4. *Save and print your work often.* One way to court disaster is to write without saving or printing often enough. One wrong keystroke might cause pages of writing to disappear forever—unless you have saved them beforehand. Try to save each paragraph and print each page as you complete it.

5. *Make a backup disk.* A single electrical surge or malfunction can wipe out an entire file, floppy disk, or hard disk!

6. *Consider the benefits of revising from hard copy.* Many writers find they edit more productively by scribbling on the page. The printed page provides the whole text, in front of you.

7. *Print final copies on the right type of paper for your printer.* Ink jet printers, for example, require different paper from laser printers to get the same high quality. Also, use colour only for special effects, in special circumstances.

8. *Decide how to design and transmit your document.* Should the document be primarily verbal (text), visual, or some combination? Should it travel by conventional mail, inter-office mail, fax, or e-mail? Would your reader(s) prefer the solid feel of paper or the lure of the computer screen? Research indicates that younger readers prefer flashy graphics, older readers prefer traditional text, and readers in general trust text more than visual images (Horton, "Mix Media" 781).

offered by an electronic thesaurus may not accurately convey your intended meaning; the spell checker cannot differentiate between incorrectly used words such as "they're," "their," and "there," or "it's" and "its." And although spell and grammar checkers help, they cannot evaluate those subtle choices of phrasing that can be very important—no automated checker will tell you whether "you can reach me at …" or "call me at …" is more appropriate in a given situation.

THE PROCESS IN ACTION

The following situation illustrates how a busy person, who must balance the technical and management components of his job, uses his time efficiently to write a proposal to his supervisor.

Process to Write a Proposal

The company is MMT Consulting, an engineering firm with its headquarters in Calgary and with branch offices in Sudbury, Winnipeg, Edmonton, Kelowna, and Prince George. MMT specializes in feasibility studies, design projects, and construction management for the mining and petroleum sectors.

Art Basran manages MMT's Kelowna branch. He is responsible for MMT's contract to help the Jackson Mining Company choose a method of hauling coal from Jackson Mining's projected new mine site in the mountains north of Grand Forks, British Columbia. As Art and his staff start to investigate the project, they learn that they need to use specialized accounting methods to evaluate four alternatives for hauling the coal to the railhead.

Art's team, at Jackson Mining's request, examines the proposed route and calculates grade resistances and energy requirements along the route. The team then researches capital costs for constructing road beds and a railroad. Capital costs are also calculated for purchasing diesel trucks, a diesel train, an electric train, and a fleet of electric trains. Finally, annual operating costs are computed for the four options.

After gathering all of the data, Art realizes that he does not have a uniform method of applying the three main evaluation criteria (capital costs, annual operating costs, and potential for expanding the delivery system) to all four transportation options. He consults Brendan Winters, a chartered accountant who works with mining companies. Brendan recommends an accounting vehicle called Equivalent Uniform Annual Costs (EUAC), which combines all the variables and thus allows a uniform comparison of the four haul methods.

Now, Art faces the task of convincing his regional manager, Brenda Backstrom, to accept the EUAC method. Backstrom, a conservative thinker, tends to resist new ways of doing things. ▲

Art gets started by:

- checking notes of his meeting with Brendan Winters
- jotting questions that Brenda Backstrom will have
- brainstorming a list of points to make in explaining EUAC

Time to get started: 10 minutes.

Art's next step is to choose **an approach.** He decides to write a direct proposal because he knows that his reader prefers direct communication, not a lengthy analysis that eventually leads to a recommendation. Art also knows that his reader prefers short messages, so Art chooses a memo that will not exceed two pages.

With the format decision made, Art turns to the **choice of content**. His word-processed list includes:

- situation more complex than usual—financial considerations
- *solution;* EUAC (Brendan's idea)
- process of preparing EUAC (see my notes)
- *note:* we already do 5 of the 6 steps
- advantages of EUAC
  (1) relatively easy—Brendan's software
  (2) qualitative *and* quantitative
- our qualifications: Peter, John, Brendan
- why our "normal" method won't work: too complex, not reliable, too diffuse
- request authorization

Before thinking about the order of his material, Art **edits the content** he's listed. He decides that discussing the inadequacies of the "normal" method could be too negative and that it would take the memo beyond two pages, so he deletes that part. He also thinks of a third EUAC advantage (using the EUAC method makes MMT look good), and he adds that point, so his list now includes:

- situation more complex than usual—financial considerations
- *solution;* EUAC (Brendan's idea)
- process of preparing EUAC (see my notes)
- *note:* we already do 5 of the 6 steps
- advantages of EUAC.
  (1) relatively easy—Brendan's software
  (2) qualitative *and* quantitative
  (3) using this innovative method gives us a competitive advantage
- our qualifications: Peter, John, Brendan
- request authorization

Time to choose and edit content: 8 minutes.

At this point, Art's writing is interrupted by a phone call, which leads to other conversations between Art and two technologists in the office. After 40 minutes, Art returns to organize the content of his proposal. As he looks at his edited contents list, Art realizes that he has jotted down his points in the order he has successfully used for proposals in the past:

1. reader connection and statement of problem
2. proposed solution
3. description of proposal
4. supporting arguments
5. request for authorization

However, Art also realizes that it would be better to place his team's qualifications *before* his supporting arguments, to establish that his Kelowna team can handle the proposed approach.

Using the cut-and-paste function on his word processor, Art places his list of points as follows:

- situation more complex than usual—financial considerations
- *solution:* EUAC (Brendan's idea)
- process of preparing EUAC (see my notes)
- *note:* we already do 5 of the 6 steps
- our qualifications: Peter, John, Brendan
- advantages of EUAC
  (1) relatively easy—Brendan's software
  (2) merges qualitative *and* quantitative methods
  (3) using this innovative method gives us a competitive advantage
- request authorization

Re: Recommending a Coal-haul Method to Jackson Mining

As you requested, I have begun to prepare a recommendation for Jackson Mining's consideration. Its situation, however, is considerably much more complex than other transportation issues we have analyzed in the past. In this case, we must consider the factor of fluctuating interest rates as well as varying amortization periods for carrying the capital debt. This problem is additionally compounded by not knowing how long Jackson Mining expects to operate the proposed Othello mine and haul coal from it.

Consulting with Brendan Winters, a Penticton C.A. who has a special interest in mining projects, we should use Equivalent Uniform Annual Costs (EUAC) as the main method of comparing Jackson Mining's four-alternative transportation methods. EUAC provides a comprehensive, as well as a clear, way of comparing the four alternatives.

The Process

Preparing a comprehensive set of EUAC requires a six-step process. You will note from the following list of steps that we would normally preform the first five steps in this kind of analysis, the sixth step, the actual EUAC calculations, uses data generated during the first five steps of the process. These, then, are the steps:

1. Gather field data about the proposed road and railroad routes. This step will require doing our own surveys as well as gathering data from existing maps and surveys.

Figure 3.3 Draft Document

2. Research specifications of the diesel trucks and train options that are suitable for the type of terrain found on the proposed routes. We will need physical and mechanical specifications as well as the capital costs for this equipment.

3. Determine the fuel consumption figures for each of the four options. This set of calculations will be quite extensive, particularly in this case, because of the length of the haul road and because of the grade variances on that road.

4. Combine the fuel costs with projected labour and maintenance costs to determine the annual operating costs for each alternative.

5. Determine the capital costs for the four alternatives.

6. Combine the annual operating costs and the capital costs with varying interest rates to produce an EUAC for each of the study periods of one to 20 years. Please see the attached sample table and sample figure which show the results for three of the 20 years that could be considered. These samples are based on roughly estimated data.

Qualifications

I believe our team is well equipped to handle this innovative method of assessing Jackson Mining's transportation needs. Peter Bondra, the engineer who will lead the research team, has degrees in both civil and mechanical engineering, and useful experience in railbed construction. His main assistant will be John Housley who graduated two years ago from the Civil Engineering Technology program here in Kelowna. John has 15 years' experience in road construction and knows the area north of Grand Forks because of his frequent hunting trips in that area. In addition, John is well versed in a variety of surveying techniques. He'll be able to deal with the area's rough terrain.

After the data has been collected, Brendan Winters has indicated his willingness to be available to direct the cost analysis. He has particular expertise in calculating and presenting EUAC data.

Advantages of Using the EUAC Comparisons

The main advantage of using EUAC comparisons is that it provides qualitative comparisons as well as quantitative data. In effect, an EUAC creatively merges enginerring analysis with the maximal kind of accounting techniques to help our client see what the data really means.

Providing this kind of inovative analysis will position our firm as a creative, advanced engineering firm. Brendan Winters and I have recently surveyed a variety of resource-based companies, and not one of them had used the EUAC method. I believe its use will give us a competitive advantage in this bid and in others.

Figure 3.3   Draft Document   *(continued)*

Furthermore, the complicated EUAC caclations will only take an extra day of work on this project because Brendan Winters has modified an accounting software program to help do the computations. His contribution will cost our firm an additional $1,200.00, a relatively small part of our budget for this project.

Conclusion

I need to talk to you about using Equivalent Uniform Annual Costs in the Jackson Mining project before the end of this week. If I'm out of the office when you call, call me at my new cellular phone number, (250) 863-2999. If you would like a more comprehensive view of how EUAC can be used, I can prepare such a document in three hours and fax it to you immediately.

Art Basran

Figure 3.3   Draft Document

Before using this outline to compose his memo, Art **reviews its structure.** He realizes that he should start his supporting arguments with his strongest point, so he rearranges the EUAC advantages as follows: (1) merges qualitative and quantitative methods; (2) gives us a competitive advantage; (3) Brendan's software—easy to use and inexpensive.
Time to organize content: 3 minutes.

Feeling in control of the writing process, Art **composes the first draft**, working quickly from his outline and detailed notes. He composes the 757-word document in 33 minutes. At that point, Art has to leave for a meeting, which takes the rest of the afternoon. This is Art's draft:

Art returns to the memo the next morning. As he reads through his draft, he finds a dangling modifier ("Consulting with Brendan Winters, . . . , we") and three spelling errors ("preform," "enginerring," "inovative"), even though he is primarily checking the readability and tone of the document. As he **edits the draft** of his memo, Art asks himself:

- Does the content explain my ideas clearly?
- Have I established the need for my proposed action?
- Is everything accurate, complete, and correct?
- Will my supporting arguments appeal to my reader?
- Is the style readable? (concise? direct? natural?)
- Are the sentences and paragraphs about the right length?
- Have I used too much jargon? Have I used the right descriptive words?

MMT Consulting   Inter-office MEMORANDUM

To: Brenda Backstrom   **Date:** December 3, 2004
            Western Regional Manager

From: Art Basran
            Kelowna Branch Manager

Re: Method of Comparing Transportation Alternatives for Jackson Mining's
            Proposed Othello Mine

As you requested, I'm preparing a proposal for Jackson Mining's consideration. Its situation, however, is more complex than other transport issues we have analyzed. In this case, we must consider fluctuating interest rates and varying amortization periods for carrying the capital debt. Also, we don't know how long Jackson Mining expects to operate the proposed Othello mine.

After consulting with Brendan Winters, a Penticton C.A. who has a special interest in mining projects, I propose that we use Equivalent Uniform Annual Costs (EUAC) as the main method of comparing Jackson Mining's four alternative transportation methods. EUAC provides a complete, clear comparison of the four options.

The Process

Preparing a comprehensive set of EUAC requires a six-step process, the first five of which we would normally perform in this kind of analysis; the sixth step, the actual EUAC calculations, uses data generated during the first five steps of the process. Here are the steps:

1. *Gather field data about the proposed road and railroad routes.* This step will require doing our own surveys as well as gathering data from existing maps and surveys.

2. *Research specifications of the diesel trucks and train options* that are suitable for the type of terrain found on the proposed routes. We will need physical and mechanical specifications as well as the capital costs of this equipment.

3. *Determine the fuel consumption figures* for each of the four options. This set of calculations will be quite extensive for the Othello Mine road, which is about 80 kilometres and which has many grade variances.

4. *Combine the fuel costs* with projected labour and maintenance costs to *determine the annual operating costs for each alternative.*

5. *Determine the capital costs* for the four alternatives.

6. *Combine the annual operating costs and the capital costs with varying interest rates* to produce an EUAC for each of the study periods of one to 20 years. Please see the attached sample table and figure that show the results for three of the 20 years which could be considered. These samples are based on estimated data.

Qualifications

Our team is well equipped to handle this innovative method of assessing Jackson Mining's transportation needs. Peter Bondra, the engineer who will lead the research team, has degrees in both civil and mechanical engineering and useful experience in railbed construction. His main assistant will be John Housley who graduated two years ago from the Civil Engineering Technology program here in Kelowna.

...2

Figure 3.4   An Efficiently Produced Document

Brenda Backstrom
December 3, 2004
Page 2

John has 15 years' experience in road construction and knows the area north of Grand Forks because of his frequent hunting trips in that area. He'll be able to deal with the area's rough terrain. In addition, John is well versed in a variety of surveying techniques.

After the data has been collected, Brendan Winters will be available to direct the cost analysis. He's an expert in calculating and presenting EUAC data.

Advantages of Using the EUAC Comparisons

The main advantage of using EUAC comparisons is that they provide qualitative comparisons as well as quantitative data. EUAC combines engineering and accounting ideas to help our client see what the data really means.

Providing this kind of innovative analysis will position MMT as a creative, advanced engineering firm. Brendan Winters and I have recently surveyed 12 mining and forestry companies, not one of which has used the EUAC method. I believe its use will give us a competitive advantage in this and future bids.

Furthermore, the complicated EUAC calculations will take only an extra day of work on this project because Brendan Winters has modified some accounting software to do the computations. His work will cost our firm an additional $1, 200, a small part of our project budget.

Conclusion

May we discuss using EUAC for Jackson Mining by this Friday? If I'm out of the office when you call, you can reach me at my new cellular phone, (250) 863-2999. If you would like a fuller explanation of how EUAC can be used, I could fax one to you.

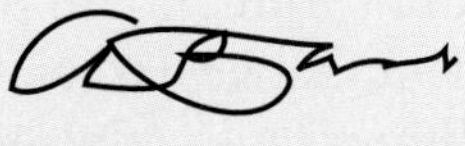

Art Basran

Attachments: Sample Calculations

Figure 3.4   An Efficiently Produced Document

His evaluation leaves him satisfied with the content of the proposal, but he sees wordy phrases and some phrases whose tone needs changing. Some of these are shown in the following table:

Paragraph	Phrasing	Concern/problem	Improvements
1	Its situation, however, is *considerably much* more complex than other transportation issues we have analyzed in the past. This problem is *additionally compounded by not knowing* . . . operate the proposed Othello mine *and haul coal from it.*	wordy, pompous

wordy, pompous | Its situation, however, is more complex than other transport issues we have analyzed.

Also, we don't know . . . operate the proposed Othello mine. |
2	EUAC provides a comprehensive, *as well as* a clear, way of comparing the four alternatives.	wordy	EUAC provides a complete, clear comparison of the four options.
5	. . . Brendan Winters has indicated his willingness to be available . . .	wordy	. . . Brendan Winters will be available . . .
7	. . . surveyed a variety of resource-based companies . . .	inexact and therefore not persuasive	. . . surveyed 12 mining and forestry companies . . .
9	I need to talk to you about using Equivalent Uniform Annual Costs . . . before the end of this week.	the tone is too aggressive for addressing one's supervisor	May we discuss using EUAC . . . by this Friday?

Art's paring of unnecessary words reduces the word count from 757 to 663 and makes several sentences easier to read. Then, also confident that he has established the right tone in the memo, Art puts the document through his software's spell checker, which detects one error ("calclations"). But, not trusting either his own proofreading or the spell checker, Art gives the document to engineering technologist Cathy Haldane, who finds a comma splice in paragraph 3 and a pronoun agreement error (". . . advantage of using EUAC comparisons is that *it* provides. . .") in paragraph 6.

Art's revision, editing, and proofreading time: 18 minutes.

Finally, Art takes four minutes to apply some final touch-ups and print the proposal.

Art's finished document is shown in Figure 3.4 on pages 50 and 51. His time to produce the memo, including time in discussion with Cathy Haldane, totals 84 minutes. Every minute of that time has been productive.

EXERCISES

1. Assume that you are a training manager for XYZ Corporation. After completing this first section of the text and the course, what advice would you have for a beginning writer who will frequently need to write reports on the job? In a one- or two-page, single-spaced memo to new employees, explain the writing process briefly, and give a list of guidelines these writers should follow.

2. Use the following form (Figure 3.5) to audit your writing habits as you produce your next few assignments. The time log should give you an idea of how efficiently you write, and where you could begin to save time in the writing process.

Stage of Writing Process	Time Spent	Methods and Results
1. Getting started		
2. Gathering/choosing content: information, ideas, and analysis		
3. Organizing content		
4. Phrasing first draft		
5. Revising content and structure Editing style and paragraphs Finding and correcting errors		

Figure 3.5   **Time Audit**

COLLABORATIVE PROJECT

Reviewing, Editing, and Revising a Document

After several weeks in a technical writing class, you have a good sense of what material is covered and of how the material is taught. Imagine that at a recent meeting your school's Advanced Writing Committee passed this motion:

> Because of the popularity of the technical writing course, and the 200 percent enrollment increase within two years, many new sections have been added. To ensure a unified program, we suggest that all sections follow a standard syllabus and similar teaching approaches.

For help in developing a standard model, the committee has decided to survey students about to complete the technical writing course. Each student has been asked to submit a memo evaluating the section, with suggestions for improvement. The responses will form a database for the committee's decisions about a course model that meets students' needs.

As a guide for evaluation, the committee has provided this question:

> How well do the content and teaching approach in your technical writing course fulfill your needs and expectations? How can this course best be taught, and what material should be covered? Be specific in your evaluation of strengths and weaknesses. Along with suggestions for improvement, explain how a specific change or improvement would benefit you.

Assume that Fran White, a student in some other section, has responded with this memo:

To: The Advanced Writing Committee
From: Fran White, Technical Writing Student
Subject: *Section 1499: Evaluation and*
         Recommendation

This course is providing me with a great deal of useful information. Also, the teacher usually manages to hold the attention of the class very well. Learning about writing can be a pretty boring experience, but this class hardly ever is bored because the teacher does such a good job of making the material interesting. Also, the teacher is a very nice person. I'm happy to say that I've learned a number of approaches that have helped me improve my writing.

Writing skills are important in just about anyone's career, so I'm glad we have the chance to take a technical writing course before graduation. Having the course as a requirement is a good idea, because many students (myself included) probably would avoid any course that requires this much writing. It isn't until we get there that we realize how worthwhile (and difficult!) this course is for everyone. I'd like to see every section taught the way this one is: by a teacher who knows how to get a tough job done.

The only complaint I have against this teacher is the fact that he talks too much about computers. I know that computers are important, but I'd like to learn to use one for my writing instead of just being exposed to computers in general. And too many writing assignments have been saved for the latter part of the course. Everything is just stacking up, leaving me buried and confused.

Also, the class is much too large, and the layout is awful. With so many students, the teacher has no way of providing individual attention, and so too many students get lost in the crowd. We need a classroom layout that would make group editing easier.

I really enjoyed the audiovisual presentations in class and would like to see more of them. All in all, the concrete stuff was always the most useful.

In general, the material has been covered well, except for the material on oral communication.

With a few minor changes, this course would be excellent.

Before Fran can submit this memo, it will need heavy revision for content that is informative, persuasive, and ethical; organization that is easy to follow; and style that is readable. Working in teams, review, edit, and revise Fran's memo.

a. As a first step, complete an Audience/Purpose Profile Sheet based on the following data, as well as on the details given earlier.

Primary audience: The writing committee and the department chair. Several committee members also are on the tenure committee, and they are likely to use this information for an additional purpose: to evaluate Fran's instructor for tenure. These are the decision makers. Above all, Fran wants to convey a positive impression of her instructor.

Secondary audience: Fran's instructor (whom Fran likes, but who deserves an honest and detailed evaluation). Fran wants to be fair to her instructor, but also wants to make some realistic suggestions for improving a course so important to everyone's career.

b. Assume you helped Fran prepare the brainstorming list she should have prepared *before* writing the previous draft. To visualize what it was that specifically pleased or displeased Fran, imagine that these items appeared on her list:

- ◆ instructor is always willing to help students individually
- ◆ instructor always takes time in class to answer all questions thoroughly
- ◆ emphasis on planning, drafting, and revising for a specific audience is helpful
- ◆ instructor spends a lot of time encouraging us
- ◆ instructor spends too much time talking about computers and automated offices—what I need is to develop strong writing skills by *using* the computer
- ◆ now and then we should have a guest lecturer from business and industry

- ◆ we should spend more time discussing the documents *we* have prepared
- ◆ instructor spends too much time emphasizing mistakes—not enough time on the positive
- ◆ instructor should spend more time on writing, and less than the present four weeks on oral communication
- ◆ when they are used (which is not often enough), the overhead, slide, and data projectors make things more vivid and interesting
- ◆ instructor gives a lot of feedback on our papers
- ◆ the class should have the same type of computers that are in our campus micro labs so that we have hands-on knowledge of how automation can affect the writing and revising process
- ◆ we should have a class with several round tables that seat 4–6 students for editing groups
- ◆ class size should be reduced from 30 to no more than 20 students
- ◆ tutoring should be available on a regularly scheduled basis for students with any type of writing problem
- ◆ we should spend at least one full week on job applications
- ◆ too few editing assignments early in the course; thus, too many at the end
- ◆ instructor should begin discussing the long report (term project) as early as the first or second week

From Fran's original draft and from the brainstorming list, select only that which is useful and appropriate for *this* audience and purpose. Compose a final draft of Fran's memo. Appoint one team member to present the document in class.

CHAPTER 4

Writing Persuasively

ASSESS THE POLITICAL REALITIES

EXPECT READER RESISTANCE

KNOW HOW TO CONNECT WITH READERS

ASK FOR A SPECIFIC DECISION

NEVER ASK FOR TOO MUCH

RECOGNIZE ALL CONSTRAINTS

SUPPORT YOUR CLAIMS CONVINCINGLY

CONSIDER THE CULTURAL CONTEXT

In Brief Questions for Analyzing Cross-cultural
Audiences

Guidelines for Persuasion

Checklist for Cross-cultural Documents

Chapter 1 explained how writers face the *information challenge: How can I make readers understand exactly what I mean?* But writers face a persuasion challenge as well, outlined in Figure 4.1. Persuasion means trying to influence people's thinking or win their cooperation. In the workplace, persuasive efforts often are aimed at building group consensus. The size of your persuasion challenge depends on who your readers are, how you want them to respond, and how firmly individuals are committed to their position.

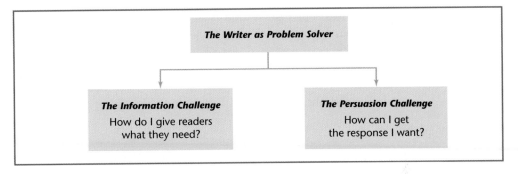

Figure 4.1 Two Challenges Confronted by Writers

You face a persuasion challenge whenever you express a viewpoint readers might dispute. Viewpoints ordinarily are expressed in a thesis or a *claim* (a statement of the point you are trying to prove). For instance, you might want readers to *recognize* facts they've ignored:

A claim about what the facts are

> The O-rings in the space shuttle's booster rockets have a serious defect that could have disastrous consequences.

Or you might want to influence how they *evaluate* the facts:

A claim about what the facts mean

> Taking time to redesign and test the O-rings is better than taking unacceptable risks to keep the shuttle program on schedule.

Or you might want readers to *take immediate action:*

A claim about what should be done

> We should call for a delay of tomorrow's shuttle launch because the risks simply are too great.

Whenever an audience disagrees about what things mean or what is better or worse or what should be done, you face a persuasion challenge.

Your own letters, memos, and reports will be asking readers to accept and act on claims like these:[1]

- ◆ We cannot meet this production deadline without sacrificing quality.
- ◆ We're doing all we can to correct your software problem.
- ◆ This hiring policy is discriminatory.
- ◆ Our software is superior to the competing brand.
- ◆ We all need to work harder.
- ◆ I deserve a raise.

---

1. This list of claims was inspired by Gilsdorf, "Executives' and Academics' Perception."

Your goal might be to convert readers to a different way of thinking, to reinforce one particular way of thinking, or to create a new way of thinking. In any event, you need to make the best case for seeing things *your* way.

ASSESS THE POLITICAL REALITIES

Besides their varied backgrounds, different readers have different attitudes. On one level, your readers are consumers of information; on another they are human beings, who react on the basis of their personality and feelings, and who create their own meanings for what they have just read (Littlejohn and Jabusch 5).

Any document can evoke different reactions—depending on a reader's temperament, preferences, interests, fears, biases, preconceptions, misconceptions, ambitions, or general attitude. Whenever readers feel their views are being challenged, they respond with questions like these:

TYPICAL READER QUESTIONS ABOUT A DOCUMENT THAT ATTEMPTS TO PERSUADE

- Says who?
- So what?
- Why should I?
- Why rock the boat?
- What's in it for me?

- What are you up to?
- What's in it for you?
- What does this *really* mean?
- Will it mean more work for me?
- Will it make me look bad?

Some readers might be impressed and pleased by your suggestions for increasing productivity; some might feel offended or threatened; others might think you are trying to make yourself look good or make them look bad. People can read much more between the lines than what actually is on the page. Such are the political realities of writing in any organization.

If you have worked with others, you already know something about office politics: how some people seek favour, influence, status, or power; how some resent, envy, or intimidate others; how some are easily threatened. Writing consultant Robert Hays sums up the writer's political situation this way:[2] "A writer must labour under political pressures from boss, peers, and subordinates. Any conclusion affecting other people can arouse resistance" (19). Some readers might resist your suggestion for shortening lunch breaks, cutting expenses, or automating the assembly line. Or your document might be seen as an attempt to undermine your supervisor.

No-one wants bad news; some people prefer to ignore it (as the Walkerton water crisis demonstrated). If you know something is wrong, that a project or product is unsafe, inefficient, or worthless, you have to decide whether "to try to change company plans; to keep silent; to 'blow the whistle'; or to quit" (Hays 19).

---

2. Hays' "Political Realities" has excellent suggestions for analyzing and addressing political realities faced by writers.

Does your organization encourage or discourage outspokenness and constructive criticism? Find out—preferably before you accept the job. Ignoring political realities, you might write something that violates expectations and hurts your career.

EXPECT READER RESISTANCE

People who haven't made up their minds about what to do or think are more likely to be receptive to persuasive influence:

We rely on persuasion to help us make up our minds

> *We are all consumers as well as providers of persuasion. Daily, we open OURSELVES to the persuasion of others. We need others' arguments and evidence. We're busy. We can't and don't want to discover and reason out everything for ourselves. We look for help, for short cuts, in making up our minds.* (Gilsdorf, "Write Me" 12)

In a world overwhelmed by information, persuasion can help people "process" the information and decide on its meaning.

People who already have decided what to do or think, however, don't like to change their minds without good reason. Sometimes, even for the best reasons, people refuse to budge. Whenever you question people's stand on an issue or try to change their behaviour, expect resistance. One researcher explains why persuasion is so difficult:

Once our minds are made up, we tend to hold stubbornly to our views

> *By its nature, informing "works" more often than persuading does. While most people do not mind taking in some new facts, many people do resist efforts to change their opinions, attitudes, or behaviours.* (Gilsdorf, "Executives' and Academics' Perception" 61)

The bigger the readers' stake in the issue, the more personal their involvement will be, and the more resistance you can expect. Research indicates that inducing permanent change in behaviour is especially difficult because people tend to revert to the familiar patterns and activities that are part of their lifestyle or work habits (Perloff 321).

Some ways of yielding to persuasion are better than others

When people do yield to persuasion, they yield either grudgingly, willingly, or enthusiastically (as in Figure 4.2). Researchers categorize these responses as *compliance, identification,* or *internalization* (Kelman 51–60):

- ◆ *Compliance:* "I'm yielding to your demand in order to get a reward or to avoid punishment. I really don't accept it, but I feel pressured and so I'll go along to get along."
- ◆ *Identification:* "I'm yielding to your appeal because I like and believe you, I want you to like me, and I feel we have something in common."
- ◆ *Internalization:* "I'm yielding because what you're saying makes good sense and it fits my goals and values."

Although compliance is sometimes a necessary response (as in military orders or workplace safety regulations), nobody likes to be coerced. Effective persuasion relies on identification or internalization. If readers merely comply because they feel they have no choice, then you probably have lost their loyalty and goodwill—and as soon as the threat or reward disappears, you will lose their compliance as well.

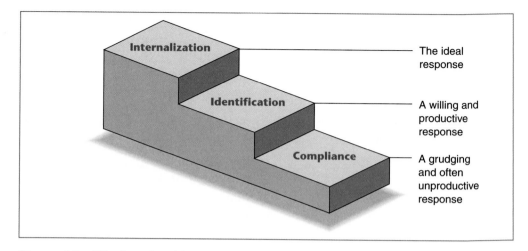

Figure 4.2   The Levels of Response to Persuasion

KNOW HOW TO CONNECT WITH READERS

Persuasive people know when to merely declare what they want, when to reach out and create a relationship, when to appeal to reason—or when to employ some combination of these strategies. These three strategies for connecting have been categorized as *hard, soft,* and *rational* (Kipnis and Schmidt 40–46). Let's call them the *power connection,* the *relationship connection,* and the *rational connection,* as shown in Figure 4.3.

For an illustration of these different connections, picture the following situation:

> Your Company, XYZ Engineering, has just developed a fitness program, based on findings that healthy employees work better, take fewer sick days, and cost less to insure. This program offers clinics for smoking, stress reduction, and weight loss, along with group exercise. In your second month on the job you receive this notice through e-mail:

POWER CONNECTION

To: All Employees   1/20/04

From: G. Maximus, Human Resources Director

Subject: Physical Fitness

Orders readers to show up

On Monday, June 10, all employees will report to the company gymnasium at 8:00 a.m. for the purpose of choosing a walking or jogging group. Each group will meet 30 minutes three times weekly during lunch time.

How would you react to this memo—and to the person who wrote it? Here the writer seeks nothing more than compliance. Although the writer speaks of "choosing," you are given no real choice but simply ordered to show up. This kind of *power connection* is typically used by supervisors and others in power. Although it

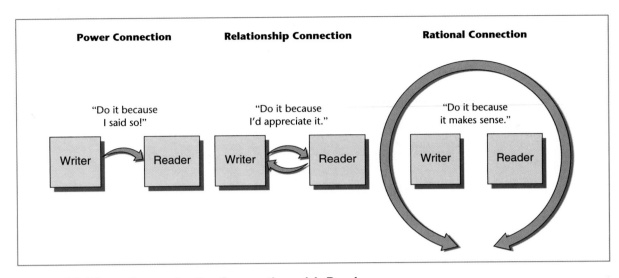

Figure 4.3 Three Strategies for Connecting with Readers

may or may not achieve its goal, the power connection almost surely will alienate its readers.

Now assume instead that you received the following version of our memo. How would you react to this message and its writer?

RELATIONSHIP CONNECTION

To: All Employees   1/20/04

From: G. Maximus, Human Resources Director

Subject: An Invitation to Physical Fitness

Invites readers to participate

I realize most of you spend lunch hour playing cards, reading, or just enjoying a bit of well-earned relaxation in the middle of a hectic day. But I'd like to invite you to join our lunchtime walking/jogging club.

Leaves the choice to readers

We're starting this club in hopes that it will be a great way for us all to feel better. Why not give it a try?

This version evokes a sense of identification, of shared feelings and goals. Instead of being commanded, readers are invited—they are given a real choice. This *relationship connection* establishes goodwill.

Often, the biggest factor in persuasion is a readers' perception of the writer. Readers are more receptive to people they like, trust, and respect. Of course, you would be unethical in appealing to the relationship or in faking the relationship merely to hide the fact that you had no evidence to support your claim (R. Ross 28). Audiences need to find the claim believable ("Exercise will help me feel better") and relevant ("I personally need this kind of exercise"). The relationship connection, moreover, might strike some readers as too "chummy" to carry any real authority—and so the request might be ignored.

Here is a third version of our memo. As you read, think about the ways it makes a persuasive case.

RATIONAL CONNECTION

To: All Employees   1/20/04

From: G. Maximus, Human Resources Director

Subject: Invitation to Join One of Our Jogging or Walking Groups

Presents authoritative evidence

I want to share a recent study from the *New England Journal of Medicine,* which reports that adults who walk two miles a day could increase their life expectancy by three years.

Other research shows that 30 minutes of moderate aerobic exercise, at least three times weekly, has a significant and long-term effect in reducing stress, lowering blood pressure, and improving job performance.

Offers alternatives

As a first step in our exercise program, XYZ Engineering is offering a variety of daily jogging groups: The One-Milers, Three-Milers, and Five-Milers. All groups will meet at designated times on our brand new, quarter-mile, rubberized clay track.

For beginners or skeptics, we're offering daily two-mile walking groups. For the truly resistant, we offer the option of a Monday-Wednesday-Friday two-mile walk.

Offers a compromise

Coffee and lunch breaks can be rearranged to accommodate whichever group you select.

Leaves the choice to readers

Offers incentives

Why not take advantage of our hot new track? As small incentives, XYZ will reimburse anyone who signs up as much as $100 for running or walking shoes, and will even throw in an extra 15 minutes for lunch breaks. And with a consistent turnout of 90 percent or better, our company insurer may be able to eliminate everyone's $200 yearly deductible in medical costs.

Here the writer shows willingness to compromise ("If you do this, I'll do that"). This *rational connection* communicates respect for the reader's intelligence *and* for the relationship by presenting good reasons, a variety of alternatives, and attractive incentives—all framed as an invitation. Whenever an audience is willing to listen to reason, the rational connection stands the best chance of succeeding.

Keep in mind that each kind of connection (or some combination) can work in particular situations. But no cookbook formula exists, and in many situations, your persuasive attempts may fail.

ASK FOR A SPECIFIC DECISION

Unless you are giving an order, diplomacy is essential in persuasion. But don't be afraid to ask for the specific decision you want, preferably at the end of the message:

Let people know exactly what you want

> *Studies show that the moment of decision is made easier for people when we show them what the desired action is, rather than leaving it up to them. . . . Without this directive, people may misunderstand or lose interest in the entire message. No one likes to make decisions: there is always a risk involved. But if the writer asks for the action, and makes it look easy and urgent, the decision itself looks less risky and the entire persuasive effort has a better chance of succeeding.* (Cross 3)

Let readers know what you want them to do or think.

NEVER ASK FOR TOO MUCH

No amount of persuasion will move people to accept something they consider unreasonable. And the definition of *reasonable* depends on the individual. Employees at XYZ, for example, will differ as to which walking/jogging option they might accept.

To the jock writing the memo, a daily five-mile jog might seem perfectly reasonable, but some employees would think it outrageous. XYZ's program therefore has to offer something most of its audience (except, say, couch potatoes and those in poor health) accept as reasonable. Any request that exceeds its audience's "latitude of acceptance" (Sherif 39–59) is doomed (see Figure 4.4).

Identify your audience's latitude of acceptance

Options offered by XYZ

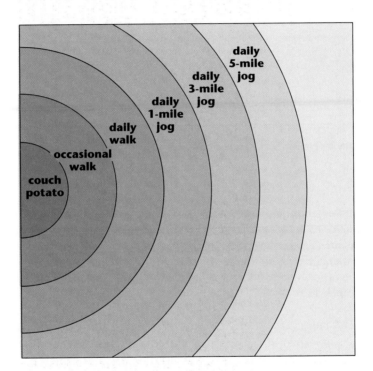

Figure 4.4 Options and Latitude of Acceptance

RECOGNIZE ALL CONSTRAINTS

Persuasive communicators observe certain limits or restrictions imposed by their situation. These are the *constraints,* and they govern what should or should not be said, who should say it and to whom, when and how it should be said, and through which medium (printed document, on-line, telephone, face to face, and so on). In the workplace you need to account for constraints like those in Figure 4.5.

ORGANIZATIONAL CONSTRAINTS

Organizations often have their own official constraints such as schedules, deadlines, budget limitations, writing style, the way a document is organized and formatted, and its chain and medium of distribution throughout the organization. But writers also face unofficial constraints:

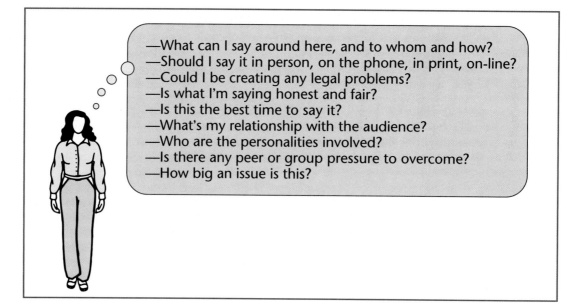

Figure 4.5 Every Writing Situation Poses Its Own Constraints

Decide carefully
when to say what
to whom

Most organizations have clear rules for interpreting and acting on (or responding to) statements made by colleagues. Even if the rules are unstated, we know who can initiate interaction, who can be approached, who can propose a delay, what topics can or cannot be discussed, who can interrupt or be interrupted, who can order or be ordered, who can terminate interaction, and how long interaction should last. (Littlejohn and Jabusch 143)

The exact rules vary among organizations, depending on whether communication channels are open and flexible or closed and rigid, on whether employee participation in decision making is encouraged or discouraged.

Although the rules of the game mostly are unspoken, anyone who ignores them (for example, going over a supervisor's head with a complaint or suggestion) invites disaster.

Airing even the most legitimate gripe in the wrong way through the wrong medium to the wrong person can be fatal to your work relationships and your career. The following memo, for instance, is likely to be interpreted by the executive officer as petty and whining behaviour, and by the maintenance director as a public attack.

Wrong way to the
wrong person

Dear Chief Executive Officer:

Please ask the Maintenance Director to get his people to do their job for a change. I realize we're all short-staffed, but I've received 50 complaints this week about the filthy restrooms and overflowing wastebaskets in my department. If he wants us to empty our own wastebaskets, why doesn't he let us know?

cc: Maintenance Director

Instead, why not address the memo directly to the key person—or better yet, phone the person?

Dear Maintenance Director:

I wonder if we could meet to exchange some ideas about how our departments might be able to help one another during these staff shortages.

Can you identify the unspoken rules where you have worked? What happens when such rules are ignored?

LEGAL CONSTRAINTS

Sometimes what you can say is limited by contract or by laws protecting confidentiality or customers' rights, or laws affecting product liability. For example, in a collection letter for non-payment, you can threaten legal action but cannot threaten any kind of violence or to publicize the refusal to pay, or pretend to be a lawyer (Varner and Varner 31–40). If someone requests information on one of your employees, you can "respond only to specific requests that have been approved by the employee. Further, your comments should relate only to job performance which is documented" (Harcourt 64). When writing sales literature or manuals, you and your company are liable for faulty information that leads to injury or damage.

Know how the law applies to any document you prepare. Suppose, for instance, an employee drops dead while participating in the new jogging program you've marketed so persuasively. Could you and your company be liable? Perhaps you should require physical exams and stress tests (at company expense) for participants.

ETHICAL CONSTRAINTS

While legal constraints are defined by federal and provincial laws, ethical constraints are defined by good conscience, honesty, and fair play. For example, it may be perfectly legal to promote a new pesticide by emphasizing its effectiveness, while downplaying its carcinogenic effects; whether such action is *ethical*, however, is another issue entirely. To earn people's trust, you will find that "saying the right thing" involves more than legal considerations.

Persuasive skills carry tremendous potential for abuse. There is a difference between honestly presenting your best case and using deception to manipulate the reader. (Chapter 5 is devoted to various ethics problems in communication.)

TIME CONSTRAINTS

Persuasion often is a matter of good timing. Do you have a deadline? If not, should you delay your message, release it immediately, or what? Let's assume you're trying to "bring out the vote" among members of your professional society on some hotly debated issue; for example, whether to refuse work on any project related to biological warfare. You might want to wait until you have all of the information you need or until you've analyzed the situation and planned a strategy. But you don't want to delay so long that rumours, misinformation, or paranoia cause people to harden their position *and* their resistance to your appeals. If delay might place the situation beyond your control, you might have to speak out sooner than you would like.

SOCIAL AND PSYCHOLOGICAL CONSTRAINTS

Too often, what we say can be misunderstood or misinterpreted. Here are just a few of the human constraints routinely encountered by communicators.

◆ *Relationship between communicator and audience:* Are you writing to a superior, a subordinate, or an equal? (Try not to appear dictatorial to subordinates nor to shield superiors from bad news.) How well do you and your audience know each other? Can you joke around or should you be serious? Do you get along or have a history of conflict? Do you trust and like one another? What you say and how you say it—and how it is interpreted—will be influenced by the relationship.

◆ *Audience's personality:* Researchers claim that "some people are easier to persuade than others, regardless of the topic or situation" (Littlejohn 136). Any reader's ability to be persuaded might depend on such personality traits as confidence, optimism, self-esteem, willingness to be different, desire to conform, open- or closed-mindedness, or regard for power (Stonecipher 188–89). The less your audience is open to persuasion, the harder you have to work. If you sense that your audience is totally resistant, you may want to back off—or give up altogether.

◆ *Audience's sense of identity and affiliation as a group:* How close-knit is the group? Does it have a strong sense of identity (as, for example, union members, conservationists, or engineering majors)? Will group loyalty or pressure to conform prevent certain appeals from working? Address the group's collective concerns.

◆ *Perceived size and urgency of the problem or issue:* In the audience's view, how big is this issue or problem? Has it been understated or overstated? Big problems are more likely to cause people to exaggerate their fears, anxieties, loyalties, and resistance to change—or to desperately seek some quick and easy solution. Assess the problem realistically. You don't want to downplay it, but you don't want to cause panic, either.

Writers who can assess a situation's constraints avoid serious blunders and can develop their message for greatest effectiveness.

SUPPORT YOUR CLAIMS CONVINCINGLY

The persuasive argument is the one that makes the best case in the audience's view. The strength of your case depends on the *reasons* you offer to support your claims.

> *When we seek a project extension, argue for a raise, interview for a job, justify our actions, advise a friend, speak out on issues of the day . . . we are involved in acts that require good reasons. Good reasons allow our audience and ourselves to find a shared basis for cooperating. . . . In speaking and writing, you can use marvellous language, tell great stories, provide exciting metaphors, speak in enthralling tones, and even use your reputation to advantage, but what it comes down to is that you must speak to your audience with reasons they understand.* (Hauser 71)

Persuasive claims are backed up by reasons that have meaning for the reader

To see how reasons support persuasive claims, imagine yourself in the following situation: As documentation manager for Bemis Software, a rapidly growing company, you supervise preparation and production of all user manuals. The present system for producing manuals is inefficient because three respective departments are involved in (1) assembling the required material, (2) word processing and designing, and (3) publishing the manuals. As a result, much time and energy are wasted as a manual goes back and forth among software specialists, communication specialists, and the art and printing department. After studying the problem and calling in a consultant, you decide that greater efficiency could be achieved if desktop publishing software were installed on all computers. This way, all employees involved could contribute to all three phases of the process. To sell this plan to supervisors and co-workers you will need good reasons, in the form of *evidence* and *appeals to readers' needs and values* (Rottenberg 104–06).

OFFER CONVINCING EVIDENCE

Evidence is any factual support from an outside source. Evidence is a powerful element in persuasion—as long as it measures up to readers' standards. Discerning readers evaluate evidence by using these criteria (Perloff 157–58):

- ◆ *The evidence has quality.* Instead of sheer quantity, readers expect evidence that is strong, specific, new, or different.
- ◆ *The sources are credible.* Readers want to know where the evidence comes from, how it was collected, and who collected it.
- ◆ *The evidence is considered reasonable.* It falls within a reader's "latitude of acceptance" (Sherif 39–59).

Common types of evidence include factual statements, statistics, examples, and expert testimony.

Factual Statements. A *fact* is something whose existence can be demonstrated by observation, experience, research, or measurement.

Offer the facts

Many of our competitors already have desktop publishing networks in place.

When your space and your reader's tolerance are limited—as they usually are—be selective. Decide which facts best support your case.

Statistics. Numbers can be highly convincing. Before considering other details of your argument, many workplace readers are interested in the "bottom line": costs, savings, profits (Goodall and Waagen 57).

Give the numbers

After a cost/benefit analysis, our accounting office estimates that an integrated desktop publishing network will save Bemis 30 percent in production costs and 25 percent in production time—savings that will enable the system to pay for itself within one year.

Numbers also can be highly misleading. Any statistics you present have to be accurate, trustworthy, and easy for readers to understand and verify. (See pages 141–144 for ways to avoid faulty statistical reasoning.) Always cite your source.

Examples. By showing specific instances of your point, examples help audiences *visualize* the idea or the concept. For example, the best way to explain what you mean by "inefficiency" in your company is to show one or more instances of it:

Show what you
mean

The following figure illustrates the inefficiency of Bemis's present system for producing manuals:

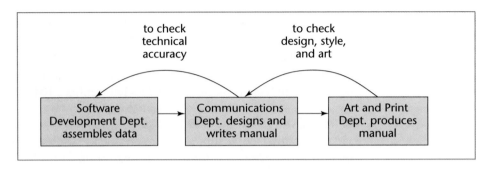

A manual typically goes back and forth through this cycle three or four times, wasting time and effort in all three departments.

Good examples have persuasive force; they give readers something solid, a way of understanding even the most surprising or unlikely claim. Use examples that the audience can identify with and that fit the point they are designed to illustrate.

Expert Testimony. Expert testimony lends authority and credibility to any claim.

Cite the experts

Ron Catabia, nationally recognized networking consultant, has studied our needs and strongly recommends we move ahead with the integrated network.

To be credible, however, an expert has to be unbiased and considered reliable by the audience.

Although solid evidence can be persuasive, evidence alone isn't always enough to influence the reader. At Bemis, for example, the bottom line might be very persuasive for company executives but could mean little to some managers and employees who will be asking: *Does this threaten my authority? Will I have to work harder? Will I fall behind? Is my job in danger?* These readers will have to perceive some benefit beyond company profit.

APPEAL TO COMMON GOALS AND VALUES

Audiences expect a writer to share their goals and values. If you hope to create any kind of consensus, you have to identify at least one goal you and your audience have in common: "What do we all want most?"

Bemis employees, like most people, share these goals: job security, a sense of belonging, control over their jobs and destinies, a growing and fulfilling career. Any persuasive recommendation will have to take these goals into account. For example:

Appeal to shared
goals

I'd like to show how desktop publishing skills, instead of threatening anyone's job, would only increase career mobility for all of us.

Our goals are shaped by our values (qualities we believe in, ideals we stand for): friendship, loyalty, ambition, honesty, self-discipline, equality, fairness, achievement, among others (Rokeach 57–58). Beyond appealing to common goals, you can appeal to shared values.

At Bemis, you might appeal to the commitment to quality and achievement shared by the company and by individual employees:

Appeal to shared goals values

> None of us needs reminding of the fierce competition in the software industry. The improved collaboration among networking departments will result in better manuals, keeping us on the front line of quality and achievement.

Give your audience reasons that have real meaning for *them* personally. For example, in a recent study of teenage attitudes about the negative consequences of smoking, respondents listed these reasons for not smoking: bad breath, difficulty concentrating, loss of friends, and trouble with adults. None of them listed dying of cancer—presumably because this last reason carries little meaning for young people personally (Baumann et al. 510–30).

CONSIDER THE CULTURAL CONTEXT

Reaction to persuasive appeals can be influenced by a culture's customs and values.[3] Cultures might differ in their willingness to debate, criticize, or express disagreement or emotion. They might differ in their definitions of "convincing support," or they might observe special formalities in communicating. Expressions of feelings and concern for one's family might be valued more than logic, fact, statistics, research findings, or expert testimony. Some cultures consider the *source* of a message as important as its content. Establishing trust and building a relationship might weigh more heavily than proof and might be an essential prelude to getting down to business.

Cultures might differ in their attitudes toward the environment, big business, technology, or competition. They might value delayed gratification more than immediate reward, stability more than progress, time more than profit, politeness more than candour, age more than youth.

One essential element of reader expectations in all cultures is the primacy of *face saving:*

Face saving is every reader's bottom line

> *Face saving [is] the act of preserving one's prestige or outward dignity. People of all cultures, to a greater or lesser degree, are concerned with face saving. Yet . . . [its] importance . . . varies significantly from culture to culture. . . . Indirectness in high face-saving cultures is viewed as consideration for another's sense of dignity; in low face-saving cultures, indirectness is seen as dishonesty.* (Victor 159–61)

Figure 4.6 illustrates how our guidelines are employed in an actual persuasive situation. This letter is from a company that distributes systems for generating electrical power from recycled steam (cogeneration). General manager Manson Harding writes a persuasive answer to a potential customer's question: "Why should I invest in a geothermal heat pump system you are prosing for my home?" As you read the letter, notice the kinds of evidence and appeals that support the opening claim. Notice also how the writer focuses on reasons important to the reader.

---

3. Adapted from Beamer 293–95; Gestelend 24; Hulbert,"Overcoming" 42; Jameson 9–11; Kohl et al. 65; Martin and Chaney 271–77; Nydell 61; Thrush 276–77; Victor 159–66.

In Brief QUESTIONS FOR ANALYZING CROSS-CULTURAL AUDIENCES[4]

What Is Accepted Behaviour?

- Formalities for making requests, expressing disagreement, criticism, or praise
- Preferred form for greetings or introductions (first or family names, titles)
- Willingness to criticize or request clarification
- Willingness to argue, debate, or express disagreement
- Willingness to be contradicted
- Willingness to express emotion (pleasure, gratitude, anger)
- Importance of trust and relationship building
- Importance of politeness and euphemism and leaving certain things unsaid
- Preference for casual or formal interaction
- Preference for directness and plain talk or for indirectness and ambiguity
- Preference for rapid decision making or for extensive analysis of a topic

How Do the Social and Legal Systems Work?

- Social/political system inflexible or open
- Class distinctions
- Democratic, egalitarian ideals or rank-conscious, authoritarian system
- Relative importance of the law versus interpersonal trust
- Formality of the contract process: a mere handshake or extensive legal documents
- Extent of lawyer involvement
- Extent to which the legal system enforces contractual agreements
- Role of gift giving (viewed as bribery or as a display of respect)

What Are the Values and Attitudes?

- Attitude toward the environment, big business, technology, competition, risk taking, status, youth versus age, rugged individualism versus group loyalty
- Preference for immediate reward or delayed gratification, progress or stability
- Importance of gender equality, interaction, and differences in the workplace
- Importance of the personal relationship in a business transaction
- Importance of time ("Time is money!" or "Never rush!")
- Importance of feelings versus logic and facts, results versus relationships
- Importance of candour versus saving face and sparing other people's feelings
- Extent of belief in fate, luck, or destiny
- View of our culture (admiration, contempt, envy, fear)

4. Adapted from Beamer 293–95; Gestelend 24; Hulbert, "Overcoming" 42; Jameson 9–11; Kohl et al. 65; Martin and Chaney 271–77; Nydell 61; Thrush 276–77; Victor 159–66. 5. This list was largely adapted from Caswell-Cowardo 265; Weymouth 144; Beamer 293–95; Martin and Chaney 271–77; Victor 159–61.

GUIDELINES

for Persuasion

Later chapters offer specific guidelines for various persuasive documents. But beyond attending to specific requirements of a particular document, remember this principle:

No matter how brilliant, any argument rejected by its audience is a failed argument.

If readers find cause to dislike you or conclude that your argument has no meaning for them personally, they usually reject *anything* you say. Connecting with an audience means being able to see things from their perspective. The following guidelines can help you make that connection.

1. *Assess the political climate.* Can you be outspoken? Who will be affected by your document? How will they react? How will your motives be interpreted? Will the document enhance your reputation or damage it? The better you assess readers' political feelings, the less likely your document will backfire. Do what you can to earn confidence and goodwill:

   ◆ Be diplomatic; try not to make anyone look bad or lose face.
   ◆ Be aware of your status in the organization; don't overstep.
   ◆ Don't expect anyone to be perfect—including yourself.
   ◆ Ask your intended readers to review early drafts.

   When reporting company negligence, dishonesty, stupidity, or incompetence, expect political fallout. Decide beforehand whether you want to keep your job or your dignity (more in Chapter 5).

2. *Learn the unspoken rules.* Know the constraints on what you can say, to whom you can say it, and how and when you can say it.

3. *Be clear about what you want.* Diplomacy is important, but people won't like having to guess about your purpose.

4. *Never make a claim or ask for something you know readers will reject outright.* Be sure readers can live with whatever you're requesting or proposing. Offer a genuine choice.

5. *Anticipate your audience's reaction.* Will they be defensive, surprised, annoyed, angry, or what? Try to address their biggest objections beforehand. Express your judgments ("We could do better") without making people defensive ("It's your fault").

GUIDELINES

for Persuasion

(continued)

6. *Decide on a connection (or combination of connections).* Does the situation call for you to merely declare your position, appeal to the relationship, or appeal to common sense and reason?

7. *Avoid an extreme persona.* Persona is the image or impression of the writer's personality suggested by a document's tone. Resist the urge to "sound off," no matter how strongly you feel, because audiences tune out aggressive people, no matter how sensible the argument. Try to be likable and reasonable. Admit the imperfections in your case—a little humility never hurts. Don't hesitate to offer praise when it's deserved.

8. *Find points of agreement with your audience.* Focusing early on a shared value, goal, or concern can reduce conflict and help win agreement on later points.

9. *Never distort the opponent's position.* A sure way to alienate people is to cast the opponent as more of a villain or simpleton than the facts warrant.

10. *Try to concede something to the opponent.* Surely the opposing case is based on at least one good reason. Acknowledge the merits of that case before arguing for your own. Instead of seeming like a know-it-all, show some empathy and willingness to compromise.

11. *Use only your best material.* Not all your reasons or appeals will have equal strength or significance. Decide which material—from your *audience's* view—best advances your case.

12. *Make no claim or assertion unless you can support it with good reasons.* "Just because" does not constitute adequate support!

13. *Use your skills responsibly.* Persuasive skills are easily abused. People who feel they have been bullied, manipulated, or deceived most likely will become your enemies. Know when to back off.

14. *Seek a second opinion of your document before you release it.* Ask someone you trust and who has no stake in the issue at hand.

15. *Decide on the appropriate medium.* Given the specific issue and audience, should you communicate in person, in print, by phone, e-mail, fax, newsletter, bulletin board, or what? Should all recipients receive your message via the same medium?

Harding Heartland HVAC

Bay 17, 1699 Larose Avenue St. Boniface, Manitoba
R4C 2Y1 (204) 948-6662

May 12, 2004

Mr. Pascual di Santos
245 Carleton Way
Winnipeg, Manitoba R3T 6L7

Dear Mr. di Santos:

The writer states his claim

Bill Neville has asked me to suggest a heating/cooling system for the new house he has contracted to build for you. I understand you had planned to install a conventional natural gas furnace and standard air conditioner, but I recommend a geothermal heat pump system as the best long-term system for your needs.

Provides background, in case the reader has little previous knowledge

Geothermal heat pumps use electricity to move heat; thus, the same pumps would take heat from the earth in winter and dissipate heat from the house to the earth in summer. Enclosure A describes a horizontal-loop geothermal system that would best suit your house design and your 6-acre location.

States major reason and gives supporting figures

A geothermal system would result in major savings for your proposed residence. We've calculated heat loss for your home design and we've projected the costs for both a conventional system and a geothermal system. Over the next 10 years, your 4950 square-foot residence would average $3,500 annually in heating costs and $2,450 in annual air conditioning costs. By contrast, a geothermal system would cost $1,500 annually. All calculations are listed in Enclosure B.

Acknowledges potential reader resistance, and refutes reason for reader concern

A geothermal system would cost $14,250 more to install than a conventional system. However, I understand that you have a contingency fund for "necessary modifications." More to the point, at an annual operating saving of $4,450, the geothermal system's additional cost would be regained in less than four years.

States a second "reason"—appeals to emotion

Geothermal systems also offer increased comfort and a healthy environment. My clients comment about the air feeling cleaner and they've observed that the temperature doesn't vary like it does with gas furnaces. Allergy sufferers tell me that they notice fewer symptoms in winter, now that they've switched to a geothermal system. If you wish, I can put you in contact with clients who've reported feeling better about the air they breathe with a geothermal system.

Offers additional proof

Further, geothermal systems contribute to a cleaner **external** environment, because fossil fuels aren't burned and because geothermal systems use less electricity than conventional air conditioners.

Figure 4.6 Supporting a Claim with Good Reasons

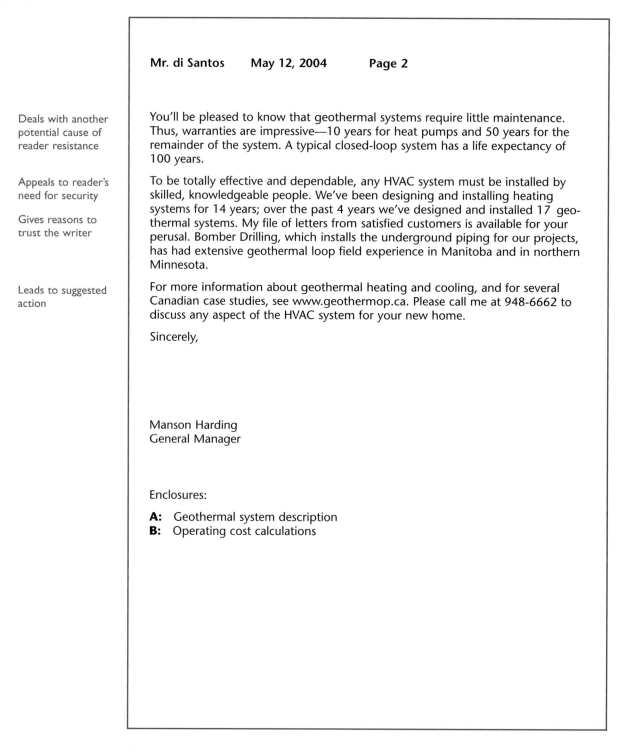

Deals with another
potential cause of
reader resistance

Appeals to reader's
need for security

Gives reasons to
trust the writer

Leads to suggested
action

Mr. di Santos   May 12, 2004   Page 2

You'll be pleased to know that geothermal systems require little maintenance. Thus, warranties are impressive—10 years for heat pumps and 50 years for the remainder of the system. A typical closed-loop system has a life expectancy of 100 years.

To be totally effective and dependable, any HVAC system must be installed by skilled, knowledgeable people. We've been designing and installing heating systems for 14 years; over the past 4 years we've designed and installed 17 geothermal systems. My file of letters from satisfied customers is available for your perusal. Bomber Drilling, which installs the underground piping for our projects, has had extensive geothermal loop field experience in Manitoba and in northern Minnesota.

For more information about geothermal heating and cooling, and for several Canadian case studies, see www.geothermop.ca. Please call me at 948-6662 to discuss any aspect of the HVAC system for your new home.

Sincerely,

Manson Harding
General Manager

Enclosures:

A: Geothermal system description
B: Operating cost calculations

Figure 4.6 **Supporting a Claim with Good Reasons**

Checklist FOR CROSS-CULTURAL DOCUMENTS

Use this checklist[5] to verify that your documents respect audience diversity.

- Does the document enable everyone to save face?
- Is the document sensitive to the culture's customs and values?
- Does the document conform to the safety and regulatory standards of the province and/or country?
- Does the document provide the expected level of detail?
- Does the document avoid possible misinterpretation?
- Is the document organized in a way that readers will consider appropriate?

- Does the document observe interpersonal conventions important to the culture (accepted forms of greeting or introduction, politeness requirements, first names, family names, titles, and so on)?
- Does the document's tone reflect the appropriate level of formality or casualness one would expect?
- Is the document's style appropriately direct or indirect?
- Is the document's format consistent with the culture's expectations?
- Does the document embody universal standards for ethical communication?

EXERCISES

1. Assume that you work for a technical marketing firm proud of its reputation for honesty and fair dealing. A handbook being prepared for new personnel includes a section titled "How to Avoid Abusing Your Persuasive Skills." All employees have been asked to contribute to this section by preparing a written response to the following:

   Share a personal experience in which you or a friend were the victim of persuasive abuse in a business transaction. In a one- or two-page memo, describe the situation and explain exactly how the intimidation, manipulation, or deception occurred. Write the memo and be prepared to discuss it in class.

2. Find an example of an effective persuasive letter. In a memo to your instructor, explain why and how the message succeeds. Base your evaluation on the persuasion guidelines listed on pages 71–72. Attach a copy of the letter to your evaluation memo. Be prepared to discuss your evaluation in class.

   Now, evaluate a poorly written document, explaining how and why it fails.

3. Think about some change you would like to see on your campus or at your part-time job. Perhaps you would like to make something happen, such as a campus-wide policy on plagiarism, changes in course offerings or requirements, more access to computers, a policy on sexist language, or a day-care centre. Or perhaps you would like to improve something, such as the grading system, campus lighting, the system for student evaluation of teachers, or the promotion system at work. Or perhaps you would like to stop something from happening, such as noise in the library or sexual harassment at work.

   Decide whom you want to persuade, and write a memo to that audience. Anticipate carefully your audience's implied questions, such as:

   - Do we really have a problem or need?
   - If so, should we care enough about it to do anything?
   - Can the problem be solved?
   - What are some possible solutions?
   - What benefits can we anticipate? What liabilities?

   Can you envision additional audience questions? Complete an audience/purpose profile (see page 34).

   Don't think of this memo as the final word, but as a consciousness-raising introduction that gets the reader to acknowledge that the issue deserves attention. At this early stage, highly specific recommendations would be premature and inappropriate.

4. Challenge an attitude or viewpoint that is widely held by your audience. Maybe you want to persuade your colleagues that the time required to earn a bachelor's degree should be extended to five years or that grade inflation is watering down your school's education. Maybe you want to claim that the campus police should (or should not) carry guns. Or maybe you want to ask students to support a 10 percent tuition increase in order to make more computers and software available.

Complete an audience/purpose profile (see page 34). Write specific answers to the following questions: What are the political realities? What kind of resistance could you anticipate? How would you connect with readers? What about their latitude of acceptance? Are there any other constraints? What reasons could you offer to support your claim?

In a memo to your instructor, submit your plan for presenting your case. Be prepared to discuss your plan in class.

COLLABORATIVE PROJECT

Assume that you work for an environmental consulting firm that is under contract with various countries for a range of projects, including these:

- ◆ a plan for rain forest regeneration in Latin America and Sub-Saharan Africa
- ◆ a plan to decrease industrial pollution in Eastern and Western Europe
- ◆ a plan for "clean" industries in developing countries
- ◆ a plan for organic agricultural development in Africa and India
- ◆ a joint Canadian/American plan to decrease acid rain
- ◆ a plan for developing alternative energy sources in Southeast Asia

Each project will require environmental impact statements, feasibility studies, grant proposals, and a legion of other documents. These are often prepared in collaboration with members of the host country, and in some cases prepared by your company for audiences in the host country: from political, social, and industrial leaders to technical experts and so on.

For such projects to succeed, people from different cultures have to communicate effectively and sensitively, creating goodwill and cooperation.

Before your company begins work in earnest with a particular country, your co-workers will need to develop a degree of cultural awareness. Your assignment is to select a country and to research that culture's behaviours, attitudes, values, and social system in terms of how these variables influence the culture's communication preferences and expectations. What should your colleagues know about this culture in order to communicate effectively and diplomatically? Do the necessary research using the questions from In Brief on page 70 as a guide.

CHAPTER 5

Writing Ethically

RECOGNIZE UNETHICAL COMMUNICATION

EXPECT SOCIAL PRESSURE TO PRODUCE UNETHICAL COMMUNICATION

NEVER CONFUSE TEAMWORK WITH GROUPTHINK

RELY ON CRITICAL THINKING FOR ETHICAL DECISIONS

ANTICIPATE SOME HARD CHOICES

NEVER DEPEND ONLY ON LEGAL GUIDELINES

UNDERSTAND THE POTENTIAL FOR COMMUNICATION ABUSE

KNOW YOUR COMMUNICATION GUIDELINES

DECIDE WHERE AND HOW TO DRAW THE LINE

Checklist **for Communicators**

Chapters 2 and 4 explained how audience analysis helps us tailor informative and persuasive communication (so we can complete the project, win the contract, or the like). But an *effective* message (one that achieves its purpose) isn't necessarily an *ethical* message. Think of examples from advertising: "Our artificial sweetener is composed of proteins that occur naturally in the human body (amino acids)" or "Our potato chips contain no cholesterol." Such claims are technically accurate but misleading: amino acids in certain sweeteners can alter body chemistry to cause headaches, seizures, and possibly brain tumours; potato chips contain saturated fat—which produces cholesterol. While the advertisers' facts may be accurate, they often are incomplete and imply misleading conclusions.

Whether the miscommunication occurs deliberately or through neglect, a message is unethical when it leaves readers at a disadvantage or prevents readers from making their best decision. Therefore, writers ultimately face the threefold challenge outlined in Figure 5.1. Ethical communication is measured by standards of honesty, fairness, and concern for everyone involved (Johannesen 1).

RECOGNIZE UNETHICAL COMMUNICATION

Thousands of people are injured or killed yearly in avoidable accidents—the result of faulty communication that prevented intelligent decision making. Following are descriptions of tragedies caused ultimately by unethical communication.

The Challenger Accident

Unethical communication has consequences

On January 28, 1986, the space shuttle *Challenger* exploded 43 seconds after launch, killing all seven crew members. The explosion was traced to two rubber O-ring seals in a booster rocket; these seals permitted hot exhaust gases to escape and ignite the adjacent fuel tank. The problem could have been prevented: although the O-ring hazard had been recognized since 1977 and documented by engineers, management had largely ignored the problem. (Managers had claimed that the O-ring system was safe because it was "redundant"—each primary O-ring was backed up by a secondary O-ring.)

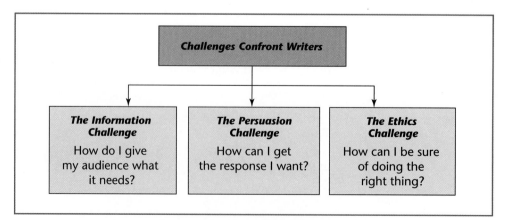

Figure 5.1   Three Challenges Confronted by Writers

In the hours preceding the launch, engineers argued against launching because that day's low temperature would drastically increase the danger of *both* the primary and secondary O-rings failing. But, under pressure to meet schedules and deadlines, managers chose to relay only a highly downplayed version of those warnings to NASA decision makers who were to make *Challenger's* fatal launch decision (Winsor; Pace; Rowland; and Gouron et al.).

For a chronology of the event, see the website maintained by the Online Ethics Center for Engineering and Science, at **onlineethics.org**. Click on "Cases," then "Detailed Cases of Exemplary Behavior–Moral Leaders," and finally, "Roger Boisjoly's Attempts to Avert the *Challenger* Disaster."

The Westray Coal Mining Accident

On May 9, 1992, 26 men were killed in an explosion at the Westray coal mining operation in Plymouth, Nova Scotia. The mine, employing the latest mining technology, had been operating for about eight months when the explosion occurred. In a subsequent report of the Westray Mine Public Inquiry, Justice K. Peter Richard said that the Westray disaster was a "complex mosaic of actions, omissions, mistakes, incompetence, apathy, cynicism, stupidity, and neglect."[1] As the evidence emerged during the enquiry, it became clear that many people and entities had not communicated appropriately, had not acted on clear directives, and had not applied or enforced coal mine and occupational health and safety regulations. While criminal negligence and manslaughter charges were withdrawn late in 1998, in 1999 the families of the victims of the disaster commenced a negligence lawsuit against the federal and Nova Scotia governments, former mine officials, and equipment manufacturers. ▲

These catastrophes make for dramatic headlines, but more routine examples of deliberate miscommunication rarely are publicized. Messages like the following succeed by saying whatever works.

◆ A person lands a great job by exaggerating his credentials, experience, or expertise.
◆ A marketing specialist for a chemical company negotiates a huge bulk sale of its powerful new pesticide by downplaying its carcinogenic hazards.
◆ To meet the production deadline on a new auto model, the test engineer suppresses, in her final report, data indicating that the fuel tank could explode upon impact.
◆ A manager writes a strong recommendation to get a friend promoted, while overlooking someone more deserving.

Can you recall some instances of deliberate miscommunication from your personal experience or that you learned about in the news? What were some of the effects of this miscommunication? How might the problems have been avoided?

To save face, escape blame, or get ahead, anyone might be tempted to say what people want to hear or to suppress bad news or make it seem "rosier." Some of these decisions are not simply black and white. Here is one engineer's description of the grey area in which issues of product safety and quality often are decided:

---

[1]The Westray Story: A Predictable Path to Disaster. Executive Summary (www.gov.ns.ca/labr/westray/summary.htm).

The company must be able to produce its products at a cost low enough to be competitive. . . . To design a product that is of the highest quality and consequently has a high and uncompetitive price may mean that the company will not be able to remain profitable, and be forced out of business. (Burghardt 92)

Do you emphasize to a customer the need for extra careful maintenance of a highly sensitive computer—and risk losing the sale? Or do you downplay maintenance requirements, focusing instead on the computer's positive features? Do you tell a white lie so as not to hurt a colleague's feelings, or do you "tell it like it is" because you're convinced that lying is wrong in any circumstance? The decisions we make in these situations often are influenced by the pressures we feel.

EXPECT SOCIAL PRESSURE TO PRODUCE UNETHICAL COMMUNICATION

Pressure to get the job done can cause normally honest people to break the rules. At some point in your career you might have to choose between doing what your employer wants ("just follow orders" or "look the other way") and doing what you know is right. Maybe you will be pressured to ignore a safety hazard in order to meet a project deadline:

Just as your automobile company is about to unveil its hot, new pickup truck, your safety engineering team discovers that the reserve gas tanks (installed beneath the truck but *outside* the frame) can explode on impact in a side collision. The company has spent a small fortune developing and producing this new model and doesn't want to hear about this problem.

Companies often face the contradictory goals of *production* (which means *making* money on the product) and *safety* (which means *spending* money to avoid accidents that may or may not happen). When productivity receives exclusive priority, safety concerns may suffer (Wickens 434–36). Thus it seems no surprise that well over 50 percent of managers studied nationwide feel "pressure to compromise personal ethics for company goals" (Golen et al. 75). These pressures come in varied forms (Lewis and Reinsch 31):

◆ the drive for profit
◆ the need to beat the competition (other organizations or co-workers)
◆ the need to succeed at any cost, as when superiors demand more productivity or savings without questioning the methods
◆ an appeal to loyalty—to the organization and to its way of doing things

Figure 5.2 depicts how such pressures can add up.
Here is a tragic example of how organizational pressure and/or personal agendas can help create a climate where important information is not reported or not heeded by those who receive the reports.

On May 17, 2000, residents of Walkerton, Ontario, began to display symptoms of bacterial poisoning—bloody diarrhea, vomiting, cramps, and fever. Eventually, more than 2300 of the town's more than 4000 citizens became seriously ill. Seven died. The tragic results were preventable, had various public officials fulfilled their obligations.

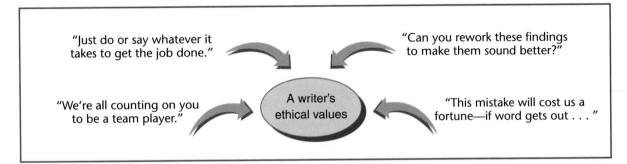

Figure 5.2 How Workplace Pressures Can Influence Ethical Values

The outbreak was primarily caused by *Escherichia coli 0157:H7* (*E. coli*) and by *Campylobacter jejuni,* traced to bacteria from cattle manure that was washed into No. 5 well, which supplied most of Walkerton's water supply during the time that supply became contaminated.

Sadly, many of Walkerton's residents might have escaped the illness if Stan Koebel, Walkerton's water manager, had notified the public or the Ministry of the Environment or the public health office that tests of water sampled May 15 had revealed *E. coli* contamination. Koebel sat on these results from May 18 to 23. Indeed, when health officials called on May 19 and 20 to enquire about Walkerton's water supply, he lied; he said that the water was safe to drink. So, the health officials looked for other possible causes of contamination.

On May 21, in response to more and more cases of *E. coli* poisoning, and with increasing likelihood of a water contamination, medical officer Dr. Murray Mc Quigge issued a boil water advisory. Dr. McQuigge's unit took samples on May 21 and on May 23, and learned on May 23 that Walkerton's water was indeed *E. coli* contaminated. When confronted with the health unit findings, Mr. Koebel admitted that he had known about the contamination since May 18 and also that another well's chlorination equipment had not been working for several days (Evenson A9).

Mr. Justice Dennis O'Connor was commissioned to head an enquiry into the disaster. His report concluded that "the outbreak would have been prevented by the use of continuous chlorine residual and turbidity monitors at Well 5. The failure to use continuous monitors at Well 5 resulted from shortcomings in the approvals and inspections programs of the Ministry of the Environment" (Summary 3). The Walkerton Inquiry, in May and June 2001, uncovered a number of possible contributing factors:

◆ From 1997 through 1999, a London, Ontario, lab found *E. coli* bacteria in Walkerton water on five separate occasions; the lab also noticed unusually low levels of chlorine disinfectant. Microbiologist Gary Palmateer faxed the results to Walkerton officials and the Ministry of the Environment, but no corrective actions were taken (Canadian Press, October 20, 2000).

◆ Despite knowing about water safety issues in Walkerton and elsewhere, the Ministry of the Environment did not "release [water safety] reports and test results documenting the severity of the situation" (Canadian Press, May 26, 2001).

◆ Dr. Richard Schabas was Ontario's chief medical officer of health in May 1997 when he tried to warn Premier Mike Harris, among others (at a meeting of the provincial cabinet's policy and priorities committee), that eliminating provincial funding for regional health boards would allow small-town politicians to bully or circumvent local medical health officers. Ignoring such warnings, the government downloaded public health boards to municipalities.

◆ Also, at Dr. Schabas's request, Health Minister Jim Wilson sent, on August 20, 1997, a letter to Norm Sterling, Minister of the Environment. In that letter, Mr. Wilson urgently recommended that the Ontario Water Resources Act be amended to require local waterworks to inform medical health officers of tests showing water contamination. In the 2001 enquiry, Mr. Sterling said he didn't recall receiving the letter; at the time, he was "distracted by other issues" (Mittelstaedt, June 28, 2001). But Mr. Sterling said he would accept some blame for the Walkerton tragedy if the enquiry establishes that his failure to react to the letter contributed to the *E. coli* outbreak.

◆ Previous to Norm Sterling's tenure at the Ministry of the Environment, Brenda Elliott was the minister responsible for chopping the provincial environmental-protection budget by nearly 50 percent from June 1995 to August 1996. Under Elliott's leadership, the ministry moved to privatize water testing services in two months, not in the three years recommended in reports submitted by ministry staffers. At the Walkerton Inquiry, Ms. Elliott said that she had been "acting as part of a team" and that she should not be held personally responsible for the Walkerton *E. coli* outbreak (Mittelstaedt, June 27, 2001).

◆ For years, Stan Koebel had not reported any test results that would reflect badly on his municipal water operation. And, along with his brother Frank (waterworks maintenance foreman), Mr. Koebel for years had "routinely falsified water samples, test results, and chlorination records" (Canadian Press, January 8, 2001). Indeed, in his annual reports, Stan Koebel had described a conscientious, competent staff and a safe secure water system (Blatchford A1).

◆ At the apex of the pyramid of responsible officials, Premier Mike Harris initially claimed at the Walkerton Inquiry that he didn't know of risks associated with eliminating government water testing labs. Later at the Inquiry, he admitted that he did know of such risks but considered them "manageable" and that therefore the public didn't need to be informed.

  Justice O'Connor's report concluded, however, that "there is no evidence that the specific risks, including the risks arising from the fact that the notification protocol was a guideline rather than a regulation, were properly assessed or addressed" (Summary 35). In effect, concluded O'Connor, the government's drive to reduce "red tape" led to the Walkerton tragedy because privatizing lab testing of water samples led directly to the events of May 2000, and reducing the Ministry of the Environment's approvals and inspections indirectly helped cause the outbreak by reducing the chance of reporting negative test results (Full Report 406–08).

◆ Dr. Murray McQuigge, who blew the whistle on Stan Koebel, faced intense criticism from Koebel's lawyer at the Walkerton Inquiry. Also, enquiry witnesses suggested that McQuigge's office knew about "problems with Walkerton's water system two years earlier but did nothing about it" (Blackwell A3) and therefore McQuigge was not above reproach himself. Such criticism illustrates one of the potential difficulties faced by whistle blowers. ▲

NEVER CONFUSE TEAMWORK WITH GROUPTHINK

Any successful organization relies on teamwork, everyone cooperating to get the job done. But teamwork is not the same as *groupthink* (Janis 9).

  Groupthink occurs when group pressure prevents individuals from questioning, criticizing, or "making waves." Group members feel a greater need for acceptance and a sense of belonging than for critically examining the issues. In a conformist climate, critical thinking is impossible. Anyone who has lived through adolescent peer pressure has already experienced a version of groupthink.

Yielding to pressure can be especially tempting in a large company or on a complex project, where individual responsibility is easy to hide in the crowd:

How some corporations evade responsibility for their actions

> *Lack of accountability is deeply embedded in the concept of the corporation. Shareholders' liability is limited to the amount of money they invest. Managers' liability is limited to what they choose to know about the operation of the company. The corporation's liability is limited by governments, . . . by insurance, and by laws allowing corporations to duck liability by altering their . . . structure.* (Mokhiber 16)

All kinds of people work at all levels on a major project. Countless decisions at any level have far-reaching effects on the whole project. But with so many people collaborating, identifying those responsible for an error often is impossible—especially when the error is one of omission, that is, of *not* doing something that should have been done (Unger 137).

After completing their assigned task, employees too often assume their job is done. Figure 5.3 depicts the kind of thinking that enables people to deny personal responsibility for the consequences of their communication.[2]

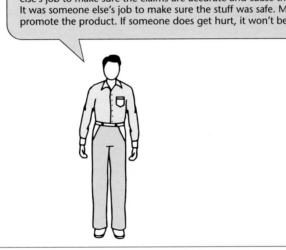

My job is to put together a persuasive ad for this brand of diet pill. It is someone else's job to make sure the claims are accurate and cause the customer no harm. It was someone else's job to make sure the stuff was safe. My job is only to promote the product. If someone does get hurt, it won't be my problem!

Figure 5.3   Groupthink Can Provide a Handy Hiding Place

RELY ON CRITICAL THINKING FOR ETHICAL DECISIONS

Because of their impact on people and on your career, ethical decisions challenge your critical thinking skills:

---

2. Our thanks to Judith Kaufman for this idea.

- ◆ How can I know the "right thing" in this situation?
- ◆ What are my obligations, and to whom, in this situation?
- ◆ What values or ideals do I want to stand for in this situation?
- ◆ What is likely to happen if I do X, or Y?

Can you rely on more than intuition or conscience in navigating the grey areas of ethical decisions? How will you make a convincing case against danger or folly to a roomful of people caught up in groupthink? Although ethical issues resist simple formulas, you can avoid two major fallacies that obstruct good judgment: the fallacy of "doing one's thing" and the fallacy of "one rule fits all."

THE FALLACY OF "DOING ONE'S THING"

One way to oversimplify a complex ethical issue is through a misguided notion known as *ethical relativism*: "Since we have no way to agree on right or wrong, it all depends on personal preference. *Right*, then, is what I think is right!" Such denial of any reasonable criteria of course legitimizes *any* action, including the atrocities of Hitler, Milosevic, Osama bin Laden, or other fanatics who claim the "right" intentions.

THE FALLACY OF "ONE RULE FITS ALL"

The opposite extreme of ethical relativism is *absolutism*, the inflexible notion that one set of criteria governs all ethical decisions. If, for example, you abide absolutely by the rule "Thou shall not lie," you would violate this rule under no circumstance (even to save your family from harm).

REASONABLE CRITERIA FOR ETHICAL JUDGMENT

Somewhere between the extremes of relativism and absolutism are *reasonable criteria* (standards of measurement that most people would consider acceptable). These criteria for ethical judgment take the form of *obligations, ideals,* and *consequences* (Ruggiero 55–56; Christians et al. 17–18).

Obligations are the responsibilities we have to everyone involved:

- ◆ *obligation to ourselves,* to act in our own self-interest and according to good conscience
- ◆ *obligation to clients and customers,* to stand by the people to whom we are bound by contract—and who pay the bills
- ◆ *obligation to our company,* to advance its goals, respect its policies, protect confidential information, and expose misconduct that would harm the organization
- ◆ *obligation to co-workers,* to promote their safety and well-being
- ◆ *obligation to the community,* to preserve the local economy, welfare, and quality of life
- ◆ *obligation to society,* to consider the national and global impact of our actions

When the interests of these parties conflict—as they often do—we have to decide very carefully where our primary obligations lie.

Ideals are "notions of excellence" (Ruggiero 55), the positive values that we believe in or stand for: loyalty, friendship, courage, compassion, dignity, fairness, and whatever qualities that make us who we are.

Consequences are the beneficial or harmful results of our actions. Consequences may be immediate or delayed, intentional or unintentional, obvious or subtle (Ruggiero 56). Some consequences are easy to predict; some aren't so easy; some are impossible.

Figure 5.4 depicts the relationship among these three criteria.

The above criteria help us understand why even good intentions can produce bad judgments, as in the following situation:

Someone observes . . . that waste from the local mill is seeping into the water table and polluting the water supply. This is a serious situation and requires a remedy. But before one can be found, extremists condemn the mill for lack of conscience and for exploiting the community. People get upset and clamour for the mill to be shut down and its management tried on criminal charges. The next thing you know, the plant does close, 500 workers are without jobs, and no solution has been found for the pollution problem. (Hauser 96)

What seems like the "right thing" might be the wrong thing

Because of their zealous dedication to the *ideal* of a pollution-free environment, the extremist protestors failed to anticipate the *consequences* of their protest or to respect their *obligation* to the community's economic welfare.

ETHICAL DILEMMAS

Ethics decisions are especially frustrating when no single answer seems acceptable:

[An ethical] dilemma exists whenever the conflicting obligations, ideals, and consequences are so very nearly equal in their importance that we feel we cannot choose among them, even though we must. (Ruggiero 91)

Ethical questions often resist easy answers

In private and public ways, such dilemmas are inescapable. For example, political candidates speak of drastic plans to eliminate the federal deficit in five years. One could argue that such a dedication to the *consequences* (or results) would violate our *obligations* (to the poor, the sick, etc.) and our *ideals* (of compassion, fairness, etc.). On the basis of our three criteria, how else might the deficit issue be considered?

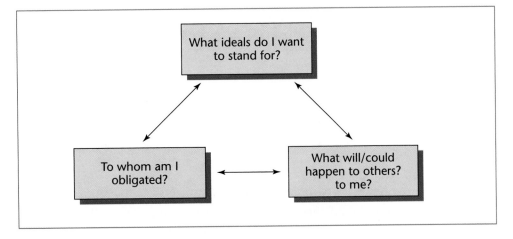

Figure 5.4 Reasonable Criteria for Ethical Judgment

ANTICIPATE SOME HARD CHOICES

Communicators' ethical choices basically are concerned with honesty in choosing to reveal or conceal information:

- ◆ What exactly do I report, and to whom?
- ◆ How much do I reveal or conceal?
- ◆ How do I say what I have to say?
- ◆ Could misplaced obligation to one party be causing me to deceive others?

The following scenario illustrates a hard choice at a workplace.

A Hard Choice

You may have to choose between the goals of your organization and what you know is right

You are an assistant structural engineer working on the construction of a nuclear power plant near a northern city. After years of construction delays and cost overruns, the plant finally has received its limited operating licence from the Atomic Energy Control Board (AECB).

During your final inspection of the nuclear core containment unit on February 15, you discover a 10-foot-long hairline crack in a section of the reinforced concrete floor, within 20 feet of the area where the cooling pipes enter the containment unit. (The especially cold and snowless winter likely has caused a frost heave under a small part of the foundation.) The crack has either just appeared or was overlooked by AECB inspectors on February 10.

The crack could be perfectly harmless, caused by normal settling of the structure; and this is, after all, a "redundant" containment system (a shell within a shell). But then again, the crack could signal some kind of serious stress on the entire containment unit, which furthermore could damage the entry and exit cooling pipes or other vital structures.

You phone your supervisor, who is just about to leave on a ski vacation, and who tells you, "Forget it; no problem," and hangs up.

You know that if the crack is reported, the whole start-up process scheduled for February 16 will be delayed indefinitely. More money will be lost; excavation, reinforcement, and further testing will be required—and many people with a stake in this project (from company executives to construction officials to shareholders) will be furious—especially if your report turns out to be a false alarm. All segments of plant management are geared up for the final big moment. Media coverage will be widespread. As the bearer of bad news—and bad publicity—you suspect that, even if you turn out to be right, your own career could be hurt by what some people will see as your overreaction that has made them look bad.

On the other hand, ignoring the crack could compromise the system's safety, with unforeseeable consequences. Of course, no one would ever be able to implicate you. The AECB has already inspected and approved the containment unit, leaving you, your supervisor, and your company in the clear. You have very little time to decide. Start-up is scheduled for tomorrow, at which time the containment system will become intensely radioactive. ▲

Working professionals commonly face similar choices, the product of conflicting goals and expectations, the pressure to meet deadlines and achieve results, to be a "loyal" employee, a "team player," to consider "the bottom line." Often, these choices have to be made alone or on the spur of the moment, without the luxury of meditation or consultation.

NEVER DEPEND ONLY ON LEGAL GUIDELINES

Can the law tell you how to communicate ethically? Sometimes. If you stay within the law, are you being ethical? Not always. Legal standards "sometimes do no more than delineate minimally acceptable behaviour." In contrast, ethical standards "often attempt to describe ideal behaviour, to define the best possible practices for corporations" (Porter 183). Consider these legal actions:

- ◆ The investigative TV show *20/20* exposed the common and legal practice for trucks to haul garbage or toxic substances (such as formaldehyde) one way and then haul food products (such as juice concentrates) on the return trip—without ever informing the customer.

Deception often is legal

- ◆ It is perfectly legal to advertise a cereal made with oat bran (which allegedly lowers cholesterol) without mentioning that another ingredient in the cereal is coconut oil (which raises cholesterol).

Lying is rarely illegal, except in cases of lying under oath or breaking a contractual promise (Wicclair and Farkas 16). But putting aside these and other illegal lies, such as defamation of character or lying about a product so as to cause injury, we see plenty of room for the kinds of legal lies depicted in Figure 5.5. Later chapters cover other kinds of legal lying, such as page design that distorts the real emphasis or words that are deliberately unclear, misleading, or ambiguous.

What then are a communicator's legal guidelines? Besides obscenity laws (not especially relevant here), workplace writing is regulated by the types of laws described below.

- ◆ *Laws against libel* prohibit any false written statement that maliciously attacks or ridicules anyone. A statement is considered libelous when it damages someone's reputation, character, career, or livelihood or when it causes humiliation or mental suffering. Material that is damaging but *truthful* would not be considered libelous unless it were used intentionally to cause harm. In the event of a libel suit, a writer's ignorance is no defence; even when the damaging material has been obtained from a source presumed reliable, the writer (and publisher) are legally accountable.[3]

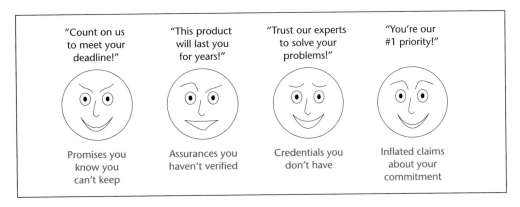

Figure 5.5 Some Legal Lies in the Workplace

---

3. Thanks to Peter Owens for the material on libel.

- ◆ *Copyright laws* protect the ownership rights of authors—or of their employers, in cases where the writing was done as part of one's employment. You will find a copy of the Copyright Act posted at **insight.mcmaster.ca/org/efc/pages/law/ canada/copyright/html.**
- ◆ *Laws protecting software* provide penalties for illegally duplicating copyrighted software. (Guidelines for ethical, legal use of software are provided through CAN-COPY's website: **www.cancopy.com**). Also, in Canada and internationally, the Software and Information Industry Association fights software piracy through education and enforcement measures.
- ◆ *Laws against deceptive or fraudulent advertising* make it illegal, for example, to falsely claim or imply that a product or treatment will cure cancer, or to represent and sell a used product as new. Fraud can be defined as "lying that causes another person monetary damage" (Harcourt 64).
- ◆ *Liability laws* define the responsibilities of authors, editors, and publishers for damages resulting from the use of incomplete, unclear, misleading, or otherwise defective information. The misinformation might be about a product (failure to warn about the toxic fumes from a spray-on oven cleaner) or a procedure (misleading instructions in an owner's manual for using a tire jack). Even if misinformation is given out of ignorance, the writer is liable (Walter and Marsteller 164–65).

    Legal standards with which product literature must comply vary from country to country. A document must satisfy the legal standards for safety, health, accuracy, language, or other issues for any country in which it will be distributed. For example, instructions for any product requiring assembly or operation have to carry warnings as stipulated by the laws of the country in which the product will be sold. Inadequate documentation, as judged by that country's standards, can result in a lawsuit (Caswell-Coward 264–66; Weymouth 145).

Laws regulating communication practices are few because such laws traditionally have been seen as threats to our freedom of speech (Johannesen 86). Many companies, however, have legal departments you can turn to with questions about a document's legality.

Recognizing that ethical behaviour is a matter of commitment, and not of legislation, most professions have developed their own ethics guidelines. If your field has its own formal code, obtain a copy.

UNDERSTAND THE POTENTIAL FOR COMMUNICATION ABUSE

On the job, you write in the service of your employer. Your effectiveness is judged by how well your documents speak for the company and advance its interests and agendas (Ornatowski 100–01). You walk the proverbial line between telling the truth and doing what your employer expects (Dombrowski 79).

Workplace writing influences the thinking, actions, and welfare of different people: customers, investors, co-workers, the public, policy-makers—to name a few. These people are victims of communication abuse whenever we give them information that is less than the truth as we know it. Following are some examples of such abuses.

SUPPRESSING KNOWLEDGE THE PUBLIC DESERVES

Except for disasters that make big news (Bhopal, Walkerton, Chernobyl, Westray mine), people hear plenty about the *wonders* of technology: how fluoride eradicates tooth decay, how smart bombs and cruise missiles never miss their targets, how nuclear power will solve our energy problems—but they rarely hear about the failures or the dangers (Staudenmaier 67).

In fact, the pressure to downplay failure sometimes results in censorship. For instance, some prestigious science journals have refused to publish studies linking chlorine and fluoride in drinking water with cancer risk, and fluorescent lights with childhood leukemia. The papers allegedly were rejected as part of widespread suppression of news about dangers of technological products (Begley 63).

Another example of censorship is what occurred at Toronto's Hospital for Sick Children (HSC) when University of Toronto clinician Dr. Nancy Olivieri discovered unexpected risks in drug trials sponsored by Apotex Inc. Apotex terminated the trials but warned Dr. Olivieri not to inform her HSC patients about the risks, or to publish her findings. Fulfilling her ethical obligations, she ignored the warnings and subsequently faced legal actions from Apotex. Also, she was disciplined and publicly attacked by HSC and the University of Toronto for jeopardizing millions of dollars of Apotex grants.

A comprehensive report commissioned by the Canadian Association of University Teachers (CAUT) was released in October 2001. This report vindicates Dr. Olivieri and warns of the conflicts of interest faced by researchers and research facilities when they are funded by corporations such as Apotex. The 500-page report and its 48-page Overview was placed on CAUT's website, **www.caut.ca**.

EXAGGERATING CLAIMS ABOUT TECHNOLOGY

Organizations that have a stake in a particular technology are especially tempted to exaggerate its benefits, potential, or safety.

Unwarranted claims help technology sell

> *An entrepreneur needs financiers. Scientists in a large corporation need advocates high enough in the hierarchy to allocate funds. And government-supported researchers at universities and national labs have an obvious incentive to overstate their progress and understate the problems that lie ahead: the better the chances for success, the more money an agency is willing to shell out.* (Brody 40)

If your organization depends on outside funding (as in the defence or space industry), you might find yourself pressured to make unrealistic promises.

STEALING OR DIVULGING PROPRIETARY INFORMATION

Proprietary information is any document or idea that can be considered the exclusive property of the company in which it originated. Proprietary documents may include company records, test and experiment results, surveys paid for by clients, market research, minutes of meetings, plans, and specifications (Lavin 5). In theory, such information is legally protected,[4] but it remains vulnerable to sabotage or theft. Rapid developments in technology create fierce competition among rival companies for the very latest intelligence, giving rise to measures like these:

---

4. Although various Trade Secrets Acts around the world are designed to curb such abuses, these laws are subject to broad and varied interpretation by individual courts.

Examples of
corporate
espionage

Companies have been known to use business school students to garner information on competitors under the guise of conducting "research." Even more commonplace is interviewing employees for slots that don't exist and wringing them dry about their current employer. (Gilbert 24)

Moreover, employees within a company can leak confidential information to the press or to anyone else who has no legal right to know.

MISMANAGING ELECTRONIC INFORMATION

With so much information stored in databases (by schools, employers, government agencies, mail order retailers, credit bureaus, banks, credit card companies, insurance companies, pharmacies, etc.), questions of how we combine, use, and disseminate the information become increasingly important (Finkelstein 471). Moreover, a database is easier to alter than its printed equivalent; one simple command can wipe out or transform the facts.

WITHHOLDING INFORMATION PEOPLE NEED TO DO THEIR JOBS

Nowhere is the adage that "information is power" more true than among coworkers. One sure way to sabotage a colleague is to withhold vital information about the task at hand.

Beyond these deliberate communication abuses is this reality: *all* information is a matter of personal or social interpretation (Dombrowski 79); therefore, "objective reporting," practically speaking, is impossible. What we say on the job and how we say it are influenced by the expectations of our employer and by our own self-interest.

EXPLOITING CULTURAL DIFFERENCES

Cross-cultural documents carry great potential for communication abuse. Based on its level of business experience, technological development, or financial need, a particular culture might be especially vulnerable to manipulation or deception. Some countries, for instance, are persuaded to purchase, from North American companies, pesticides and other chemicals whose use is banned in North America. Other countries witness depletion of their natural resources or exploitation of their labour force by more developed countries—all in the name of progress. All communication in all cultural contexts should embody universal standards of honesty and fairness.

KNOW YOUR COMMUNICATION GUIDELINES

How do you balance self-interest with the interests of others—the organization, the public, your customers? How can you be practical and responsible at the same time? Here are two practical guidelines for ethical communication (Clark 194):

1. *Give the audience everything it needs to know.* To see things as clearly as you do, people need more than just a partial view. Don't bury readers in needless details, but do make sure they get all of the facts and get them straight.

2. *Give the audience a clear understanding of what the information means.* Even when all of the facts are known, they can be misinterpreted. Do all you can to ensure that your readers understand the facts as you do.

The Checklist for Communicators on page 93 incorporates additional guidelines from various chapters. Use the checklist for any document you prepare or for which you are responsible. Also, see "On-line Science Ethics Resources" at **www.chem.vt.edu/ ethics/vinny/ethxonline.html.**

DECIDE WHERE AND HOW TO DRAW THE LINE

Suppose your employer asks you to do something unethical—for example, alter data to cover up a violation of federal pollution standards. If you decide to resist, your choices seem limited: resign or go public (i.e., blow the whistle).

Walking away from a job isn't easy, however, and whistle-blowing can spell career disaster (Rubens, *Reinventing* 330). Many organizations refuse to hire anyone blacklisted as a whistle-blower (Wicclair and Farkas 19). Even if you aren't fired, expect your job to become hellish. Here is one communicator's gloomy assessment, based on personal experience:

Know what to expect

Most of the ethical infractions [you] witness will be so small that blowing the whistle will seem fruitless and self-destructive. And leaving one company for another may prove equally fruitless, given the pervasiveness of the problem. (Bryan 86)

Anglo-Canadian law tends to protect employees who blow the whistle and are then dismissed. An employee has the right to bring a wrongful dismissal lawsuit and, in certain circumstances, to include human rights and/or labour relations claims. The outcome of these claims varies.

Employees covered by a contract or collective agreement normally seek advice from their shop steward and/or union executive as well as a personal and/or union lawyer. Employees not covered by a contract or collective agreement often seek initial advice from local departments of labour to establish employment standards law provisions. It is quite common for these people to also seek advice from their personal lawyer.

◆ Anyone who reports employer violations to a regulatory agency such as Occupational Health and Safety (OHS) and who is punished can request a Department of Labour investigation. Employees whose claims are ruled valid can win reinstatement and reimbursement for back pay and legal expenses.[5]

Even with such protections, an employee who takes on a company without the backing of a labour union or other powerful group can expect lengthy court battles, high legal fees (which may or may not be recouped), and disruption of life and career. Exactly where you draw the line (on having your integrity or health instead of your job) will be strictly your own decision.

If you do decide to take a stand, be reasonable and cautious, and follow these suggestions (Unger 127–30):

---

5. Although employees legally are entitled to speak confidentially with OHS and WCB inspectors about violations of health and safety in their company, a survey revealed that the inspectors themselves feel such laws offer employees little actual protection against company retribution (Kraft 5).

- *Get your facts straight, and get them on paper.* Don't blow matters out of proportion, but do keep a paper trail in case of legal proceedings.
- *Appeal your case in terms of the company's interests.* Instead of being pious and judgmental ("This is a racist and sexist policy, and you'd better get your act together"), focus on what the company stands to gain or lose ("Promoting too few women and minorities makes us vulnerable to legal action").
- *Aim your appeal toward the right person.* If you have to go to the top, find someone who knows enough to appreciate the problem and who has enough clout to make something happen.
- *Get professional advice.* Contact a lawyer and your professional society for advice about your legal rights.

Before accepting a job offer, do some discreet research about the company's ethical reputation. (Of course you can learn only so much about a company before actually working there.) Some companies have ombudspersons, who help employees lodge complaints. Others offer hotlines for advice on ethics problems or for reporting violations. Also, realizing that good ethics is good business, companies increasingly are developing codes for personal and organizational behaviour. Without such supports, don't expect to last long as an ethical employee in an unethical organization.

Remember that very few employers tolerate any public statement, no matter how truthful, that makes the company look bad.

A final note: Sometimes the right choice is obvious, but often not so obvious. No one has any sure way of always knowing what to do. This chapter is only an introduction to the inevitable hard choices that, throughout your career, will be yours to make and to live with. For further guidance, see the website maintained by the Online Ethics Center for Engineering and Science, established by the Institute of Electrical and Electronics Engineers (IEEE) at **onlineethics.org**. This website covers a wide range of ethics issues for technical people trying to cope with difficult situations. In particular, see the "Famous Cases" and "Recent Engineering Cases" secondary sites.

Checklist FOR COMMUNICATORS

Use this checklist[6] to help your documents reflect reasonable, ethical judgment.

◆ Do I avoid exaggeration, understatement, sugar-coating, or any distortion or omission that leaves readers at a disadvantage?

◆ Do I make a clear distinction between "certainty" and "probability"?

◆ Am I being honest and fair?

◆ Have I explored all sides of the issue and all possible alternatives?

◆ Are my information sources valid, reliable, and unbiased?

◆ Do I actually believe what I'm saying, instead of being a mouthpiece for group-think or advancing some hidden agenda?

◆ Would I still advocate this position if I were held publicly accountable for it?

◆ Do I provide enough information and interpretation for readers to understand the facts as I know them?

◆ Am I reasonably sure this document will harm no innocent persons or damage their reputation?

◆ Am I respecting all legitimate rights to privacy and confidentiality?

◆ Do I inform readers of the consequences or risks (as I am able to predict) of what I'm advocating?

◆ Do I state the case clearly, instead of hiding behind jargon and generalities?

◆ Do I give candid feedback or criticism, if it is warranted?

◆ Am I distributing copies of this document to every person who has the right to know about it?

◆ Do I credit all contributors and sources of ideas and information?

EXERCISES

1. Prepare a memo (one or two pages) for distribution to first-year college students in which you introduce the ethical dilemmas they will face in college. For instance:

◆ If you receive a final grade of A by mistake, would you inform your instructor or professor?

◆ If the library loses the record of books you've signed out, would you return them anyway?

◆ Would you plagiarize—and would that change in your professional life?

◆ Would you support lowering standards for student athletes if the team's success was important for the school's funding and status?

◆ Would you allow a friend to submit a paper you've written for some other course?

What other ethical dilemmas can you envision? Tell your audience what to expect, and give them some *realistic* advice for coping. No sermons, please.

2. In your workplace communications, you may end up facing hard choices concerning what to say, how much to say, how to say it, and to whom. Whatever your choice, it will have definite consequences. Be prepared to discuss the following cases in terms of the obligations, ideals, and consequences involved. Can you think of similar choices you or someone you know has already faced? What happened?

◆ While travelling on assignment that is being paid for by your employer, you visit an area in which you would really like to live and work, an area in which you have lots of contacts but never can find time to visit on your own. You have five days to complete your assignment, and then you must report on your activities. You complete the assignment in three days. Should you spend the remaining two days checking out other job possibilities, without reporting this activity?

---

6. Adapted from Brownell and Fitzgerald 18; Bryan 87; Johannesen 21–22; Larson 39; Unger 39–46; Yoos 50–55.

◆ As a marketing specialist, you are offered a lucrative account from a cigarette manufacturer; you are expected to promote the product. Should you accept the account? Suppose instead the account were for beer, junk food, suntanning parlours, or ice cream. Would your choice be different? Why, or why not?

◆ You have been authorized to hire a technical assistant, and so you are about to prepare an advertisement. This is a time of threatened cutbacks for your company. People hired as "temporary," however, have never seemed to work out well. Should your ad include the warning that this position could be only temporary?

◆ You are one of three employees being considered for a yearly production bonus, which will be awarded in six weeks. You've just accepted a better job, at which you can start anytime in the next two months. Should you wait until the bonus decision is made before announcing your plans to leave?

◆ You are marketing director for a major importer of coffee beans. Your testing labs report that certain African beans contain roughly twice the caffeine of South American varieties. Many of these African varieties are big sellers, from countries whose coffee bean production helps prop otherwise desperate economies. Should your advertising of these varieties inform the public about the high caffeine content? If so, how much emphasis should this fact be given?

◆ You are research director for a biotechnology company working on an AIDS vaccine. At a national conference, a researcher from a competing company secretly offers to sell your company crucial data that could speed discovery of an effective vaccine. Should you accept the offer?

3. Review the scenario entitled "A Hard Choice," on page 86. What would you do? Come to the class prepared to justify your decision on the basis of the obligations, ideals, and consequences involved.

COLLABORATIVE PROJECT

After dividing into groups, study the following scenario and complete the assignment: You belong to the forestry management division in a province whose year-round economy depends almost totally on forest products (lumber, paper, etc.), but whose summer economy is greatly enriched by tourism, especially from fishing, kayaking, and other outdoor activities. The province's poorest area is also its most scenic, largely because of the virgin stands of hardwoods. Your division has been facing growing political pressure from this area to allow logging companies to harvest the trees. Logging here would have good and bad consequences: for the foreseeable future, the area's economy would benefit greatly from the jobs created; but traditional logging practices would erode the soil, pollute waterways, and decimate wildlife, including several endangered species—besides posing a serious threat to the area's tourist industry. Logging, in short, would give a desperately needed boost to the area's standard of living, but would put an end to many tourist-oriented businesses and would change the landscape forever.

Your group has been assigned to weigh the economic and environmental impacts of logging, and prepare recommendations (to log or not to log) for your managers, who will use your report in making their final decision. To whom do you owe the most loyalty here: the unemployed or underemployed residents, the tourist businesses (mostly owned by residents), the wildlife, the land, future generations? The choices are by no means simple. In cases like this, it isn't enough to say that we should "do the right thing," because we are sometimes unable to predict the consequences of a particular action—even when it seems the best thing to do. In a memo to your supervisor, outline what action you would recommend and explain why. Be prepared to defend your group's ethical choice in class on the basis of the obligations, ideals, and consequences involved.

PART II

Information Gathering, Analysis, and Manipulation

Garth Homer, Okanagan University College (OUC), Kelowna

He jokingly calls himself "Garth Homer, Boy Librarian," but he's very serious about the art and science of research. Besides guiding OUC students and professors in their research, Garth (with Ross Tyner) has organized OUC's library website into a full array of research tools. And, as a freelance researcher, he mines data for business and government clients.

Garth says the "new electronic library" has "changed our access to the world." But he cautions that "only 8 percent of published material finds its way onto the Web . . . It's hard to tell what's relevant and valid on the Internet: the Web is an open-ended garbage can of information. The shift has gone from 'Can I find it?' to 'Is it any good?' and that's a big shift." He further cautions that Web material is hard to document and is ephemeral.

Garth has three secrets for successful data searches:

1. In your research strategy, identify possible paper *and* electronic sources to answer your question.

2. Use your library's on-line catalogue. Universities compete to make their holdings and electronic gateways the most comprehensive and the easiest to use.

3. Look for websites that provide "one-stop shopping," with links to many related sites.

He says that all of us must be able to research effectively: "Twenty years ago, most people in a given field had all the necessary information in their heads, or they could phone a colleague, or they could read two or three journals to keep current. Not any more!"

CHAPTER 6

Gathering Information

THINKING CRITICALLY ABOUT THE RESEARCH PROCESS

EXPLORING SECONDARY SOURCES

EXPLORING PRIMARY SOURCES

In Brief **Guidelines for Internet Research**

Guidelines **for Informative Interviews**

Major decisions in the workplace typically are based on careful research, with the findings recorded in a written report. The report's readers expect current information that can help answer their specific questions.

Research is classified as *primary* or *secondary*. Primary research involves an original, first-hand study of your topic or problem: observations, interviews, questionnaires, enquiry letters, personal experiments, analysis of samples, fieldwork, or company records. Secondary research includes materials published by other researchers: journal articles, books, encyclopedias, reports, textbooks, handbooks, on-line articles, electronic databases, government documents, Internet sites, and material held by public agencies and special interest groups.

Research strategies and resources differ widely among disciplines. This chapter focuses on research for preparing a formal report.

THINKING CRITICALLY ABOUT THE RESEARCH PROCESS

Research is a deliberate form of enquiry, a process of problem solving, in which certain procedures follow a recognizable sequence, as shown in Figure 6.1.

But research does not simply follow a numbered set of procedures ("First, do this; then, do that"). The procedural stages depend on the many decisions that accompany any legitimate enquiry (see Figure 6.2).

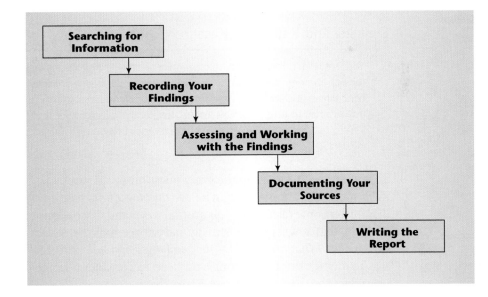

Figure 6.1 Procedural Stages of the Research Process

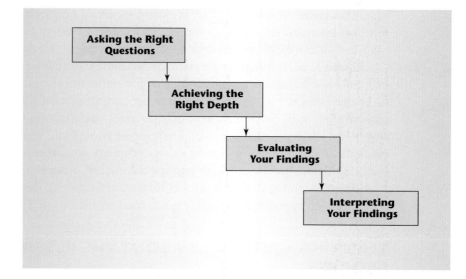

Figure 6.2 The Enquiry Stages of the Research Process

Asking the Right Questions

The answers you uncover will depend on the questions you ask. Assume, for instance, that you are faced with the following scenario:

Defining and Refining a Research Question

You are the public health manager for a small New Brunswick town in which high-tension power lines run within 30 metres of the elementary school. Parents are concerned about the danger from electromagnetic radiation (EMR) emitted by these power lines, in energy waves known as electromagnetic fields (EMFs). Town officials ask you to research the issue and prepare a report to be distributed at the next town meeting in six weeks. ▲

First, you need to identify the exact question or questions you want answered. Initially, the major question might be: *Do the power lines pose any real danger to our children?* After some telephone calls around town and discussions at the coffee shop, you discover that townspeople actually have three major questions about electromagnetic fields: *What are they? Do they endanger our children? If so, what can be done?*

To answer these questions, you need to consider a range of subordinate questions, like those in the Figure 6.3 tree chart.

As research progresses, this chart will grow. For instance, after some preliminary reading, you learn that electromagnetic fields radiate not only from power lines but also from *all* electrical equipment, and even from the earth iteself. So you face this additional question: *Do power lines present the greatest hazard as a source of EMFs?*

You now wonder whether the greater hazard comes from power lines or from other sources of EMF exposure. Critical thinking, in short, has enabled you to define and refine the essential questions.

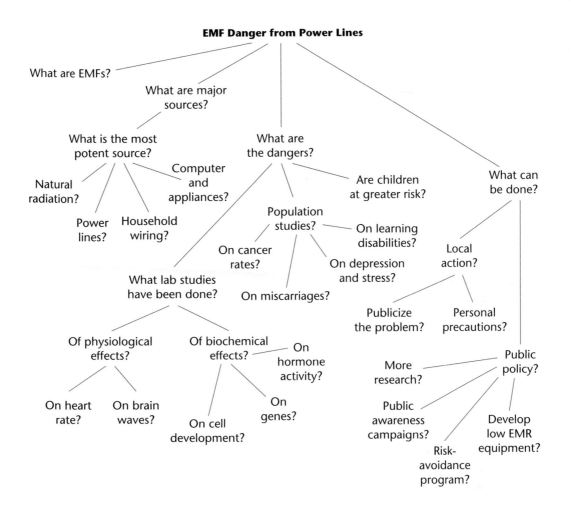

Figure 6.3   How the Right Questions Help Define a Research Problem

ACHIEVING ADEQUATE DEPTH IN YOUR SEARCH

Balanced research examines a broad *range* of evidence; thorough research, however, examines that evidence at an appropriate *depth*. As depicted in Figure 6.4, different types of secondary information about any topic occupy different levels of detail and dependability.

1. At the surface level are items from the popular media (newspapers, radio, TV, general magazines). Designed for general consumption, this layer of information often contains more journalistic interpretation than factual detail.

2. At the next level are trade and business publications (*Frozen Food World, Publisher's Weekly,* etc.). Designed for readers who range from moderately informed to highly specialized, this level of information focuses more on practice than on theory, on items considered newsworthy to group members, on issues affecting the field, on public relations, on viewpoints that tend to reflect the particular biases of that field.

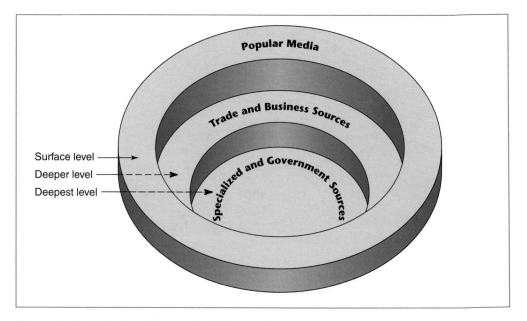

Figure 6.4   Effective Research Achieves Adequate Depth

3. At a deeper level is the specialized literature (journals from professional associations: medical, legal, engineering, etc.). Designed for practising professionals, this level of information focuses on theory as well as practice, on descriptions of the latest studies—written by the researchers themselves and scrutinized by others for accuracy and objectivity, on debates among scholars and researchers, and on reviews and critiques and refutations of prior studies and publications.

Also at this deeper level are government sources and corporate documents available through the Freedom of Information Act. Designed for anyone willing to investigate its complex resources, this layer of information offers hard facts and highly detailed and (in many instances) *relatively* impartial views of virtually any issue or topic in any field.

Web pages, of course, offer links to increasingly specific levels of detail. But the actual "depth" and quality of a website's information depends on the sponsorship and reliability of that site.

How deep is deep enough? This depends on your purpose, your audience, and your topic. But the real story and the hard facts more likely reside at the deeper levels of information.

EVALUATING YOUR FINDINGS

Once you have collected all of the essential evidence about your topic, you need to decide how much of it is legitimate and then decide what it means.

Not all findings have equal value. Some information might be distorted, incomplete, or misleading. Information might be tainted by *source bias.* With an emotional issue involving children, a source might understate or overstate certain facts, depending on whose interests that source represents (power company, government agency, parent organization, etc.).

> ## QUESTIONS FOR EVALUATING A PARTICULAR FINDING
>
> - Is this information accurate, reliable, and relatively unbiased?
> - Can the claim be verified by the facts?
> - How much of the information is useful?
> - Is this the whole or the real story?
> - Does something seem to be missing?
> - Do I need more information?

Ethical researchers rely on evidence that represents a fair balance of views. They don't merely emphasize findings that support their own biases or assumptions.

INTERPRETING YOUR FINDINGS

Once you have decided which of your findings seem legitimate, you need to decide what they all mean.

> ## QUESTIONS FOR INTERPRETING YOUR FINDINGS
>
> - What do all these facts or observations mean?
> - Do any findings conflict?
> - Are other interpretations possible?
> - Should I reconsider the evidence?
> - What are my conclusions?
> - What, if anything, should be done?

Perhaps you will reach a definite conclusion; for example, "The evidence about EMF dangers seems persuasive enough for us to be concerned and to take the following actions"—perhaps you will not.

Even the best research can produce contradictory or indefinite conclusions. For instance, some scientists show that studies linking electromagnetic radiation to health hazards are flawed. They point out that some studies indicate increased cancer risk while others indicate beneficial health effects. Other scientists claim that stronger EMFs are emitted by natural sources, such as earth's magnetic field, than by electrical sources (McDonald 5). An accurate conclusion would have to come from your analyzing all views and then deciding that one outweighs the others—or that only time will tell.

Never force a simplistic conclusion on a complex issue. Sometimes the best you can come up with is an indefinite conclusion: "Although controversy continues over the extent of EMF hazards, we all can take simple precautions to reduce our exposure." A wrong conclusion is far worse than no definite conclusion at all.

EXPLORING SECONDARY SOURCES

Although electronic searches for information are becoming the norm, a *thorough* search often requires careful examination of hard copy sources as well.

HARD COPY VERSUS ELECTRONIC SOURCES

Advantages and drawbacks of each search medium (Table 6.1) often provide good reason for exploring both.

Table 6.1   Hard Copy verus Electronic Sources: Benefits and Drawbacks

	Benefits	Drawbacks
Hard Copy Sources	◆ discovered and organized by librarians ◆ easier to preserve and keep secure	◆ time-consuming and inefficient to search ◆ difficult to update
Electronic Sources	◆ more current, efficient, and accessible ◆ searches can be narrowed or broadened ◆ can offer material that has no hard copy equivalent	◆ access to recent material only ◆ not always reliable: Sources may be very biased ◆ user might get lost ◆ material may disappear

Benefits of Hard Copy Sources.   Hard copy libraries offer the judgment and expertise of librarians who organize and search for information. Compared with electronic files (on disks, tapes, and hard drives), hard copy is easier to protect from tampering and to preserve from aging. An electronic file's life-span can be as brief as 10 years.

For many electronic searches, a manual search of hard copy usually is needed as well. One recent study found greater than 50 percent inconsistency among database indexers. Thus, even an electronic search by a trained librarian can miss improperly indexed material (Lang and Secic 174–75). In contrast, a manual search provides the whole "database" (the bound index or abstracts). As you browse, you often randomly discover something useful.

Drawbacks of Hard Copy Sources.   Manual searches (flipping pages by hand), however, are time-consuming and inefficient: books can get lost; relevant information has to be pinpointed and retrieved, or "pulled" by the user. Also, hard copy cannot be updated easily (Davenport 109–111).

Benefits of Electronic Sources.   Compared with hard copy, electronic sources are more current, efficient, and accessible. Sources are updated rapidly. Ten or 15 years of an index can be reviewed in minutes. Searches can be customized; for example, narrowed to specific dates or topics. They also can be broadened: A keyword search can uncover material that a hard copy search may have overlooked; Web pages can provide links to material of all sorts—much of which exists in no hard copy form.

Drawbacks of Electronic Sources.   Drawbacks of electronic sources include the fact that databases rarely contain entries published before the mid-1960s and that material, especially on the Internet, can change or disappear overnight or be highly unreliable. Also, given the potential for getting lost in cyberspace, a thorough electronic search calls for a preliminary conference with a trained librarian.

TYPES OF HARD COPY SOURCES

Where you begin your hard copy search depends on whether you are searching background and basic facts or the latest information. Library sources are shown in Figure 6.5.

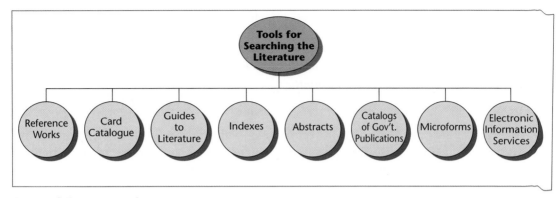

Figure 6.5 Library Sources

If you are an expert in the field, you might simply do a computerized database search or browse through specialized journals and listservs. If you have limited knowledge or you need to focus your topic, you probably will want to begin with general reference sources.

REFERENCE WORKS

Reference sources, which provide background information, include encyclopedias, almanacs, handbooks, dictionaries, histories, and biographies. These provide background and bibliographies that can lead to more specific information. Make sure the work is current by checking the last copyright date.

Bibliographies. These comprehensive lists of publications about a subject are issued yearly, or even weekly. However, some can quickly become dated. Annotated bibliographies (which include an abstract for each entry) are most helpful. The following are sample listings (with annotations):

Bibliographies

> *Bibliographic Index.* A list (by subject) of bibliographies that contain at least 50 citations; to see which bibliographies are published in your field, begin here.
>
> *A Guide to Canadian Government Scientific and Technical Resources.* A list of everything published by the government in these broad fields.
>
> *Health Hazards of Video Display Terminals: An Annotated Bibliography.* One of many bibliographies focused on a specific subject.

Dictionaries. Dictionaries can be generalized or they can focus on specific disciplines or give biographical information. The following are sample listings:

Dictionaries

> *Dictionary of Engineering and Technology*
> *Dictionary of Telecommunications*
> *Dictionary of Scientific Biography*

Handbooks. These research aids amass key facts (formulas, tables, advice, examples) about a field in condensed form. The following are sample listings:

Handbooks

> *Business Writer's Handbook*
> *Civil Engineering Handbook*
> *The McGraw-Hill Computer Handbook*

Almanacs.   Almanacs contain factual and statistical data. The following are sample listings:

> *World Almanac and Book of Facts*
> *Almanac for Computers*
> *Almanac of Business and Industrial Financial Ratios*

Directories.   In directories you will find updated information about organizations, companies, people, products, services, or careers, often including addresses and phone numbers. The following are sample listings:

> *The Career Guide: Dun's Employment Opportunities Directory*
> *The Internet Directory*

Reference works increasingly are accessible by computer. Some, such as the Free Online Dictionary of Computing, are wholly electronic.

THE CARD CATALOGUE

All books, reference works, indexes, periodicals, and other materials held by a library are usually listed in its card catalogue under three headings: author, title, and subject.

Most libraries have automated their card catalogues. These electronic catalogues offer additional access points (beyond *author, title,* and *subject*) including:

- ◆ *descriptor:* for retrieving works on the basis of a keyword or phrase (for example, "electromagnetic" or "power lines and health") in the subject heading, in the work's title, or in the full text of its bibliographic record (its catalogue entry or abstract)
- ◆ *document type:* for retrieving works in a specific format (videotape, audiotape, compact disk, motion picture)
- ◆ *organizations and conferences:* for retrieving works produced under the name of an institution or professional association (for example, Brookings Institute or Canadian Heart and Stroke Foundation)
- ◆ *publisher:* for retrieving works produced by a particular publisher (for example, Prentice Hall Canada Inc.)
- ◆ *combination:* for retrieving works by combining any available access points (a book about a particular subject by a particular author or institution)

Figure 6.6 displays the first three computer screens you might encounter in an automated search using the descriptor ELECTROMAGNETIC. Through the Internet, a library's electronic catalogue can be searched from anywhere in the world.

INDEXES

Indexes are lists of books, newspaper articles, journal articles, or other works, as shown in Figure 6.7. They are excellent sources for current information. Because different indexes list sources in different ways, always read the introductory pages for instructions, or ask a librarian for help.

You begin by pressing any key, and the computer responds with the screen:

This first screen lists your options for getting help or for searching the catalogue from various access points

```
Type of searches:                          Press HELP key for Help

1  AU  =  Author                     8    PU  =  Publisher
2  OC  =  Organization or conference 9    SH  =  Subject heading
3  TI  =  Title                      10   DT  =  Document type
4  UT  =  Uniform or collective title 11         Combination
5  DE  =  Descriptor                 12         ISBN
6  CN  =  Call number                13         ISBN
7  SE  =  Series                     14         Numeric

Enter the NUMBER of your search request and press RETURN:
```

After selecting the DE search mode, you enter your keyword (ELECTRO MAGNETIC), and then press RETURN. This next screen appears (the first of several with all 108 entries):

A journal that might be of interest for this topic

Conference proceedings

A recent and relevant title for this topic

The number of items in your library that contains your keyword in their subject headings

Publication dates

```
            #1 = Keyword: ELECTROMAGNETIC (108 records)

1:  The classical theory of fields   Landau, L. D. 1908–1968.   1989
2:  The Journal of microwave po...   International Microwave P   1985–
3:  Basic electromagnetic theory   Paris, Demetrius T., 1928   1969
4:  The theory of electrons and...   Lorentz, H. A. 1853–1928.   1952
5:  Principles of optics: elect...   Born, Max, 1882–1970.   1964
6:  Field theory of guided waves   Collin, Robert E.   1991
7:  Electromagnetic theory and...   Walsh, John B.   1960
8:  Electrodynamics of continuo...   Landau, L. D. 1908–1968.   1960
9:  Electromagnetic fields and...   Becker, Richard, 1887–1955   1964
10: IEEE 1990 International Sym...   IEEE International Sympos   1990
11: CPEM '90 digest: Conferenc...   Conference on Precision E   1990
12: Weak and electromagnetic in...   Depommier, Pierre.   1989
13: The EMF book: what you...   Pinsky, Mark A.   1995
14: Wave transmission and fiber...   Daiment, Paul.   1990
15: Electromagnetic wave propag...   Dearholt, Donald W.   1973
```

You select entry #13 and then press RETURN. The computer responds with detailed bibliographic information on your selected item.

This screen shows the electronic equivalent of the printed catalogue entry

```
Selection:
01-0211132

AUTHOR       Pinsky, Mark A.
TITLE        The EMF book: what you should know about electromagnetic
             fields, electromagnetic radiation, and your health
PUBLISHER    New York: Warner
DATE         1995
PHYS. FEAT.  246 p.; 20 cm.
SUBJECTS     Electromagnetic fields—Health aspects.
             ELF electromagnetic fields—Health aspects.
             Electric lines—Health aspects.
             Computer terminals—Health aspects.
```

Figure 6.6   Searching an Electronic Card Catalogue

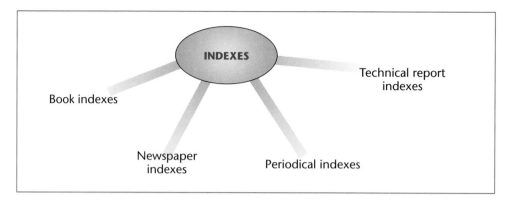

Figure 6.7   Useful Indexes for Technical Disciplines

Book Indexes.   All books currently published (up to a set date) are listed in book indexes by author, title, or subject. The following are sample indexes (shown with annotations):

Book indexes

> *Books in Print.* An annual listing of all books published in Canada.
> *Cumulative Book Index.* A monthly worldwide listing of books in English.
> *Forthcoming Books.* A listing every two months of Canadian books to be published.
> *Scientific and Technical Books and Serials in Print.* An annual listing of literature in science and technology.
> *New Technical Books: A Selective List with Descriptive Annotations.* Issued ten times yearly.
> *Technical Book Review Index.* A monthly listing (with excerpts) of book reviews.
> *Medical Books and Serials in Print.* An annual listing of works from medicine and psychology.

In research on electromagnetic radiation, you might check the current issue of *New Technical Books in Print* or *Scientific and Technical Books and Serials in Print.* But no book is likely to offer the very latest information because of the time required to publish a book manuscript (from several months to over one year).

Newspaper Indexes.   Most newspaper indexes list articles by subject. The *New York Times Index* is best known, but other major newspapers have their own indexes. The following are sample titles:

Newspaper indexes

> *The Globe and Mail Index* (www.theglobeandmail.com)
> *Christian Science Monitor Index*
> *Wall Street Journal Index*

For research on electromagnetic radiation, you might check recent editions of the *Canadian Index* available monthly in print or every six months on CD-ROM.

Periodical Indexes.   For recent information in magazines and journals, consult periodical indexes. To find useful indexes, first decide whether you seek general or specialized information.

One general index is the *Magazine Index,* a subject index (on microfilm) of 400 general periodicals. A popular index is the *Readers' Guide to Periodical Literature,* listing articles from 150 general magazines and journals. Because the *Readers' Guide* is updated every few weeks, you can locate current material. In research on electromagnetic radiation, you would find numerous entries under the subject heading, "Electromagnetic waves."

For specialized information, consult indexes that list journal articles in specific disciplines, such as *Ulrich's International Periodicals Directory.* Another comprehensive source of specialized information, the *Applied Science and Technology Index* carries a monthly listing, by subject, of articles in more than 200 scientific and technical journals. In research on electromagnetic radiation, you would find multiple entries under the heading "Electromagnetic fields" in recent issues of the *AS&T Index.*

Other broad indexes that cover specialized fields in general include the *General Science Index.* The *Statistical Reference Index* lists statistical works not published by the government.

Along with these broad indexes, some disciplines have their own specific indexes. The following are sample listings:

Periodical indexes

Agricultural Index
Biological Sciences Index
Canadian Business and Current Affairs Index
Canadian Periodical Index
Energy Index
Environment Index
F&S Index of Corporations and Industries
GEOBASE—Earth Sciences and Geography
GEOSCAN—Geological Survey of Canada
International Nursing Index
PubMed (MEDLINE)

Ask your librarian about the best indexes for your topic and about the many indexes that can be searched by computer.

Technical Report Indexes.   Countless government and private-sector reports written worldwide offer specialized and highly current information. (Proprietary or security restrictions, of course, restrict public access to certain corporate or government documents.) The following are sample indexes for these reports:

Technical report indexes

Canadian Research Index
CISTI (The National Research Council of Canada's "Canada's Institute for Scientific and Technical Information")
Monthly Catalog of United States Government Publications

ABSTRACTS

Beyond indexing various works, abstracts summarize each article. The abstract can save you from going all the way to the journal in order to decide whether to read the article.

Collection of abstracts

Abstracts usually are titled by discipline. The following is a sample list:

Biological Abstracts
Computer Abstracts
Environment Abstracts
Excerpta Medica
Forestry Abstracts
International Aerospace Abstracts
Metals Abstracts

In researching electromagnetic radiation, you might consult *Energy Research Abstracts* under the subject heading "Electromagnetic Fields." Abstracts (such as *Energy Research Abstracts* and *Pollution Abstracts*) increasingly are searchable by computer. Check with your librarian.

For some current research, you might consult abstracts of doctoral dissertations in *Dissertation Abstracts International*.

Locating the Source.   If your library does not hold the article you need, you can search an on-line database to identify a holding library, then request the article through interlibrary loan.

ACCESS TOOLS FOR GOVERNMENT PUBLICATIONS

The Canadian and U.S. federal governments publish maps, periodicals, books, pamphlets, manuals, monographs, annual reports, research reports, and a bewildering array of other information. The *Canadian Research Index* lists a wide range of Canadian federal and provincial government publications.

Types of Canadian information available to the public include ministerial and government proclamations, government bills and reports, judiciary rulings, and publications from all other government agencies (Departments of Agriculture, Transportation, etc.). A few of the countless titles available are:

Government publications

Electromagnetic Fields in Your Environment
Economic Report of the President
Major Oil and Gas Fields of the Free World
Decisions of the Federal Trade Commission
Journal of Research of the National Bureau of Standards
Siting Small Wind Turbines

Much of this information can be searched on-line as well as in printed volumes. Your best bet for tapping this valuable but complex resource is to request assistance from the librarian in charge of government documents. If your library does not hold the publication you seek, it can be obtained through electronic access or interlibrary loan.

Here are the basic access tools for documents issued or published at government expense and access tools for privately sponsored documents. Though most of these access tools charge fees, you can use many of them free at your college library.

- ◆ *Micromedia's Canadian News and Periodical Reference Services* provides access to the *Canadian Business and Current Affairs Index CBCA, Canadian Index, Canadian Serialism Microform, Canadian Education Index, Canadian NewsDisc,* and the *Canadian Research Index* and its microlog collection.
- ◆ *Canadian Research Index.* This comprehensive research guide includes all depository publications of research value issued by federal and provincial government agencies and departments. It also indexes scientific and technical reports issued by research institutes and government laboratories. It even lists policy, social, economic, and political reports, and theses and dissertations from Canadian universities.
- ◆ *The Monthly Catalog of the United States Government,* the major access to government publications and reports, is indexed by author, subject, and title.
- ◆ *Government Reports Announcements & Index* is a listing published every two weeks by the National Technical Information Service (NTIS), a U.S. federal clearinghouse for scientific and technical information—all stored in a database. The collection has summaries of over 1 million federally sponsored research reports published and patents issued since 1964. About 70 000 new summaries are added annually in 22 subject categories, from aeronautics to medicine and biology. Full copies of reports are available from NTIS.

A growing body of government information is posted to the Internet. One starting point is the Government of Canada's federal departments and agencies home page: **canada.gc.ca/depts/major/depind_e.html.**

Examples of government agency postings include:

1. The Canadian Council for Tobacco Control (a Health Canada agency) and the British Columbia Ministry of Health combined in December 1998 to post the results of testing constituents of tobacco smoke.
2. The U.S. Food and Drug Administration's electronic bulletin board lists information on experimental drugs to fight AIDS, drug and device approvals, recalls and litigations involving drugs or devices, health fraud, and a host of related items.

MICROFORMS

Microform technology enables vast quantities of printed information to be reproduced and stored on rolls of microfilm or packets of microfiche. (This material is read on machines that magnify the reduced image. Ask your librarian for assistance.) Among the growing array of microform products are government documents, technical reports, newspapers, business directories, and translated documents from around the world (Lavin 12).

INTERNET SOURCES

Today's Internet connects computer users by the tens of millions. Websites and addresses (uniform resource locators, or URLs) numbering in the hundreds of millions continue to multiply across the globe.

Internet Service Providers (ISPs), including *Sympatico, Compuserve,* and *Microsoft Network,* provide Internet access via "gateways," along with aids for navigating their many resources (see Figure 6.8).

E-mail Enquiries. The global e-mail network is excellent for contacting knowledgeable people in any field worldwide. E-mail addresses are increasingly accessible via locator programs that search various local directories listed on the Internet. However, unsolicited and indiscriminate enquiries might offend the recipient.

World Wide Web ("the Web"). The Web is a global network of databases, documents, images, and sounds. All types of information from anywhere in the Web network can be accessed and explored through navigation programs such as *Netscape Navigator* or *Microsoft Internet Explorer,* known as "browsers." Hypertext links among Web resources enable users to explore information along different paths by clicking on key words or icons that reveal additional paths for browsing and discovery.

Web Usage. Each website has its own *home page* that serves as an introduction to the site and is linked to additional "pages" that individual users can explore according to their information needs.

Through your campus or office network, or using a modem, a phone line, and an Internet Service Provider, you can search Internet files, databases, and home pages. You can participate in various newsgroups, subscribe to discussion lists, send e-mail enquiries, and gain access to publications that exist only in electronic form. Using a browser, you can explore sites on the Web, locate experts in all types of specialties, read the latest articles in journals such as *Nature* or *Science,* or review the latest listings of jobs in your specialty. For on-line databases in your field, ask your librarian.

Assume that any material obtained from the Internet is protected by copyright. Before using such material any place other than in a college paper (properly documented), obtain written permission from its owner.

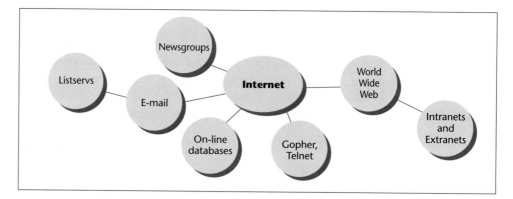

Figure 6.8   Various Parts of the Internet

Intranets and Extranets. An *intranet* is an in-house computer network that employs Internet technology for information access within a company.

Invaluable for on-the-job research, training, and collaboration, a customized intranet provides authorized users access to the company's document library, price lists, on-line discussions, and progress reports—even the data files of colleagues. Fast-food chains and other franchises increasingly use intranets to respond to franchisee questions and to distribute advice, industry/company news, sales figures, and other timely messages (Wallace 12). An intranet puts the company's knowledge and expertise at everyone's fingertips. Some organizations have their company "yellow pages," listing the expertise and information possessed by each employee.

An *extranet* integrates a company's intranet with the global Internet. External users (customers, subcontractors, outside vendors) with an Internet connection and a password can browse non-restricted areas of a company's website and download selected information, including customized reports. Extranets eliminate the need for the traditional printing and mailing of information to clients or suppliers (Stedman 49). They also enable collaboration among organizations. At Caterpillar Tractors, Inc., for example, when a customer's equipment breaks down, the entire company can mobilize immediately, contact experts inside and outside the organization, access records of previous solutions to similar problems, and collaborate on a solution—for example, designing an improved mechanical part for the equipment (Haskin 57–60).

The extranet blend of in-house information and Internet access raises security issues for any organization, leaving its network vulnerable to hackers, spies, or saboteurs (Meyerson 35). Therefore, each extranet site has its own *firewall* (software that keeps out uninvited users and that controls the data they can access) that includes password protection and *encryption* (coding) of sensitive information (Haskin 59).

Keyword Searches Using Boolean Operators. Most search engines that retrieve by keyword allow the use of Boolean[1] operators (commands such as "AND," "OR," "NOT") to define relationships among various keywords. Table 6.2 shows how these commands can expand a search or narrow it by generating fewer "hits."

Table 6.2 Using Boolean Operators to Expand or Limit a Search

If you enter these terms ...	The computer searches for ...
◆ electromagnetism AND health	◆ only entries that contain both words
◆ electromagnetism OR health	◆ all entries that contain either word
◆ electromagnetism NOT health	◆ only entries that contain Word 1 and do not contain Word 2
◆ electromag*	◆ all entries that contain this root within other words

---

1. British mathematician and logician George Boole (1815–1864) developed the system of symbolic logic (Boolean logic) now widely used in electronic information retrieval.

Boolean commands also can be combined, as in

(electromagnetic **or** radiation) **and** (fields **or** tumours)

The results (hits) produced from this query would contain any of these combinations:

electromagnetic fields, electromagnetic and tumours, radiation and fields, radiation and tumours

Using *truncation* (cropping a word to its root and adding an asterisk), as in *electromag**, would produce a broad array of hits, including these:

electromagnet, electromagnetic energy, electromagnetic pulse, electromagnetic wave . . .

In Brief GUIDELINES FOR INTERNET RESEARCH [2]

1. *Try to focus your search beforehand.* The more precisely you identify the information you seek, the lower your chance of wandering aimlessly through cyberspace.

2. *Select keywords or search phrases that are varied and technical, rather than general.* Some search terms generate better hits than others. In addition to "electromagnetic radiation," for example, try "electromagnetic fields," "power lines and health," or "electrical fields." Specialized terms (for example, *vertigo* versus *dizziness*) offer the best access to sites that are reliable, professional, and specific. Always check your spelling.

3. *Look for websites that are specific.* Compile a *hotlist* of sites that are most relevant to your needs and interests. (Specialized newsletters and trade publications are good sources for site listings.)

4. *Set a time limit for searching.* It's no secret that Internet searching ("surfing") can be addictive. Recent surveys indicate that employees spend sizable amounts of time surfing for personal instead of business-related information. As you begin a search set a 10–15 minute time limit, and avoid tangents that, no matter how engaging, are irrelevant to your search.

5. *Expect limited results from any search.* Each search engine (*Alta Vista, Excite, Hot Bot, Infoseek, WebCrawler, Yahoo,* etc.) has its own strengths and weaknesses. Some are faster and more thorough while others yield more targeted and updated hits. Some search titles only—instead of the full text—for keywords. In addition, studies show that "Web content is increasing so rapidly that no single search engine indexes more than about one-third of it" (Peterson 286). Broaden your coverage by using multiple search engines.

6. *Use bookmarks and hot lists for quick access to favourite websites.* Mark a useful site with a bookmark, which you then add to your hot list.

7. *Expect material on the Internet to have a brief lifespan.* Site addresses can change overnight; material is rapidly updated or discarded. If you find something of special value, save or print it before it changes or disappears.

8. *Be selective about what you download.* Download only what you need. Unless they are crucial to your research, omit graphics, sound, and video files because these consume time and disk space. Focus on text files only.

9. *Never download copyrighted material without written authorization from the copyright holder.* According to the 1997 No Electronic Theft Act (NET), you commit a federal crime if you possess or distribute "unauthorized electronic copies of copyrighted material valued over $1,000, even when no profit is involved" (Grossman 37). Only material in the public domain (page 132) is exempted. Such crimes are punishable by heavy fines and/or prison sentences.

Before downloading *anything* from the Internet, ask yourself: "Am I violating someone's privacy (as in forwarding an e-mail or a newsgroup entry)? or "Am I decreasing, in any way, the value of this material for the person who owns it?" Obtain permission beforehand and cite the source.

10. *Consider using information retrieval services.* An electronic service such as *Inquisit* or *Dialog* protects copyright holders by selling access to all materials in its database. For a monthly fee and/or per-page fee, users can download full texts of articles from countless periodicals, ranging from general to highly specialized. Subscribers to these Internet-accessible databases include companies and educational institutions.

Although these services effectively filter out a lot of Internet junk, they do not catalogue material that exists only in electronic form (e-zines, newsgroup and listserv entries, etc.). Therefore, these databases exclude potentially valuable material (such as research studies not yet available in hard copy) accessible only through a general Web search.

Note: Rick Broadhead and Jim Carroll publish several titles dealing with the Internet in Canada:

Canadian Internet Directory
Canadian Internet Handbook
Canadian Internet New Users Handbook
Canadian Internet Access Kit
Canadian Internet Advantage

To learn more about these titles (and their updates), see your bookseller.

---

2. Adapted from Baker 57; Branscrum 78; Busiel and Maeglin 39–40, 76; Fugate 40–41; Kawasaki 156; Matson 249–52.

Different search engines use Boolean operators in slightly different ways; many include additional options (such as NEAR, to search for entries that contain search terms within 10 or 20 words of each other). Click on the HELP option of your particular search engine to see which strategies it supports.

EXPLORING PRIMARY SOURCES

Work-related research is often based on primary research, an original, first-hand study of the topic, involving sources like those in Figure 6.9.

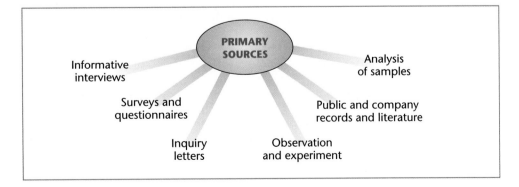

Figure 6.9 Sources for Primary Research

THE INFORMATIVE INTERVIEW

An excellent primary source of information unavailable in any publication is the personal interview. Much of what an expert knows may never be published (Pugliano 6). Also, a respondent might refer you to other experts or sources of information.

Of course, an expert opinion can be just as mistaken or biased as anyone else's. Like patients who seek second opinions about serious medical conditions,

---

GUIDELINES

for Informative Interviews[3]

Purpose statement

Planning the Interview

1. *Focus on your purpose.* Determine exactly what you hope to learn from this interview. Write out your purpose.

   > I will interview Anne Hector, chief engineer at Northport Electric, to ask her about the company's approaches to EMF risk avoidance—within the company as well as in the community.

2. *Do your homework.* Learn all you can about the topic beforehand. The more you know, the better your chance of getting the facts straight. If the respondent has published anything relevant, read it before the interview. Be sure the information this person might provide is unavailable in print.

3. *Contact the intended respondent.* Do this by telephone, letter, or e-mail, and be sure to introduce yourself and your purpose.

4. *Request the interview at your respondent's convenience.* Give the respondent ample notice and time to prepare, and ask whether she or he objects to being quoted or taped. If you use a tape recorder, insert fresh batteries and a new tape, and set the recording volume loud enough.

Preparing the Questions

1. *Make each question clear and specific.* Vague, unspecific questions elicit vague, unspecific answers.

A vague question

   > How is this utility company dealing with the problem of electromagnetic fields?

   > Which problem—public relations, potential liability, danger to electrical workers, to the community, or what?

Clear and specific question

   > What safety procedures have you developed for risk avoidance among electrical work crews?

2. *Avoid questions that can be answered with a mere "yes" or "no."*

An unproductive question

   > In your opinion, can technology find ways to decrease EMF hazards?

---

3. Several guidelines are adapted from Blum 88; Dowd 13–14; Kotulak 147; McDonald 190; Rensberger 15; Hopkins-Tanne 23, 29; Young 114, 115, 177.

GUIDELINES
for Informative Interviews
(continued)

A productive
question

A loaded question

An impartial
question

Clarifying questions

Instead, phrase your question to elicit a detailed response:

> Of the various technological solutions being proposed or considered, which do you consider most promising?

This is one instance in which your earlier homework pays off.

3. *Avoid loaded questions.* A loaded question invites or promotes a particular bias:

> Wouldn't you agree that EMF hazards have been overstated?

An impartial question does not lead the interviewee to respond in a certain way.

> In your opinion, have EMF hazards been accurately stated, overstated, or understated?

4. *Save the most difficult, complex, or sensitive questions for last.* Leading off with your toughest question might annoy respondents, making them uncooperative.

Conducting the Interview

1. *Make a good start.* Dress appropriately and arrive on time. Thank your respondent: restate your purpose; explain why you believe she or he can be helpful; explain exactly how you will use the information.

2. *Be sensitive to cultural differences.* If the respondent belongs to a culture different from your own, then consider the level of formality, politeness, directness, relationship building, and other behaviours considered appropriate in that culture.

3. *Let the respondent do the most talking.* Keep opinions to yourself.

4. *Be a good listener.* Don't doodle or let your eyes wander. People reveal more when the listener seems genuinely interested.

5. *Stick to your interview plan.* If the respondent wanders, politely nudge the conversation back on track (unless the added information is useful).

6. *Ask for clarification or explanation whenever necessary.* If you don't understand an answer, say so. Request an example, an analogy, or a simplified version—and keep asking until you understand.

> ◆ Would you go over that again please?
> ◆ Is there a simpler explanation?

Science writer Ronald Kotulak argues that, "No question is dumb if the answer is necessary to help you understand something. . . . Don't pretend to know more than you do" (144).

Follow-up questions

7. *Keep checking on your understanding.* Repeat major points in your own words and ask if the technical details are accurate and if your interpretation is correct.

8. *Be ready with follow-up questions.* Some answers may lead to additional questions.

   ◆ Why is it like that?
   ◆ Could you say something more about that?
   ◆ What more needs to be done?
   ◆ What happened next?

9. *Keep note-taking to a minimum.* Record statistics, dates, names, and other precise data, but not every word. Jot key terms or phrases that later can refresh your memory.

Concluding the Interview

1. *Ask for closing comments.* Perhaps the respondent can lead you to additional information.

Concluding
questions

   ◆ Have I missed anything?
   ◆ Would you care to add anything?
   ◆ Is there anything I've neglected to ask?
   ◆ Is there anyone else I should talk to?
   ◆ Is there anyone who has a different point of view?

2. *Invite the respondent to review your version.* If the interview will be published, ask the respondent to check your final draft (for misspelled names, inaccurate details, misquotations, and so on) and to approve it. Offer to provide copies of any document in which this information appears.

3. *Thank your respondent and leave promptly.*

4. *As soon as you leave the interview, write a complete summary (or record one verbally).* Do this while responses are fresh in your memory.

researchers seek a balanced range of expert opinions about a complex problem or controversial issue—not only from a company engineer and environmentalist, for example, but also from an independent and presumably more objective third party such as a professor or journalist who has studied the issue.

Selecting an Interview Medium.   Once you decide whom to interview about what, select your medium carefully:

   ◆ *In-person interviews* are most productive because they allow human contact (Hopkins-Tanne 24).

- *Phone interviews* can be convenient and productive, but they lack the human contact of in-person interviews—especially when the interviewer and respondent have not met.
- *E-mail interviews* are convenient and inexpensive, and they allow plenty of time for respondents to consider their answers.
- *Fax interviews* are highly impersonal, and using them is generally a bad idea.

Whatever your medium, obtain a respondent's approval *beforehand*—instead of waylaying this person with an unwanted surprise.

A Sample Interview

Figure 6.10 shows the partial text of an interview on persuasive challenges in the workplace. Notice how the interviewer probes, seeks clarification, and follows up on certain responses.

Surveys and Questionnaires

Surveys help us develop profiles and estimates about the concerns, preferences, attitudes, beliefs, or perceptions of a large, identifiable group (a *target population*) by studying representatives of that group (a *sample group*).

Surveys help us make assessments like these:

- Do consumers prefer brand A or brand B?
- What percentage of students feel safe on our campus?
- Is public confidence in technology increasing or decreasing?

The questionnaire is the tool for conducting surveys. While interviews allow for greater clarity and depth, questionnaires offer an inexpensive way to survey a large group. Respondents can answer privately and anonymously—and often more than in an interview.

Questionnaires carry certain limitations:

Limitations of survey research

- *A low rate of response (often less than 30 percent).* People refuse to respond to a questionnaire that seems too long, complicated, or in some way threatening. They might be embarrassed by the topic or afraid of how their answers could be used.
- *Responses that might be non-representative.* A survey will get responses from the people who want to respond, but you will know nothing about the people who didn't respond. Those who responded might have extreme views, a particular stake in the outcome, or some other motive that represents inaccurately the population being surveyed (Plumb and Spyridakis 625–26).
- *Lack of follow-up.* Survey questions do not allow for the kind of follow-up and clarification possible with interview questions.

Even surveys by professionals carry potential for error. As consumers of survey research, we need to understand how surveys are designed, administered, and interpreted, and what can go wrong in the process. The following is an introduction to creating surveys and to avoiding pitfalls along the way.

Defining the Survey's Purpose. Why is this survey being done? What, exactly, is it measuring? How much background research is needed? How will the survey findings be used?

Probing and following up

Q. *Would you please summarize your communication responsibilities?*
A. The corporate relations office oversees three departments: customer service (which handles claims, adjustments, and queries), public relations, and employee relations. My job is to supervise the production of all documents generated by this office.

Q. *Isn't that a lot of responsibility?*
A. It is, considering we're trying to keep some people happy, getting others to cooperate, and trying to get everyone to change their thinking and see things in a positive light. Just about every document we write has to be persuasive.

Seeking clarification

Q. *What exactly do you mean by "persuasive"?*
A. The best way to explain is through examples of what we do. The customer service department responds to problems like these: some users are unhappy with our software because it won't work for a particular application, or they find a glitch in one of our programs, or they're confused by the documentation, or someone wants the software modified to meet a specific need. For each of these complaints or requests we have to persuade our audience that we've resolved the problem or that we're making a genuine effort to resolve it quickly. The public relations department works to keep up our reputation through links outside the company. For instance, we keep in touch with this community, with consumers, the general public, government, and educational agencies. . . .

Seeking clarification

Q. *Can you be more specific? "Keeping in touch" doesn't sound much like persuasion.*
A. Okay, right now we're developing programs with colleges and universities, in which we offer heavily discounted software, backed up by an extensive support network (regional consultants, an 800 phone hotline, and workshops). We're hoping to persuade them that our software is superior to our well-entrenched competitor's. And locally we're offering the same kind of service and support to business clients.

Following up

Q. *What about employee relations?*
A. Day to day we face the usual kinds of problems: trying to get 100 percent employee contributions to the United Way, or persuading employees to help out in the community, or getting them to abide by new company regulations restricting smoking or to limit personal phone calls. Right now, we're facing a real persuasive challenge. Because of market saturation, software sales have flattened across the board. This means temporary layoffs for roughly 28 percent of our employees. Our only alternative is to persuade *all* employees to accept a 10 percent salary and benefit cut until the market improves.

Probing

Q. *How, exactly, do you persuade employees to accept a cut in pay and benefits?*
A. Basically, we have to make them see that by taking the cut, they're really investing in the company's future—and, of course, in their own.

(The interview continues.)

Figure 6.10   Partial Text of an Informative Interview

Defining the Target Population.   Who is the exact population being studied ("the chronically unemployed," "part-time students," "computer users")? For example, in its research on science and technology activity, the *1997 Statistical Abstract of the United States* differentiates "scientists and engineers" from "technicians":

Target populations
clearly defined

Scientists and engineers are defined as persons engaged in scientific and engineering work at a level requiring a knowledge of sciences equivalent at least to that acquired through completion of a 4-year college course. Technicians are defined as persons engaged in technical work at a level requiring knowledge acquired through a technical institute, junior college, or other type of training less extensive than 4-year college training. Craftspersons and skilled workers are excluded. (603)

Identifying the Sample Group.   How will intended respondents be selected? How many respondents will there be? Generally, the larger the sample surveyed, the more dependable the results (assuming a well-chosen and representative sample). Will the sample be randomly chosen? In the statistical sense, "random" does not mean "chosen haphazardly": a random sample means that any member of the target population stands an equal chance of being included in the sample group.

Even a sample that is highly representative of the target population carries a measure of *sampling error*.

A type of survey
error

The particular sample used in a survey is only one of a large number of possible samples of the same size that could have been selected using the same sampling procedures. Estimates derived from the different samples would, in general, differ from each other. (U.S. Department of Commerce 949)

The larger the sampling error (usually expressed as the *margin of error*, page 143) the less dependable the survey findings.

Defining the Survey Method.   What type of data (opinions, ideas, facts, figures) will be collected? Is timing important? How will the survey be administered—in person, by mail, by telephone? How will the data be collected, recorded, analyzed, and reported (Lavin 277)?

Telephone, e-mail, and in-person surveys yield fast results, but respondents consider telephone surveys annoying and, without anonymity, people tend to be less candid. They generate high response rates, but mail surveys are less expensive and more confidential. Computerized surveys create the sense of a video game: the program analyzes each response and automatically designs the next question. Respondents who dislike being quizzed by a human researcher seem more comfortable with this automated format (Perelman 89–90).

Guidelines for Developing a Questionnaire.
1. *Choose the types of questions.* (Adams and Schvaneveldt 202–12; Velotta 390). Questions can be *open-ended* or *closed-ended*. Open-ended questions allow respondents to express exactly what they're thinking or feeling in a word, phrase, sentence, or short essay:

Open-ended
questions

How much do you know about electromagnetic radiation at our school?
What do you think should be done about electromagnetic fields (EMFs) at our school?

Since one never knows what people will say, open-ended questions are a good way to uncover attitudes and obtain unexpected information. But essay-type questions are difficult to answer and tabulate.

When you want to measure where people stand on an issue, choose close-ended questions:

Are you interested in joining a group of concerned parents?
YES_____   NO_____

Rate your degree of concern about EMFs at our school.
HIGH _____   MODERATE _____   LOW _____   NO CONCERN _____

Circle the number that indicates your view about the town's proposal to spend $20,000 to hire its own EMF consultant.

1.......... 2.......... 3.......... 4.......... 5.......... 6.......... 7

Strongly   No   Strongly
Approve   Opinion   Disapprove

Respondents may be asked to *rate* one item on a scale (from high to low, best to worse), to *rank* two or more items (by importance, desirability), or to select items from a list. Other questions measure percentages or frequency.

How often do you . . . ?
ALWAYS _____ OFTEN _____ SOMETIMES _____ RARELY_____ NEVER_____

Although closed-ended questions are easy to answer, tabulate, measure, and analyze, they might elicit biased responses. Some people, for instance, automatically prefer items near the top of a list or the left side of a rating scale (Plumb and Spyridakis 633). Also, people are prone to agree rather than disagree with assertions in a questionnaire (Sherblom, Sullivan, and Sherblom 61).

2. *Design an engaging introduction and opening questions.* Persuade respondents that the survey relates to their concerns, that their answers matter, and that their anonymity is assured. Explain how respondents will benefit from your findings, or offer an incentive (say, a copy of your final report).

Your answers will enable our school board to speak accurately for your views at our next town meeting. Results of this survey will appear in our campus newspaper. Thank you.

Researchers often include a cover letter with the questionnaire.

Begin with the easiest questions. Once respondents commit to these, they are likely to complete more difficult questions later.

3. *Make each question unambiguous.* All respondents should be able to interpret identical questions identically. An ambiguous question leaves room for misinterpretation.

Do you favour weapons for campus police?   YES _____   NO _____

"Weapons" might mean tear gas, clubs, handguns, all three, or two out of three. Consequently, responses to the above question would produce a misleading statistic, such as "Over 95 percent of students favour handguns for campus police," when the accurate conclusion might be "Over 95 percent of students favour some form of weapon." Moreover, the limited choice ("yes/no") reduces an array of possible opinions to an either/or choice.

A clear, incisive question

Do you favour (check all that apply):

_____ Having campus police carry mace and a club?

_____ Having campus police carry non-lethal "stun guns"?

_____ Having campus police store handguns in their cruisers?

_____ Having campus police carry small-calibre handguns?

_____ Having campus police carry large-calibre handguns?

_____ Having campus police carry no weapons?

_____ Don't know

To ensure a full range of possible responses, include options such as "Other_____," "Don't know," "Not Applicable," or an "Additional Comments" section.

4. *Make each question unbiased.* Avoid *loaded questions* that invite or advocate a particular viewpoint or bias:

A loaded question

Should our campus tolerate the needless endangerment of innocent students by lethal weapons?

YES _____   NO _____

Emotionally loaded and judgmental words ("endangerment," "innocent," "tolerate," "needless," "lethal") in a survey are unethical because their built-in judgments manipulate people's responses (Hayakawa 40).

5. *Make it brief, simple, and inviting.* Try to limit questions and response space to two sides of a single page. Include a stamped, return-addressed envelope, and give a specific return date. Address each respondent by name, sign your letter or your introduction, and give your title.

A SAMPLE QUESTIONNAIRE

The student-written questionnaire in Figure 6.11 sent to presidents of local companies is designed to elicit responses that can be tabulated easily.

Written reports of survey findings usually include an appendix that contains a copy of the questionnaire as well as the tabulated responses.

ENQUIRIES

Letters, phone calls, or e-mail enquiries to experts listed in Web pages are handy for obtaining specific information from government agencies, legislators, private companies, university research centres, trade associations, and research foundations.

OFFICE FILES

Organization records (reports, memos, computer printouts, etc., are good primary sources. Most organizations also publish pamphlets, brochures, annual reports, or prospectuses for consumers, employees, investors, or voters. But be alert for bias in company literature. If you were evaluating the safety measures at a local nuclear power plant, you would want the complete picture. Along with the company's literature, you would also want studies and reports from government agencies and publications from environmental groups.

PERSONAL OBSERVATION AND EXPERIMENT

If possible, amplify and verify your findings with a first-hand look. Observation should be your final step because you now know what to look for. Have a plan. Know how, where, and when to look, and jot down observations immediately. You might even take photos or make drawings.

Informed observations can pinpoint real problems. Here is an excerpt from a report investigating low morale at an electronics firm. This researcher's observations and interpretation are crucial in defining the problem:

Direct observation is often essential

> Our on-site communications audit revealed that employees were unaware of any major barriers to communication. Over 75 percent of employees claimed they felt free to talk to their managers, but the managers, in turn, estimated that fewer than 50 percent of employees felt free to talk to them.
>
> The problem involves misinterpretation. Because managers don't ask for complaints, employees are afraid to make them, and because employees never ask for an evaluation, they never get one. Each side has inaccurate perceptions of what the other side expects, and because of ineffective communications, each side fails to realize that its perceptions are wrong.

Even direct observation is not foolproof; for instance, you might be biased about what you see (focusing on the wrong events or ignoring something important), or, instead of behaving normally, people being observed might behave in ways they think they expect (Adams and Schvaneveldt 244).

An experiment is a controlled form of observation designed to verify an assumption (e.g., the role of fish oil in preventing heart disease) or to test something untried (the relationship between background music and worker productivity). Each specialty has its own guidelines for experiment design.

ANALYSIS OF SAMPLES

Workplace research can involve collecting and analyzing samples: water, soil, or air, for contamination and pollution; foods, for nutritional value; ore, for mineral value; or plants, for medicinal value. Investigators analyze material samples to find the cause of an airline accident. Engineers analyze samples of steel, concrete, or other building materials to determine their load-bearing capacity. Medical specialists analyze tissue samples for disease.

Communication Questionnaire

1. Describe your type of company (e.g., manufacturing, high tech) _____
   _____

2. Number of employees? (Please check one.)

   _____ 5–25   _____ 51–100   _____ 151–300

   _____ 26–50   _____ 101–150   _____ 301–450

3. What types of written communication occur in your company? (Label by frequency:
   never, rarely, sometimes, often.)

   _____ memos   _____ letters   _____ advertising
   _____ manuals   _____ reports   _____ newsletters
   _____ procedures   _____ proposals   _____ other (Specify.)
   _____ e-mail   _____ catalogues   _____

4. Who does most of the writing? (Pls. give titles.) _____
   _____

5. Please characterize your employees' writing effectiveness.

   _____ good   _____ fair   _____ poor

6. Does your company have formal guidelines for writing?

   _____ no   _____ yes (Pls. describe briefly.) _____
   _____

7. Do you offer in-house communication training?

   _____ no   _____ yes (Pls. describe briefly.) _____
   _____

8. Please rank the usefulness of the following areas in communication training (from 1 through 10).

   _____ organization information   _____ audience awareness

   _____ summarizing information   _____ persuasive writing

   _____ editing for style   _____ grammar

   _____ document design   _____ researching

   _____ e-mail etiquette   _____ Web page design

   _____other   (Pls. specify.) _____

9. Please rank these skills in order of importance (from 1–6).

   _____ reading   _____ listening   _____ speaking to groups
   _____ writing   _____ collaborating   _____ speaking face-to-face

10. Do you provide tuition reimbursement for employees?

    _____ no   _____ yes

11. Would you consider having UMD communication interns work for you part-time?

    _____ no   _____ yes

12. Should UMD offer Saturday seminars in communication?

    _____ no   _____ yes

    Additional comments/suggestions: _____
    _____

Figure 6.11   **A Sample Survey**

EXERCISES

1. Begin researching for the analytical report (Chapter 19) due at semester's end. Complete these steps. (Your instructor might establish a timetable.)

   Phase One: Preliminary Steps

   a. Choose a topic of *immediate practical importance*, something that affects you or your community directly.

   b. Identify a specific audience and its intended use of your information. Complete an audience/purpose profile (page 34).

   c. Narrow your topic, and check with your instructor for approval.

   d. Make a working bibliography to ensure sufficient primary and secondary resources. Don't delay this step!

   e. List things you already know about your topic.

   f. Write a clear statement of purpose and submit it in a proposal memo (pages 47–49) to your instructor.

   g. Develop a tree chart of possible questions (as on page 99).

   h. Make a working outline.

   Phase Two: Collecting Data (Read Chapter 7 in preparation for this phase.)

   a. In your research, move from general to specific; begin with general reference works for an overview.

   b. Skim your material, looking for high points.

   c. Take selective notes. Don't write everything down! Use notecards.

   d. Plan and administer questionnaires, interviews, and enquiry letters.

   e. Whenever possible, conclude your research with direct observation.

   f. Evaluate and interpret your findings.

   g. Use the checklist on page 146 to reassess your research methods and reasoning.

   Phase Three: Organizing Your Data and Writing Your Report

   a. Revise and adjust your working outline, as needed.

   b. Compose an audience/purpose analysis, like the sample on page 34.

   c. Fully document all sources of information.

   d. Proofread carefully and add all needed supplements (title page, letter of transmittal, abstract, summary, appendix, glossary).

   Due Dates: To Be Assigned by Your Instructor

   List of possible topics due:

   Final topic due:

   Proposal memo due:

   Working bibliography and working outline due:

   Notecards due:

   Copies of questionnaires, interview questions, and enquiry letters due:

   Revised outline due:

   First draft of report due:

   Final draft with supplements and documentation due:

2. Using the printed or electronic card catalogue, locate and record the full bibliographic data for five books in your field or on your semester report topic, all published within the past year.

3. List the title of each of these specialized reference works in your field or on your topic: a bibliography, an encyclopedia, a dictionary, a handbook, an almanac (if available), and a directory.

4. Identify the major periodical index in your field or on your topic. Locate a recent article on a specific topic (e.g., use of artificial intelligence in medical diagnosis). Photocopy the article (get CANCOPY clearance) and write an informative abstract.

5. Consult the appropriate librarian and identify two databases you would search for information on the topic in Exercise 1.

6. Using technical report indexes, locate abstracts of three recent reports on one specific topic in your field. Provide complete bibliographic information.

7. Most Web browsers allow you to do keyword searches (page 111) by using a search engine such as Yahoo, Lycos, Alta Vista, or Infoseek. Each engine has its own guidelines and peculiarities; these usually are explained in a "help" file or user guide. Learn to use at least one search engine; for your colleagues, write instructions for designing and conducting a Web search using that engine.

   URLs: **www.yahoo.com**
   www.lycos.com
   www.altavista.com
   www.infoseek.com

8. Using Netscape Navigator, Internet Explorer, or a similar browser, search websites to locate resources for your report topic.

9. If your library belongs to a consortium of electronically networked libraries, search the holdings of other libraries on the network for topic resources not available in your library. Prepare a list of promising possibilities.

10. Revise these questions to make them appropriate for inclusion in a questionnaire:

    a. Would a female prime minister do the job as well as a male?

    b. Don't you think that euthanasia is a crime?

    c. Do you oppose increased government spending?

    d. Do you think welfare recipients are too lazy to support themselves?

    e. Are teachers responsible for the decline in literacy among students?

    f. Aren't humanities studies a waste of time?

    g. Do you prefer Rocket Cola to other leading brands?

    h. In meetings, do you think men are more interruptive than women?

11. Arrange an interview with someone in your field. Decide on general areas for questioning: job opportunities, chances for promotion, salary range, requirements, outlook for the next decade, working conditions, job satisfaction, etc. Compose specific interview questions; conduct the interview, and summarize your findings in a memo to your instructor.

COLLABORATIVE PROJECT

Divide into small groups, and decide on a campus or community issue or some other topic worthy of research. Elect a group manager to assign and coordinate tasks. At project's end, the manager will provide a performance appraisal by summarizing, in writing, the contribution of each team member. Assigned tasks will include planning, information gathering from primary and secondary sources, document preparation (including visuals) and revision, and classroom presentation. (See pages 11–19 for collaboration guidelines.)

Do the research, write the report, and present your findings to the class. (In conjunction with this project, your instructor may assign Chapter 23.)

Recording and Reviewing Research Findings

As you discover material during research, you confront questions like these: How much is worth keeping? How should I record it? Can I trust this information? What, exactly, does it mean? How will I credit the source? These latter stages of the research process require the same quality of critical thinking as the earlier stages discussed in Chapter 6.

RECORDING THE FINDINGS

Findings should be recorded in ways that enable you to easily locate, organize, and control the material as you work with it. Record primary research findings by using a laptop computer, photographs, drawings, tape recorder, videotape, or whichever medium suits your purpose. Record secondary research findings in the form of notes.

TAKING NOTES

Notecards are convenient because they are easy to organize and reorganize. In place of notecards, many researchers take notes on a laptop computer, using information or database management software that allows notes to be filed, rearranged, and retrieved by author, title, topic, date, etc.

Follow these suggestions when using notecards:

1. Make a separate bibliography card for each work you plan to consult (Figure 7.1). Record the complete entry, using the identical citation format that will appear in your document (see Chapter 8 for sample entries). When searching an on-line catalogue, you often can print out the full bibliographic record for each work, thereby ensuring accurate citation.
2. Skim the entire work to locate relevant material.
3. Go back and decide what to record. Use a separate card for each item.
4. Decide how to record the item: as a quotation or a paraphrase. When quoting others directly, be sure to record words and punctuation accurately. When restating or adapting material in your own words, be sure to preserve the original meaning and emphasis.

Record each bibliographic citation exactly as it will appear in your final report

> Pinsky, Mark A. *The EMF Book: What You Should Know About Electromagnetic Fields, Electromagnetic Radiation, and Your Health.* New York: Warner, 1995.

Figure 7.1   Bibliography Card

QUOTING THE WORK OF OTHERS

When you borrow exact wording, whether the words were written or spoken (as in an interview or presentation) or whether they appeared in electronic form, you must place quotation marks around all borrowed material. Even a single borrowed sentence or phrase, or a single word used in a special way, needs quotation marks, with the exact source properly cited.

If your notes fail to identify quoted material accurately, you may forget to credit the source in your report. Even when this omission is unintentional, writers face the charge of *plagiarism* (misrepresenting as one's own the words or ideas of someone else). Possible consequences of plagiarism include expulsion from school, the loss of your job, or a lawsuit.

In recording a direct quotation, copy the selection word for word (Figure 7.2) and include the page numbers.

If your quotation omits parts of a sentence, use an *ellipsis* (three periods: . . .) to indicate each part that you have omitted from the original. If your quotation omits the end of a sentence, the beginning of the subsequent sentence, or whole sentences or paragraphs, show the ellipsis with four periods (. . . .).

Ellipsis within and between sentences

If your quotation omits parts . . . use an ellipsis. . . . If your quotation omits the end. . . .

Be sure that your elliptical expression is grammatical and that the omitted material in no way distorts the original meaning.

If you insert your own words within the quotation, place them inside brackets to distinguish them from those of your source:

Brackets setting off personal comments within quoted material

"This profession [aircraft ground controller] requires exhaustive attention."

(For more on brackets, see the Appendix.)

Sentences and paragraphs that include quotations must be clear and understandable. Read your sentences aloud to be sure they make sense and they read smoothly and grammatically. Generally, integrated quotations are introduced by

Place quotation marks around all directly quoted material

Pinsky, Mark A. pp. 29–30.

"Neither electromagnetic fields nor electromagnetic radiation cause cancer per se, most researchers agree. What they may do is promote cancer. Cancer is a multistage process that requires an "initiator" that makes a cell or group of cells abnormal. Everyone has cancerous cells in his or her body. Cancer—the disease as we think of it—occurs when these cancerous cells grow uncontrollably."

Figure 7.2   Notecard for a Quotation

phrases such as "Wong argues that," "Dupuis suggests that," so that readers know who said what. More importantly, readers must see the relationship between the quoted idea and the sentence that precedes it. Use a transitional phrase that emphasizes this relationship by looking back as well as ahead:

> After you decide to develop a program, "the first step in the programming process . . . ."

Besides showing how each quotation helps advance the main idea you are developing, your integrated sentences should be grammatical:

> "The agricultural crisis," Marx acknowledges, "resulted primarily from unchecked land speculation."
>
> "She has rejuvenated the industrial economy of our region," Smith writes of Berry's term as regional planner.

(For quoting long passages and for punctuating at the end of a quotation, see the Appendix.)

Use a direct quotation only when precision, clarity, or emphasis requires the exact words from the original. Avoid excessively long quoted passages. Research writing is more a process of independent thinking, in which you work with the ideas of others in order to reach your own conclusions; you should therefore paraphrase, instead of quoting, much of your borrowed material.

PARAPHRASING THE WORK OF OTHERS

We paraphrase not only to preserve the original idea but also to express it in a clear, simple, direct, or emphatic way—without distorting the idea. *Paraphrasing* means more than changing or shuffling a few words; it means restating the original idea in your own words and giving full credit to the source.

To borrow or adapt someone else's ideas or reasoning without properly documenting the source is plagiarism. To offer as a paraphrase an original passage only slightly altered—even when you document the source—also is plagiarism. It is equally unethical to offer a paraphrase, although documented, that distorts the original meaning.

An effective paraphrase generally displays all or most of the following elements (Weinstein 3):

- reference to the author early in the paraphrase, to indicate the beginning of the borrowed passage
- keywords retained from the original, to preserve the meaning
- original sentences restructured and combined, for emphasis and fluency
- needless words from the original deleted, for conciseness
- your own words and phrases that help explain the author's ideas, for clarity
- a citation (in parentheses) of the exact source, to mark the end of the borrowed passage and to give full credit
- preservation of the author's original intent

Figure 7.3 shows an entry paraphrased from the passage in Figure 7.2. Paraphrased material does not have quotation marks, but you must acknowledge your debt to

Signal the beginning of the paraphrase by citing the author, and the end by citing the source

> Pinsky, Mark A.
>
> Pinsky explains that electromagnetic waves probably do not directly cause cancer. However, they might contribute to the uncontrollable growth of those cancer cells normally present—but controlled—in the human body (29–30).

Figure 7.3   Notecard for a Paraphrase

the source. Failing to acknowledge ideas, findings, judgments, lines of reasoning, opinions, facts, or insights not considered *common knowledge* (page 128) is plagiarism—even when these are expressed in your own words.

EVALUATING AND INTERPRETING INFORMATION

Not all information is equal. Not all interpretations are equal. Whether you work with your own findings or the findings of other researchers, you need to decide if the information is valid and reliable. Then you need to decide what your information means. Figure 7.4 outlines this challenge.

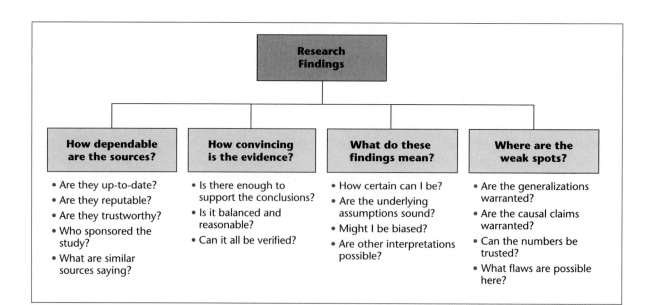

Figure 7.4   Decisions in Reviewing Research Findings

EVALUATING THE SOURCES

Not all data sources are equally dependable. A source might offer information that is out of date, inaccurate, incomplete, mistaken, or biased.

"How current is the information?"

Is the Source Up-to-date? Newly published books contain information that can be more than one year old, and journal articles often undergo a lengthy process of peer review.

Certain types of information become outdated more quickly than others. For topics that focus on *technology* (superconductivity, multi-media law, Internet censorship, alternative cancer treatments), information more than a few months old may be outdated. Except for historical or background research, sources in those areas generally should offer the most recent information available. For topics that focus on *people* (business ethics, management practices, workplace gender equality, employee motivation), information several decades old might offer valuable perspective on present situations.

"What is the source's reputation?"

Is the Source Reputable? Some sources enjoy better reputations than others. For research on alternative cancer treatments, you could depend more on reports in the *Canadian Medical Association Journal* or *Scientific American* than on those in scandal sheets or movie magazines. Even researchers with expert credentials, however, can disagree or be mistaken.

One way to assess a publication's reputation is to check its copyright page for information. Is the work published by a university, professional society, museum, or respected news organization? Do members of the editorial and advisory board have distinguished titles and degrees? Is the publication *refereed* (all submissions reviewed by experts prior to acceptance)?

One way to assess an author's reputation is to check citation indexes to see what others have said about this research. Many periodicals also provide brief biographies or descriptions of authors' earlier publications and other achievements.

"Can the source be trusted?"

Is the Source Trustworthy? The Internet offers information that may never appear in other sources, for example, from listservs and newsgroups. But much of this information may reflect the bias of the special interest groups that provide it. Moreover, anyone can publish almost anything on the Internet—including a great deal of misinformation—without having it verified, edited, or reviewed for accuracy (Snyder 89–90).

Even in a commercial database, decisions about what to include and what to leave out depend on the biases, priorities, or interests of those who assemble that database. In general, try not to rely on any single information source.

"Who sponsored the study?"

Is the Information Biased? Much of today's research is paid for by private companies or special-interest groups that have their own social, political, or economic agendas (Crossen 14, 19). Medical research may be sponsored by drug or tobacco companies; nutritional research, by food manufacturers; and environmental research, by oil or chemical companies. Public policy research (on gun control, school prayer, seat-belt laws, endangered species) may be sponsored by opposing groups (environmentalists versus the logging industry), producing

In Brief COPYRIGHT PROTECTION AND FAIR USE OF PRINTED INFORMATION

Copyright Law

A copyright is the exclusive legal right to reproduce, publish, and sell a literary, dramatic, musical, or artistic work. The law grants the copyright owner the exclusive rights to do and to authorize any of the following:

1. To reproduce the copyrighted work.
2. To prepare derivative works.
3. To distribute copies of the copyrighted work to the public by sale, rental, lease, or lending.
4. In certain cases, such as for literary and musical works, to perform the copyrighted work publicly.
5. In certain cases, such as for graphics, images, or other audiovisuals, to display the copyrighted work publicly.

You must obtain written permission to use all copyrighted material. Works are copyrighted for the author's life plus 50 years. If the author is unknown, the copyright lasts for 50 years from the document's publication date.

Public Domain

Public domain refers to material on which copyright has expired or material that is not protected by copyright. Most government publications and commonplace information such as height and weight charts are in the public domain. These works occasionally contain copyrighted material used with permission and properly acknowledged. **However, a new translation or version of a work in the public domain can be protected by copyright;** if you are not sure whether something is in the public domain, obtain permission.

Fair Dealing

Fair dealing allows quotes from, or reproduction of, minor excerpts of a copyrighted work, if the quotation or reproduction is for bona fide private study, research, criticism, or newspaper sum-

mary. But there's a problem: It's difficult to tell the difference between fair dealing and copyright infringement. The limits of fair dealing copying have not yet been defined in the Canadian Copyright Act or by case law.

Copyright law provides the following criteria to be considered in the determination of fair dealing:

1. The purpose and character of the use, including whether such use is of a commercial nature or is for non-profit educational purposes.
2. The nature of the copyrighted work.
3. The amount and substantiality of the portion used in relation to the whole.
4. The effect of the use upon the potential market for or value of the copyrighted work.

When the quoted material forms the core, distinguishable creative effort of the work being cited, use of the material without permission isn't considered fair.

Fair dealing ordinarily does not apply to use of poetry, musical lyrics, dialogue of a play, entries in a diary, case studies, charts and graphs, author's notes, private letters, testing materials, or quotations for use as epigraphs.

Copying Under the CANCOPY Licence

Many Canadian college, university, high school, and public libraries have signed licence agreements with CANCOPY, a non-profit organization representing Canadian and foreign authors and publishers. Under such licences, library users can copy certain portions of certain kinds of documents without permission. HOWEVER, EACH LICENCE IMPOSES CLEAR LEGAL RESTRICTIONS, so ask your librarian for details.

Based on *HarperCollins Author's Guide.* Copyright ©1995, and on Canada's Copyright Act, as amended by Bill C-32.

opposing results. Instead of a neutral and balanced enquiry, this kind of "strategic research" is designed to support one special interest group or another. Those who pay for strategic research are not likely to publicize findings that contradict their original claims, opinions, or beliefs (profits lower than expected, losses or risks

In Brief COPYRIGHT PROTECTION AND FAIR DEALING OF ELECTRONIC INFORMATION

The Problem

Copyright and fair dealing law is quite specific for printed works or works in other tangible form (paintings, photographs, music). But how do we define "fair dealing" (above) of intellectual property in electronic form? How does copyright protection apply (Dyson 137)? How do fair-dealing restrictions apply to material used in multi-media presentations or to text or images that have been altered or reshaped to suit the user's specific needs (Steinberg, "Travels" 30)?

Information obtained via e-mail or discussion groups presents additional problems: Sources often do not wish to be quoted or named or to have early drafts made public. How do we protect source confidentiality? How do we avoid infringing on works in progress that have not yet been published? How do we quote and cite this material without violating ownership and privacy rights (Howard 40–41)?

Present Status of Electronic Copyright Law

Subscribers to commercial on-line databases pay fees, and copyholders in turn receive royalties (Communication Concepts, Inc. 13). But as of this writing, few specific legal protections exist for non-commercial types of electronic information. Since April 1989, however, most works are considered copyrighted as soon as they are produced—even if they carry no copyright notice. Fair dealing of electronic information generally is limited to brief excerpts that serve as a basis for response—for example, in a discussion group.

Except for certain government documents, no Internet posting is in the public domain unless it is expressly designated as such by its author (Templeton).

Until specific laws are enacted, the following examples can be considered violations of copyright (Communication Concepts, Inc. 13; Templeton)

- downloading a work from the Internet and forwarding copies to other readers
- editing, altering, or incorporating an original work as part of your own document or multi-media presentation
- putting someone else's printed work on-line without the author's written permission
- copying and forwarding an e-mail message without the sender's authorization. The e-mail *text* is copyrighted, but its *content* legally may be revealed—except for proprietary information.

Penalties for Copyright Infringement

Individuals, businesses, or organizations that hold copyright may sue those who infringe copyright. Canadian legislation stipulates minimum penalties for copyright infringements. Violations of copyright on printed or electronic works may exceed the boundaries of civil law and may be prosecuted as summary convictions under criminal law. When in doubt, assume the work is copyrighted and obtain written permission for its use.

greater than expected). As consumers of research, we should try to determine exactly what the sponsors of a particular study stand to gain or lose from the results. The following hints may further help you evaluate Internet sites:

1. Look at the website's commercial component. If a company has published a site to promote its products, be suspicious. In particular, notice what information the company does *not* provide.
2. Don't be satisfied with generalities. Look for specific facts, examples, or statistics. Then, check to see which of these "facts" can be verified by other sources.
3. Try to verify the accuracy of quoted facts and statistics by contacting other sources or by checking for similar results in other studies or experiments.

4. Look at your own attitudes and beliefs to see if you're predisposed to automatically accept certain things that you read.

5. Exercise extreme caution in using anything picked up from usenet discussion groups.

GUIDELINES

for Evaluating Websites[1]

1. *Consider the site's domain type and sponsor.* In the typical address **www.umass.edu**, the site or domain information follows the *www.* The *.edu* signifies the type of organization from which the site originates. Standard domain types in North America are:

.com	=	business/commercial organization
.edu	=	educational institution
.gov	=	government organization
.mil	=	military organization
.net	=	any group or individual with simple software and Internet access
.org	=	non-profit organization

The domain type might signal a certain bias or agenda that could skew the data. For example, at a *.com* site, you might find accurate information, but also some type of sales pitch. At an *.org* site, you might find a political or ideological bias (say, The Heritage Foundation's conservative ideology versus the Brookings Institution's more liberal slant). Knowing about a site's sponsor can help you evaluate the credibility of its postings.

2. *Identify the purpose of the page or message.* Decide whether the message is intended to merely relay information, to sell something, or to promote a particular ideology or agenda.

3. *Look beyond the style of a site.* Fancy graphics, video, and sound do not always translate into dependable information. Sometimes the most reliable material resides in the less attractive, text-only sites. People can design flashy-looking pages without necessarily knowing what they are talking about. Even something written in clear, plain English instead of in difficult scientific terms (as in medical information) might be inaccurate.

4. *Assess the site's/material's currency.* An up-to-date site should clearly indicate when the material was created or published, and when it was posted and updated.

5. *Assess the author's credentials.* Learn all you can about the author's reputation, level of expertise on the topic, and institutional affiliation (a university, a Fortune 500 company, a reputable environmental group). Do this by following links to other sites that mention the author or by using search

1. Adapted from Barnes; Busiel and Maeglin 39; Elliot; Facklemann 397; Grassian; Hall 60–61; Hammett; Harris; Stemmer.

GUIDELINES

for Evaluating
Websites

(continued)

engines to track the author's name. Newsgroup postings often contain "a *signature file* that includes the author's name, location, institutional or organizational affiliation, and often a quote that suggests something of the writer's personality, political leanings, or sense of humour" (Goubil-Gambrell).

6. *Compare the site with other sources.* Check related sites, publications, and other sources to compare the quality of information and to discover what others might have said about this site or author. Comparing many similar sites helps you create a *benchmark*, a standard for evaluating any particular site (based on the criteria in these guidelines). Ask a librarian for help.

7. *Decide whether the assertions/claims make sense.* Based on what you know about the issue, decide where, on the spectrum of informed opinion and accepted theory, this author's position resides. How well is each assertion supported? Never accept any claim that seems extreme without verifying it through other sources, such as a professor, a librarian, or a specialist in the field.

8. *Look for other indicators of quality.*

   ◆ *Worthwhile content:* The material displays a clear and sharply focused main point, technical accuracy, opinions based on fact or good sense, and assertions supported by evidence that is documented (citing all sources of data presented as "factual").

   ◆ *Sensible organization:* The material is organized for the reader's understanding, with a clear line of reasoning.

   ◆ *Readable style:* The material is well written (clear, concise, and easy to understand) and free of typos, misspellings, and other errors.

   ◆ *Objective coverage:* Debatable topics are addressed in a balanced and impartial way, with fair, accurate representation of opposing views. The tone is reasonable, with no "sounding off."

   ◆ *Expertise:* The author refers to related theory and other work in the field and uses specialized terminology accurately and appropriately.

   ◆ *Peer review:* The material has been evaluated and verified by related experts.

   ◆ *Links to reputable sites:* The site offers a gateway to related sites that meet quality criteria.

   ◆ *Follow-up option:* The material includes a signature block or a link for contacting the author or organization for clarification or verification.

6. When considering material provided in a listserv academic discussion group, look at the author's credentials: Does the author have a solid track record of publications or other contributions to the field discussed by that group?

7. In general, maintain a suspicious mindset: Demand evidence for any assertions you read.

EVALUATING THE EVIDENCE

Evidence is any finding used to support or refute a particular conclusion. While evidence can serve the truth, it also can create distortion, misinformation, and deception. For example, how much money, material, or energy does recycling really save? How good for your heart is oat bran? How well are public schools educating our children? Which investments or automobiles are safest? Conclusions about such matters are based on evidence that often can be manipulated in support of one view or another. As consumers of research, we have to assess for ourselves the quality of the evidence presented.

We assess the quality of evidence by examining it critically to understand its limitations, to see if findings conflict; to discover connections, similarities, trends, or relationships; to determine the need for further enquiry; and to raise new questions.

"Is there enough evidence?"

Is the Evidence Sufficient?   Evidence is sufficient when it enables us to reach an accurate judgment or conclusion. A study of the stress-reducing benefits of low-impact aerobics, for example, would require a broad survey sample: people who have practised aerobics for a long time; people of different genders, different ages, different occupations, and different lifestyles before they began aerobics, etc. Even responses from hundreds of practitioners might constitute insufficient evidence unless those responses were supported by laboratory measurements of metabolism, heart rates, and blood pressure.

Personal experience usually offers insufficient evidence from which to generalize. You cannot tell whether your experience is representative, no matter how long you might have practised aerobics. Although anecdotal evidence ("This worked great for me!") might offer a good starting point for an investigation, personal experience should be evaluated within the broader context of *all* available evidence.

"Is the evidence hard or soft?"

Can the Evidence Be Verified?   *Hard evidence* consists of factual statements, expert opinion, or statistics that can be verified (shown to be true). *Soft evidence* consists of uninformed opinion or speculation, data obtained or analyzed unscientifically, and findings that have not been replicated or reviewed by experts. Reputable news organizations employ fact-checkers to verify information before it appears in print.

Evidence that seems scientific can turn out to be soft. For example, information obtained from polling often is reported in fancy charts, graphs, and impressive statistics—but it is based on public opinion, which is almost always changing (Crossen 104).

Base your conclusions on hard evidence. For example, suppose an article makes positive claims about low-impact aerobics but provides no data on measurements of pulse, blood pressure, or metabolic rates. Although these claims might coincide with your own experience, your evidence so far consists of only two opinions: yours and the author's—without scientific support. Any conclusion at this point would rest on soft evidence. Only after carefully assessing dependable sources can you decide which conclusions are supported by the bulk of the evidence.

INTERPRETING THE EVIDENCE

Interpreting means trying to reach the truth of the matter: an overall judgment about what the evidence means and what conclusion or action it suggests. Unfortunately, research does not always yield answers that are conclusive or about which we can be certain. Instead of settling for the most *convenient* answer, we should pursue the most *reasonable* answer by examining critically a full range of possible meanings.

What Level of Certainty Is Warranted?   As possible outcomes of research, we can identify three distinct and very different levels of certainty:

1. The definitive truth—the *conclusive answer:*

A practical definition of "truth"

> *Truth is* what is so *about something, the reality of the matter, as distinguished from what people wish were so, believe to be so, or assert to be so. From another perspective, in the words of Harvard philosopher Israel Scheffler, truth is the view "which is fated to be ultimately agreed to by all who investigate." The word* ultimately *is important. Investigation may produce a wrong answer for years, even for centuries. . . . Does the truth ever change? No. . . . One easy way to spare yourself any further confusion about truth is to reserve the word* truth *for the final answer to an issue. Get in the habit of using the words,* belief, theory, *and* present understanding *more often.* (Ruggiero 21–22)

We often are mistaken in our certainty about the *truth*. For example, in the second century A.D., Ptolemy's view of the universe concluded that the earth was its centre. Though untrue, this judgment was based on the best information available at that time. Ptolemy's view survived for 13 centuries, even after new information had discredited this belief. When Copernicus and Galileo proposed more truthful views in the fifteenth century, they were labelled heretics.

Conclusive answers are the research outcome we seek, but often we have to settle for answers that are less than certain.

2. The *probable answer:* the answer that stands the best chance of being true or accurate—given the most we can know at this particular time. Probable answers are subject to revision in the light of new information.

3. The *inconclusive answer:* the realization that the truth of the matter is far more elusive, ambiguous, or complex than we expected.

"Exactly how certain are we?"

To ensure an accurate outcome, we must decide what level of certainty the findings warrant. For example, we are *highly certain* about the perils of smoking or sunburn, *reasonably certain* about the benefits of fruits and vegetables and moderate exercise, but *far less certain* about the perils of coffee drinking, or electromagnetic waves, or the benefits of vitamin supplements.

Are the Underlying Assumptions Sound?   *Assumptions* are notions we take for granted, things we accept without proof. The research process rests on assumptions like these: that a sample group accurately represents a larger target group, that survey respondents remember certain facts accurately, and that mice and humans share enough biological similarities for meaningful research. For a particular study to be valid, the underlying assumptions must be accurate.

Consider this example: You are an education consultant evaluating the accuracy of IQ testing as a predictor of academic performance. Reviewing the evidence, you perceive an association between low IQ scores and low achievers. You then check your statistics by examining a cross-section of reliable sources. Can you then conclude that IQ tests do predict performance accurately? This conclusion might be invalid unless you could verify the following assumptions:

1. That neither parents, teachers, nor children had seen individual test scores, which could produce biased expectations.
2. That, regardless of score, each child had completed an identical curriculum at an identical pace, instead of being "tracked" on the basis of his or her score.

Do I Have a Personal Bias?   To support a particular version of the truth, our own bias might cause us to overestimate (or deny) the certainty of our findings.

Expect yourself to be biased, and expect your bias to affect your efforts to construct arguments. Unless you are perfectly neutral about the issue, an unlikely circumstance, at the very outset . . . you will believe one side of the issue to be right, and that belief will incline you to . . . present more and better arguments for the side of the issue you prefer. (Ruggiero 134)

Because personal bias is hard to transcend, *rationalizing* often becomes a substitute for *reasoning:*

Reasoning versus rationalizing

You are reasoning if your belief follows the evidence—that is, if you examine the evidence first and then make up your mind. You are rationalizing if the evidence follows your belief—if you first decide what you believe and then select and interpret evidence to justify it. (Ruggiero 44)

Personal bias often is unconscious until we examine our own value systems, attitudes long held but never analyzed, notions we've inherited from our own backgrounds, and so on. Recognizing our own biases is a crucial first step in managing them.

"What else could this mean?"

Are Other Interpretations Possible?   Perhaps other researchers would disagree with the meaning of these findings. Some controversial issues (the need for defence spending or causes of inflation) will never be resolved. Although we can get verifiable data and can reason persuasively on some subjects, no close reasoning by any expert and no supporting statistical analysis will prove anything about a controversial subject to everyone's satisfaction. For instance, one could only *argue* (more or less effectively) that federal funds will or will not alleviate poverty or unemployment.

Settling on a final meaning can be difficult. For example (Ledermen 5): What does a reported increase in violent crime on North American college campuses mean—especially in light of national statistics that show violent crime decreasing?

- That college students are becoming more violent?
- That some drugs and guns in high schools end up on campuses?
- That off-campus criminals see students as easy targets?

Or could these findings mean something else entirely?

- That increased law enforcement has led to more campus arrests—and thus, greater recognition of the problem?
- That crimes haven't increased, but more are being reported?

Depending on our interpretation, we might conclude that the problem is worsening—or improving!

AVOIDING ERRORS IN REASONING

Finding the truth, especially in a complex issue or problem, often is a process of elimination, of ruling out or avoiding errors in reasoning. As we interpret, we make *inferences:* We derive conclusions about what we don't know by reasoning from what we do know (Hayakawa 37). For example, we might infer that a drug that boosts immunity in laboratory mice will boost immunity in humans, or that a rise in campus crime statistics is caused by the fact that young people have become more violent. Whether a particular inference is on target or dead wrong depends largely on our answers to one or more of these questions:

- To what extent can these findings be generalized?
- Is Y really caused by X?
- How much can the numbers be trusted, and what do they mean?

Following are three major reasoning errors that can distort our interpretations.

"How much can we generalize?"

Faulty Generalizations. When we accept research findings uncritically and jump to conclusions about their meaning, we commit the error of *hasty generalization*. When we overestimate the extent to which the findings reveal some larger truth, we commit the error of *overstated generalization*.

A study in Greece on the role of fruits, vegetables, and olive oil in lowering breast cancer risk was widely publicized in 1995 because of the alleged benefits of olive oil for women who consume it twice or more a day. Subsequent analysis of this study revealed that data about the women's food consumption covered only one year and were based on a single questionnaire asking respondents to estimate their previous year's diet. (Estimates of this type tend to be highly inaccurate.) Also, the study did not identify the quantities of olive oil individual users consumed. In this instance, the study's generalization about olive oil was shown to be *hasty* (based on insufficient evidence).

Further analysis revealed that only 99 respondents (of the nearly 2500 surveyed) claimed to have consumed olive oil twice or more a day ("Olive Oil" 1). In this instance, the study's generalization was shown to be *overstated* (a limited generalization made to apply to all cases). Something true in one instance need not be true in other instances.

Although this particular study was flawed, many other studies support the generalization that fruits and vegetables do help lower the risk of cancer. Generalizing is vital and perfectly legitimate—when it is warranted.

Faulty Causal Reasoning. Causal reasoning tries to explain *why* something happened or *what* will happen: often very complex questions. Faulty causal reasoning oversimplifies or distorts the cause-effect relationship through errors like these:

Ignoring other causes	Investment builds wealth. [Ignores the role of knowledge, wisdom, timing, and luck in successful investing.]
Ignoring other effects	Running improves health. [Ignores the fact that many runners get injured, and that some even drop dead while running.]
Inventing a cause	Right after buying a rabbit's foot, Felix won the 649 lottery. [Posits an unwarranted causal relationship merely because one event follows another.]
Confusing correlation with causation	Poverty causes disease. [Ignores the fact that disease, while highly associated with poverty, has many causes unrelated to poverty.]
Rationalizing	My grades were poor because my exams were unfair. [Denies the possible causes of one's failures.]

Because of bias or impatience, we can be tempted to settle for a hasty cause or to confuse possible, probable, and definite causes.

"Did X possibly, probably, or definitely cause Y?"

Sometimes a definite cause is apparent (e.g., "The engine's overheating is caused by a faulty radiator cap"), but usually much analysis is needed to isolate a specific cause. Suppose you want to answer this question: Why does our local college not have daycare facilities? Brainstorming yields these possible causes:

◆ lack of need among students
◆ lack of interest among students, faculty, and staff
◆ high cost of liability insurance
◆ lack of space and facilities on campus
◆ lack of trained personnel
◆ prohibition by law
◆ lack of government funding for such a project

Say you proceed with interviews, questionnaires, and research into provincial laws, insurance rates, and availability of personnel. You begin to rule out some items, and others appear as probable causes. Specifically, you find a need among students, high campus interest, an abundance of qualified people for staffing, and no provincial laws prohibiting such a project. Three probable causes remain: lack of funding, high insurance rates, and lack of space. Further inquiry shows that lack of funding and high insurance rates *are* issues. These obstacles, however, could be eliminated through new sources of revenue: charging a fee for each child, soliciting donations, or diverting funds from other campus organizations.

Finally, after examining available campus space and speaking with school officials, you arrive at one definite cause: lack of space and facilities. One could argue that lack of space and facilities is somehow related to funding, and the college's being unable to find funds or space may be related to student need, which is not sufficiently acute or interest sufficiently high to exert real pressure. Lack of space and facilities, however, appears to be the *immediate* cause.

When you report on your research, be sure readers can draw conclusions identical to your own on the basis of the evidence. The process might be diagrammed like this:

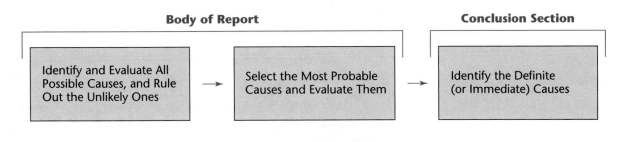

Body of Report / Conclusion Section

Identify and Evaluate All Possible Causes, and Rule Out the Unlikely Ones → Select the Most Probable Causes and Evaluate Them → Identify the Definite (or Immediate) Causes

Initially you might have based your conclusions hastily on soft evidence (an opinion—buttressed by a newspaper editorial—that the campus was apathetic). Now you base your conclusions on solid, factual evidence. You have moved from a wide range of possible causes to a narrow range of probable causes, and finally to one definite cause.

Sometimes, finding a single cause is impossible, but this reasoning process can be tailored to most problem-solving analyses. Anything but the simplest effect is likely to have more than one cause. By narrowing the field, you can focus on the real issues.

Faulty Statistical Reasoning. The purpose of statistical analysis is to determine the meaning of a collected set of numbers. In primary research, our surveys and questionnaires often lead to some kind of numerical interpretation ("What percentage of respondents prefer X?" "How often does Y happen?"). In secondary research, we rely on numbers collected by primary researchers. Numbers seem more precise, more objective, more scientific and less ambiguous than words. They are easier to summarize, measure, compare, and analyze. But numbers can be misleading. For example, radio or television phone-in surveys produce grossly distorted data: Although 90 percent of callers might express support for a particular viewpoint, callers tend to be those with the greatest anger or the strongest feelings about the issue—representing only a fraction of overall attitudes (Fineman 24). Mail-in surveys can produce similar distortion because only people with certain attitudes might choose to respond.

Before relying on any set of numbers, we need to know exactly where they came from, how they were collected, and how they were analyzed (Lavin 275–76). Can the numbers be trusted, and if so, what do they mean?

Faulty statistical reasoning produces conclusions that are unwarranted, inaccurate, or downright deceptive. The following are some common statistical fallacies:

♦ *The sanitized statistic*: Numbers are manipulated (or "cleaned up") to obscure the facts. For instance, a recently revised formula enables the government to exclude from its unemployment figures an estimated 5 million people who remain unemployed after one year—thus creating a far rosier economic picture than the facts warrant. Similar formulas allow for all sorts of sugarcoating in reports of wages, economic growth, inflation, and other statistics that affect the political climate (Morgenson 54). The College Board's recentring of SAT scores has raised the average math score from 478 to 500 and the average verbal score from 424 to 500 (a boost of almost 5 and 18 percent, respectively) although actual student performance remains unchanged (Samuelson 44).

(margin note) "How much can we trust these numbers?"

(margin note) "Exactly how well are we doing?"

◆ *The meaningless statistic:* Exact numbers are used to quantify something so inexact or vaguely defined that it should only be approximated (Huff 247; Lavin 278): "Only 38.2 percent of college graduates end up working in their specialty." "Toronto has 3 247 561 rats." "Zappo detergent makes laundry 10 percent brighter." An exact number looks impressive, but certain subjects (child abuse, cheating in college, virginity, drug and alcohol abuse on the job, eating habits) cannot be quantified exactly because respondents don't always tell the truth (because of denial, embarrassment, or merely guessing). Or they respond in ways they think the researcher -expects.

◆ *The undefined average:* The mean, median, and mode are confused in determining an average (Huff 244; Lavin 279). The *mean* is the result of adding up the value of each item in a set of numbers, then dividing by the number of items. The *median* is the result of ranking all the values from high to low, then choosing the middle value (or the 50th percentile, as in calculating SAT scores). The *mode* is the value that occurs most often in a set of numbers.

"How many rats was that?"

Each of these three measurements represents some kind of average, but unless we know which average is being presented, we cannot interpret the figures accurately. Assume that we are computing the average salary among female vice presidents at XYZ Corporation:

Vice President	Salary
A	$90,000
B	90,000
C	80,000
D	65,000
E	60,000
F	55,000
G	50,000

In the above example, the mean salary (total salaries divided by people) equals $70,000; the median salary (middle value) equals $65,000; the mode (most frequent value) equals $90,000. Each is legitimately an average, and each could be used to support or refute a particular assertion (for example, "Women vice presidents are paid too little" or "Women vice presidents are paid too much").

"Why is everybody griping?"

Research expert Michael Lavin sums up the potential for bias in reporting averages:

Depending on the circumstances, any one of these measurements [mean, median, or mode] may describe a group of numbers better than the other two. . . . [But] people typically choose the value which best presents their case, whether or not it is the most appropriate to use. (279)

Although the mean is the most commonly computed average, this measurement can be misleading when values on either end of the scale are extremely high or low. Suppose, for instance, that vice president A received a $200,000 salary. Because this figure deviates so far from the normal range of salary figures for B through G, it distorts the average for the whole group—increasing the mean salary by more than 20 percent (Plumb and Spyridakis 636).

◆ *The distorted percentage figure:* Percentages are reported without explanation of the original numbers used in the calculation (Adams and Schvaneveldt 359; Lavin 280): "Seventy-five percent of respondents prefer our brand over the competing brand"—without mention that only four people were surveyed. Or "Sixty-six percent of employees we hired this year are women and minorities, compared to the national average of 40 percent"—without mention that only three people have been hired this year, by a company that employs 300 (mostly white males).

Another fallacy in reporting percentages occurs when the *margin of error* is ignored. This is the margin within which the true figure lies, based on estimated sampling errors in a survey. For example, a claim that most people surveyed prefer brand X might be based on the fact that 51 percent of respondents expressed this preference; but if the survey carried a 2 percent margin of error, the true figure could be as low as 49 percent or as high as 53 percent. In a survey with a high margin of error, the true figure may be so uncertain that no definite conclusion can be drawn.

◆ *The bogus ranking:* Items are compared on the basis of ill-defined criteria (Adams and Schvaneveldt 212; Lavin 284): "Last year, the Batmobile was the number-one selling car in Canada"—without mention that some competing car makers actually sold *more* cars to private individuals, and that the Batmobile figures were inflated by hefty sales to rental-car companies and corporate fleets. Unless we know how the ranked items were chosen and how they were compared (the criteria), a ranking can produce a scientific-seeming number based on a completely unscientific method.

◆ *The fallible computer model:* Computer models process complex assumptions to produce impressive but often inaccurate statistical estimates about costs, benefits, risks, or probable outcomes.

Assumptions are notions we take for granted, things we often accept without proof. The research process rests on assumptions like these: that a sample group accurately represents a larger target group, that survey respondents remember certain facts accurately, that mice and humans share enough biological similarities for meaningful research. For a particular study to be valid, the underlying assumptions have to be accurate.

Computer models to predict global warming levels, for instance, are based on differing assumptions about wind and weather patterns, cloud formations, ozone levels, carbon dioxide concentrations, sea levels, or airborne sediment from volcanic eruptions. Despite their seemingly scientific precision, different global warming models generate 50-year predictions of sea-level rises that range from a few inches to several feet (Barbour 121). Other models suggest that warming effects could be offset by evaporation of ocean water and by clouds reflecting sunlight back to outer space (Monastersky 69). Still other models suggest that the 0.56°C (1°F) warming over the last 100 years may not be the result of the greenhouse effect at all, but of "random fluctuations in global temperatures" (Stone 38). The estimates produced by any model depend on the assumptions (and data) programmed in.

Choice of assumptions might be influenced by researcher bias or the sponsor's agenda. For example, a prediction of human fatalities from a nuclear plant meltdown might rest on assumptions about availability of safe shelter, evacuation routes, time of day, season, wind direction, and structural integrity of the containment unit. But the assumptions could be manipulated to produce an overstated or understated estimate of risk (Barbour 228). For computer-modelled estimates of accident risk (oil spill, plane crash) or of the costs and benefits of a proposed project or policy

"Is 51 percent really a majority?"

"Which car should we buy?"

"Garbage in, garbage out."

(a space station, welfare reform), consumers rarely know the assumptions behind the numbers. We wonder, for example, about the assumptions underlying NASA's pre-*Challenger* risk assessment, in which a 1985 computer model reportedly showed an accident risk of less than 1 in 100 000 shuttle flights (Crossen 54).

◆ *Confusion of correlation with causation: Correlation* is the measure of association between two variables (between smoking and increased lung cancer risk, or between education and income). *Causation* is the demonstrable production of a specific effect (smoking causes lung cancer). Correlations between smoking and lung cancer or education and income signal a causal relationship that has been proven by studies of all kinds. But not every correlation implies causation. For instance, a recently discovered correlation between moderate alcohol consumption and decreased heart disease risk offers insufficient proof that moderate drinking *causes* less heart disease.

"Does a beer a day keep the doctor away?"

Many highly publicized correlations are the product of "data dredging." In this process, computers randomly compare one set of variables (various eating habits) with another set (a range of diseases). From these countless comparisons, certain relationships are revealed (say, between coffee drinking and pancreatic cancer risk). As dramatic as such isolated correlations may be, they constitute no proof of causation and often lead to hasty conclusions (Ross 135).

"Is this good news or bad news?"

◆ *Misleading terminology:* The terms used to interpret statistics sometimes hide their real meaning. For instance, the widely publicized figure that people treated for cancer have a "50 percent survival rate" is misleading in three ways: (1) *survival* to laypersons means "staying alive," but to medical experts, staying alive for only five years after diagnosis qualifies as survival; (2) the "50 percent" survival figure covers *all* cancers, including certain skin or thyroid cancers that have extremely high cure rates, as well as other cancers (such as lung or ovarian) that rarely are curable and have extremely low survival rates; (3) more than 55 percent of all cancers are skin cancers with a nearly 100 percent survival rate, thereby greatly inflating survival statistics for other types of cancer ("Are We" 5; *Facts and Figures* 2).

Even the most valid and reliable statistics require that we interpret the reality behind the numbers. For instance, the overall cancer rate today is higher than it was in 1910. What this may mean is that people are living longer and thus are more likely to die of cancer and that cancer today rarely is misdiagnosed—or mislabelled because of stigma ("Are We" 4). The finding that rates for certain cancers double after prolonged exposure to electromagnetic waves may really mean that cancer risk actually increases from 1 in 10 000 to 2 in 10 000.

These are only a few examples of statistics and interpretations that seem highly persuasive but that in fact cannot always be trusted. Any interpretation of statistical data carries the possibility that other, more accurate, interpretations have been overlooked or deliberately excluded (Barnett 45).

In Brief   HOW STANDARDS OF PROOF VARY FOR DIFFERENT AUDIENCES AND CULTURAL SETTINGS

How much evidence is enough to "prove" a particular claim? The answer often depends on whether the enquiry occurs in the science lab, the courtroom, or the boardroom—as well as on the specific cultural setting:

♦ The scientist demands evidence that indicates at least 95 percent certainty. A scientific finding must be evaluated and replicated by other experts. Good science looks at the entire picture. Findings are reviewed before they are reported. Enquiries and answers in science are never "final," but open-ended and ongoing: What seems probable today may be shown improbable by tomorrow's research.

♦ The juror demands evidence that indicates only 51 percent certainty (a "preponderance of the evidence").

   Jurors are not scientists. Instead of the entire picture, jurors get only the information made available by lawyers and witnesses. A jury bases its opinion on evidence that exceeds "reasonable doubt" (Monastersky, *Courting,* 249; Powell 32+). Courts have to make decisions that are final.

♦ The corporate executive demands immediate (even if insufficient) evidence. In a global business climate of overnight developments (in world markets, political strife, military conflicts, natural disasters), important business decisions often are made on the spur-of-the-moment, often on the basis of incomplete or unverified information—or even hunches—in order to react to crises and capitalize on opportunities (Seglin 54).

♦ Specific cultures may have their own standards for authentic, reliable, and persuasive evidence. "For example, African cultures rely on story telling for authenticity. Arabic persuasion is dependent on universally accepted truths. And Chinese value ancient authorities over recent empiricism."

REASSESSING THE ENTIRE RESEARCH PROCESS

Chapters 6 and 7 show that the research process is a minefield of potential errors, in what we do and how we reason: We might ask the wrong questions; we might rely on the wrong sources; we might collect or record data incorrectly; we might analyze or document data incorrectly. We therefore need to critically examine our methods and our reasoning before reporting findings and conclusions. The following research checklist helps guide our assessment.

Checklist FOR THE RESEARCH PROCESS

Use this checklist to assess your research process.

Method

◆ Did I ask the right questions?
◆ Are the sources appropriately up-to-date?
◆ Is each source reputable, trustworthy, and relatively unbiased?
◆ Does the evidence clearly support all of the conclusions?
◆ Can all of the evidence be verified?
◆ Is a fair balance of viewpoints presented?
◆ Has the research achieved adequate depth?
◆ Has the entire research process been valid and reliable?
◆ Is all quoted material clearly marked throughout the text?
◆ Are direct quotations used sparingly and appropriately?
◆ Are all quotations accurate and integrated grammatically?
◆ Are all paraphrases accurate and clear?
◆ Have I documented all sources not considered common knowledge?
◆ Is the documentation consistent, complete, and correct?

Reasoning

◆ Am I reasonably certain about the meaning of these findings?
◆ Does my final answer seem definitive, only probable, or inconclusive?
◆ Am I reasoning instead of rationalizing?
◆ Is this the most reasonable conclusion (or merely the most convenient)?
◆ Can I rule out other possible interpretations or conclusions?
◆ Have I accounted for all sources of bias, including my own?
◆ Are my generalizations warranted by the evidence?
◆ Am I confident that my causal reasoning is correct?
◆ Can all of the numbers and statistics and interpretations be trusted?
◆ Have I resolved (or at least acknowledged) any conflicts among my findings?
◆ Should the evidence be reconsidered?

EXERCISES OR COLLABORATIVE PROJECTS

1. Assume you are an assistant communications manager for a new organization that prepares research reports for decision makers worldwide. (A sample topic: "What is the expected long-term impact of the North American Free Trade Agreement on the Canadian computer industry?") These clients expect answers based on the best available evidence and reasoning.

    Although your recently hired co-workers are technical specialists, few have experience in the kind of wide-ranging research required by your clients. Training programs in the research process are being developed by your communications division but will not be ready for several weeks.

    Meanwhile, your manager directs you to prepare a one- to two-page memo that introduces employees to major procedural and reasoning errors that affect validity and reliability in the research process. Your manager wants this memo to be comprehensive but not vague.

2. Assume the scenario from Exercise 1. In a memo to colleagues, offer guidelines for avoiding unintentional plagiarism in quoting, paraphrasing, and citing the work of others. Explain what to document and how, using MLA style for a parenthetical reference and a works cited entry. (Illustrate with examples, but not those from the book!)

3. From print or broadcast media or from personal experience, identify an example of each of the following sources of distortion or of interpretive error:

- a study with questionable sponsorship or motives
- reliance on soft evidence
- overestimating the level of certainty
- biased interpretation
- rationalizing
- faulty causal reasoning
- hasty generalization

- overstated generalization
- sanitized statistic
- meaningless statistic
- undefined average
- distorted percentage figure
- bogus ranking
- fallible computer model
- misinterpreted statistic

Submit your examples to your instructor, along with a memo explaining each error, or be prepared to discuss your material in class.

CHAPTER 8

Documenting Research Findings

WHY YOU SHOULD DOCUMENT

WHAT YOU SHOULD DOCUMENT

HOW YOU SHOULD DOCUMENT

MLA DOCUMENTATION STYLE

APA DOCUMENTATION STYLE

CBE NUMERICAL DOCUMENTATION STYLE

Documenting research findings means acknowledging one's debt to each information source. Proper documentation satisfies professional requirements for ethics, efficiency, and authority.

WHY YOU SHOULD DOCUMENT

Documentation is *ethical* in that the originator of borrowed material deserves full credit and recognition. Moreover, all published material is protected by copyright law. Failure to credit a source could make you liable to legal action, even if your omission was unintentional.

Documentation also is *efficient*. It provides a network for organizing and locating the world's recorded knowledge. If you cite a particular source correctly, your reference will enable interested readers to locate that source themselves.

Finally, documentation *provides authority*. In making any claim—for example, "A Honda Accord is more reliable than a Ford Taurus"—you invite challenge: "Says who?" Data on road tests, frequency of repairs, resale value, workmanship, and owner comments can help validate your claim by showing its basis in *fact*. A claim's credibility increases in relation to the expert references supporting it. For a controversial topic, you may need to cite several authorities who hold various views, as in this next example, instead of forcing a simplistic conclusion on your material:

Citing a balance of views

> Opinion is mixed as to whether a marketable quantity of oil rests beneath Great Slave Lake. Edmonton geologist Mo Rajabully feels that extensive reserves are improbable ("Geologist Dampens Hopes" 3). Oil geologist Marta Silverlaug is uncertain about the existence of any oil in quantity at this location ("Northern Oil Drilling" 2). But the Canadian Geological Survey reports that the lake bed may overlay 3.5 billion barrels of oil (Ruston 8).

Readers of your research report expect the *complete* picture.

WHAT YOU SHOULD DOCUMENT

Document any insight, assertion, fact, finding, interpretation, judgment, or other "appropriated material that readers might otherwise mistake for your own" (Gibaldi and Achtert 155)—whether the material appears in published form or not. Specifically, you must document:

Sources that require documentation

- ◆ any source from which you use exact wording
- ◆ any source from which you adapt material in your own words
- ◆ any visual illustration: chart, graph, drawing, or the like

You don't need to document anything considered *common knowledge:* material that appears repeatedly in general sources. In medicine, for instance, it is common knowledge that foods high in fat correlate with higher incidences of cancer, so in a report on fatty diets and cancer, you probably would not need to document that well-known fact. But you would document information about how the fat/cancer connection was discovered, subsequent studies (e.g., the role of saturated versus unsaturated fats), and any information for which some other person could claim specific credit. If the borrowed material can be found in only one specific source and not in multiple sources, document it. When in doubt, document the source.

HOW YOU SHOULD DOCUMENT

Borrowed material has to be cited twice: at the exact place that you use the material, and at the end of your document. Documentation practices vary widely, but all systems work almost identically: A brief reference in the text names the source and refers readers to the complete citation, which enables the source to be retrieved.

This chapter illustrates citations and entries for three styles widely used for documenting sources in respective disciplines:

- Modern Language Association (MLA) style, for the humanities
- American Psychological Association (APA) style, for social sciences
- Council of Biology Editors (CBE) style, for natural and applied sciences

Unless your audience has a particular preference, any of these three styles can be adapted to most research writing. Another widely used format is that of the *Chicago Manual of Style*, which covers documentation in the humanities, related fields, and natural sciences. Whichever style you select, use it consistently throughout the document.

MLA DOCUMENTATION STYLE

Until recently, MLA documentation of sources used superscript numbers (like this: [1]) in the text, followed by the full reference at the bottom of the page (footnotes) or at the end of the document (endnotes) and, finally, by a bibliography. A more current form of documentation appears in the *MLA Handbook for Writers of Research Papers*. In the new MLA style, an in-text parenthetical reference briefly identifies the source. Full documentation then appears in a "Works Cited" section at the end of the document. Footnotes or endnotes now are used only to comment on material in the text.

A parenthetical reference usually includes the author's surname and the exact page number of the borrowed material:

Parenthetical reference in the text

A recent study indicates an elevated risk of leukemia among children exposed to certain types of electromagnetic fields (Bowman et al. 59).

Readers seeking the complete citation for Bowman can move easily to "Works Cited," listed alphabetically by author:

Full citation at document's end

Bowman, J. D., et al. "Hypothesis: The Risk of Childhood Leukemia Is Related to Combinations of Power-Frequency and Static Magnetic Fields." *Bioelectromagnetics* 16.1 (1995): 48–59.

This complete citation includes page numbers for the entire article.

MLA PARENTHETICAL REFERENCES

For clear and informative parenthetical references, observe these guidelines:

- ◆ If your discussion names the author, do not repeat the name in your parenthetical reference; simply list the page number:

Citing page numbers only

Bowman et al. explain how their recent study indicates an elevated risk of leukemia for children exposed to certain types of electromagnetic fields (59).

- ◆ If you cite two or more works in a single parenthetical reference, separate the citations with semicolons:

Three works in a single reference

(Jones 32; Leduc 41; Gomez 293–94)

- ◆ If you cite two or more authors with the same last name, include the first initial in your parenthetical reference to each author:

Two authors with identical last names

(R. Jones 32)

(S. Jones 14–15)

- ◆ If you cite two or more works by the same author, include the first significant word from each work's title, or a shortened version:

Two works by one author

(Lamont, *Biophysics* 100–01)

(Lamont, *Diagnostic Tests* 81)

- ◆ If the work is by an institutional or corporate author or if it is unsigned (that is, author unknown), use only the first few words of the institutional name or the work's title in your parenthetical reference:

Institutional, corporate, or anonymous author

(American Medical Assn. 2)

("Distribution Systems" 18)

To avoid distracting your readers, keep each parenthetical reference as brief as possible. One method is to name the source in your discussion and to place only the page number in parentheses.

Where to place a parenthetical reference

For a paraphrase, place the parenthetical reference *before* the closing punctuation mark. For a quotation that runs into the text, place the reference *between* the final quotation mark and the closing punctuation mark. For a quotation set off (indented) from the text, place the reference two spaces *after* the closing punctuation mark.

As you will see in the following examples, documenting electronic sources is often trickier than documenting printed sources. Most often than not, Internet sources do not name authors, so special identification techniques are required. Also, electronic documents seldom number pages or provide other types of reference numbers. Therefore, MLA format recommends avoiding parenthetical references to electronic sources. Instead, the *MLA Handbook* suggests direct references in the text.

Susan Ireland's on-line *Resume Guide* shows how to format a résumé for electronic submission. Also, the *Proven Resumes* website provides workshops for producing a full variety of electronic résumés.

MLA WORKS CITED ENTRIES

The Works Cited list includes each source you have paraphrased or quoted. Double space the list for academic papers in the humanities. In all other situations, single space within each entry and double space between entries. Key the first line of each entry flush with the left margin. Indent the second and subsequent lines five spaces (1.25 cm [$^1/_2$"]). Use a one-character space after any period, comma, or colon.

How to space and indent entries

Following are examples of complete citations as they would appear in the Works Cited section of your document. Shown italicized below each citation is its corresponding parenthetical reference as it would appear in the text. Note capitalization, abbreviations, spacing, and punctuation in sample entries.

INDEX TO SAMPLE MLA WORKS CITED ENTRIES

Books

1. Book, single author
2. Book, two or three authors
3. Book, four or more authors
4. Book, anonymous author
5. Multiple books, same author
6. Book, one or more editors
7. Book, indirect source
8. Anthology selection or book chapter

Periodicals

9. Article, magazine
10. Article, journal with new pagination each issue
11. Article, journal with continuous pagination
12. Article, newspaper

Other Sources

13. Encyclopedia, dictionary, other alphabetic reference
14. Report
15. Conference presentation
16. Interview, personally conducted
17. Interview, published
18. Letter, unpublished

19. Questionnaire
20. Brochure or pamphlet
21. Lecture
22. Government document
23. Document with corporate authorship
24. Map or other visual
25. Unpublished dissertation, report, or miscellaneous items

Electronic Sources

26. On-line database source
27. Computer software
28. CD-ROM
29. Internet (bulletin board, discussion list)
30. E-mail
31. Web source (on-line article or posting)
32. Web source (home page— personal site)
33. Web source (home page— professional site)
34. Web source (secondary page)
35. Magazine article, on-line
36. Journal article, on-line
37. Newspaper article, on-line

MLA Works Cited Entries for Books.   Book citations should contain the following information (found on the book's title page and copyright page): author, title, and facts about publication (city, publisher, date). In some cases, other information will be available: editor or translator, edition, and volume number. MLA format underlines or italicizes the titles of books (and other publications).

1. Book, Single Author—MLA

> Bender, Peter Urs. *Secrets of Power Presentations.* 5th ed. Toronto: The Achievement Group, 1991.

Parenthetical reference: (Bender 29–30)

Identify the province or state of publication by Canada Post or U.S. Postal Service abbreviations. If the city of publication is well known (Toronto, Vancouver, etc.), omit the province or state abbreviation. If several cities are listed on the title page, give only the first. For other countries, except the U.S., include an abbreviation of the country name.

2. Book, Two or Three Authors—MLA

> Aronson, Linda, Roger Katz, and Candide Moustafa. *Toxic Waste Disposal Methods.* New Haven: Yale UP, 1996.

Parenthetical reference: (Aronson, Katz, and Moustafa 121–23)

Shorten publishers' names, as in "Simon" for Simon & Schuster or "Yale UP" for Yale University Press. For page numbers having more than two digits, give only the final two digits for the second number.

3. Book, Four or More Authors—MLA

> Beebe, Morton, et al. *Cascadia: A Tale of Two Cities, Seattle and Vancouver, B.C.* New York: Henry N. Abrams, 1996.

Parenthetical reference: (Beebe et al. 14)

"Et al." is the abbreviated form of the Latin "et alia," meaning "and others."

4. Book, Anonymous Author—MLA

> *A Primer on Freshwater: Questions and Answers.* Ottawa: Environment Canada, 2000.

Parenthetical reference: (Primer 33)

5. Multiple Books, Same Author—MLA

> Chang, John W. *Biophysics.* Boston: Little, 1997.
>
> ---. *Diagnostic Techniques.* New York: Radon, 1994.

Parenthetical references: (Chang, *Biophysics* 123–26) (Chang, *Diagnostic* 87)

When citing more than one work by the same author, do not repeat the author's name; simply key three hyphens followed by a period. List the works alphabetically by title.

6. Book, One or More Editors—MLA

Gunn, John. M., ed. *Restoration and Recovery of an Industrial Region: Sudbury.* New York: Springer-Verlag, 1995.

Parenthetical reference: (Gunn 34)

For more than three editors, name only the first, followed by "et al."

7. Book, Indirect Source—MLA

Kline, Thomas. *Automated Systems.* Boston: Rhodes, 1992.

Stubbs, John. *White-Collar Productivity.* Miami: Harris, 1996.

Parenthetical reference: (qtd. in Stubbs 116)

When your source (as in Stubbs, above) has quoted or cited another source, include each source in its appropriate alphabetical place in your Works Cited list. Use the name of the original source (here, Kline) in your text and begin the parenthetical reference with "qtd. in"—or "cited in" for a paraphrase.

8. Anthology Selection or Book Chapter—MLA

Bowman, Joel P. "Electronic Conferencing." *Communication and Technology: Today and Tomorrow.* Ed. Al Williams. Denton, TX: Assn. for Business Communication, 1994. 123–42.

Parenthetical reference: (Bowman 129)

Page numbers in the entry cover the selection cited from the anthology.

MLA Works Cited Entries for Periodicals.   Give all available information in this order: author, article title, periodical title, volume and issue, date (day, month, year), and page numbers for the entire article—not just pages cited.

9. Article, Magazine—MLA

Jenish, D'Arcy. "A Car That Just May Fly." *Maclean's* 21 June, 1999: 46–47.

Parenthetical reference: (Jenish 46)

No punctuation separates the magazine title and date. Nor is the abbreviation "p." or "pp." used to designate page numbers. If no author is given, list all other information:

"Too Much of a Good Thing." *Canada and the World Backgrounder* May 2001: 10–15.

Parenthetical reference: ("Too Much" 10)

When an article does not appear on consecutive pages, give only the number of the first page, followed immediately by a plus sign.

10. Article, Journal with New Pagination Each Issue—MLA

Ackerman, Nancy. "Landfill Landscape." *Canadian Geographic* 119.4 (1999): 56–63.

Parenthetical reference: (Ackerman 56–63)

Because each issue for that year will have page numbers beginning with "1," readers need the number of the issue. The "119" denotes the volume number; the "4" denotes the issue number. Omit "The" or "A" or any other introductory article from a journal or magazine title.

11. Article, Journal with Continuous Pagination—MLA

Norcliffe, Glen. "John Cabot's Legacy in Newfoundland." *Geography: an International Journal* 84 (1999): 97–109.

Parenthetical reference: (Norcliffe 104)

When page numbers continue from issue to issue for the full year, readers do not need the issue number, because no other issue in that year repeats these same page numbers. (Include the issue number if you think it will help readers retrieve the article more easily.) The "84" denotes the volume number.

12. Article, Newspaper—MLA

Chase, Steven. "Spam Under Attack." *Globe and Mail* 26 Mar. 2001, natl. ed., T1.

Parenthetical reference: (Chase 1)

When a daily newspaper has more than one edition, cite the specific edition after the date. Omit any introductory article in the newspaper's name (not *The Globe and Mail*). If no author is given, list all of the other information. If the newspaper's name does not contain the city of publication, insert it, using brackets: "*Northern Miner* [LaRonge, SK]."

MLA Works Cited Entries for Other Sources. Miscellaneous sources range from unsigned encyclopedia entries to conference presentations to government publications. A full citation should give this information (as available): author, title, city, publisher, date, and page numbers.

13. Encyclopedia, Dictionary, Alphabetic Reference—MLA

"Communication." *The Business Reference Book.* 1999 ed.

Parenthetical reference: ("Communication")

Begin a signed entry with the author's name. For any work arranged alphabetically, omit page numbers in the citation and the parenthetical reference. For a well-known reference book, only an edition (if stated) and a date are needed. For other reference books, give the full publication information.

14. Report—MLA

MacHutchon, Arthur, S. Himmer, and C.A. Bryden. *Khatzeywateen Valley Grizzly Bear Study: Final Report.* Victoria: B.C. Ministry of Forests, Oct. 1993.

Parenthetical reference: (MacHutchon, Himmer, and Bryden 29)

If no author is given, begin with the organization that sponsored the report; e.g., Canadian Professional Sales Association (CPSA). For any report or other document with group authorship, include the group's abbreviated name in your first parenthetical reference; e.g., (Canadian Professional Sales Association [CPSA] 49), and then use only that abbreviation in any subsequent reference; e.g., (CPSA 78).

15. Conference Presentation—MLA

Smith, Abelard A. "Radon Concentrations in Molded Concrete." *First British Symposium in Environmental Engineering. London, 11–13 Oct. 1995.* Ed. Anne Hodkins. London: Harrison, 1996. 106–21.

Parenthetical reference: (Smith 109)

The previous example shows a presentation that has been included in the published proceedings of a conference. For an unpublished presentation, include the presenter's name; the title of the presentation; and the conference title, location, and date, but do not underline or italicize the conference information.

16. Interview, Personally Conducted—MLA

Turner, Dan. Operations Manager for Prairie Power. Personal Interview. Winnipeg. 4 Mar. 2004.

Parenthetical reference: (Turner)

17. Interview, Published—MLA

Lescault, James. "The Future of Graphics." *Executive Views of Automation.* Ed. Karen Prell. Miami: Haber, 1997. 216–31.

Parenthetical reference: (Lescault 218)

The interviewee's name is placed in the entry's author slot.

18. Letter, Unpublished—MLA

Singh, Jopal. Letter to the author. 15 May 2004.

Parenthetical reference: (Singh)

19. Questionnaire—MLA

Sakamoto, Yoshi. Questionnaire sent to 612 Quebec business executives. 14 Feb. 2003.

Parenthetical reference: (Sakamoto)

20. Brochure or Pamphlet—MLA

> *Nissan-01 Maxima.* Mississauga, Ontario: Nissan Canada Inc. June 2000.

Parenthetical reference: (Nissan)

If the work is signed, begin with its author.

21. Lecture—MLA

> Jack, David. "Energy Levels and Spectrum of the Hydrogen Atom." Lecture. Concordia University, Montreal, 7 Nov. 2001.

Parenthetical reference: (Jack)

If the lecture title is not known, write Address, Lecture, or Reading but do not use quotation marks. Include the sponsor and the location if available.

22. Government Document—MLA

> British Columbia. Ministry of Highways. *Standard Specifications for Bridge Maintenance.* Victoria: B.C. Ministry of Highways, 1999.

Parenthetical reference: (B.C. Ministry of Highways 49)

If the author is unknown (as shown), list the information in this order: name of the government, name of the issuing agency, document title, place, publisher, and date.

23. Document with Corporate Authorship—MLA

> Canada Post Corporation. *The Canadian Addressing Standard.* Ottawa: Canada Post Corporation, 1995.

Parenthetical reference: (Canada Post 5)

24. Map or Other Visual—MLA

> *Deaths Caused by Breast Cancer, by Country.* Map. *Scientific American* Oct. 1995: 32D.

Parenthetical reference: (Deaths Caused)

If the creator of the visual is listed, list that name first. Identify the type of visual ("Map," "Graph," "Table," "Diagram") immediately following its title.

25. Unpublished Dissertation, Report, or Miscellaneous Items—MLA

> Author (if known), title (in quotes), sponsoring organization or publisher, date, page numbers.

For any work that has group authorship (corporation, committee, task force), cite the name of the group or agency in place of the author's name.

MLA Works Cited Entries for Electronic Sources.   In general, citation for an electronic source with a printed equivalent should begin with that publication information (see relevant sections above). But any citation should enable readers to retrieve the material electronically whether or not a printed equivalent exists.

26. On-line Database Source—MLA

> Sahl. J. D. "Power Lines, Viruses, and Childhood Leukemia." *Cancer Causes Control* 6.1 (Jan. 1995): 83. <u>MEDLINE.</u> On-line. DIALOG. 7 Nov. 1995.
>
> *Parenthetical reference:* (Sahl 83)

For entries with a printed equivalent, begin with complete publication information, then the database title (underlined), the "On-line" designation to indicate the medium, service provider, and date of access. The access date is important because frequent updatings of databases can produce different versions of the material. For entries with no printed equivalent, give the title and date of the work in quotation marks, followed by the electronic source information:

> Argent, Roger R. "An Analysis of International Exchange Rates for 2003." <u>Accu-Data.</u> On-line. Dow Jones News Retrieval. 10 Jan. 2004.
>
> *Parenthetical reference:* (Argent 4)

If the author is not known, begin with the work's title.

27. Computer Software—MLA

> *Virtual Collaboration.* Diskette. New York: Harper, 1994.
>
> *Parenthetical reference:* (Virtual)

Begin with the author's name, if known.

28. CD-ROM—MLA

> Cavanaugh, Herbert A. "EMF Study: Good News and Bad News." *Electrical World* Feb. 1995: 8. <u>ABI/INFORM.</u> CD-ROM. Proquest. Sept. 1995.
>
> *Parenthetical reference:* (Cavanaugh 8)

If the material is also available in print, begin with complete publication information, followed by the name of the database (underlined), "CD-ROM" designation, vendor's name, and electronic publication date. If the material has no printed equivalent, list its author (if known) and its title (in quotation marks), followed by the electronic source information.

For CD-ROM reference works and other material that is not routinely updated, give the work title followed by the "CD-ROM" designation, place, electronic publisher, and date:

> <u>*Time Almanac.*</u> CD-ROM. Washington: Compact, 1994.
>
> *Parenthetical reference:* (Time Almanac 74)

Begin with the author's name, if known.

29. Internet (Bulletin Board, Discussion List)—MLA

Templeton, Brad. "10 Big Myths about Copyright Explained." 29 Nov. 1994. On-line posting. Listserv law/copyright-FAQ/myths/part 1. BITNET. 6 May 1995.

Parenthetical reference: (Templeton)

Begin with the author's name (if known), followed by the title of the work (in quotation marks), publication date, "On-line posting" designation, name of discussion group, name of network, and date of your access. If appropriate, include the on-line address at the end of the entry, after the word "Available." The parenthetical reference should not include a page number as none is given in an on-line posting.

30. E-mail—MLA

Konecsni, Jerome. "E-mail Survey Questions." E-mail to Don Klepp. 21 Oct. 2001.

Parenthetical reference: (Konecsni)

Cite personal e-mail as you would print correspondence. If the document has a subject line or title, enclose it in quotation marks. For publicly posted e-mail (for a newsgroup or discussion list), include the address and the date of access.

31. Web Source (On-line Article or Posting)—MLA

Dumont, R. A. "An Online Course in Technical Writing." 10 Dec. 1995. On-line posting. http://www.umassd.edu/englishdepartment (6 Jan. 1996).

Parenthetical reference: (Dumont 7–9)

Begin with the author's name (if known), followed by title of the work (in quotation marks), posting date, "On-line" designation, Web address, and date of access. In place of (or in addition to) the Web address, include the name of the website (underlined), if available:

Rogers, S. E. "Chemical Risk Assessment Guidelines." 12 Feb. 1996. OTA Online. http://www.ota.gov (10 Mar. 1996).

"OTA" stands for Office of Technology Assessment.

32. Web Source (Home Page, Personal Site)—MLA

Ireland, Susan. *The Resume Guide.* Copyright 2000. Susan Ireland.com. Retrieved 12 Nov. 2001. *http://susanireland.com/*

Parenthetical reference: (Ireland)

Name the site owner, then the title of the site. (If no title is available, add a description, such as *Home page*, neither underlined nor in quotation marks.) Provide the date of latest posting, if that's available. Name the organization that sponsors the site, if named. Provide the access date and the URL.

33. Web Source (Home Page, Professional Site)—MLA

> *WCWWA.* Home page. Last updated 21 Sept. 2001. Western Canada Waste Water Association. Retrieved 15 Nov. 2001. http://www.wcwwa.ca

Parenthetical reference: (WCWWA)

Name the site author or creator, then the title of the site. (If no title is available, provide a descriptive phrase and indicate *Home page*, neither underlined nor in quotation marks.) Provide the date of latest posting, if that's available. Name any organization or institution associated with the site. Provide the access date and the URL.

34. Web Source (Secondary Page)—MLA

> Coneybeare, Steve. "Application of Chlorine Capping Kits." *Papers 2001.* Alberta Water and Wastewater Operators Association Seminar. Last updated 24 Oct. 2001. Retrieved 28 Oct. 2001. http://www.awwoa.ab.ca/2001%20Papers.htm

Parenthetical reference: (Coneybeare 2)

Name the site author or creator, if available. Place the title of the topic or article in quotation marks. Underline the page title, if it's named. Name any organization or institution associated with the site. Provide the latest update, if that's available. Provide the access date and the URL.

35. Magazine Article, On-line—MLA

> Gatehouse, Jonathon. "Terror By Mail." *Maclean's* on-line. 5 Nov. 2001. Retrieved 30 Oct. 2001. http://www.macleans.ca/xta-doc/2001/11/05/World/59288.shtml

Parenthetical reference: (Gatehouse)

Name the author. Place the article title in quotation marks. Underline the magazine's name. List the date and page, if indicated. Provide the access date and the URL.

36. Journal Article, On-line—MLA

> Kitts, Kara. "Technical Communication Provides Vital Link." *Employment Review.Com.* Mar. (2001). Retrieved 5 Aug. 2001. www.employmentreview.com/2001-03/features/Cnfeat04.asp

Parenthetical reference: (Kitts)

Name the author. Place the article title in quotation marks. Underline the journal's name. Provide the issue or other identifying number. (The year of publication goes in brackets.) Give the page numbers, if known. Provide the access date and the URL.

37. Newspaper Article, On-line—MLA

> Garwood-Jones, Alison. "Cattle Battle." *Globe and Mail.* 23 June 2001. *Globe and Mail on-line.* Retrieved 28 June 2001. http://www.globeandmail.com

Parenthetical reference: (Garwood-Jones)

Name the author. Place the article title in quotation marks. Underline the newspaper's name. List the date, edition, and section and page, if indicated. Name the database, if applicable. Give the access date and the URL.

MLA Sample Works Cited Page

Place your Works Cited section on a separate page at the end of the document. Arrange entries alphabetically by author surname. When the author is unknown, list the title alphabetically according to its first word (excluding introductory articles). For a title that begins with a digit ("5," "6," etc.), alphabetize the entry as if the digit were spelled out.

See the list of works cited in Figure 8.1. In the left margin, numbers refer to the elements discussed on the page facing Figure 8.1. Bracketed labels on the right identify different types of sources the first time a particular type is cited.

1   **WORKS CITED**

2 Broad, William J. "Cancer Fear Is Unfounded, Physicists Say." *New York Times*
14 May 1995, sec. A: 19.   *[newspaper article]*

3 Brodeur, Paul, "Annals of Radiation: The Cancer at Slater School."
New Yorker 7 Dec. 1992: 86+.   *[magazine article]*

4 Castleman, Michael. "Electromagnetic Fields." *Sierra* Jan./Feb. 1992: 21–22.

5 Cavanaugh, Herbert A. "EMF Study: Good News and Bad News." *Electrical
World* Feb. 1995: 8. *ABI/INFORM.* CD-ROM. Proquest. Sept. 1995.
    [trade-magazine article from CD-ROM database]

6 Dana, Amy, and Tom Turner. "Currents of Controversy." *Amicus Journal*
Summer 1993: 29–32.   *[alternative press]*

de Jager, L., and L. deBruyn. "Long-Term Effects of a 50 HZ Electric Field on
the Life-Expectancy of Mice." *Review of Environmental Health* 10.3
7 (1994): 221–24. *MEDLINE.* On-line. DIALOG. 8 Mar. 1996.
    [journal article from on-line database]

8 Des Marteau, Kathleen. "Study Links Sewing Machine Use to Alzheimer's
Disease." *Bobbin* Oct. 1994: 36–38. *ABI/INFORM*. CD-ROM. Proquest.
Aug. 1995.

"Electrophobia: Overcoming Fears of EMFs." *University of California Wellness
Letter* Nov. 1994: 1.   *[newsletter]*

Goodman, E. M., B. Greenebaum, and M. T. Marron. "Effects of
Electromagnetic Fields on Molecules and Cells." *International Review of
Cytology* 158 (1995): 279–338. *MEDLINE.* On-line. DIALOG. 8 Mar. 1996.

9 Halloran-Barney, Marianne B. Energy Service Advisor for County Electric.
E-mail to author. 3 Apr. 1996.   *[E-mail enquiry]*

10 Jauchem, J. "Alleged Health Effects of Electromagnetic Fields: Misconceptions in
the Scientific Literature." *Journal of Microwave Power and Electromagnetic
Energy* 26.4 (1991): 189–95.   *[journal article from print source]*

Kirkpatrick, David. "Can Power Lines Give You Cancer?" *Fortune* 31 Dec.
1990: 80–85.

11 Lee, J. M., Jr., et al. *Electrical and Biological Effects of Transmission Lines: A Review.*
U.S. Dept. of Energy. NTIS no. PC Ao6/MF A01. Washington: GPO, 1989.

Figure 8.1 A List of Works Cited (MLA Style)   *(continued)*

Discussion of Figure 8.1

1. Centre the Works Cited title at the top of the page. Use 2.5 cm (1") margins. Single space *within* the entries, double space *between* the entries. Order the entries alphabetically. For numbering Works Cited pages, follow numbering of text pages.

2. Indent five spaces (1.25 cm [½"]) for the second and subsequent lines of an entry.

3. Place quotation marks around article titles. Underline or italicize periodical or book titles. Capitalize the first letter of key words in all titles (also articles, prepositions, and conjunctions, only if they come first or last).

4. Do not cite a magazine's volume number, even if it is given.

5. For a CD-ROM database that is updated often (such as Proquest), conclude your citation with the date of electronic publication.

6. For additional perspective beyond "establishment" viewpoints, examine "alternative" publications (such as *The Amicus Journal* and *In These Times,* in this list).

7. Conclude an on-line database citation with the date you accessed the source.

8. Use a period and one space to separate a citation's three major items (author, title, publication data). Leave one space after a comma or colon. Use no punctuation to separate magazine title and date.

9. Alphabetize hyphenated surnames according to the name that appears first.

10. Include the issue number for a journal with new pagination in each issue. For page numbers of more than two digits, give only the final digits in the second number.

11. For government reports, name the sponsoring agency and include all available information for retrieving the document. Use the first author's name and "et al." for works with four or more authors or editors.

Maugh, Thomas H. "Studies Link EMF Exposure to Higher Risks of Alzheimer's." *Los Angeles Times* 31 July 1994, sec A: 3.

12 Mevissen, M., M. Keitzmann, and W. Loscher. "In Vivo Exposure of Rats to a Weak Alternating Magnetic Field Increases Ornithine Decarboxylase Activity in the Mammary Gland by a Similar Extent as the Carcinogen DMBA." *Cancer Letter* 90.2 (1995): 207–14. *MEDLINE.* On-line. DIALOG. 8 Mar. 1996.

Miller, M. A., et al. "Variation in Cancer Risk Estimates for Exposure to Powerline Frequency Electromagnetic Fields: A Meta-analysis Comparing EMF Measurement Methods." *Risk Analysis* 15.2 (1995): 281–87. *MEDLINE.* On-line. DIALOG. 8 Mar. 1996.

Miltane, John. Chief Engineer for County Electric. Personal Interview. Adams, MA. 5 Apr. 1996.   *[personal interview]*

Moore, Taylor. "EMF Health Risks: The Story in Brief." *EPRI Journal* Mar./Apr. 1995: 7–17.

13 Moulder, John. "Power Lines and Cancer" 6 Oct. 1995. On-line posting. Newsgroup powerlines.cancer.FAQ: USENET. 10 Mar. 1996.
                                                    [World Wide Web newsgroup]

Palfreman, Jon. "Apocalypse Not." *Technology Review* 24 April 1996: 24–33.

Pinsky, Mark A. *The EMF Book: What You Should Know about Electromagnetic Fields, Electromagnetic Radiation, and Your Health.* New York: Warner, 1995.
                                                    [book—one author]

14 Schneider, David. "High Tension: Researchers Debate EMF Experiments on Cells." *Scientific American* Oct. 1995: 26+.

Taubes, Gary. "Fields of Fear." *Atlantic Monthly* Nov. 1994: 94–108. On-line. U of Virginia Electronic Text Center. Internet. 15 Mar. 1996. Available http://www:etext. libvirginia.edu/english.html.   *[Internet source]*

15 United States Environmental Protection Agency. *EMF in Your Environment.* Washington: GPO, 1992.

16 White, Peter. "Bad Vibes." *In These Times* 28 June 1993: 14–17.

Figure 8.1   A List of Works Cited (MLA Style)

Discussion of Figure 8.1 *Continued* ▬▬▬▬▬

12. Use three-letter abbreviations for months with five or more letters.

13. Because an on-line conference source such as a listserv or newsgroup provides no page numbers, you can eliminate the in-text parenthetical reference by referring directly to that source in your discussion ("Dr. Jones of Harvard points out that").

14. When an article skips pages in a publication, give only the first page number followed by a plus sign.

15. When the privacy of the electronic source is not an issue (e.g., a library versus an e-mail correspondent), consider including its electronic address in your entry, after the word *Available.*

16. Shorten publishers' names (as in "Simon" for Simon & Schuster; "Knopf" for Alfred A. Knopf, Inc.; "GPO" for Government Printing Office; or "U of T Press" for University of Toronto Press).

APA DOCUMENTATION STYLE

One popular alternative to MLA style appears in the *Publication Manual of the American Psychological Association*. APA style is useful when writers wish to emphasize the publication dates of their references. A parenthetical reference in the text briefly identifies the source, date, and page number:

Reference cited in the text

> In a recent study, mice continuously exposed to an electromagnetic field tended to die earlier than mice in the control group (de Jager & de Brun, 1994, p. 224).

The full citation then appears in the alphabetic listing of "References," at the end of the report:

Full citation at the end of the document

> de Jager, L., & de Brun, L. (1994). Long term effects of a 50 Hz electric field on the life-expectancy of mice. *Review of Environmental Health, 10* (3–4), 221–224.

APA style (or some similar author-date style) is preferred in the sciences and social sciences, where information quickly becomes outdated.

APA PARENTHETICAL REFERENCES

APA's parenthetical references differ from MLA's as follows: the citation includes the publication date; a comma separates each item in the reference; and "p." or "pp." precedes the page number (this is optional in the APA system). When a subsequent reference to a work follows closely after the initial reference, the date need not be included. Here are specific guidelines:

- If your discussion names the author, do not repeat the name in your parenthetical reference; simply give the date and page number:

Author named in the text

> Researchers de Jager and de Brun explain that experimental mice exposed to an electromagnetic field tended to die earlier than mice in the control group (1994, p. 224).

When two authors of a work are named in your text, their names are connected by "and," but in a parenthetical reference their names are connected by an ampersand (&).

Two or more works in a single reference

- If you cite two or more works in a single reference, list the authors in alphabetical order and separate the citations with semicolons:

> (Jones, 2003; Gomez, 1999; Leduc, 2004)

- If you cite a work with three to five authors, try to name them in your text, to avoid an excessively long parenthetical reference:

A work with three to five authors

> Franks, Oblesky, Ryan, Jablar, and Perkins (1993) studied the role of electromagnetic fields in tumour formation.

In any subsequent references to this work, name only the first author, followed by "et al."

◆ If you cite two or more works by the same author published in the same year, assign a different letter to each work:

Two or more works by the same author in the same year

(Lamont, 2002a, p. 135)

(Lamont, 2002b, pp. 67–68)

Other examples of parenthetical references appear with their corresponding entries in the following discussion of the list of references.

APA REFERENCE LIST ENTRIES

How to space and indent entries

The APA reference list includes each source that you cited in your document. In preparing the list, which you should single space, key the first line of each entry flush with the left margin. Indent the second and subsequent lines five spaces 1.25 cm [$\frac{1}{2}$"]). Use one character space after any period, comma, or colon.

INDEX TO SAMPLE ENTRIES FOR APA REFERENCES

Books

1. Book, single author
2. Book, two to five authors
3. Book, six or more authors
4. Book, anonymous author
5. Multiple books, same author
6. Book, one or more editors
7. Book, indirect source
8. Anthology selection or book chapter

Periodicals

9. Article, magazine
10. Article, journal with new pagination for each issue
11. Article, journal with continuous pagination
12. Article, newspaper

Other Sources

13. Encyclopedia, dictionary, other alphabetic reference
14. Report
15. Conference presentation
16. Interview, personally conducted
17. Interview, published

18. Personal correspondence
19. Brochure or pamphlet
20. Lecture
21. Government document
22. Computer software or manual
23. Miscellaneous items

Electronic Sources

24. E-mail
25. Message posted to Internet group
26. Magazine or journal article from a database
27. CD-ROM reference work
28. Internet article based on a print source
29. Article in Internet-only magazine or journal
30. Newspaper article, searchable electronic version
31. Internet report
32. Website document, author named
33. Website document, author not named
34. Website document (non-periodical), multi-page, not dated, author not named
35. On-line encyclopedia article

Following are examples of complete citations as they would appear in the "References" section of your document. Shown immediately below each entry is its corresponding parenthetical reference as it would appear in the text. Note the capitalization, abbreviation, spacing, and punctuation in the sample entries.

APA Entries for Books. Any citation for a book should contain all applicable information in the following order: author, date, title, editor or translator, edition, volume number, and facts about publication (city and publisher).

1. Book, Single Author—APA

Bender, P.U. (1991). *Secrets of power presentations* (5th Ed.). Toronto: The Achievement Group.

Parenthetical reference: (Bender, 1991, pp. 29–30)

Use only initials for an author's first and middle name. Capitalize only the first words of a book's title and subtitle and any proper names. Identify a later edition in parentheses.

2. Book, Two to Five Authors—APA

Aronson, L., Katz, R., & Moustafa, C. (1996). *Toxic waste disposal methods.* New Haven: Yale University Press.

Parenthetical reference: (Aronson, Katz, & Moustafa, 1996)

Use an ampersand (&) before the name of the final author listed in an entry. As an alternative parenthetical reference, name the authors in your text and include date (and page numbers, if appropriate) in parentheses.

3. Book, Six or More Authors—APA

Fogle, S. T., et al. (1995). *Hyperspace technology.* Boston: Little, Brown.

Parenthetical reference: (Fogle, et al., 1995, p. 34)

For more than five authors, name only the first, followed by "et al."

4. Book, Anonymous Author—APA

A Primer on Freshwater: Questions and Answers. (2000). Ottawa: Environment Canada.

Parenthetical reference: (Primer, 2000, p. 33)

In your list of references, place an anonymous work alphabetically by the first keyword (not The, A, or An) in its title. In your parenthetical reference, capitalize all key words in a book, article, or journal title. But in your list of references, capitalize only journal titles in this way.

5. Multiple Books, Same Author—APA

Chang, J. W. (1997a). *Biophysics.* Boston: Little, Brown.

Chang, J. W. (1997b). *MindQuest.* Chicago: John Pressler.

Parenthetical references: (Chang, 1997a) (Chang, 1997b)

Two or more works by the same author not published in the same year are distinguished by their respective dates alone, without the added letter.

6. Book, One or More Editors—APA

Gunn, J. M. (Ed.). (1995). *Restoration and recovery of an industrial region: Sudbury.* New York: Springer-Verlag.

Parenthetical reference: (Gunn, 1995, p. 34)

For more than five editors, name only the first, followed by "et al."

7. Book, Indirect Source

Stubbs, J. (1996). *White-collar productivity.* Miami: Harris.

Parenthetical reference: (cited in Stubbs, 1996, p. 47)

When your source (as in Stubbs, above) cites another source, list your source in the References section, but name the original source in your text: "Kline's study (cited in Stubbs, 1996, p. 47) supports this conclusion."

8. Anthology Selection or Book Chapter—APA

Bowman, J. (1994). Electronic conferencing. In A. Williams (Ed.), *Communication and technology: Today and tomorrow.* (pp. 123–142). Denton, TX: Association for Business Communication.

Parenthetical reference: (Bowman, 1994, p. 126)

The page numbers in the complete reference are for the selection cited from the anthology.

APA Entries for Periodicals. A citation for an article should give this information (as available) in this order: author, publication date, article title (without quotation marks), volume or number (or both), and page numbers for the entire article—not just the page cited.

9. Article, Magazine—APA

Jenish, D. (1999, 21 June). A car that just may fly. *Maclean's,* 112, 46–47.

Parenthetical reference: (Jenish, 1999, p. 46)

If no author is given, provide all other information. Capitalize the first word in an article's title and subtitle, and any proper nouns. Capitalize all keywords in a periodical title. Italicize the periodical title. Then, list the volume number and page numbers, using commas to separate title, volume number, and page numbers (as above).

10. Article, Journal with New Pagination for Each Issue—APA

> Ackerman, N. (1999). Landfill landscape. *Canadian Geographic,* 119 (4), 56–63.

Parenthetical reference: (Ackerman, 1999, 56–58)

Because each issue for a given year has page numbers that begin at "1," readers need the issue number ("1"). The "119" denotes the volume number, which is underlined or italicized.

11. Article, Journal with Continuous Pagination—APA

> Norcliffe, G. (1999). John Cabot's legacy in Newfoundland. *Geography: An international journal, 84*

Parenthetical reference: (Norcliffe, 1999, p. 104)

The "84" denotes the volume number. When page numbers continue from issue to issue for the full year, readers do not need the issue number, because no other issue in that year repeats these same page numbers. (You can include the issue number if you think it will help readers retrieve the article more easily.)

12. Article, Newspaper—APA

> Chase, Steven. (2001, Mar. 26) Spam under attack. *The Globe and Mail,* natl. ed., p. T1.

Parenthetical reference: (Chase, 2001, p. T1)

In addition to the year of publication, include the month and day. If the newspaper's name begins with "The," include it in your citation. Include "p." or "pp." before page numbers. For an article on non-consecutive pages, list each page, separated by a comma.

APA Entries for Other Sources.   Miscellaneous sources range from unsigned encyclopedia entries to conference presentations to government documents. A full citation should give this information (as available): author, publication date, work title (and report or series number), page numbers (if applicable), city, and publisher.

13. Encyclopedia, Dictionary, Alphabetic Reference—APA

> Communication. (1993). In *The business reference book.* Boston: Business Resources Press.

Parenthetical reference: ("Communication," 1993)

For a signed entry, begin with the author's name and publication date.

14. Report—APA

MacHutchon, A. Himmer, S., and Bryden, C. A. (1993). *Khatzeywateen Valley grizzly bear study: final report.* Victoria: B.C. Ministry of Forests, Oct. 1993.

Parenthetical reference: (MacHutchon, Himmer, & Bryden, 1993, p. 29)

If the authors are named, list them first, followed by the publication date. When citing a group author; e.g., Canadian Professional Sales Association, include the group's abbreviated name in your first parenthetical reference; e.g., (Canadian Professional Sales Association [CPSA]), and use only that abbreviation in any subsequent reference; e.g., (CPSA). When the agency (or organization) and publisher are the same, list "Author" in the publisher's slot.

15. Conference Presentation—APA

Smith, A. A. (1996). Radon concentrations in molded concrete. In A. Hodkins (Ed.), *First British Symposium on Environmental Engineering* (pp. 106–121). London: Harrison Press.

Parenthetical reference: (Smith, 1995, p. 109)

The example shows a presentation included in the published proceedings of a conference. The name of the symposium is a proper name and so is capitalized. For an unpublished presentation, include the presenter's name, year and month, title of the presentation (underlined or italicized), and all available information about the conference or meeting: "Symposium held at . . ." Do not underline or italicize this information.

16. Interview, Personally Conducted—APA

Parenthetical reference: (D. Turner, personal interview, March 4, 2008)

This material is considered a non-recoverable source, and so is cited in the text only, as a parenthetical reference. If you name the interviewee in your text, do not repeat the name in your parenthetical reference.

17. Interview, Published—APA

Jable, C. K. (1997). The future of graphics [Interview with James Lescault]. In K. Prell (Ed.), *Executive Views of Automation* (pp. 216–231). Miami: Haber Press.

Parenthetical reference: (Jable, 1997, pp. 218–223)

Begin with the name of the interviewer, followed by the publication date, title, designation (in brackets), and publication information.

18. Personal Correspondence—APA

Parenthetical reference: (L. Nguyen, personal correspondence, May 15, 2006)

This material is considered non-recoverable data, and so is cited in the text only, as a parenthetical reference. If you name the correspondent in your text, do not repeat the name in your citation.

19. Brochure or Pamphlet—APA

This material follows the citation format for a book entry. After the title of the work, include the designation "Brochure" in brackets.

20. Lecture—APA

Jack, D. (2001, November 7). *Energy levels and spectrum of the hydrogen atom.* Lecture presented at Concordia University.

Parenthetical reference: (Jack, 2001)

If you name the lecturer in your text, do not repeat the name in your citation.

21. Government Document—APA

British Columbia Ministry of Highways. (1999). *Standard specifications for bridge maintenance.* Victoria: Author.

Parenthetical reference: (British Columbia Ministry of Highways, 1999, p. 49)

If the author is unknown, present the information in this order: name of the issuing agency, publication date, document title, place, and publisher. When the issuing agency is both author and publisher, list "Author" in the publisher's slot.

22. Computer Software or Manual—APA

Rotellan, P. (2002) Forest Inventory Mapping Control (Version 2.0) [Computer software]. Wakaw, Saskatchewan: Branch Services.

Parenthetical reference: (Rotellan, 2002, p. 17)

Do not reference standard software and programming languages such as Microsoft Word and Excel, Java, or AutoCAD. In the text, name the software and its version number ("this spring, AutoCAD 16 will be updated"). For specialized software, in limited distribution, use the above format.

23. Miscellaneous Items (Unpublished Manuscripts, Dissertations, etc.)—APA

Author (if known), date of publication, title of work, sponsoring organization or publisher, page numbers.

For any work that has group authorship (corporation, committee, etc.), cite the name of the group or agency in place of the author's name.

APA Entries for Electronic Sources.   APA documentation standards for electronic sources continue to be refined and defined. A sampling of currently preferred formats follows. Any citation for electronic media should enable the reader to identify the original source (printed or electronic) and provide an electronic path for retrieving the material.

Begin with the publication information for the printed equivalent. Then in brackets name the electronic source ([On-line], [CD-ROM], [Computer software]), the protocol[1] (Bitnet, Dialog, FTP, Telnet), and any other items that define a clear path (service provider, database title, access code, retrieval number, or site address).

24. E-mail—APA

Jerome Konecsni (personal communication, Oct. 21, 2001) comments on the use of e-mail for messages sent locally, nationally, and internationally.

Provide a parenthetical reference for this personal communication; do not list the entry in the References section.

25. Message Posted to Internet Group (Newsgroup, On-line Forum, or Discussion)—APA

Lewis, S. (2001, December 30). Checking out the geology when buying a house [Msg1] Message posted to http://groups.google.com/groups?h1=en&group-sci.geo.geology

Parenthetical reference: (Lewis, 2001, para. 1)

The basic form is the same for all types: Author(s). (Date of posting). Message subject line [Message ID]. Message posted to [group address].

26. Magazine or Journal Article from a Database—APA

Blackburn-Brockman, E. & Belanger, K. (2001, January). One page or two? A national study of CPA recruiters' preferences for resume length. *The Journal of Business Communication, 38*(1), 29. Retrieved June 20, 2001, from InfoTracCollege Edition database, Article No. A71327300.

Parenthetical reference: (Blackburn-Brockman, 2001, p. 29)

Name the author. In brackets, give the publication date. Then, name the title of the article, followed by the italicized journal name and volume number, in title case (with the issue number in brackets), and the page number(s). Provide the retrieval date and the database (with an article number, if given).

27. CD-ROM Reference Work—APA

Time almanac. (2001). Washington: Compact, 1999.

Parenthetical reference: (Time almanac, 2001)

If the work on CD-ROM has a print equivalent, APA prefers that it be cited in its print form.

---

1. A protocol is a body of standards that ensures compatibility among the different products designed to work together on a particular network.

28. Internet Article Based on a Print Source—APA

> Gatehouse, J. (2001, November 5). Terror by mail [Electronic version]. *Maclean's,* 114 (45), 22–24.

Parenthetical reference: (Gatehouse, 2001, November 5, p. 22)

Name the author or editor, if known. List the publication date (or write "n.d." for "no date"—if the date is unknown.) Name the article's title and the publication medium. Name the magazine or journal, its volume and number, and the page numbers of the print version. (Add the date of retrieval and the URL only if you believe the electronic version differs from the print version.)

29. Article in Internet-only Magazine or Journal—APA

> Kitts, K. (2001, March). Technical communication provides vital link. *Employment Review.Com,* Vol.: March 2001. Retrieved November 7, 2001 from www. employmentreview.com/2001-03/features/Cnfeat04.asp

Parenthetical reference: (Kitts, 2001, March, para. 6)

Name the author and the publication date, then the title of the article. Place the journal's name in italics. Name the publication medium. Provide the volume and issue, if they're given. Provide the access date and the URL.

30. Newspaper Article, Searchable Electronic Version—APA

> Garwood-Jones, A. (2001, June 23). Cattle battle. *The Globe and Mail.* Retrieved June 28, 2001, from http://www.globeandmail.com

Parenthetical reference: (Garwood-Jones, 2001, para. 3)

Name the author and the publication date, then the article title. Italicize the newspaper's name. Give the access date and the URL.

31. Internet Report—APA

> Thompson, J., Baird, P., & Downie, J. (2001, October). Report of the committee of inquiry on the case involving Dr. Nancy Olivieri, the Hospital for Sick Children, the University of Toronto, and Apotex Inc. Retrieved June 28, 2001, from http://222.caut/ca/english/issues/acadfreedom/olivieri.asp

Parenthetical reference: (Thompson, Baird, & Downie, 2001, p. 45)

Name the author(s) and the publication date, then the report title. (If no author is named, name the sponsoring organization.) Give the access date and the URL.

32. Website Document, Author Named—APA

> Coneybeare, S. (2001, October 24), Papers 2001: Application of chlorine capping kits. [On-line]. Alberta Water and Wastewater Association. Retrieved October 28, 2001, from http://www.awwoa.ab.ca/2001%20Papers.htm

Parenthetical reference: (Coneybeare, 2001, para. 6)

Name the author or editor, if known. Then give the revision or copyright date, if available. Next, give the page's title. Name the page publisher or site sponsor, list the access date, provide the URL.

NOTE   *APA recommends using a paragraph indicator (e.g., for parenthetical references when page numbers are not available for electronic sources).*

33. Website Document, Author Not Named—APA

> National Sciences and Engineering Research Council. (Updated 2001, December 17). *Truck research hit the road.* Retrieved through the NSERC website November 15, 2001, from http://www.nserc.ca/news/features/truck_e.htm

Parenthetical reference: (NSERC, 2001, para 6)

Name the website sponsor. Provide the latest revision date or copyright date, if either is available. Give the title of the hyper-linked document. Provide the access date and the URL.

34. Website Document (Non-periodical), Multi-page Not Dated, Author Not Named—APA

> Tire Rack. (n.d.) *Tire test results: Building max performance that lasts.* Retrieved January 2, 2002, from within Tire Rack's website, http://222.tirerack.com/tires/tests/bs_s03_rd.jsp

Parenthetical reference: (Tire Rack tests, n.d., para. 4)

Name the sponsoring organization. Indicate that the article is not dated. Name the document or page title. Provide the access date and the URL.

35. On-line Encyclopedia Article—APA

> Bartleby.com. (2001). *Phospholipid.* Found in *The Columbia Encyclopedia,* 6th edition [On-line]. Retrieved January 11, 2002, from http://www.bartleby.com/65/ph/phosphol.html

Parenthetical reference: (Columbia, 2001, para. 1)

Name the author, if known. If not, name the sponsoring organization—in the above example, Bartleby.com is the Internet arm of Columbia University Press, which publishes the printed version. (Give the date or indicate that the article is not dated.) Name the hyperlinked document, the encyclopedia, and the edition, if known. Provide the access date and the URL.

APA SAMPLE LIST OF REFERENCES

APA's References section is an alphabetic listing (by author) equivalent to MLA's Works Cited section. Like Works Cited, the reference list includes only those works actually cited. (A bibliography usually would include background works or works consulted as well.) One notable difference from MLA style is that the APA style calls for only "recoverable" sources to appear in the reference list. Therefore, personal interviews, e-mail messages, and other unpublished materials are cited in the text only.

REFERENCES

Anson, D. (1999, March 12). *Engineering graduates and the job market.* Lecture presented at the University of Massachusetts at Dartmouth. *[lecture]*

Basta, N. (1984, September). Take a good look at sales engineering. *Graduating Engineer, 32,* 84–87.   *[journal article]*

College Placement Council. (1998). *CPC Annual* (42nd ed.). Bethlehem, PA: Author.

                    [reference book—author/organization as publisher]

Cornelius, H., & Lewis, W. (1983). *Career guide for sales and marketing* (2nd ed.). New York: Monarch Press.   *[book with two authors]*

Electronic sales positions. (1996). *The national job bank.* Holbrook, MA: Bob Adams, Inc.   *[directory entry—no author named]*

Jones, B. (1997, December). Giving women the business. *Harper's Magazine, 296* (1772), 47–58.

Occupational employment. (1997–98, Winter) *Occupational Outlook Quarterly, 41*(4), 6–24.   *[govt. periodical—no author named]*

Shelley, K. J. (1997, Fall). A portrait of the M.B.A. *Occupational Outlook Quarterly, 41*(3), 26–33.   *[govt. periodical—author named]*

Solomon, S. D. (1996, January). An engineer goes to Wall Street. *Technology Review,* 99(1). Retrieved March 17, 2001 from http://www.web.mit.edu/techreview/www/   *[online article]*

Technology Marketing Group. (n.d.). *TMG Services.* Retrieved March 24, 2001 from http://www.technology-marketing.com   *[website]*

Tolland, M. (1999, April). *Alternate careers in marketing.* Presentation at Electro '99 Conference in Boston.   *[unpublished conference presentation]*

U.S. Department of Labor. (1997) *Tomorrow's jobs.* Washington, DC: Author. *[govt. publication—no author named]*

Young, J. (1995, August). Can computers really boost sales? *Forbes ASAP,* 84–101.   *[magazine article—no vol. or issue number]*

Figure 8.2 A List of References (APA Style)   *(continued)*

Discussion of Figure 8.2

1. Centre the References title at the top of the page. Number reference pages consecutively with text pages. Include only recoverable data (material that readers could retrieve for themselves); cite personal interviews, unpublished lectures, electronic discussion lists, and e-mail and other personal correspondence parenthetically in the text only. See also item 9 in this list.

2. Single space *within* entries and double space *between* entires. Order the entries alphabetically by author's last name (excluding *A, An,* or *The*). List initials only for authors' first and middle names. Write out names of all months. In student papers, indent the second and subsequent lines of an entry five spaces. In papers submitted for publication in an APA journal, the *first* line is indented instead.

3. Do not enclose article titles in quotation marks. Underline or italicize periodical titles. Capitalize the first word in article or book titles and subtitles, and any proper nouns. Capitalize all keywords in magazine or journal titles.

4. Identify the edition of a book in parentheses. If the author is also the publisher, use the word "Author" after the place of publication. Otherwise, write out the publisher's name in full.

5. For more than one author or editor, use ampersands instead of spelling out "and."

6. Use the first keyword in the title to alphabetize works whose author is not named.

7. Use italics or a continuous underline for a journal article's title, volume number, and the comma. Give the issue number in parentheses only if each issue begins on page 1. Do not include "p." or "pp." before journal page numbers (only before page numbers from a newspaper).

8. Omit punctuation from the end of an electronic address.

9. Treat an unpublished conference presentation as a recoverable source; include it in your list of references instead of only citing it parenthetically in your text.

The list of references in Figure 8.2 accompanies the report on technical marketing, pages 442–51. In the left margin, blue numbers denote elements discussed on the page facing Figure 8.2. Bracketed labels on the right identify different types of sources the first time a particular type is cited.

CBE NUMERICAL DOCUMENTATION STYLE

The Council of Science Editors (known as the Council of Biology Editors prior to January 1, 2000) accepts a "name-year" system similar to the APA format, but prefers a numerical "citation-sequence" system. In this numerical system, each work is assigned a number the first time it is cited. This same number then is used for any subsequent reference to that work. Numerical documentation often is used in the physical sciences (astronomy, chemistry, geology, physics) and the applied sciences (mathematics, medicine, engineering, computer science).

Preferred documentation styles for particular disciplines are defined in style manuals such as:

- American Chemical Society. *The ACS Style Guide: A Manual for Authors and Editors* (1986)
- American Institute of Physics. *AIP Style Manual* (1990)
- American Mathematical Society. *A Manual for Authors of Mathematical Papers* (1990)
- Engineering. *How to Write and Publish Engineering Papers and Reports* (1986)

One widely consulted guide for numerical documentation is *Scientific Style and Format: The CBE Manual for Authors, Editors, and Publishers.* 6th ed., 1994, from the Council of Science Editors. Descriptions of CBE formatting can also be found at these websites:

- **www.bedfordstmartins.com/online/cite8.html**
- **www.lib.ucdavis.edu/citing/**
- **www.columbia.edu/cu/cup/cgos/basic.html**
- **www.wisc.edu/writing/Handbook/DocCBE6.html**

CBE NUMBERED CITATIONS

In one version of CBE style, a citation in the text appears as a raised number immediately following the source to which it refers:

Numbered citations in the text

A recent study[1] indicates an elevated leukemia risk among children exposed to certain types of electromagnetic fields. Related studies[2-3] tend to confirm the EMF/cancer hypothesis.

When referring to two or more sources in a single superscripted note (as in "2–3" above), separate the numbers with a hyphen if they are in sequence and by commas but no space if they are out of sequence: ("2,6,9").

The full citation for each source then appears in the numerical listing of references at the end of the document:

REFERENCES

Full citations at the end of the document

1. Bowman JD, et al. Hypothesis: the risk of childhood leukemia is related to combinations of power-frequency and static magnetic fields. Bioelectromagnetics 1995; 16(1): 48–59.

2. Feychting M, Ahlbom A. Electromagnetic fields and childhood cancer: meta-analysis. Cancer Causes Control 1995 May; 6(3): 275–277.

To refer again to any of these sources later in your document, use the same number.

CBE REFERENCE LIST ENTRIES

CBE's References section lists each source in the numerical order in which it was first cited. In preparing the list, which should be single spaced, begin each entry on a new line. Key the number flush with the left margin, followed by a period and a space. Align subsequent lines directly under the first word of line one.

Following are examples of complete citations as they would appear in the References section for your document.

INDEX TO SAMPLE CBE ENTRIES

1 Book, single author

2. Book, multiple authors

3. Book, anonymous author

4. Book, one or more editors

5. Anthology selection or book chapter

6. Article, magazine

7. Article, journal with new pagination each issue

8. Article, journal with continuous pagination

9. Article, newspaper

10. Article, on-line source

11. Web source (home page—personal site)

12. Web source (home page—professional site)

13. Web source (secondary page)

14. Magazine article, on-line

15. Journal article, on-line

16. Newspaper article, on-line

CBE Entries for Books.   Any citation for a book should contain all available information in the following order: number assigned to the entry, author or editor, work title (and edition), facts about publication (place, publisher, date), and number of pages. Note the capitalization, abbreviation, spacing, and punctuation in the following sample entries.

1. Book, Single Author—CBE

1. Bender PU. 1991. Secrets of power presentations. 5th ed. Toronto: Achievement Press; 1991.

2. Book, Multiple Authors—CBE

2. Aronson L, Katz R, Moustafa C. Toxic waste disposal methods. New Haven: Yale Univ. Pr.; 1996. 316p.

3. Book, Anonymous Author—CBE

3. [Anonymous]. A primer on freshwater: questions and answers. Ottawa: Environment Canada; 2000.

4. Book, One or More Editors—CBE

4. Morris AJ, Pardin-Walker LB, editors. Handbook of new information technology. New York: Harper; 1996. 345p.

5. Anthology Selection or Book Chapter—CBE

5. Bowman JP. Electronic conferencing. In: Williams A, editor. Communication and technology: today and tomorrow. Denton, TX: Assn. for Business Communication; 1994. p. 123–42.

CBE Entries for Periodicals. Any citation for an article should contain all available information in the following order: number assigned to the entry, author, article title, periodical title, date (year, month), volume and issue number, and inclusive page numbers for the article. Note the capitalization, abbreviation, spacing, and punctuation in the sample entries.

6. Article, Magazine—CBE

6. Jenish D. A car that just might fly. Maclean's 1999 June 21: 46–47.

7. Article, Journal with New Pagination Each Issue—CBE

7. Thackman-White JR. Computer-assisted research. American Library Journal 1997; 51(1): 3–9.

8. Article, Journal with Continuous Pagination—CBE

8. Barnstead MH. The writing crisis. Journal of Writing Theory 1994; 12: 415–433.

9. Article, Newspaper—CBE

9. Chase S. Spam under attack. Globe and Mail 2001 Mar 26; Sect T:1

10. Article, On-line Source—CBE

10. Alley RA. Ergonomic influences on worker satisfaction. Industrial Psychology [serial on-line] 1995 Jan; 5(11). Available from: ftp. pub/journals/industrial_psychology/1995 via the INTERNET. Accessed 1996 Feb 10.

11. Web Source (Home Page, Personal Site)—CBE

11. Ireland S. 2000. The Resume Guide. http://susanireland.com Accessed 2001 Nov 12.

12. Web Source (Home Page, Professional Site)—CBE

12. [WCWWA]. Western Canada Waste Water Association. 2001 Sept. 21.
    WCWWA home page. http://www.wcwwa.ca
    Accessed 2001 Nov. 15.

13. Web Source (Secondary Page)—CBE

13. Coneybeare S. Application of chlorine capping kits. 2001 Sept. 21. Papers 2001:
    Alberta Water and Wastewater Operators Association Seminar.
    http://www.awwoa.ab.ca/ 2001%20Papers.htm
    Accessed 2001 Oct. 28.

14. Magazine Article, On-line—CBE

14, Gatehouse J. (2001 Nov. 5). Terror by mail. Maclean's. [On-line].
    http://www.macleans.ca/xta-doc/2001/11/05/World/59288.shtml
    Accessed 2001 Oct. 30.

15. Journal Article, On-line—CBE

15. Kitts K. (2001 March). Technical communication provides vital link. Employment
    Review.Com. [On-line]. (Volume not indicated.)
    http://www.employmentreview. com/2001-03/features/Cnfeat04.asp
    Accessed 2001 Nov. 7.

16. Newspaper Article, On-line—CBE

16. Garwood-Jones A. (2001 June 23). Cattle battle. The Globe and Mail. [On-line].
    http://www.globeandmail.com
    Accessed 2001 June 28.

For more detailed guidelines on CBE style, consult the CBE manual and websites such as those listed on the next page.

EXERCISES

1. Locate the style manual for your discipline. (Ask the faculty in your major or a librarian). Redesign Figure 8.1 according to the guidelines in this manual. Submit your document along with a memo outlining the main differences in the two documentation styles. If your discipline stipulates no particular style, use the *APA Manual* for this assignment.

2. Locate the latest updates for MLA and APA documentation of electronic sources at the following websites:

   ◆ www.wisc.edu/writing/Handbook/handbook.html
   ◆ www.westwords.com/guffey/students.html
   ◆ www.apastyle.org/elecref.html

COLLABORATIVE PROJECT

Work in groups of three. Examine the documentation format each of you has used in your major report assignment to confirm that you have correctly used the documentation system stipulated by your project supervisor.

CHAPTER 9

Summarizing Information

PURPOSE OF SUMMARIES

ELEMENTS OF A SUMMARY

CRITICAL THINKING IN THE SUMMARY PROCESS

A SAMPLE SITUATION

FORMS OF SUMMARIZED INFORMATION

Checklist **for Summaries**

A summary is a short version of a longer document. An economical way to communicate, a summary saves time, space, and energy.

PURPOSE OF SUMMARIES

Chapter 6 shows how abstracts (a type of summary) aid our research by providing an encapsulated glimpse of an article or other long document. As we record our research findings, we summarize and paraphrase to capture the main ideas in a compressed form. In addition to this dual role as a research aid, summarized information is vital in day-to-day workplace transactions.

On the job, you have to write concisely about your work. You might report on meetings or conferences, describe your progress on a project, or propose a money-saving idea. A routine assignment for many new employees is to provide superiors (decision makers) with summaries of the latest developments in their field.

Given today's pace and volume of information, summaries are more vital than ever. Some reports and proposals can be hundreds of pages long. Those who must act on this information need to rapidly identify what is most important in a document. From a good summary, busy readers can get enough information to decide whether they should read the entire document, parts of it, or none of it.

Whether you summarize someone else's document or your own, your job is to communicate the *essential message* accurately and in the fewest words. The essential message in any well-written document is easy enough to identify, as in the following passage:

The original passage

> The lack of technical knowledge among owners of television sets leads to their suspicion about the honesty of television repair technicians. Although television owners might be fairly knowledgeable about most repairs made to their automobiles, they rarely understand the nature and extent of specialized electronic repairs. For instance, the function and importance of an automatic transmission in an automobile are generally well known; however, the average television owner knows nothing about the flyback transformer in a television set. The repair charge for a flyback transformer failure is roughly $150—a large amount to a consumer who lacks even a simple understanding of what the repairs will accomplish. In contrast, a $450 repair charge for the transmission on the family car, though distressing, is more readily understood and accepted.

Three significant ideas make up the essential message: (1) television owners lack technical knowledge and are suspicious of repair technicians; (2) an owner usually understands even the most expensive automobile repairs; and (3) owners do not understand or accept expenses for television repairs. A possible summary might read like this:

A summarized version

> Because television owners lack technical knowledge about their sets, they are often suspicious of repair technicians. Although consumers may understand expensive automobile repairs, they rarely understand or accept repair and parts expenses for their television sets.

This summary is almost 30 percent of the original length because the original itself is short. With a longer original, a summary might be 5 percent or less. But length is less important than informative value: An effective summary gives readers just what they need and no more. For letters, memos, or other short documents that can be read quickly, the only summary needed is usually an *opening thesis* or *topic sentence* that previews the contents.

Summaries are vital to some key executives who do not have time to read everything that crosses their desks. For example, some executives ask that all significant world news of the last 24 hours be condensed into one page and placed on their desks, first thing each morning. Others employ writers who summarize articles from relevant and reputable magazines.

ELEMENTS OF A SUMMARY

All effective summaries display the following elements.

- *The essential message:* The essential message is the significant material from the original: controlling ideas (thesis and topic sentences); major findings; important names, dates, statistics, and measurements; and conclusions or recommendations. Significant material does not include background; the author's personal comments or conjectures; introductions; long explanations, examples, or definitions; visuals; or data of questionable accuracy.
- *Non-technical style:* More people generally read the summary than any other part of a document. So, write at the lowest level of technicality. Translate technical data into plain English. "The patient's serum glucose measured 240 mg%" can be translated: "The patient's blood sugar remained critically high." Of course, if you know all your readers are experts, you don't need to simplify.
- *Independent meaning:* In meaning as well as style, your summary should stand alone as a self-contained message. Readers should have to read the original only for more detail—not to make sense of the basic ideas.
- *No personal assessment:* Avoid personal comments ("This interesting report" or "The author is correct in assuming"). Add nothing to the original except for a brief clarifying definition, if needed.
- *Conciseness:* Conciseness is vital, but never at the expense of clarity and accuracy. Make the summary short enough to be economical, but long enough to be clear and comprehensive.

CRITICAL THINKING IN THE SUMMARY PROCESS

Follow these guidelines for summarizing your own writing or another's.

1. *Read the entire original.* When summarizing another's work, get a complete picture before writing a word.
2. *Reread and underline.* Reread the original, underlining essential material. Focus on the thesis and topic sentences.
3. *Edit the underlined information.* Reread the underlined material and cross out whatever does not advance the meaning.
4. *Rewrite in your own words.* Include all essential material in the first draft, even if it's too long; you can trim later.
5. *Edit your own version.* When you have everything readers need, edit for conciseness.
   a. Cross out all needless words without harming clarity or grammar. Use complete sentences.

The summer internship in journalism gives the ~~journalism~~ student ~~first-hand~~ experience ~~at what goes~~ on ~~within~~ a ~~real~~ newspaper.

    b.   Cross out needless prefaces such as "The writer argues" or "Also discussed is."

    c.   Use numerals for numbers, except to begin a sentence.

    d.   Combine related ideas in order to emphasize relationships (page 229).

6. *Check your version against the original.* Verify that you have preserved the essential message and added no comments.

7. *Rewrite your edited version.* Add transitional expressions to reinforce the connection between related ideas.

8. *Document your source.* If summarizing another's work, cite the source immediately below the summary, and place directly quoted statements within quotation marks. (See Chapter 8 for documentation formats.)

Although the summary is written last by the writer, it is read *first* by the reader. Take time to do a good job.

A SAMPLE SITUATION

Imagine that you work in the information office of your province's Ministry of the Environment. In a coming election, citizens will vote on a referendum proposal for constructing municipal trash incinerators. Referendum supporters argue that incinerators would help solve the growing problem of waste disposal in highly populated parts of the province. Opponents argue that incinerators cause air pollution.

To clarify the issues for voters, the Ministry will mail a newsletter to each registered voter. You have been assigned the task of researching the recent data and summarizing it. Here is one of the articles. You have underlined key phrases. (The margin notes reflect your critical thinking as you prepare to summarize the article.)

INCINERATING TRASH: A HOT ISSUE GETTING HOTTER

Combine as orienting sentence (controlling idea)

Include Definition

Alarmed by the tendency of landfills to contaminate the environment, both public officials and citizens are vocally seeking alternatives. The most commonly discussed alternative is something called a resource recovery facility. Nearly 40 Canadian cities have built such in the last 15 years, and another 50 or so are in various stages of planning.

These recovery facilities are a new form of an old technology. Basically, they're incinerators. But, unlike the incinerators of old, they don't just burn waste. They also recover energy. The energy is sold as steam to an industrial customer, or it is converted to electricity and sold to the local utility. (A few facilities, but not many, also recover metals or other materials before using the waste as fuel.)

Include Major Fact

Include Major Statistics

Include Major Fact

Include Major Fact

Delete Explanation

A ton of trash possesses the energy content of a barrel and a half of oil. This is not a trivial amount. Canada discards 50 million tons of municipal refuse a year. If all of it were converted to energy, we could replace the equivalent of 12 percent of our oil imports.

At the local level, selling energy or materials not only replaces non-renewable resources; but it also provides a source of income that partly offsets the cost of operating the facility.

The new facilities are, on average, much cleaner than the municipal incinerators of old. Many have two-stage combustion units, in which the second-stage burns exhaust gases at high temperature, converting many potential organic pollutants to less harmful emissions such as carbon dioxide. Some, especially the larger and newer facilities, also come equipped with the latest in pollution control devices.

Delete questionable point

The environmental community is uneasy with this new technology. Environmentalists have argued for many years that the best method of handling municipal trash is to recycle it—i.e., to separate the glass, metal, paper, and other materials and use them again, either without reprocessing or as raw materials in producing new products. The thought of the potential resources in municipal solid waste simply being burned, even with energy recovery, has made many environmentalists opponents of resource recovery.

Include key finding and explanation

More recently, opponents have found a stronger reason to oppose <u>burning waste: dioxin in the plants' emissions</u>. The <u>amounts</u> present are <u>extremely small</u>, measured in trillionths of a gram per cubic metre of air. <u>But dioxin can be deadly, at least to animals, at very low levels.</u>

What is Dioxin?

Include definition

Delete technical Details

Include major fact

<u>Dioxin</u> is a <u>generic term for any of 75 chemical compounds</u>, the technical name for which is poly-chlorinated dibenzo-p-dioxins (PCDDs). A related group of 135 chemicals, the PCDFs, or furans, are often found in association with PCDDs.

<u>The most infamous of these substances</u>, 2,3,7,8-TCDD, is often <u>referred to as the "most toxic chemical known."</u> This judgment is <u>based on animal test data</u>. In laboratory tests, 2,3,7,8-TCDD is lethal to guinea pigs at a concentration of *500 parts per trillion*. A part per trillion is roughly equivalent to the thickness of a human hair compared to the distance across Canada.

Include major point

Delete long explantation

Include continuation of major point

Delete long example

<u>The effects on humans are less certain</u>, for many reasons: it is difficult to measure the amounts to which humans have been exposed; difficult to isolate the effects of dioxin from the effects of other toxic substances on the same population; and the latency period for many potential effects, such as cancer, may be as long as 20 to 30 years.

<u>Nevertheless</u>, because of the <u>extreme effects</u> of this substance <u>on animals</u>, known releases of dioxin have generated <u>considerable public alarm</u>. One of the most publicized releases occurred at Seveso, Italy, in July 1976, where a pharmaceutical plant explosion resulted in the contamination of at least 700 acres of fields and affected more than 5000 people. Dioxin was found in the soil in concentrations of 20 to 55 parts per billion.

The immediate effects on humans were nausea, headaches, dizziness, diarrhea, and an acute skin condition called chloracne, which causes burn-like sores. The effects on animals were more severe: birds, rabbits, mice, chickens, and cats died by the hundreds, within days of the explosion. In response to the explosion, the Italian provincial authorities evacuated 730 people from the zone nearest the plant, and sealed off an area containing another 5 000 people from contact with non-residents.

Delete long example

In North America, perhaps the best known dioxin contamination incident occurred at Times Beach, Missouri, where used oil, contaminated with dioxin, was sprayed on roads as a dust suppressant. Soil samples showed dioxin at levels exceeding 100 parts per billion. While no human health effects were documented at Times Beach, a flood in December 1982 led to widespread dispersal of the contamination, as a result of which the entire town was condemned, the population evacuated, and over $30 million of the U.S.'s Superfund money used to purchase the condemned property.

Include the most striking and familiar example

Include key findings

<u>Dioxin was among the substances</u> of concern <u>at Love Canal</u>. And it was <u>the major contaminant</u> in the chemical defoliant <u>Agent Orange</u>, the subject of a <u>lawsuit by 15 000 Vietnam veterans</u> and dependents and an out-of-court settlement of those complaints valued at $180 million.

<u>As early as 1978, trace amounts of dioxin were found in the routine emissions of a municipal incinerator. Virtually every incinerator tested since</u> that date <u>has shown traces of dioxin.</u>

The Meaning of it All

Include major fact

At the request of the U.S. Congress, the Environmental Protection Agency began in 1994 a major research effort on dioxin, the National Dioxin Study. The study is intended to provide a context in which to place mounting concerns about dioxin. Research for the study was organized into seven "tiers," each tier including a group of sites at which dioxin contamination may be present.

Condense list

- Tier 1, production sites, includes the 10 sites at which 2,4,5-TCP, a pesticide known to have been contaminated by dioxin, was produced, and additional sites where waste materials from its production were disposed.
- Tier 2, precursor sites, includes 9 sites where 2,4,5-TCP was used as a precursor to make other chemical products, and related waste disposal sites. The chemical products included the herbicides 2,4,5-T and silvex, and hexachlorophene, a disinfectant that was widely used in soaps and deodorants, but was banned from non-prescription uses by the U.S. Food and Drug Administration in 1981.
- Tier 3 includes 60 to 70 sites at which 2,4,5-TCP and its derivatives were formulated into herbicide products, and associated waste disposal sites.
- Tier 4 includes a wide range of combustion sources, including internal combustion engines, wood stoves, fireplaces, forest fires, oil burners and other sources burning waste oil, and many others.
- Tier 5 includes 20 to 30 of the thousands of sites at which dioxin-contaminated pesticides have been used; for example, power line rights-of-way, forests, and rice and sugar cane fields.
- Tier 6 includes about 20 of the chemical and pesticide production facilities where improper quality control may have led to the accidental production of dioxin.
- Tier 7 includes samples from sites where the Agency least expected to find dioxin, to determine whether there are background levels of dioxin in the environment. Soil samples have been taken at 500 randomly selected locations across the country—200 in rural areas, 300 in urban areas—and fish have been sampled from over 400 locations.

Include key finding

While the study is not yet complete, data from a variety of sources have already produced disturbing—yet, perhaps, in an odd way, reassuring—results. Dioxin in trace amounts appears to be widely present in the environment, even in remote locations where industrial activity and waste combustion are unlikely to be the source.

Delete long example

Environment Canada, the Canadian EPA, also has an extensive dioxin testing program underway. One of the more startling findings of its research, conducted at a resource recovery facility on picturesque Prince Edward Island, is that the garbage delivered to the plant contained more dioxin than the plant's emissions. The source of the

Delete speculation

dioxin in this case is not known, though it could include pesticide residues or other products contaminated with dioxin during manufacturing processes.

Include key conclusion

In short, dioxin is not just a problem created by burning municipal waste. It is not clear at this time whether municipal waste combustion is even the major source of dioxin in the environment.

include conclusion

Ultimately, the dioxin problem is like other toxic substance issues. We know less than we need to know to thoroughly evaluate the risk. The more we find out, the more complex the issues tend to become. There is no risk-free solution, since all the potential disposal methods may result in some release of toxic substances to the environment.

Delete personal comment

Yet those who counsel delay, to allow the collection of more data, are met with the suspicion that their real agenda is to prevent action entirely.

Include
recommendations

What we do know at present <u>does not seem</u> to suggest <u>that we should stop planning</u> to build <u>resource recovery facilities</u>. What it does suggest is that we <u>proceed cautiously, inform the public</u> of both what is known and what is unknown, <u>install pollution controls</u> if the plants' uncontrolled emissions are significant or if the exposed population wants added protection, and <u>hope that continued examination</u> of all the sources and effects of dioxin <u>will eventually produce a consensus</u>.

Adapted from James E. McCarthy, *Congressional Research Service Review* Apr. 1986: 19–21.

Assume that in two early drafts of your summary, you rewrote and edited; for coherence and emphasis, you then inserted transitions and combined related ideas. Here is your final draft.

INCINERATING TRASH: A HOT ISSUE GETTING HOTTER
(A Summary)

Because landfills often contaminate the environment, trash incinerators (resource recovery facilities) are becoming a popular alternative. Nearly 40 are operating in Canadian cities, and 50 more are planned. Besides their relatively clean burning of waste, these incinerators recover energy, which can be sold to offset operating costs. One ton of trash has roughly the energy content of 1.5 barrels of oil. Converting all Canadian refuse to energy could reduce oil imports by 12 percent.

However, incinerator emissions contain very small amounts of dioxin (a generic name for any of 75 related chemicals). Even low dioxin levels can be deadly to animals. In fact, animal tests have helped label one dioxin substance "the most toxic chemical known." Although effects on human beings are less certain, news of dioxin in the environment creates public alarm, as evidenced at Love Canal and by the successful Agent Orange lawsuit by 15 000 Vietnam veterans. Almost every municipal incinerator tested since 1978 has shown traces of dioxin.

At the U.S. Congress's request, the Environmental Protection Agency in 1994 began the National Dioxin Study. Sites included herbicide and pesticide production and waste-disposal facilities, combustion sources such as woodstoves and forest fires, areas of dioxin-contaminated pesticide use such as rice and sugarcane fields, and random soil and fish samples nationwide. Findings indicate that trace amounts of dioxin are widely present in the environment, even in areas remote from industry and waste combustion. Waste incineration is by no means the only source of environmental dioxin—and may not even be the major source.

At this stage, we know too little to evaluate the risk. No risk-free waste-disposal solution exists. But the evidence so far does not suggest that we stop planning incinerators. We should, however, move cautiously, fully informing the public, installing pollution controls as needed, and searching for a better solution.

Adapted from James E. McCarthy, *Congressional Research Service Review* Apr. 1986: 19–21.

The version above is trimmed, tightened, and edited: word count is reduced to roughly 20 percent of the original. A summary this long serves well in many situations, but other audiences might want a briefer and more compressed summary—say, 10 to 15 percent of the original:

A More Compressed Summary

Because landfills often contaminate the environment, trash incinerators (resource recovery facilities) are becoming a popular alternative across Canada. Besides their relatively clean burning of waste, these incinerators recover energy, which can be sold to offset operating costs.

However, incinerator emissions contain very small amounts of dioxin, a chemical proven so deadly to animals, even at low levels, that it has been labelled the most toxic chemical known. Although its effects on human beings are less certain, news of dioxin in the environment creates public alarm. Almost every municipal incinerator tested since 1978 has shown traces of dioxin.

Findings of a U.S. study begun in 1994 suggest that trace amounts of dioxin are widely present in the environment, even in areas remote from industry and waste combustion. The dioxin source is by no means only waste incineration.

We lack risk-free disposal solutions and know too little to evaluate risks. But nothing so far suggests that we should stop planning incinerators. We should, however, move cautiously, fully informing the public, installing pollution controls as needed, and continuing to search for better solutions.

Adapted from James E. McCarthy, *Congressional Research Service Review* Apr. 1986: 19–21.

Notice that the essential message is still intact; related ideas are again combined and fewer supporting details are included. Clearly, length is adjustable according to your audience and purpose.

FORMS OF SUMMARIZED INFORMATION

In preparing a report, proposal, or other document, you might summarize others' material as part of your presentation. But you will often summarize your own material as well. For instance, if your document extends to several pages, it might include three forms of summary, in different locations, with different levels of detail: *closing summary, informative abstract,* or *descriptive abstract.*[1] Figure 9.1 depicts these forms.

---

1. Adapted from Vaughan. Although we take liberties with his classification, Vaughan's insightful article helped clarify our thinking about the overlapping terminology that perennially seems to confound discussions of these distinctions.

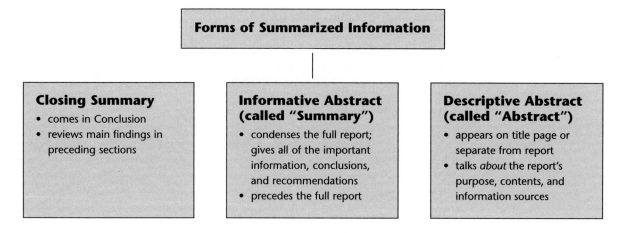

Figure 9.1   **Summarized Information Assumes Various Forms**

CLOSING SUMMARY

A *closing summary* refers to summarized information at the beginning of a Conclusion section or at the end of a report's central Body sections. It helps readers review and remember the preceding major findings. This look back at the "big picture" helps readers appreciate the conclusions and recommendations that follow.

INFORMATIVE ABSTRACT ("SUMMARY")

Many report readers appreciate condensed versions of reports. Some of these readers like to see a capsule version of a report before reading the complete document; others simply want to know basically what a report says without having to read the full document.

In order to meet reader needs, the *informative abstract* appears just after the title page. This summary tells the reader essentially what the full version says: It identifies the need or issue that has prompted the report; it describes the report's analytical method; it reviews the main facts and findings; and it condenses the report's conclusions and recommendations.

Actually, the term "Informative Abstract" is not used much these days. You are more likely to encounter the term "Summary." The heading, "Executive Summary," is used for material summarized for managers who may not understand all the technical jargon a report might contain. By contrast, a "Technical Summary" is written for readers at the same technical level as the author of the report. It's possible that you may need two or three levels of summary for report readers who have different levels of technical expertise.

See Chapter 20 for more discussion of the Summary section in a report.

DESCRIPTIVE ABSTRACT ("ABSTRACT")

A *descriptive abstract* talks about a report; it doesn't give the report's main points. Such an abstract helps potential readers decide whether to read the report. Thus, a descriptive abstract conveys only the nature and extent of a document. It presents the broadest view and offers no major facts from the original. It indicates whether conclusions and recommendations are included, but doesn't list them.

Descriptive abstracts usually appear in special publications or in electronic databases, both of which name and briefly describe hundreds of reports. One such publication is the widely used *Fisheries Abstracts*. However, some reports place a one- to three-sentence abstract on the report's title page. In all these placements, the term "Abstract" will signal a brief description, not all of the report's highlights.

Checklist FOR SUMMARIES

Use this checklist to refine your summaries.

Content

◆ Does the summary contain only the essential message?
◆ Does the summary make sense as an independent piece?
◆ Is the summary accurate when checked against the original?
◆ Is the summary free of any additions to the original?
◆ Is the summary free of needless details?
◆ Is the summary economical yet clear and comprehensive?
◆ Is the source documented?
◆ Does the descriptive abstract tell what the original is about?

Organization

◆ Is the summary coherent?
◆ Are there enough transitions to reveal the line of thought?

Style

◆ Is the summary's level of technicality appropriate for its audience?
◆ Is the summary free of needless words?
◆ Are all sentences clear, concise, and fluent?
◆ Is the summary written in correct English?

EXERCISES

1. Read each of these two paragraphs, and then list the significant ideas comprising each essential message. Write a summary of each paragraph.

    In recent years, ski-binding manufacturers, in line with consumer demand, have redesigned their bindings several times in an effort to achieve a non-compromising synthesis between performance and safety. Such a synthesis depends on what appear to be divergent goals. Performance, in essence, is a function of the binding's ability to hold the boot firmly to the ski, thus enabling the skier to change rapidly the position of his or her skis without being hampered by a loose or wobbling connection. Safety, on the other hand, is a function of the binding's ability both to release the boot when the skier falls, and to retain the boot when subjected to the normal shocks of skiing. If achieved, this synthesis of performance and safety will greatly increase skiing pleasure while decreasing accidents.

    Contrary to public belief, sewage-treatment plants do not fully purify sewage. The product that leaves the plant to be dumped into the leaching (sievelike drainage) fields is secondary sewage containing toxic contaminants such as phosphates, nitrates, chloride, and heavy metals. As the secondary sewage filters into the ground, this conglomeration is carried along. Under the leaching area develops a contaminated mound through which groundwater flows, spreading the waste products over great distances. If this leachate reaches the outer limits of a well's drawing radius, the water supply becomes polluted. And because all water flows essentially toward the sea, more pollution is added to the coastal regions by this secondary sewage.

2. Attend a campus lecture on a topic of interest and take notes on the significant points. Write a summary of the lecture's essential message.
3. Find an article about your major field or area of interest and write both an informative abstract and a descriptive abstract of the article.
4. Select a long paper that you have written for one of your courses; write an informative abstract and a descriptive abstract of the paper.

COLLABORATIVE PROJECT

Organize into small groups and choose a topic for discussion: an employment problem, a campus problem, plans for an event, suggestions for energy conservation, or the like. (A possible topic: Should employers have the right to require lie detector tests, drug tests, or AIDS tests for their employees?) Discuss the topic for one class period, taking notes on significant points and conclusions. Afterward, organize and edit your notes in line with the directions for writing summaries. Next, write a summary of the group discussion in no more than 200 words. Finally, as a group, compare your individual summaries for accuracy, emphasis, conciseness, and clarity.

Structural, Style, and Format Elements

Sue-Tina Kong, Toronto

When Sue-Tina Kong was young, she didn't ever think of becoming a technical writer: "You think of growing up to be a firefighter or a rocket scientist, not a technical writer." Following her English degree at Carleton University, "drawn to writing," she took Seneca College's Technical Communications program. Now she writes software user manuals and training programs at Amdocs Corporate Training, for clients mainly in the IP (Internet provider) broadband telephone market.

Ms. Kong's writing style has developed from her college education but also from extensive reading and from "learning new things each project." She's learned that a writer "must marry her understanding of English with the special technical nature of each project."

To choose exactly the right style, she talks to clients and to software designers. She learns the range of users for a manual or training program, from "new people to very technical people—the phrasing can't be overly technical, but it must be exact!" And, she says, "Visuals can be just as important as words: people remember pictures, and some visuals—such as flow charts—can convey ideas better than words alone. Also, pictures break up the text, to improve accessibility."

Page design, then, is also important at Amdocs: "For our document design, we've developed standards clients can count on. Consistency is important to them." She adds, laughing, "Standard page designs help us meet our deadlines!"

CHAPTER 10

Organizing for Readers

TOPICAL ARRANGEMENT

OUTLINING

PARAGRAPHING

SEQUENCING

One of your biggest writing challenges is to transform your material into manageable form. First, you need to unscramble information to make sense of it for yourself; then you need to shape it for the reader's understanding.

In order to follow your thinking, readers need a message organized in a way that makes sense to *them*. But data rarely materializes or thinking rarely occurs in neat, predictable sequences. You cannot merely report ideas or data in the same random order they occur. Instead, you must *shape* this material into an organized unit of meaning. In trying to organize, you will face questions like these:

TYPICAL QUESTIONS IN ORGANIZING FOR READERS

- What major question am I answering for my reader(s)?
- What secondary questions will help answer the main question?
- What relationships and answers do the gathered data suggest?
- What should I emphasize?
- What belongs where?
- What do I place first? Why?
- What comes next?
- How do I end?

Writers rely on the following strategies for organizing material: topical arrangement, outlining, paragraphing, and sequencing.

TOPICAL ARRANGEMENT

Whenever you analyze something, you break it down to discover constituents, connections, similarities, trends, associations, correlations, relationships, and perspectives. You break the topic into sub-topics (that's commonly called *partition*). You also see which things share similarities and should therefore be discussed at the same level in a certain category (that's called *classification*). The following example explains how this can be achieved.

A CD collection could be divided (partitioned) into categories, to discuss the collection in an orderly way: jazz, rock, blues, country, classical, new age, world music, etc. The rock CDs might further be sub-divided by the decade of the albums' release or by other descriptions—heavy metal, progressive rock, country rock, rock fusion, etc. The heavy metal bands could then be grouped together (classification) to identify the distinctive features of this rock genre.

However, someone might use a different approach to discuss a CD collection. The collection might be grouped according to how various albums have influenced selected current, popular musicians. Why use this topical arrangement instead of the arrangement described in the previous paragraph? Your choice depends on your audience and purpose—see the above list of questions.

The following case study illustrates how authors arrange and develop topics for a report.

In December 1998, the Canadian Council for Tobacco Control and British Columbia Ministry of Health released a report relating to cigarette additives and cigarette smoke constituents. The report had a persuasive purpose—to discourage smoking by releasing facts about cigarette ingredients and additives, as well as the detailed chemical analysis of the smoke of each brand of cigarette sold in British Columbia.

The information in the report emanated from the tobacco companies themselves, in response to a July 1998 British Columbia government requirement. The writers of the report were faced with two challenges: (1) to arrange the information so that readers could easily make sense of what they read, and (2) to present a progression of facts so that readers could draw the "right" conclusions about cigarette smoke.

The authors started by preparing an Introduction to the report. After the Introduction, they developed the following four main topics:

1. *background* information about British Columbia's requirements
2. *cigarette additives and ingredients* for each brand sold in British Columbia
3. *analysis of cigarette smoke* for the 11 leading brands
4. *"light"* cigarettes' test results

The authors then further developed one of the report's more contentious parts, the light cigarettes' test results, by presenting the following groups of data:

◆ B.C. tests show little difference in substances in light and regular cigarettes
◆ labels on cigarette packages mislead consumers
◆ filter vents on cigarettes help "fool" test machines
◆ machine tests don't simulate real smoking
◆ using more accurate test methods gives more accurate results
◆ a majority of smokers are fooled by the "light" label

Then, the authors further developed the first sub-topic, "differences in substances in light and regular cigarettes." They did this in a series of tables that showed tar, nicotine, and carbon monoxide values for each of the 11 corporate brands sold in British Columbia. ▲

Table 10.1 shows the full set of relationships among the report's topics, on three levels. These levels are differentiated in Table 10.1 by decreasing levels of capitalization. As the table moves from left to right through the four main topics, there is a logical development, all leading to the report's culminating point about smokers being "fooled" by the "light cigarette" designation. Further, within the fourth main topic, the six sub-topics are presented equally, but in an order that supports the authors' main point about the problem with so-called "light" cigarettes.

In organizing their documents, writers use partition and classification routinely, in a process we know as outlining.

OUTLINING

With an outline, you move from a random listing of items to a deliberate map that will guide readers from point to point. Readers more easily understand and remember material organized in a sequence they find logical. Organize in the sequence in which you expect readers to approach the material.

Table 10.1   Relationships Within a Report's Topical Arrangement (based on B.C. Ministry of Health report)

BACKGROUND	ADDITIVES AND INGREDIENTS	SMOKE ANALYSIS	IMPACT OF "LIGHT" CIGARETTES ON SMOKERS
B.C. Regulations	Nature of Disclosure	The Test Methods ◆ test conditions ◆ methodology	Little Difference in Light and Regular Cigarettes ◆ tar values ◆ nicotine values ◆ CO values
Reporting Requirements		The Test Results ◆ our values for each brand	Misleading Labels
		Impact on Health ◆ carbon monoxide ◆ cadmium ◆ benzene	Filter Vents Lower Readings Machine vs. Real Smoking
		The Chemicals ◆ 29 toxic chemicals in cigarettes	Better Tests, Better Results Smokers' Beliefs

A DOCUMENT'S BASIC SHAPE

How should you organize the document to make it logical from your audience's point of view? Begin with the basics. Useful writing of any length—a book, chapter, news article, letter, or memo—typically follows this organizing pattern:

◆ The *introduction* provides orientation by doing any of these things: explaining the topic's origin and significance and the document's purpose; identifying briefly your intended audience and your information sources; defining specialized terms or general terms that have special meanings in your document; accounting for limitations such as incomplete or questionable data; previewing the major topics to be discussed in the body section.

Some introductions need to be long and involved; others, short and to the point. Reports too often waste readers' time with needless background information. If you don't know your readers well enough to give them only what they need, use subheadings so that they can choose what they want to read.

◆ The *body* delivers on the promise implied in your introduction ("Show me!"). Here you present your data, discuss your evidence, lay out your case, or tell readers what to do and how to do it. Body sections come in all different sizes, depending on how much readers need and expect.

Body sections are titled to reflect their specific purpose: "Description and Function of Parts," for a mechanical description; "Required Steps," for a set of instructions; "Collected Data," for a feasibility analysis; "Operating Instructions," for a user's manual.

◆ The *conclusion* of a document has assorted purposes: It might evaluate the significance of the report, re-emphasize key points, take a position, predict an outcome, offer a solution, or suggest further study. If the issue is straightforward, the conclusion might be brief and definite. If the issue is complex or controversial, the conclusion might be lengthy and open ended. Whatever the conclusion's specific purpose, readers expect a clear perspective on the whole document.

Conclusions vary with the document. You might conclude a mechanical description by reviewing the mechanism's major parts and then briefly describing one operating cycle. You might conclude a comparison or feasibility report by offering judgments about the facts you've presented and then recommending a course of action.

Most workplace documents display this basic shape

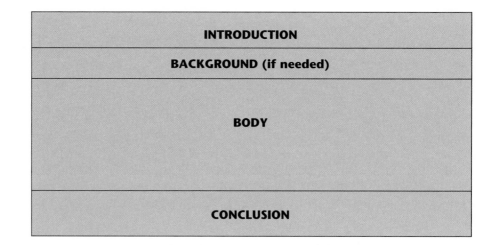

A suitable beginning, middle, and ending are essential, but alter your outline as you see fit. No single form of outline should be followed exactly by any writer. *The organization of any document ultimately is determined by its audience's needs and expectations.* In many cases, specific requirements about a document's organization and style are spelled out in a company's style guide. Structures for specific types of documents are provided in various sections of this text. In particular, see the Action Structure for short reports, in Chapter 21.

The computer is especially useful for rearranging outlines until they reflect the sequence in which you expect readers to approach your message.

THE FORMAL OUTLINE

A simple list usually suffices for organizing short documents. However, long or complex documents call for a systematic, formal outline, to mark divisions and to show how categories relate. Here, for example, is a formal outline for a report examining the health effects of electromagnetic fields:

A formal outline whose topical arrangement uses an appropriate pattern:

- causes (of problem)
- effects (nature of problem)
- interpretation of those effects
- possible solutions
- conclusions and recommendations

CHILDREN EXPOSED TO EMFS: A RISK ASSESSMENT

I. INTRODUCTION

A. Definition of electromagnetic fields

B. Background on the health issues

C. Description of the local power-line configuration

D. Purpose of this report[1]

E. Brief description of data sources

F. Scope of this report

II. DATA SECTION [Body]

A. Sources of EMF exposure

  1. home and office

    a. kitchen

    b. workshop[2] [etc.]

  2. power lines

  3. natural radiation

  4. risk factors

    a. current intensity

    b. source proximity

    c. duration of exposure

B. Studies of health effects

  1. population surveys

  2. laboratory measurements

  3. workplace links

C. Conflicting views of studies

  1. criticism of methodology in population studies

  2. criticism of overgeneralized lab findings

D. Power industry views

  1. uncertainty about risk

  2. confusion about risk avoidance

E. Risk-avoidance measures

  1. nationally

  2. locally

III. CONCLUSION

A. Summary and overall interpretation of findings

B. Recommendations

---

1. Long reports often begin directly with a statement of purpose. For the intended audience (i.e., generalists) of this report, however, the technical topic first must be defined for the readers' clear understanding of the context.
2. Note that each level of division yields at least two items. If you cannot divide a major item into at least two subordinate items, retain only your major heading.

A formal outline easily converts to a table of contents for the finished report, as shown in Chapter 20. (Because they serve mainly to guide the *writer*, omit minor outline headings, such as a and b [under II.A.1 in our example] from the table of contents or the report itself. Excessive headings make a document seem fragmented.)

In technical documents, the alphanumeric form of notation shown above often is replaced by decimal notation:

Decimal notation in a technical document

2.0 DATA SECTION
    2.1 Sources of EMF Exposure
        2.1.1 home and office
            2.1.1.1 kitchen
            2.1.1.2 workshop [etc.]
        2.1.2 power lines
        2.1.3 natural radiation
        2.1.4 risk factors
            2.1.4.1 current intensity
            2.1.4.2 source proximity
            2.1.4.3 [etc.]

The decimal outline makes it easier to refer readers to specifically numbered sections of the document: ("See 2.1.2"). While both systems achieve the same organizing objective, decimal notation usually is preferred in business, government, and industry.

In some cases, you may want to expand the above *topic outline* into a *sentence outline,* in which each sentence serves as a topic sentence for a paragraph in the report:

A sentence outline

2.0 DATA SECTION
    2.1 Although the several million miles of power lines criss-crossing North America have been the focus of the EMF controversy, potentially harmful waves also are emitted by household wiring, appliances, electric blankets, and computers.
        2.1.1 [etc.]

Sentence outlines are used mainly in collaborative projects in which various team members prepare different sections of a long document.

THE IMPORTANCE OF BEING ORGANIZED

The neat and ordered outline shown earlier represents the *product* of outlining, not the *process*. Beneath any finished outline (or document) lies *planning*. Whether you work alone or as part of a team, planning is an essential element of the writing process. Initially, planning may take the form of general discussions and brainstorming. The ideas generated are often written on a whiteboard or displayed on a computer screen as notes or flowcharts (see "Getting Started," Chapter 3).

The second stage of planning involves *organizing* the ideas into specific topics. Ideas may be discarded because they are not within the scope of the project; other ideas may be generated as the topics are being identified. When you know the main topics you need to cover for your project, you are ready to put them into an outline. If you have used a computer during your planning and organizing, you already have a draft outline. The next step is to rearrange your outline so that it flows logically (has a sequence). Various types of sequencing are covered later in this chapter.

Two rules of effective and efficient writing are:

1. NEVER start writing until you have *thoroughly* planned the topics to be included in your document and created an organized, logical outline.
2. When you start writing, follow your outline precisely; don't start planning again!

OUTLINING AND REORGANIZING ON A COMPUTER

Most word-processing programs enable you to work on your document and your outline simultaneously. An "outline view" of the document helps you to see relationships among ideas at various levels, to create new headings, to add text beneath headings, and to move headings and their subtext (Figure 10.1). You also can *collapse* the outline view to display the headings only.

Switch between "normal view" (to compose your text) and "outline view" (to examine the arrangement of material). You can add or delete headings or text and reorganize whole sections of your document. (*Microsoft Word* 504–05).

As a visual alternative to traditional outlining, many computer graphics programs enable you to display prose outlines as tree charts.

ORGANIZING FOR CROSS-CULTURAL AUDIENCES

Different cultures often have different expectations as to how information should be organized. A document considered well organized by one culture may confuse or offend another. For instance, a paragraph in English typically begins with a main idea directly expressed as a topic or orienting sentence, followed by specific support; any digression from this main idea is considered harmful to the paragraph's *unity*. Some cultures, however, consider digression a sign of intelligence or politeness. To native readers of English, the long introductions and digressions in certain Spanish or Russian documents might seem tedious and confusing, but a Spanish or Russian reader might view the more direct organization of English as overly abrupt and simplistic (Leki 151).

Expectations can differ even among same-language cultures. British correspondence, for instance, typically expresses the bad news directly up front, instead of the indirect approach preferred in North America. A bad-news letter or memo appropriate for North American readers could be considered evasive by British readers (Scott and Green 19).

Despite all our electronic communication tools, connecting with readers—especially in a global context—requires, above all, human sensitivity and awareness of audience.

Chapters 19 and 21 show structures for long and short reports. Chapter 19 also demonstrates how three levels of outlines (planning outlines, detailed working outlines, and paragraph outlines) can be used to efficiently write a complex document.

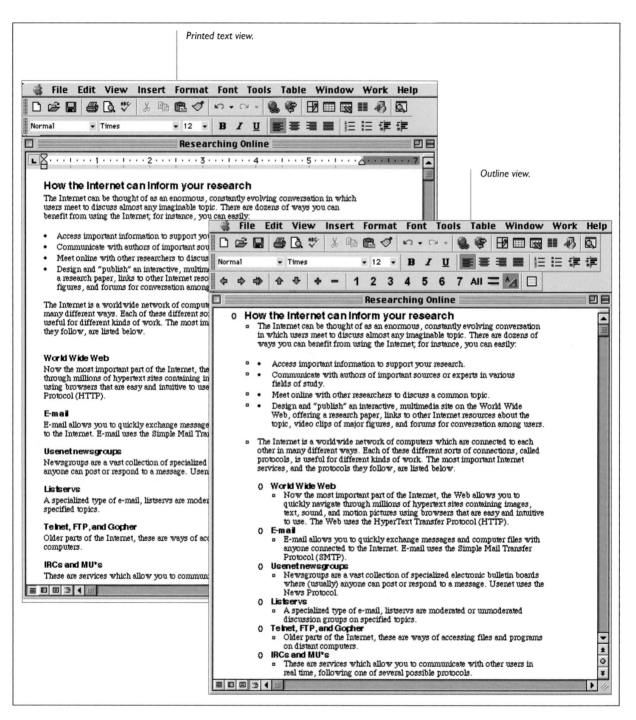

Figure 10.1   Using a Computer's Outline Features

Source: Munger, David, et al. Researching Online. 2nd ed. New York: Longman, 1998:2.

PARAGRAPHING

Readers look for orientation, for shapes they can recognize. Beyond its larger shape (introduction, body, conclusion), a document depends on the smaller shapes of each paragraph.

Although paragraphs can have various structures and purposes (paragraphs of introduction, conclusion, or transition), our focus here is on *standard support paragraphs*. While part of the document's larger design, each of these middle blocks of thought usually can stand alone in meaning and emphasis.

THE STANDARD PARAGRAPH

All the sentences in a standard paragraph relate to the main point, which is expressed as the *topic sentence:*

Topic sentences

Computer literacy has become a requirement for all "educated" people.

A video display terminal can endanger the operator's health.

Chemical pesticides and herbicides are both ineffective and hazardous.

Each topic sentence introduces an idea, judgment, or opinion. But in order to grasp the writer's exact meaning, readers need explanation. Consider the third statement:

Chemical pesticides and herbicides are both ineffective and hazardous.

Imagine you are a researcher for the Epson Electric Light Company and have been asked to determine whether the company should (1) begin spraying pesticides and herbicides under its power lines, or (2) continue with its manual (and non-polluting) ways of minimizing foliage and insect damage to lines and poles. If you simply responded with the preceding assertion, your employer would have questions:

◆ Why, exactly, are these methods ineffective and hazardous?
◆ What are the problems?
◆ Can you explain?

To answer these questions and to support your assertion, you need a fully developed paragraph:

Intro. (topic sent.)
Body (2–6)

Conclusion (7–8)

[1]**Chemical pesticides and herbicides are both ineffective and hazardous.** [2]Because none of these chemicals has permanent effects, pest populations invariably recover and need to be resprayed. [3]Repeated applications cause pests to develop immunity to the chemicals. [4]Furthermore, most of these products attack species other than the intended pest, killing off its natural predators, thus actually increasing the pest population. [5]Above all, chemical residues survive in the environment (and living tissue) for years, often carried hundreds of miles by wind and water. [6]This toxic legacy includes such biological effects as birth deformities, reproductive failures, brain damage, and cancer. [7]Although intended to control pest populations, these chemicals ironically threaten to make the human population their ultimate victims. [8]We therefore recommend continuing our present control methods.

Most standard paragraphs in technical writing have an introduction-body-conclusion structure. They begin with a clear topic (or orienting) sentence stating a generalization. Details in the body support the generalization.

THE TOPIC SENTENCE

Readers look to a paragraph's opening sentences for a framework. When they don't know exactly what the paragraph is about, readers struggle to grasp your meaning. Read this next paragraph once only, and then try answering the questions that follow.

A paragraph with its topic sentence omitted

> Besides containing several toxic metals, it percolates through the soil, leaching out naturally present metals. Pollutants such as mercury invade surface water, accumulating in fish tissues. Any organism eating the fish—or drinking the water—in turn faces the risk of heavy metal poisoning. Moreover, acidified water can release heavy concentrations of lead, copper, and aluminum from metal plumbing, making ordinary tap water hazardous.

Can you identify the paragraph's main idea? Probably not. Without the topic sentence, you have no framework for understanding this information in its larger meaning. And you don't know where to place the emphasis: on polluted fish, on metal poisoning, on tap water?

Now, insert the following opening sentence and reread the paragraph:

The missing topic sentence

> Acid rain indirectly threatens human health.

With this orientation, the exact meaning becomes obvious.

The topic sentence should appear *first* in the paragraph, unless you have a good reason to place it elsewhere. Think of your topic sentence as the one sentence you would keep if you could keep only one (U.S. Air Force Academy 11). In some instances, a paragraph's main idea may require a "topic statement" consisting of two or more sentences, as in this example:

A topic statement can have two or more sentences

> The most common strip-mining methods are open-pit mining, contour mining, and auger mining. The specific method employed will depend on the type of terrain that covers the coal.

The topic sentence or topic statement should immediately tell readers what to expect. Don't write: *Some pesticides are less hazardous and often more effective than others* when you mean: *Organic pesticides are less hazardous and often more effective than their chemical counterparts.* The first topic sentence leads everywhere and nowhere; the second helps the reader focus, telling the reader what to expect from the paragraph. Don't write: *Acid rain poses a danger,* leaving readers to decipher your meaning of danger. If you mean that *Acid rain is killing our lakes and polluting our water supplies,* say so. Uninformative topic sentences keep readers guessing.

ALTERNATIVE TOPIC SENTENCE PLACEMENT

Nearly all paragraphs in technical and business writing open with a topic sentence, but sometimes the opening sentence will provide a transition, which is followed by IV the topic in the second sentence and then by the rest of the paragraph. Here's an effective paragraph in a report's Conclusion section:

Transition sentence (1)
Topic sentence (2)
Supporting reasons
(3–5)

> As this report has demonstrated, erosion control blankets have a wide range of uses. **Therefore, these blankets should be considered for controlling run-offs on the Summit Highway.** The blankets provide a warm, moist environment for the hydro-seeded wild grass to germinate. Also, the blankets prevent erosion until the grass can provide a stable sub-soil-root system to prevent erosion. Finally, the straw and the coconut fibre in the blankets gradually decay into the soil to make the soil more fibrous and resistant to run-off erosion.

Some paragraphs open with a transition which continues the direction established by the previous paragraph(s), and then switch to the real topic of the new paragraph.

Transition sentence (1)
Switching to the real
topic (2)
Descriptive
explanation, to justify
the topic statement
(3–5)

> We recommend erosion control blankets for the majority of road banks on the Summit Highway. **However, for banks that are 2:1 steeper, we recommend heavy duty geogrid.** Skalned Engineering's studies have shown that steep banks require geogrids of 4-inch side wall mesh that can be filled with small-scale blast rock (Appendix 2). The geogrid can also be partially filled with soil, which can be hydro-seeded. Skalned's study revealed that geogrid covers of this type prevent erosion of banks as steep as 1.5 to 1.

Other paragraphs, such as the following, effectively place the main idea at the end of the paragraph.

Sentence 1
introduces the topic
area, but not the
paragraph's main idea.
List of details (2 and 3)
Main idea (4)

> Worried about heavy erosion, we used Gabion baskets to retain the 2:1 banks of Tumbledown Creek below LeBihan Falls. We observed that the baskets retained their geometric shapes despite the heavy aggregate that filled them. Also, because we placed geotextile fabric beneath the Gabion structure, the strong 2002 spring runoff didn't scour or undercut the Gabion system. **Overall, the entire system remained intact and stable despite the powerful erosive conditions that ran from early March to late May.**

PARAGRAPH UNITY

A paragraph is unified when all its material belongs there—when every word, phrase, and sentence directly supports the topic sentence.

A unified paragraph

> **Solar power offers an efficient, economical, and safe solution to eastern Canada's energy problems.** To begin with, solar power is highly efficient. Solar collectors installed on fewer than 30 percent of roofs in the East would provide more than 70 percent of the area's heating and air-conditioning needs. Moreover, solar heat collectors are economical, operating for up to 20 years with little or no maintenance. These savings recoup the initial cost of installation in only 10 years. Most important, solar power is safe. It can be transformed into electricity through photovoltaic cells (a type of storage battery) in a noiseless process that produces no air pollution—unlike coal, oil, and wood combustion. In sharp contrast to its nuclear counterpart, solar power produces no toxic waste and poses no catastrophic danger of meltdown. Thus, massive conversion to solar power would ensure abundant energy and a safe, clean environment for future generations.

One way to damage unity in the paragraph above would be to discuss the differences between active and passive solar heating, or manufacturers of solar technology, or the advantages of solar power over wind power. Although these matters do *broadly* relate to the general issue of solar energy, none directly advances the meaning of *efficient, economical,* or *safe.*

Every topic sentence has a key word or phrase that carries the meaning. In the pesticide-herbicide paragraph (page 203), the keywords are *ineffective* and *hazardous.* Anything that fails to advance their meaning throws the paragraph—and the reader—off track.

PARAGRAPH COHERENCE

In a unified paragraph, everything belongs. In a coherent paragraph, everything sticks together: Topic sentence and support form a connected line of thought, like the links in a chain. To convey precise meaning, a paragraph must be unified. To be readable, a paragraph must also be coherent.

Paragraph coherence can be damaged by (1) short, choppy sentences, (2) sentences in the wrong sequence, or (3) insufficient transitions and connectors (see Appendix A) for linking related ideas. Here is how the solar energy paragraph might become incoherent:

An incoherent paragraph

> Solar power offers an efficient, economical, and safe solution to eastern Canada's energy problems. Unlike nuclear power, solar power produces no toxic waste and poses no danger of meltdown. Solar power is efficient. Solar collectors could be installed on fewer than 30 percent of roofs in the East. These collectors would provide more than 70 percent of the area's heating and air-conditioning needs. Solar power is safe. It can be transformed into electricity. This transformation is made possible by photovoltaic cells (a type of storage battery). Solar heat collectors are economical. The photovoltaic process produces no air pollution.

Here, in contrast, is the original, coherent paragraph with sentences numbered for later discussion and with transitions and connectors shown in boldface. Notice how this version reveals a clear line of thought:

A coherent paragraph

> [1]Solar power offers an efficient, economical, and safe solution to eastern Canada's energy problems. [2]**To begin with**, solar power is highly efficient. [3]Solar collectors installed on fewer than 30 percent of roofs in the East would provide more than 70 percent of the area's heating and air-conditioning needs. [4]**Moreover**, solar heat collectors are economical, operating for up to 20 years with little or no maintenance. [5]**These savings** recoup the initial cost of installation within only 10 years. [6]**Most important**, solar power is safe. [7]**It** can be transformed into electricity through photovoltaic cells (a type of storage battery) in a noiseless process that produces no air pollution—unlike coal, oil, and wood combustion. [8]**In sharp contrast** to its nuclear counterpart, solar power produces no toxic waste and poses no danger of catastrophic meltdown. [9]**Thus,** massive conversion to solar power would ensure abundant energy and a safe, clean environment for future generations.

We easily can trace the sequence of thoughts in this paragraph.

1.   The topic sentence establishes a clear direction.
2–3. The first reason is given and then explained.
4–5. The second reason is given and explained.
6–8. The third and major reason is given and explained.
9.   The conclusion sums up and re-emphasizes the main point.

Within this line of thinking, each sentence follows logically from the one before it. Readers know where they are at any place in the paragraph. To reinforce the logical sequence, related ideas are combined in individual sentences, and transitions and connectors signal clear relationships. The whole paragraph sticks together. For a useful list of transitions and other connections, see pages 666–69 in Appendix A.

PARAGRAPH LENGTH

Paragraph length depends on the writer's purpose and the reader's capacity for understanding. Actual word count is less important than how thoroughly the paragraph makes your point. Consider these guidelines:

- ◆ In writing that carries highly technical information or complex instructions, short paragraphs (perhaps in a vertically displayed list) give the reader plenty of breathing space. A clump of short paragraphs, however, can make a document seem choppy and poorly organized.
- ◆ In writing that explains concepts, attitudes, or viewpoints, support paragraphs generally run from 100 to 300 words. But long paragraphs can be tiring and hard to follow, especially if important ideas get buried in the middle. On average, report paragraphs should be kept to 100 words, while paragraphs in letters and memos should average about 60 words.
- ◆ Long paragraphs can be broken into parts—using bullets, for example—to make the information more accessible to the reader.
- ◆ In letters, memos, or news articles, paragraphs of only one or two sentences focus the reader's attention. A short paragraph (even a single-sentence paragraph) can highlight an important idea in any document.
- ◆ Avoid long paragraphs at the beginning or end of a document because they can discourage the reader or obscure the emphasis.

SEQUENCING

Research demonstrates that readers more easily understand and remember material that is organized in a logical sequence (Felker et al. 11). In practice, these logical sequences tend to be one of three main types: *general-to-specific, specific-to-general,* and *chronological.* A single paragraph usually follows one particular sequence. A longer document may use one particular sequence or a combination of sequences.

Most technical and business writing relies heavily on general-to-specific patterns, such as

- ◆ descriptive sequence
- ◆ statement plus illustration
- ◆ emphatic sequence (statement plus detailed evidence or arguments)
- ◆ extended definition
- ◆ classification
- ◆ comparison and contrast

SPATIAL SEQUENCE

The most common *descriptive* pattern uses a spatial sequence, which begins at one location and ends at another. Such a sequence is most useful in describing a physical item or a mechanism; its parts appear in the sequence in which readers would actually view the parts or in the order in which each part functions (left to right, inside to outside, top to bottom). This description of a hypodermic needle proceeds from the needle's base (hub) to its point:

"What does it look like?"

> **A hypodermic needle is a slender, hollow steel instrument used to introduce medication into the body (usually through a vein or muscle). The instrument has three parts, all considered sterile: the hub, the cannula, and the point.** The hub is the lower, larger part of the needle that attaches to the neck-like opening on the syringe barrel. Next is the cannula (stem), the smooth and slender central portion. Last is the point, which consists of a bevelled (slanted) opening, ending in a sharp tip. The diameter of a needle's cannula is indicated by a gauge number; commonly, a 24–25 gauge needle is used for subcutaneous injections. Needle lengths are varied to suit individual needs. Common lengths used for subcutaneous injections include 0.85 cm, 1.25 cm, 1.5 cm, and 1.85 cm. Regardless of length and diameter, all needles have the same functional design. Product and mechanism descriptions almost always have some type of visual to support or amplify the verbal description.

STATEMENT PLUS ILLUSTRATION

Often, the best way to prove a point is to illustrate it with an example or story. The main point appears in the topic sentence and the illustration forms the remainder of the paragraph, as in the following example:

"Why should I believe you?"

> **Sometimes, steep erosion-prone slopes can be stabilized simply and inexpensively.** For example, our client at 458 Summit Road was unwilling to spend an estimated $38,000 for a recommended reinforced concrete retaining wall at the rear of his property. Our low-budget solution involved C350 erosion control blankets, hydro-seeded wild grass cover, and multi-flow drainage pipe, all for only $8,546. For 27 months, the slope has withstood erosion.

EMPHATIC SEQUENCE (STATEMENT PLUS EVIDENCE OR ARGUMENTS)

Reasons offered in support of a specific viewpoint or recommendation often appear in workplace writing, as in the pesticide-herbicide paragraph on page 203 or the solar energy paragraph on page 206. For emphasis, the reasons or examples usually are arranged in decreasing or increasing order of importance.

"What should I remember about this?"

> **Although strip mining is safer and cheaper than conventional mining, it damages the surrounding landscape.** Among its effects are scarred mountains, ruined land, and polluted waterways. Strip operations are altering our country's land at the rate of 506 hectares (1250 acres) per week. An estimated 7085 kilometres (4400 miles) of streams have been poisoned by silt drainage in British Columbia alone. If strip mining continues at its present rate, 14 900 square kilometres (5750 square miles) of Canadian land eventually will be stripped barren.

EXTENDED DEFINITION

Another descriptive pattern used to organize a paragraph is found in an extended definition (or "expanded definition"—see Chapter 14). Such a paragraph piles up the details to help a reader understand a term.

"What does this term really mean?"

Orlimar's Trimetal fairway woods incorporate the multi-metal technology found in all the Trimetal clubs: a steel shell, an alpha Maraging face, and copper-tungsten weights. The head's shell features stainless steel's high strength-to-weight ratio; to finish the hitting surface, Carpenter Metals has developed a way to merge ultra-hard alpha Maraging metal to the 17-4 steel to form the head's face. And, completing the technology, copper-tungsten weights are built as rails into the sole of the head. These rails promote a low centre of gravity, which helps the struck golf ball fly high and long.

CLASSIFICATION AND PARTITION

"How are the components of the issue related?"

Close examination of any complex issue requires both partition and classification. (See the previous discussion on pages 195–96.) The following paragraph shows both partition and classification at work.

While researching the health effects of electromagnetic fields, you'll encounter information about

- various radiation sources,
- the ratio of the level of risk to the level of exposure,
- workplace studies of effects on workers,
- lab studies of cell physiology, biochemistry, and behaviour,
- statistical studies of disease in certain populations,
- conflicting expert views, and
- local authorities' views.

COMPARISON-CONTRAST SEQUENCE

Evaluation of two or more items on the basis of their similarities or differences often appears in job-related writing.

"How do these items compare?"

Point-by-point comparison

The ski industry's quest for a binding that ensures good performance as well as safety has led to development of two basic types. **Although both bindings improve performance and increase the safety margin, they have different release and retention mechanisms.** The first type consists of two units (one at the toe, another at the heel) that are spring-loaded. These units apply their retention forces directly to the boot sole. Thus the friction of boot against ski allows for the kind of ankle movement needed at high speeds over rough terrain, without causing the boot to release. In contrast, the second type has one spring-loaded unit at either the toe or the heel. From this unit extends a boot plate that travels the length of the boot to a fixed receptacle on its opposite end. With this plate binding, the boot has no part in release or retention. Instead, retention force is applied directly to the boot plate, providing more stability for the recreational skier, but allowing for less ankle and boot movement before releasing. Overall, the double-unit binding performs better in racing, but the plate binding is safer.

For comparing and contrasting more specific data on these bindings, two lists would be more effective.

Block comparison

The **Rossignol Axium 100 X** plate binding offers the following features:

1. lightweight strength and control for carving skis
2. an elastomer built into the back of the plate to help maintain the ski's natural flex while eliminating vibration
3. a DIN setting range of 3 to 10, with an adjustable range for skiers 65 to 240 lbs.
4. 45 mm of lateral elastic travel, 12 mm of vertical elastic travel in the heel piece, and 24 mm of adjustment

The **Salomon A912 PS** offers the following features:

1. a beefy interface for enhanced stability
2. simultaneous wing adjustment
3. a DIN setting range of 4 to 12, with an adjustable range for skiers 92 to 265 lbs.
4. a diagonal pivot shock-absorption system that provides protection in forward and backward twisting falls
5. an adjustment range of 24 mm

Instead of this block structure (in which one binding is discussed and then the other), the writer might have chosen a point-by-point structure in which points common to both items are listed together (such as release methods in the above combined paragraph). The point-by-point comparison is favoured in feasibility and recommendation reports because it offers readers a meaningful comparison between common points.

SPECIFIC-TO-GENERAL SEQUENCE

Much less common than the general-to-specific sequences just described, *specific-to-general* sequences finish with the paragraph's main point. One such paragraph, the Gabion basket example on page 205, builds to a climax. The following paragraph leaves its main point until the end because the writer wants the reader to first see the writer's justification for an unpopular bottom line:

The unpopular bottom line needs preceding justification

Gendron Road Services has attempted several unsuccessful methods to repair the deteriorating approach apron at the north end of the LaSalle Bridge. First, Gendron tried four types of standard surface repair methods, but in each case, potholes reappeared within two months. Then, Gendron engineers tried deep surface patch techniques, but the patched pieces broke up during the spring thaw. Gendron has even experimented with a sub-surface heating grid, with poor results. **Gendron therefore recommends that the city authorize a complete reconstruction of the bridge approach, at a cost of \$1.9 million above Gendron's annual maintenance contract.**

In the next specific-to-general paragraph, several examples lead to a logical bottom-line conclusion:

Integrating a shop area's radiant heating system with an office area's forced air system appeals to fabrication shops and garages. Other enthusiastic users include nursing homes, hotels, and motels. In all these settings, radiant exchange systems can heat one area of a building while they cool another, with increased efficiency and economy. **Thus, these systems are becoming more popular.**

The bottom line

Chronological patterns include past-tense narration, process description, and instructions.

NARRATIVE SEQUENCE

Past-tense narration is easy to read because its patterns match the very structure of our language. Notice that the following paragraph uses time-based transitions to keep the progressions of events clear. (The transitions are bolded.)

"What happened?"

> In preparing the sub-base soil for the new fifth green, the crew **first** removed all soil and rock to a depth of 5 feet. **At that level,** they discovered a small seeping spring, so they **then** established the green's subterranean catch basin at the spring's rocky exit point. **Next**, with excavation complete, the crew laid a cross pattern of drainage pipe leading from the catch basin, and they covered the drain piping with 1.5 feet of pea gravel and 3.5 feet of fine sand. The green's surface **was thus ready** to be contoured.

INSTRUCTIONS

Explanations of how to do something or how something happened generally are arranged according to a strict time sequence: first step, second step, and so on.

"How is it done?"

> **Instead of breaking into a jog too quickly and risking injury, take a relaxed and deliberate approach.** Before taking a step, spend at least 10 minutes stretching and warming up, using any exercises you find comfortable. (Consult a jogging book for specialized exercises.) When you've completed your warm-up, set a brisk walking pace. Exaggerate the distance between steps, taking long strides and swinging your arms briskly and loosely. After roughly 100 metres at this brisk pace, you should feel ready to jog. So, break into a very slow trot: lean your torso forward and let one foot fall in front of the other (one foot barely leaving the ground while the other is on the pavement). Keep the slowest pace possible, just above a walk. *Do not bolt out like a sprinter!* While jogging, relax your body. Keep your shoulders straight and your head up, and enjoy the scenery.

PROCESS DESCRIPTION

Chapter 16 contains a sample of a complete process description, told in present tense. The same descriptive method can be used to organize a single paragraph, as in the following explanation of a vapour compression cycle in a geothermal heating system.

"How does this process work?"

> **All heat pumps use a vapour compression cycle to transport heat from one location to another.** In heating mode, the cycle starts as the cold liquid refrigerant within the heat pump passes through a heat exchanger (evaporator) and absorbs heat from the low temperature source (fluid circulated through the earth connection). The refrigerant evaporates into a gas as heat is absorbed. The gaseous refrigerant then passes through a compressor, where it is pressurized, raising its temperature to over 80°C. The hot gas then circulates through a refrigerant-to-air heat exchanger, where the heat is removed and sent through the air ducts. When the refrigerant loses the heat, it changes back to a liquid. The liquid refrigerant cools as it passes through an expansion valve, and the process begins again (Groundloop 2001).

PROBLEM-CAUSES-SOLUTION SEQUENCE

Another form of chronological development is used to explain how a problem was solved. The problem-solving sequence proceeds from description of the problem, through diagnosis, to solution. After outlining the cause of the problem, this next paragraph explains how the problem has been solved:

"How was the problem solved?"

> On all waterfront buildings, the unpainted wood exteriors had been severely damaged by the previous winter's high winds and sandstorms. **After repairing the damage, we took protective steps against further storms.** First, all joints, edges, and sashes were treated with water-repellent preservative to protect against water damage. Next, three coats of non-porous primer were applied to all exterior surfaces to prevent paint from blistering and peeling. Finally, two coats of wood-quality latex paint were applied over the non-porous primer. To prevent coats of paint separating, we applied the first coat within two weeks of the priming coats, and the second within two weeks of the first. Now, 14 months later, no blistering, peeling, or separation has occurred.

CAUSE-EFFECT ANALYSIS

Cause-effect analyses represent a useful variation of *chronological* paragraph sequencing. The direct version starts with causes and proceeds to effects—the sequence follows an action to its results. Below, the topic sentence identifies the causes, and the remainder of the paragraph discusses its effects.

"What will happen if I do this?"

> **Some of the most serious accidents involving gas water heaters occur when a flammable liquid is used in the vicinity.** The heavier-than-air vapours of a flammable liquid such as gasoline can flow along the floor—even the length of a basement—and be explosively ignited by the flame of the water heater's pilot light or burner. Because the victim's clothing frequently ignites, the resulting burn injuries are commonly serious and extremely painful. They may require long hospitalization, and can result in disfigurement or death. *Never, under any circumstances, use a flammable liquid near a gas heater or any other open flame* (Consumer Product Safety Commission).

EFFECT-TO-CAUSE SEQUENCE

The indirect version starts with the effects and traces back to the cause(s). This sequence identifies the main "effect" (the problem) in the topic sentence.

"How did this happen?"

> **Modern whaling techniques have brought the whale population to the threshold of extinction.** In the nineteenth century, invention of the steamboat increased hunters' speed and mobility. Shortly afterward, the grenade harpoon was invented so whales could be killed quickly and easily from the ship's deck. In 1904, a whaling station opened on Georgia Island, in South America. This station became the gateway to Antarctic whaling for the nations of the world. In 1924, factory ships were designed that enabled round-the-clock whale tracking and processing. These ships could reduce a 30-metre-long whale to its by-products in roughly 30 minutes. After World War II, more powerful boats with remote sensing devices gave a final boost to the whaling industry. The number of kills had now increased far beyond the whales' capacity to reproduce.

EXERCISES

1. Locate, copy, and bring to class a paragraph that has the following features:

   ◆ an orienting topic sentence
   ◆ adequate development
   ◆ unity
   ◆ coherence
   ◆ a recognizable sequence
   ◆ appropriate length for its purpose and audience

   Be prepared to identify and explain each of these features in a class discussion.

2. For each of the following documents, indicate the most logical sequence. (For example, a description of a proposed computer lab would follow a spatial sequence.)

   ◆ a set of instructions for operating a power tool
   ◆ a campaign report describing your progress in political fund raising
   ◆ a report analyzing the weakest parts in a piece of industrial machinery
   ◆ a report analyzing the desirability of a proposed oil refinery in your area
   ◆ a detailed breakdown of your monthly budget to trim excess spending
   ◆ a report investigating the reasons for student apathy on your campus
   ◆ a report evaluating the effects of the ban on DDT in insect control
   ◆ a report on any highly technical subject, written for the general reader
   ◆ a report investigating the success of a no-grade policy at other colleges
   ◆ a proposal for a no-grade policy at your college

COLLABORATIVE PROJECT

Organize into small groups. Choose *one* of these topics, or one your group settles on, and then brainstorm to develop a formal outline for the body section of a report. One representative from your group can write the final draft and display it (using a data projection unit, overhead projector, or similar equipment) for class revision.

◆ job opportunities in your career field
◆ a physical description of the ideal classroom
◆ how to organize an effective job search
◆ how the quality of your higher educational experience can be improved
◆ arguments for and against a formal grading system
◆ an argument for an improvement you think your college needs most

CHAPTER
11

Editing for Readable Style

You might write for a diverse or specific audience, or for experts or non-experts. But no matter how technically appropriate your document, audience needs are not served unless your style is *readable*.

A definition of style

Writing style influences reader response to a document. Your writing style is the product of:

- ◆ the words you choose
- ◆ the way in which you put a sentence together
- ◆ the length of your sentences
- ◆ the way in which you connect sentences
- ◆ the tone you convey

Readable sentences require correct grammar, punctuation, and spelling. However, correctness alone is no guarantee of readability. For example, this response to a job applicant is mechanically correct but inefficient:

Inefficient style

> We are in receipt of your recent correspondence indicating your interest in securing the advertised position. Your correspondence has been duly forwarded for consideration by the personnel office, which has employment candidate selection responsibility. You may expect to hear from us relative to your application as the selection process progresses. Your interest in the position is appreciated.

Notice how hard you have worked to extract information that could be expressed this simply:

More efficient

> Your application for the advertised position has been forwarded to our personnel office. As the selection process moves forward, we will be in touch. Thank you for your interest.

Inefficient style makes readers work harder than they should.

Style can be inefficient for many reasons, but it is especially inept when it:

- ◆ makes the writing impossible to interpret
- ◆ takes too long to make the point
- ◆ reads like a Dick-and-Jane story from primary school
- ◆ uses imprecise or needlessly long words
- ◆ sounds stuffy andxsonal

Regardless of its cause, inefficient style results in writing that is less informative and less persuasive. Moreover, inefficient style can be unethical—by confusing or misleading readers.

To help your audience spend less time reading, you must spend more time editing for a style that is *clear, concise, fluent, exact,* and *likeable.*

EDITING FOR CLARITY

A clear sentence requires no more than a single reading. It avoids ambiguous constructions, signals relationships among its parts, and emphasizes the main idea. The following guidelines will help you edit for clarity.

Avoid Ambiguous Phrasing. Workplace writing ideally has *one* meaning only and allows for *one* interpretation. Does one's "suspicious attitude" mean that one is "suspicious" or "suspect"?

Ambiguous phrasing	All managers are not required to submit reports. (*Are some or none required?*)
Revised	Managers are **not all required** to submit reports.
	or
	Managers are **not required** to submit reports.
Ambiguous phrasing	Most city workers strike on Friday.
Revised	Most city workers **are planning to strike** Friday.
	or
	Most city workers **typically strike** on Friday.

Avoid Ambiguous Pronoun References. Each pronoun you use (*he, she, it, their,* etc.) must refer to one clearly identified noun. If the pronoun's referent (or antecedent) is ambiguous, readers will be confused.

Ambiguous referent	Our patients enjoy the warm days while they last. (*Are the patients or the warm days on the way out?*)

Depending on whether the referent for *they* is *patients* or *warm days*, the sentence can be clarified.

Clear referent	While these warm days last, our patients enjoy them.
	or
	Our terminal patients enjoy the warm days.
Ambiguous referent	Jack resents his assistant because he is competitive. (*Who's the competitive one—Jack or his assistant?*)
Clear referent	Because his assistant is competitive, Jack resents him.
	or
	Because Jack is competitive, he resents his assistant.

Avoid Ambiguous Punctuation. A missing hyphen, comma, or other punctuation mark can obscure your meaning.

Missing hyphen	Replace the trailer's inner wheel bearings. (*The inner-wheel bearings or the inner wheel-bearings?*)
Missing comma	Does your company produce liquid hydrogen? If so, how[,] and where do you store it? (*Notice how the meaning changes with a comma after "how."*) Police surrounded the crowd[,] attacking the strikers. (*Without the comma, the crowd appears to be attacking the strikers.*)

Although missing hyphens and commas are prime culprits, other omissions can cause ambiguity as well. A missing colon after *kill* yields the headline "Moose Kill 200." A missing " *'s* " after *Mills* creates this gem: "Mills Remains Buried in Saskatoon." Punctuation *does* affect meaning.

Exercise 1

Revise the following sentences to eliminate ambiguities in phrasing, pronoun reference, or punctuation.

a. Call me any evening except Tuesday after 7 o'clock.
b. The benefits of this plan are hard to imagine.
c. I cannot recommend this candidate too highly.
d. Visiting colleagues can be tiring.
e. Janice dislikes working with Claire because she's impatient.
f. Despite his efforts, Joe misinterpreted Sam's message.
g. Our division needs more effective writers.
h. Tell the reactor operator to evacuate and sound a general alarm.
i. If you don't pass any section of the test, your flying days are over.
j. Dial "10" to deactivate the system and sound the alarm.

Avoid Telegraphic Writing. *Function words* show relationships between the *content words* (nouns, adjectives, verbs, and adverbs) in a sentence. Some examples of function words:

- articles (a, an, the)
- prepositions (*in, of, to*)
- linking verbs (*is, seems, looks*)
- relative pronouns (*who, which, that*)

Some writers mistakenly try to compress their writing by eliminating these function words.

Ambiguous	Proposal to employ retirees almost dead.
Revised	The proposal to employ retirees **is** almost dead.
Ambiguous	Uninsulated end pipe ruptured. (*What ruptured? The pipe or the end of the pipe?*)
Revised	**The** uninsulated end **of the** pipe ruptured.
	or
	The uninsulated pipe **on the** end ruptured.
Ambiguous	The reactor operator told management several times she expected an accident. (Did she tell them once or several times?)
Revised	The reactor operator told management several times **that** she expected an accident.
	or
	The reactor operator told management **that** several times she expected an accident.

Avoid Ambiguous Modifiers.   Modifiers explain, define, or add detail to other words or ideas. If a modifier is too far from the words it modifies, the message can be ambiguous.

> **Misplaced modifier**   Only press the red button in an emergency. (*Does **only** modify **press** or **emergency**?*)
>
> **Revised**   Press **only** the red button in an emergency.
>
> *or*
>
> Press the red button in an emergency **only.**

Another problem with ambiguity occurs when a modifying phrase has no word to modify.

> **Dangling modifier**   **Being so well known in the computer industry,** I would appreciate your advice.

The writer meant to say that the *reader* is well known, but with no word to join itself to, the modifying phrase dangles. Eliminate the confusion by adding a subject:

> **Revised**   Because **you** are so well known in the computer industry, I would appreciate your advice.

Exercise 2

Revise the following sentences to repair telegraphic writing or to clarify ambiguous modifiers.

a. The manager claimed repeatedly she reported the danger.
b. I want the final Amex report written by your division.
c. Replace main booster rocket seal.
d. The president refused to believe any internal report was inaccurate.
e. Only use this phone in a red alert.
f. After offending our best client, I am deeply annoyed with the new manager.
g. Send memo to programmer requesting explanation.
h. Do not enter test area while contaminated.

Unstack Modifying Nouns.   One noun can modify another noun (as in "software development"). But when two or more nouns modify a noun, the string of densely packed words becomes hard to read and ambiguous.

> **Stacked**   Be sure to leave enough time for a **training session participant** evaluation. (*Evaluation of the session or of the participants?*)

With no function words (articles, prepositions, verbs, relative pronouns) to break up the string of nouns, readers cannot see the relationships among the nouns. What modifies what?

Stacked nouns also deaden your style. Bring your style *and* your reader to life by using action verbs (*complete, prepare, reduce*) and prepositional phrases.

> **Revised** Be sure to leave enough time **for** participants **to evaluate** the training session.
>
> *or*
>
> Be sure to leave enough time **to evaluate** participants **in the** training session.

No such problem with ambiguity occurs when *adjectives* are stacked in front of a noun.

> **Clear** He was a nervous, angry, confused, but dedicated employee.

Readers can readily see that the adjectives modify *employee*.

Arrange Word Order for Coherence. In coherent writing, everything sticks together; each sentence builds on the preceding sentence and looks ahead to the following sentence. Sentences generally work best when the beginning looks back at familiar information and the end provides the new (or unfamiliar) information:

Familiar		Unfamiliar
My dog	has	fleas.
Our supervisor	just won	the lottery.
This company	is planning	a merger.

Exercise 3

Revise the following sentences to unstack modifying nouns.

 a. Develop on-line editing system documentation.
 b. We need to develop a unified construction automation design.
 c. Install a hazardous materials dispersion monitor system.
 d. I recommend these management performance improvement incentives.
 e. Sarah's job involves fault analysis systems troubleshooting handbook preparation.

Use Active Voice Often. A verb's *voice* signals whether a sentence's subject acts or is acted upon. The active voice ("I did it") is more direct, concise, and persuasive than the passive voice ("It was done by me"). In the active voice, the agent performing the action serves as the subject:

	Agent	*Action*	*Recipient*
Active	Joe	lost	your report.
	Subject	*Verb*	*Object*

The passive voice reverses the pattern, making the recipient of an action serve as subject.

	Recipient	*Action*	*Agent*
Passive	Your report	was lost	by Joe.
	Subject	*Verb*	*Prepositional phrase*

Sometimes the passive eliminates the agent altogether:

Passive	Your report was lost. *(Who lost it?)*

Some writers mistakenly rely on the passive voice because they think it sounds more objective and important. But the passive voice often makes writing wordy, indecisive, evasive, and unethical.

Concise and direct (active)	I underestimated labour costs for this project. *(7 words)*
Wordy and indirect (passive)	Labour costs for this project were underestimated by me. (9 words)
Evasive (passive)	Labour costs for this project were underestimated. (7 words)

Do not evade responsibility by hiding behind the passive voice:

Passive	A **mistake** was made in your shipment.
(not responsible)	**It** was decided not to hire you.
	A **layoff** is recommended.

Use the active voice when reporting errors or bad news. Readers appreciate clarity and sincerity.

The passive voice creates a weak and impersonal tone.

Weak and impersonal	An offer will be made by us next week.
Strong and personal	We will make an offer next week.

Use the active voice when you want action. Otherwise, your statement will have no power.

Weak passive	If my claim is not settled by May 15, the Better Business Bureau will be contacted, and their advice on legal action will be taken.
Strong active	If you do not settle my claim by May 15, I will contact the Better Business Bureau for advice on legal action.

Notice how this active version emphasizes the new and significant information by placing it at the end.

Use the active voice for giving instructions.

Faulty passive	The bid should be sealed. Care should be taken with the dynamite.
Correct active	**Seal** the bid. **Be careful** with the dynamite.

Avoid shifts from active to passive voice in the same sentence.

Faulty shift	During the meeting, project members spoke and presentations were given.
Correct	During the meeting, project members spoke and **gave** presentations.

Unless you have a deliberate reason for choosing the passive voice, use the *active* voice for making forceful connections like the one described here:

> *By using the active voice, you direct the reader's attention to the subject of your sentence. For instance, if you write a job-application letter that is littered with passive verbs, you fail to achieve an important goal of that letter: to show the readers the important things you have done, and how prepared you are to do important things for them. That strategy requires active verbs, with clear emphasis on you and what you have done/are doing.* (Pugliano 6)

Exercise 4

The following sentences are wordy, weak, or evasive because they are in the passive voice. Revise each sentence as a concise, forceful, and direct expression in the active voice, to identify the person or agent performing the action.

   a. The evaluation was performed by us.
   b. The report was written by our group.
   c. Unless you pay me within three days, my lawyer will be contacted.
   d. Hard hats should be worn at all times.
   e. It was decided to reject your offer.
   f. Gasoline was spilled on your Ferrari's leather seats.
   g. It is believed by us that this contract is faulty.
   h. Our test results will be sent to you as soon as verification is completed.
   i. The decision was made that your request for promotion should be denied.

Use Passive Voice Selectively.   Passive voice is appropriate in lab reports and other documents in which the agent's identity is immaterial to the message.

Use the passive voice when your audience does not need to know the agent.

Correct passive	Mr. Jones was brought to the emergency room.
	The bank failure was publicized provincewide.

Use the passive voice to focus on events or results when the agent is unknown, unapparent, or unimportant.

Correct passive	All memos in the firm are filed in a database.
	Josef's article was published last week.

Use the passive voice when you want to be indirect or inoffensive (as in requesting the customer's payment or the employee's cooperation, or to avoid blaming someone—such as your supervisor) (Ornatowski 94).

Active but offensive	**You** have not paid your bill. **You** need to overhaul our filing system.
Inoffensive passive	**This bill** has not been paid. **Our filing system** needs to be overhauled.

Use the passive voice if the person behind the action has reason for being protected.

Correct passive	The criminal was identified. The embezzlement scheme was exposed.

Exercise 5

The following sentences lack proper emphasis because of an improper use of the active voice. Revise each ineffective active as an appropriate passive, to emphasize the recipient rather than the actor.

a. Joe's company fired him.
b. Someone on the maintenance crew has just discovered a crack in the nuclear-core containment unit.
c. A power surge destroyed more than 2 000 lines of our new applications program.
d. You are paying inadequate attention to worker safety.
e. You are checking temperatures too infrequently.
f. The tornado destroyed the barn.
g. You did a poor job editing this report.

Avoid Overstuffed Sentences. A sentence crammed with ideas makes details hard to remember and relationships hard to identify:

Overstuffed	Publicizing the records of a private meeting that took place three weeks ago to reveal the identity of a manager who criticized our company's promotion policy would be unethical.

Clear things up by sorting out the relationships.

Revised	In a private meeting three weeks ago, a manager criticized our company's policy on promotion. It would be unethical to reveal the manager's identity by publicizing the records of that meeting. (*Other versions are possible here, depending on the writer's intended meaning.*)

Give your readers no more information than they can retain in one sentence.

Exercise 6

Unscramble this overstuffed sentence by making shorter, clearer sentences:

A smoke-filled room causes not only teary eyes and runny noses but also can alter people's hearing and vision, as well as creating dangerous levels of carbon monoxide, especially for people with heart and lung ailments, whose health is particularly threatened by second-hand smoke.

EDITING FOR CONCISENESS

Writing can suffer from two kinds of wordiness: one kind occurs when readers receive information they don't need (think of an overly detailed weather report during local television news). The other kind of wordiness occurs when too many words are used to convey information readers *do* need (as in saying "a great deal of potential for the future" instead of "great potential").

Every word in the document should advance your meaning.

> *Writing improves in direct ratio to the number of things we can keep out of it that shouldn't be there.* (Zinsser 14)

Concise writing conveys the most information in the fewest words. But it includes the details necessary for clarity.

Use fewer words whenever fewer will do. But remember the difference between *clear writing* and *compressed writing* that is impossible to decipher.

Impenetrable	Give new vehicle air conditioner compression cut-off system specifications to engineering manager advising immediate action.

First drafts rarely are concise. The following strategies will help you "trim the fat."

Avoid Wordy Phrases. Each phrase here can be reduced to one word:

Wordy		**Concise**
at a rapid rate	=	rapidly
due to the fact that	=	because
the majority of	=	most
on a personal basis	=	personally
give instruction to	=	instruct
would be able to	=	could
readily apparent	=	obvious
a large number	=	many
prior to	=	before
aware of the fact that	=	know
conduct an inspection of	=	inspect
in close proximity	=	near

Eliminate Redundancy. A redundant expression says the same thing twice, in different words, as in *fellow colleagues*.

a **dead** corpse	**end** result
completely eliminate	cancel **out**
basic essentials	consensus **of opinion**
enter **into**	**utter** devastation
mental awareness	**the month of** August
mutual cooperation	**utmost** perfection

Avoid Needless Repetition. Unnecessary repetition clutters writing and dilutes meaning.

| **Repetitious** | In trauma victims, breathing is restored by **artificial respiration.** Techniques of **artificial respiration** include mouth-to-mouth **respiration** and mouth-to-nose **respiration.** |

Repetition in the above passage disappears when sentences are combined.

| **Concise** | In trauma victims, breathing is restored by artificial respiration, **either mouth-to-mouth or mouth-to-nose.** |

Repetition can be useful. Don't hesitate to repeat, or at least rephrase, material (even whole paragraphs in a longer document) if you feel that readers need reminders. Effective repetition helps avoid cross-references like this: "See page 23" or "Review page 10."

Exercise 7

Revise the following wordy sentences to eliminate needless phrases, needless repetition, and redundancy.

a. I have admiration for Professor Singh.
b. Due to the fact that we made the lowest bid, we won the contract.
c. On previous occasions we have worked together.
d. She is a person who works hard.
e. We have completely eliminated the bugs from this program.
f. This report is the most informative report on the project.
g. Through mutual cooperation, we can achieve our goals.
h. I am aware of the fact that Sam is trustworthy.
i. This offer is the most attractive offer I've received.

Avoid *There is* and *There are* Sentence Openers.

Weak	**There is** a coaxial cable connecting the antenna to the receiver.
Revised	A coaxial cable connects the antenna to the receiver.
Weak	**There is** a danger of explosion in Number 2 mineshaft.
Revised	Number 2 mineshaft is in danger of exploding.

Dropping these openers places the keywords at sentence end, where they are best emphasized. Of course, in some contexts, proper emphasis would call for a *There* opener.

| **Correct** | People often have wondered about the rationale behind Boris's sudden decision. Actually, there are several good reasons for his dropping out of the program. |

Avoid Some *It* Sentence Openers. Avoid beginning a sentence with *It*—unless the *It* clearly points to a specific referent in the preceding sentence: "This document is excellent. It deserves special recognition."

Weak	**It** was his bad attitude that got him fired.
Revised	His bad attitude got him fired.
Weak	**It** is necessary to complete both sides of the form.
Revised	Please complete both sides of the form.

Delete Needless Prefaces. Instead of delaying the new information in your sentence, get right to the point.

Wordy	**I am writing this letter because** I wish to apply for the position of copy editor.
Concise	Please consider me for the position of copy editor.
Wordy	**As far as artificial intelligence is concerned,** the technology is only in its infancy.
Concise	Artificial-intelligence technology is only in its infancy.

Exercise 8

Revise the following sentences to eliminate *There* and *It* openers and needless prefaces.

a. There was severe fire damage to the reactor.
b. There are several reasons why Jenna left the company.
c. It is essential that we act immediately.
d. It has been reported by Clayton that several safety violations have occurred.
e. This letter is to inform you that I am pleased to accept your job offer.
f. The purpose of this report is to update our research findings.

Avoid Weak Verbs. Use verbs that express a definite action: *open, close, move, continue, begin.* Avoid weak verbs that express no specific action: *is, was, are, has, give, make, come, take.* In some cases, such verbs are essential to your meaning: "Dr. Yang is operating at 7:00 a.m." "Take me to the laboratory." But in other cases, weak verbs add words without advancing meaning. All forms of the linking verb *to be* (*am, are, is, was, were, will, have been, might have been*) generally are weak. This next sentence achieves conciseness because of the strong verb *consider:*

Concise	Please **consider** my offer.
Weak and wordy	Please **take into consideration** my offer.

Don't disappear behind weak linking verbs and their baggage of needless nouns and prepositions.

Weak	My recommendation **is** for a larger budget.
Strong	I **recommend** a larger budget.

Strong verbs, or action verbs, suggest an assertive, positive, and confident writer. Here are some weak verbs converted to strong:

Weak		Strong
is in conflict with	=	conflicts
has the ability to	=	can
give a summary of	=	summarize
make an assumption	=	assume
come to the conclusion	=	conclude
take action	=	act
make a decision	=	decide

Exercise 9

Revise the following wordy and vague sentences to eliminate weak verbs.

- *a.* Our disposal procedure is in conformity with federal standards.
- *b.* Please make a decision today.
- *c.* We need to have a discussion about the problem.
- *d.* I have just come to the realization that I was mistaken.
- *e.* Your conclusion is in agreement with mine.
- *f.* This manual gives instructions to end users.

Delete Needless *To Be* Constructions. The preceding section showed that forms of *to be* (*is, was, are,* etc.) are weak. Sometimes the *to be* form itself mistakenly appears behind such verbs as *appears, seems,* and *finds.*

> **Wordy**   Your product seems **to be** superior.
>
>   I consider this employee **to be** highly competent.

Avoid Excessive Prepositions.   Needless prepositions combined with forms of *to be* make wordy sentences.

> **Wordy**   The recommendation first appeared **in** the report written **by** the supervisor **in** January **about** that month's productivity.
>
> **Concise**   The recommendation first appeared in the supervisor's productivity report for January.

The following prepositional phrases can be reduced.

Excessive		Reduced
with the exception of	=	except for
in reference to	=	about (or regarding)
in order that	=	so
in the near future	=	soon
in the event that	=	if
at the present time	=	now
in the course of	=	during
in the process of	=	during (or in)

Fight Noun Addiction. Nouns manufactured from verbs (nominalizations) often accompany weak verbs and needless prepositions.

Weak and wordy	We ask for the **cooperation** of all employees.
Strong and concise	We ask that all employees **cooperate.**
Weak and wordy	Give **consideration** to the possibility of a career change.
Strong and concise	**Consider** a career change.

Besides causing wordiness, a nominalization can be vague—by hiding the agent of an action.

Wordy and vague	**A valid requirement** for immediate action exists. (*Who should take the action? We can't tell.*)
Precise	We **must act** immediately.

Here are nominalizations restored to their verb forms:

Vague		**Precise**
conduct an investigation of	=	investigate
provide a description of	=	describe
conduct a test of	=	test
make a discovery of	=	discover

Nominalizations drain the life from your style. In cheering for your favourite team, you wouldn't say "Blocking of that kick is a necessity!" instead of "Block that kick!"

Write as you would speak, but avoid slang or overuse of colloquialisms. Also avoid excessive economy. For example, "Employees must cooperate" would not be a desirable alternative to the first example in this section. But, for the final example, "Block that kick" would be.

Exercise 10

Revise the following sentences to eliminate needless prepositions and *to be* constructions, and to cure noun addiction.

a. Igor seems to be ready for a vacation.
b. Our survey found 46 percent of users to be disappointed.
c. In the event of system failure, your sounding of the alarm is essential.
d. These are the recommendations of the chairperson of the committee.
e. Our acceptance of the offer is a necessity.
f. Please perform an analysis and make an evaluation of our new system.
g. A need for your caution exists.
h. I consider Bjorn to be an excellent technician.
i. The appearance of this problem was just yesterday.
j. Power surges are associated, in a causative way, with malfunctions of computers.

Make Negatives Positive. A positive expression is easier to understand than a negative one.

Indirect and wordy	Please do not be late in submitting your report.
Direct and concise	Please submit your report on time.

Readers work even harder to translate sentences with multiple negative expressions:

Confusing and wordy	Do **not** distribute this memo to employees who have not received a security clearance.
Clear and concise	Distribute this memo only to employees who have received a security clearance.

Besides the directly negative words (*no, not, never*), some words are indirectly negative (*except, forget, mistake, lose, uncooperative*).

Confusing and wordy	**Do not neglect** to activate the alarm system. My diagnosis was not inaccurate.
Clear and concise	**Be sure** to activate the alarm system. My diagnosis was accurate.

The positive versions are more straightforward *and* persuasive.

Some negative expressions, of course, are perfectly correct, as in expressing disagreement.

Correct negatives	This is **not** the best plan. Your offer is **unacceptable.** This project **never** will succeed.

Select positives over negatives whenever your meaning allows:

Negatives		**Positives**
did not succeed	=	failed
does not have	=	lacks
did not prevent	=	allowed
not unless	=	only if
not until	=	only when
not absent	=	present

Clean Out Clutter Words.   Clutter words stretch a message without adding meaning. Here are some of the commonest: *very, definitely, quite, extremely, rather, somewhat, really, actually, currently, situation, aspect, factor.*

Cluttered	**Actually,** one **aspect** of a business **situation** that could definitely make me **quite** happy would be to have a **somewhat** adventurous partner who **really** shared my **extreme** attraction to risks.
Concise	I seek an adventurous business partner who enjoys risks.

Delete Needless Qualifiers. Qualifiers such as *I feel, it seems, I believe, in my opinion,* and *I think* soften the tone and impact of a statement. Use qualifiers to express uncertainty or to avoid seeming arrogant or overconfident.

Appropriate qualifiers	Despite Frank's poor grades last year he will, **I think,** do well in college.
	Your product **seems** to be what we need.

But when you are certain, eliminate the qualifier so as not to seem tentative or evasive.

Needless qualifiers	**It seems** that I've made an error.
	We **appear to** have exceeded our budget.
	In my opinion, this candidate is outstanding.

In communicating across cultures, keep in mind that a direct, forceful style might be considered offensive (page 201).

Exercise 11
Revise the following sentences to eliminate inappropriate negatives, clutter words, and needless qualifiers.

a. Our design must avoid non-conformity with building codes.
b. Never fail to wear protective clothing.
c. Do not accept any bids unless they arrive before May 1.
d. I am not unappreciative of your help.
e. We are currently in the situation of completing our investigation of all aspects of the accident.
f. I appear to have misplaced the contract.
g. Do not accept bids that are not signed.
h. It seems as if I have just wrecked a company car.

EDITING FOR FLUENCY

Fluent sentences are easy to read because of clear connections, variety, and emphasis. Their varied length and word order eliminate choppiness and monotony. Fluent sentences enhance *clarity,* allowing readers to see the most important ideas. Fluent sentences enhance *conciseness,* often replacing several short, repetitious sentences with one longer, economical sentence. The following strategies will help you write fluent sentences.

Combine Related Ideas.   A series of short, disconnected sentences is not only choppy and wordy, but also unclear.

Disconnected	Jogging can be healthful. You need the right equipment. Most necessary are well-fitting shoes. Without this equipment you take the chance of injuring your legs. Your knees are especially prone to injury. (*5 sentences*)
Clear, concise, and fluent	Jogging can be healthful if you have the right equipment. Shoes that fit well are most necessary because they prevent injury to your legs, especially your knees. (*2 sentences*)

Most sets of information can be combined in different relationships, depending on what you want to emphasize. Imagine that this set of facts describes an applicant for a junior-management position with your company.

◆ Roy Dupius graduated from an excellent management school.
◆ He has no experience.
◆ He is highly recommended.

Assume you are a personnel director, conveying to upper management your impression of this candidate. To convey a negative impression, you might combine the facts in this way:

Strongly negative emphasis	Although Roy Dupius graduated from an excellent mangement school and is highly recommended, **he has no experience.**

The *independent* idea (in boldface) receives the emphasis. To continue with our example: If you are undecided, but leaning in a negative direction, you might write:

Strongly negative emphasis	Roy Dupuis graduated from an excellent management school and is highly recommended, **but** he has no experience.

In the sentence above, the ideas before and after *but* are both independent. These independent ideas are joined by the coordinating word *but,* which suggests that both sides of the issue are equally important (or "coordinate"). Placing the negative idea last, however, gives it slight emphasis.

Finally, to emphasize strong support for the candidate, you could say:

Strongly positive emphasis	Although Roy Dupuis has no experience, he graduated from an excellent management school and is highly recommended.

In the above version the earlier idea is subordinated by *although,* leaving the two final ideas independent.

Caution: Combine sentences only to advance your meaning and to ease the reader's task. A sentence with too much information and too many connections can be difficult for readers to sort out.

Overstuffed	Our night supervisor's verbal order from upper management to repair the overheated circuit was misunderstood by Leslie Kidd, who gave the wrong instructions to the emergency crew, thereby causing the fire within 30 minutes.
Clearer	Upper management issued a verbal order to repair the overheated circuit. When our night supervisor transmitted the order to Leslie Kidd, it was misunderstood. Kidd gave the wrong instruction to the emergency crew, and the fire began within 30 minutes.

Vary Sentence Construction and Length. We have just seen how related ideas often need to be linked in one sentence, so that readers can grasp the connections:

Disconnected	The nuclear core reached critical temperature. The loss-of-coolant alarm was triggered. The operator shut down the reactor.
Connected	As the nuclear core reached critical temperature, triggering the loss-of-coolant alarm, the operator shut down the reactor.

But an idea that should stand alone for emphasis needs a sentence of its own:

Correct	Core meltdown seemed inevitable.

However, an unbroken string of long or short sentences can bore and confuse readers, as can a series with identical openings:

Dreary	There are some drawbacks about diesel engines. **They** are difficult to start in cold weather. **They** cause vibration. **They** also give off an unpleasant odour. **They** cause sulphur dioxide pollution.
Varied	Diesel engines have some drawbacks. Most obvious are their noisiness, cold-weather starting difficulties, vibration, odour, and sulphfur dioxide emission.

Similarly, when you write in the first person, overusing *I* makes you appear self-centred. (Some organizations require use of the third person, avoiding the first person completely, for all manuals, lab reports, specifications, product descriptions, etc.)

Do not avoid personal pronouns if they make the writing more readable (by eliminating passive constructions).

Use Short Sentences for Special Emphasis.   All this talk about combining ideas might suggest that short sentences have no place in good writing. Wrong. Short sentences show connections and clarify relationships; short sentences (even one-word sentences) provide vivid emphasis. They stick in a reader's mind.

Exercise 12

Combine each set of sentences into one fluent sentence that provides the requested emphasis. For example,

Sentence set	John is a loyal employee. John is a motivated employee. John is short-tempered with his colleagues.
Combined for positive emphasis	Even though John is short-tempered with his colleagues, he is a loyal and motivated employee.
Sentence set	This word processor has many features. It includes a spelling checker. It includes a thesaurus. It includes a grammar checker.
Combined to emphasize thesaurus	Among its many features, such as spelling and grammar checkers, this word processor includes a thesaurus.

   a. The job offers an attractive salary.
      It demands long work hours.
      Promotions are rapid.
      (Combine for negative emphasis.)
   b. The job offers an attractive salary.
      It demands long work hours.
      Promotions are rapid.
      (Combine for positive emphasis.)
   c. Our office software is integrated.
      It has an excellent database management program.
      Most impressive is its word-processing capability.
      It has an excellent spreadsheet program.
      (Combine to emphasize the word processor.)
   d. Company X gave us the lowest bid.
      Company Y has an excellent reputation.
      (Combine to emphasize Company Y.)
   e. Superinsulated homes are energy efficient.
      Superinsulated homes create a danger of indoor air pollution.
      The toxic substances include radon gas and urea formaldehyde.
      (Combine for a negative emphasis.)

FINDING THE EXACT WORDS

Too often, language can be a vehicle for *camouflage* rather than communication. People see many reasons to hide behind language, as when they:

- ◆ speak for their company but not for themselves
- ◆ fear the consequences of giving bad news
- ◆ are afraid to disagree with company policy
- ◆ make a recommendation some readers will resent
- ◆ worry about making a bad impression
- ◆ worry about being wrong
- ◆ pretend to know more than they do
- ◆ avoid admitting a mistake or ignorance

Inflated and unfamiliar words, borrowed expressions, and needlessly technical terms camouflage meaning. Whether intentional or accidental, poor word choices have only one result: inefficient and often unethical writing that resists interpretation and frustrates the reader.

Following are strategies for finding words that are *convincing, precise,* and *informative.*

Use Simple and Familiar Words. Don't replace technically precise words with non-technical words that are vague or imprecise. Don't write *a part that makes the computer run* when you mean *central processing unit.* Use the precise term, and define it in a glossary for non-technical readers:

   Correct   Central processing unit: the part of the computer that controls
                  information transfer and carries out arithmetic and logical
                  instructions.

Certain technical words may be indispensable in certain contexts, but the non-technical words usually can be simplified. Instead of *answering in the affirmative, say yes;* or instead of *endeavouring to promulgate* a new policy, *try to announce* it.

Unfamiliar words	Acoustically attenuate the food-consumption area.
Revised	Soundproof the cafeteria.

Don't use three syllables when one will do. Generally, trade for less:

Three Syllables		**One Syllable**
aggregate	=	total
demonstrate	=	show
effectuate	=	cause
endeavour	=	effort, try
eventuate	=	result
frequently	=	often
initiate	=	begin
is contingent upon	=	depends on
multiplicity of	=	many
optimum	=	best
subsequent to	=	after
utilize	=	use

Trim wherever you can. Most important, choose words you hear and use in everyday speaking—words that are universally familiar.

Don't write *I deem* when you mean *I think,* or *keep me apprised* instead of *keep me informed,* or *I concur* instead of *I agree,* or *securing employment* instead of *finding a job,* or *it is cost prohibitive* instead of *we can't afford it.*

Don't write like the author of a report from the Federal Aviation Administration, who recommended that manufacturers of the DC–10 re-evaluate *the design of the entire pylon assembly to minimize design factors which are resulting in sensitive and/or critical maintenance and inspection procedures* (25 words, 50 syllables). A plain English translation: *Redesign the pylons so they are easier to maintain and inspect* (11 words, 18 syllables).

Besides the annoyance they cause, needlessly long or unfamiliar words can be *ambiguous.*

Ambiguous	Make an improvement in the clerical situation.

Should we hire more clerical personnel or better personnel or should we train the personnel we have? Words chosen to impress readers too often confuse them instead. A plain style is more persuasive because "it leaves no one out" (Cross 6).

Of course, now and then the complex or more elaborate word best expresses your meaning. For instance, we would not substitute *end* for *terminate* in referring to something with an established time limit.

Correct	Our trade agreement terminates this month.

If a complex word can replace a handful of simpler words—and can sharpen your meaning—use the complex word.

Weak	Six rectangular grooves **around the outside edge** of the steel plate **are needed for** the pressure clamps **to fit into.**
Informative and precise	Six rectangular grooves on the steel plate **perimeter accommodate** the pressure clamps.
Weak	We need a **one-to-one exchange of ideas and opinions.**
Informative and precise	We need a **dialogue.**
Weak	Sexist language **contributes to the ongoing prevalence** of gender stereotypes.
Informative and precise	Sexist language **perpetuates** gender stereotypes.

Exercise 13

Revise the following sentences for straightforward and familiar language.

a. May you find luck and success in all endeavours.
b. I suggest you reduce the number of cigarettes you consume.
c. Within the copier, a magnetic-reed switch is utilized as a mode of replacement for the conventional microswitches that were in use on previous models.
d. A good writer is cognizant of how to utilize grammar in a correct fashion.
e. I will endeavour to ascertain the best candidate.
f. In view of the fact that the microscope is defective, we expect a refund of our full purchase expenditure.
g. I wish to upgrade my present employment situation.

Avoid Useless Jargon.   Every profession has its own "shorthand." Among specialists, technical terms are a precise and economical way to communicate. For example, *stat* (from the Latin "statim" or "immediately") is medical jargon for *drop everything and deal with this emergency*. For computer buffs, a *glitch* is a momentary power surge that can erase the contents of internal memory; a *bug* is an error that causes a program to run incorrectly. Such useful jargon conveys clear meaning to a knowledgeable audience.

Technical language, however, can be used appropriately or inappropriately. The latter is useless jargon, meaningless to insiders as well as outsiders. In the world of useless jargon, people don't *cooperate* on a project; instead, they *interface* or *contiguously optimize their efforts*. Rather than *designing a model*, they *formulate a paradigm*. Instead of *observing limits or boundaries*, they *function within specific parameters*.

A popular form of useless jargon is adding *-wise* to nouns, as shorthand for *in reference to* or *in terms of*.

Useless jargon	Expensewise and schedulewise, this plan is unacceptable.
Revised	In terms of expense and scheduling, this plan is unacceptable.

Writers create another form of jargon when they invent verbs from nouns or adjectives by adding an *-ize* ending: Don't invent *prioritize* from *priority;* instead use *to rank priorities.*

Jargon's worst fault is that it makes the person using it seem stuffy and pretentious:

> **Pretentious** Unless all parties interface synchronously within given parameters, the project will be rendered inoperative.

> **Possible translation** Unless we coordinate our efforts, the project will fail.

Beyond reacting with frustration, readers often conclude that useless jargon is camouflage for a writer with something to hide.

Before using any jargon, think about your specific readers and ask yourself: "Can I find an easier way to say exactly what I mean?" Use jargon only if it *improves* your communication.

If your employer insists on needless jargon or elaborate phrasing, then you have little choice. What is best in matters of style is not always what some people consider appropriate. Use the style your employer or organization expects, but remember that most documents that achieve superior results are in plain English.

Use Acronyms Selectively.   Acronyms are another form of specialized shorthand, or jargon. They are formed from the first letters of words in a phrase (as in *LOCA* from *loss of coolant accident*) or from a combination of first letters and parts of words (as in *bit* from *binary digit,* or *pixel* from *picture element*).

Computer technology has spawned countless acronyms, including:

Acronym		Meaning
ISDN	=	Integrated Services Digital Network
Telnet	=	Telephone Network
URL	=	Uniform Resource Locator

Acronyms *can* communicate concisely—but only when the audience knows their meaning, and only when you use the term often in your document. Whenever you first use an acronym, spell out the words from which it is derived.

An acronym defined

> **Modem** ("modulator+demodulator"): a device that converts, or "modulates," computer data in electronic form into a sound signal that can be transmitted via phone line and then reconverted, or "demodulated," into electronic form for the receiving computer.

For lay audiences, try to avoid acronyms altogether or be sure to define the terms that make up the acronym.

Avoid Triteness.   Writers who rely on tired old phrases (clichés) like the following seem too lazy or too careless to find exact ways to say what they mean.

make the grade   the chips are down
in the final analysis   not by a long shot

close the deal
hard as a rock
water under the bridge
holding the bag
up the creek

last but not least
welcome aboard
over the hill
bite the bullet
work like a dog

Exercise 14
Revise the following sentences to eliminate useless jargon and triteness.

a. For the obtaining of the X-33 printer, our firm will have to accomplish the disbursement of funds to the amount of $3,000.
b. To optimize your financial return, prioritize your investment goals.
c. The use of this product engenders a 50 percent repeat consumer encounter.
d. We'll have to swallow our pride and admit our mistake.
e. We wish to welcome all new managers aboard.
f. Not by a long shot will this plan succeed.
g. Managers who make the grade are those who can take daily pressures in stride.
h. Intercom utilization will be employed to initiate substitute employee operative involvement.

Avoid Misleading Euphemisms.   Euphemisms are expressions aimed at politeness or at making unpleasant subjects seem less offensive. Thus, we *powder our nose* or *use the boys' room* instead of *using the bathroom;* we *pass away* or *meet our Maker* instead of *dying*. Euphemisms make the truth seem less painful.

When euphemisms avoid offending or embarrassing our audience, they are perfectly legitimate. Instead of telling a job applicant he or she is *unqualified,* we might say, *your background doesn't meet our needs*. In addition, there are times when friendliness and inter-office harmony are more likely to be preserved with writing that is not too abrupt, bold, blunt, or emphatic (MacKenzie 2).

Euphemisms are unethical if they understate the truth when only the truth will serve. In the sugar-coated world of misleading euphemisms, bad news disappears:

◆ Instead of being *laid off* or *fired,* workers are *surplused* or *deselected,* or the company is *downsized*.
◆ Instead of *lying* to the public, the government *engages in a policy of disinformation*.
◆ Instead of *wars* and *civilian casualties,* we have *conflicts* and *collateral damage*.

Language loses all meaning when *criminals* become *offenders,* when *rape* becomes *sexual assault,* and when people who are just plain *lazy* become *underachievers*. Plain talk is always better than deception. If someone offers you a job *with limited opportunity for promotion,* expect a *dead-end* job.

Avoid Overstatement and Unsupported Generalizations.   Exaggerating destroys credibility. Be cautious when using words such as *best, biggest, brightest, most,* and *worst*.

Overstated   **Most** businesses have no loyalty toward their employees.

Revised	**Some** businesses have little loyalty toward their employees.
Overstated	You will find our product to be the **best.**
Revised	You will **appreciate the high quality** of our product.

Unsupported generalizations harm your credibility. Be aware of the vast differences in meaning among these words:

few	never
some	rarely
many	sometimes
most	often
all	always

Unless you specify *few, some, many,* or *most,* readers can interpret your statement to mean *all.*

Misleading	Assembly-line employees are doing shabby work.

Unless you mean *all,* qualify your generalization with *some, most*—or even better, specify *20 percent.*

Exercise 15

Revise the following sentences to eliminate euphemism, overstatement, or unsupported generalizations.

 a. I finally must admit that I am an abuser of intoxicating beverages.
 b. I was less than candid.
 c. This employee is poorly motivated.
 d. Most entry-level jobs are boring and dehumanizing.
 e. Clerical jobs offer no opportunity for advancement.
 f. Because of your absence of candour, we no longer can offer you employment.

Avoid Imprecise Words. Even words listed as synonyms carry different shades of meaning. Do you mean to say *I'm slender, you're slim, she's lean,* or *he's scrawny?* The wrong choice could be disastrous.

A single wrong word can offend readers, as in this statement by a job applicant:

Offensive	Another attractive feature of your company is its **adequate** training program.

While "adequate" might convey honestly the writer's intended meaning, the word seems inappropriate in this context (an applicant expressing a judgment about a program). Although the program may not have been highly ranked, the writer could have used any of several alternatives (*solid, respectable, growing*—or no modifier at all).

Be especially aware of similar words with dissimilar meanings, as in these examples:

affect/effect	ensure/insure
all ready/already	almost dead/dying
healthy/healthful	among/between
imply/infer	continual/continuous
invariably/inevitably	eager/anxious
uninterested/disinterested	fearful/fearsome

Don't write *skiing is healthy* when you mean that skiing promotes good health. Healthful things keep us healthy.

Be on the lookout for imprecisely phrased (and therefore illogical) comparisons.

Imprecise	Your bank's interest rate is higher than Central Bank. (*Can a rate be higher than a bank?*)
Precise	Your bank's interest rate is higher than Central Bank's.

Imprecise language can create ambiguity. For instance, is *send us more personal information* a request for *more* information that is personal or for information that is *more* personal?

Precision ultimately enhances conciseness, when one exact word replaces multiple inexact words.

Wordy and less exact	I have **put together** all the financial information. **Keep doing** this exercise for 10 seconds.
Concise and more exact	I have **assembled** all the.... **Continue** this exercise....

Be Specific and Concrete.   General words name broad classes of things, such as *job, computer,* or *person.* Such terms usually need to be clarified by more specific ones.

General		Specific
job	=	senior accountant for Softbyte Press
computer	=	Acer 9000
person	=	Sarah Chu, production manager

The more specific your words, the sharper your meaning.

General	structure
	dwelling
	vacation home
	log cabin
Specific	log cabin in Ontario
	a three-room log cabin on the banks of Green Lake

Notice how the picture becomes more vivid as we move to lower levels of generality.

Abstract words name qualities, concepts, or feelings (*beauty, luxury, depression*) whose exact meaning has to be nailed down by *concrete* words—words that name things we can know through our five senses.

Abstract		Concrete
a beautiful view	=	snowcapped mountains, a wilderness lake, pink granite ledge, 90-foot birch trees
a **luxurious** condominium	=	imported tiles, glass walls, oriental rugs
a **depressed** worker	=	suicidal urge, insomnia, feelings of worthlessness, no hope for improvement

Informative writing *tells* and *shows*.

> **General** One of our **workers** was **injured** by a **piece of equipment recently.**

The boldface words only *tell* without showing.

> **Specific** **Arlene Kowalchuk** suffered a **broken thumb** while working on a lathe **yesterday.**

Choose informative words that express exactly what you mean. Don't write *thing* when you mean *lever, switch, micrometer,* or *disk.* Instead of evaluating an employee as *good, great, disappointing,* or *terrible,* use terms that are more concrete, such as *reliable, skillful, dishonest,* or *incompetent*—further clarified by examples, such as *never late for work.*

In some instances, of course, you may wish to generalize for the sake of diplomacy. Instead of writing *Bill, Mary, and Sam have been tying up the office phones with personal calls,* you might prefer to generalize: *Some employees have been . . . .* The second version makes the point without accusing anyone in particular.

When you can, provide solid numbers and statistics that get your point across:

> **General** Transport Canada's 1998 statistics show that traffic fatalities were lower than in 1997. Indeed, the 1998 total was the lowest in years. However, deaths caused by rollovers (mostly in SUVs) increased.
>
> **Specific** Transport Canada's 1998 statistics show 2851 traffic fatalities; that figure is down 4.7% from 1997's total of 2992 and is the lowest total of traffic deaths since 1955. However, deaths caused by rollovers increased from 519 to 665. By contrast, before 1991 rollover deaths did not exceed 200 in any year. (The increase can be attributed to the growing popularity of SUVs—the U.S. National Highway Safety Administration reports that the rollover rate for SUVs is 98 per 1 000 000, double that of all vehicles, which is 47 per 1 000 000.)

Exercise 16

Revise the following sentences to make them more precise or informative.

    a. Our outlet does more business than Montreal.
    b. Anaerobic fermentation is used in this report.
    c. Confusion is in control of this office.
    d. Your crew damaged a piece of office equipment.
    e. His performance was admirable.
    f. This thing bothers me.

Use Analogies to Sharpen the Image.   Ordinary *comparison* shows similarities between two things *of the same class* (two computer keyboards, two methods of cleaning dioxin-contaminated sites). *Analogy* shows some essential similarity between two things of *different classes* (report writing and computer programming, computer memory and post office boxes).

Analogies are good for emphasizing a point (*Some rain is now as acidic as vinegar*). They are especially useful in translating something abstract, complex, or unfamiliar, as long as the easier subject is broadly familiar to readers. Analogy therefore calls for particularly careful analyses of audience.

Analogies can save words and convey vivid images. *Collier's Encyclopedia* describes the tail of an eagle in flight as "spread like a fan." The following sentence from a description of a trout feeder mechanism uses an analogy to clarify the positional relationship between two working parts:

> **Analogy**   The metal rod is inserted (and centred, **crosslike**) between the inner and outer sections of the clip.

Without the analogy *crosslike*, we would need something like this to visualize the relationship:

> **Missing analogy**   The metal rod is inserted, **perpendicular to the long plane and parallel to the flat plane,** between the inner and outer sections of the clip.

This second version is doubly inefficient: more words are needed to communicate, and more work is needed to understand the meaning.

Besides naming things vividly, analogies help *explain* things. The following analogy from the *Congressional Research Report* helps us understand something unfamiliar (dangerous levels of a toxic chemical) by comparing it to something more familiar (human hair).

> **Analogy**   A dioxin concentration of 500 parts per trillion is lethal to guinea pigs. One part per trillion is roughly equal to the thickness of a human hair compared to the distance across North America.

ADJUSTING THE TONE

Your tone is your personal stamp—the personality that takes shape between the lines. The tone you create depends on (1) the distance you impose between yourself and the reader, and (2) the attitude you show toward the subject.

Assume that a friend is going to take over a job you've held. You've decided to write your friend instructions for parts of the job. Here is your first sentence:

Informal

> Now that you've arrived in the glamorous world of office work, put on your track shoes; this is no ordinary manager-trainee job.

First, we notice that the sentence imposes little distance between the writer and the reader (it uses the direct address, "you," and the humorous suggestion to "put on your track shoes"). The ironic use of "glamorous" suggests that the writer means just the opposite: that the job holds little glamour.

For a different reader (the recipient of a company training manual, for example), the writer would have chosen some other opening:

Semi-formal

> As a manager trainee at GlobalTech, you will work for many managers. In short, you will spend little of your day seated at your desk.

The tone now is serious, no longer intimate, and the writer expresses no distinct attitude toward the job. For yet another audience (those who will read an annual report for clients or investors), the writer again might alter the tone:

Formal

> Manager trainees at GlobalTech are responsible for duties that extend far beyond desk work.

Here the businesslike shift from second- to third-person address makes the tone too impersonal for any document addressed to the trainees themselves.

We already know how tone works in speaking. When you meet someone new, for example, you respond in a tone that defines your relationship:

Tone announces interpersonal distance

> Honoured to make your acquaintance. [formal tone—greatest distance]
>
> How do you do? [formal]
>
> Nice to meet you. [semi-formal—medium distance]
>
> Hello. [semi-formal]
>
> Hi. [informal—least distance]
>
> What's happening? [informal—slang]

Each of these greetings is appropriate in some situations, inappropriate in others.

To decide on an appropriate distance from which to address a particular audience, follow these guidelines:

- Use a formal or semi-formal tone in writing for superiors, professionals, or academics (depending on what you think the reader expects).
- Use a semi-formal or informal tone in writing for colleagues and subordinates (depending on how close you feel to your readers).
- Use an informal tone when you want your writing to be conversational, or when you want it to sound like a person talking.
- Above all, find out what the preferences are in your organization.

Whichever tone you select, be consistent throughout your document.

Inconsistent tone　　My office isn't fit for a pig [*too informal*]; it is ungraciously unattractive [*too formal*].

Revised　　The shabbiness of my office makes it an unfit place to work.

In general, lean toward an informal tone without falling into slang.

In addition to setting the distance between writer and reader, your tone implies your *attitude* toward the subject and the reader:

Tone announces
attitude

> We dine at seven.
>
> Dinner is at seven.
>
> Let's eat at seven.
>
> Let's chow down at seven.
>
> Let's strap on the feedbag at seven.
>
> Let's pig out at seven.

The words we choose tell readers a great deal about where we stand.

If readers expect an impartial report, try to keep your own biases out of it. But for situations in which your opinion *is* expected, or in which you perceive some danger or ethics violation, let readers know where you stand.

Plain English needed	Avoid prolix nebulosity.
Revised	Don't be wordy and vague.

Say *I enjoyed the fibre optics seminar* instead of: *My attitude toward the fibre optics seminar was one of high approval.* Say *Let's liven up our dull relationship* instead of: *We should inject some rejuvenation into our lifeless liaison.*

In an upcoming meeting about the reader's job evaluation, does your reader expect to *discuss* the evaluation, *talk it over, have a chat,* or *chew the fat?* If the situation calls for a serious tone, don't use language that suggests a casual attitude—or vice versa. Use the following strategies for making your tone conversational and appropriate.

Use an Occasional Contraction.   Unless you have reason to be formal, use (but do not overuse) contractions. Balance an *I am* with an *I'm,* a *you are* with a *you're,* an *it is* with an *it's* (as we've done throughout this book).

Missing contraction	Do not be wordy and vague.
Revised	Don't be wordy and vague.

Generally, use contractions only with pronouns, not with nouns or proper nouns (names).

Awkward contractions	Barbara'll be here soon. Health's important
Ambiguous contractions	The dog's barking. Bill's skiing.

These ambiguous contractions could be confused with possessive constructions.

Address Readers Directly. Use the personal pronouns *you* and *your* to connect with readers.

Impersonal tone	Students at our college will find the faculty always willing to help.

Personal tone	As a student at our college, **you** will find the faculty always willing to help.

Research shows that readers relate better to something addressed directly to them.

Caution: Use *you* and *your* only to correspond *directly* with the reader, as in a letter, memo, instructions, or some form of advice, encouragement, or persuasion. By using *you* and *your* when your subject and purpose call for first or third person, you might write something wordy and awkward like this:

Wordy and awkward	**When you** are in northern Ontario, **you** can see wilderness lakes everywhere around **you.**
Appropriate	Wilderness lakes are everywhere in northern Ontario.

Exercise 17

The following sentences contain pretentious language, unclear expression of attitude, missing contractions, or indirect address. Adjust the tone.

a. Further interviews are a necessity to our ascertaining the most viable candidate.
b. This project is beginning to exhibit the characteristics of a loser.
c. We are pleased to tell you that you are a finalist.
d. Do not submit the proposal if it is not complete.
e. Employees must submit travel vouchers by May 1.
f. Persons taking this test should use the HELP option whenever they need it.
g. I am not unappreciative of your help.
h. My disapproval is far more than negligible.

Use *I* and *We* When Appropriate.   Instead of disappearing behind your writing, use *I* or *We* when referring to yourself or your organization.

Distant	The writer would like a refund.
Revised	I would like a refund.

A message becomes doubly impersonal when both the writer and the reader disappear.

Impersonal	The requested report will be sent next week.
Personal	**We** will send the report **you** requested next week.

Use the Active Voice.   Because the active voice is more direct and economical than the passive voice, it generally creates a less formal tone. (Review pages 219–221.)

Passive and impersonal	Travel expenses cannot be reimbursed unless receipts are submitted.
Active and personal	We cannot reimburse your travel expenses unless you submit receipts.

Exercise 18

The following sentences have too few *I* or *We* constructions or too many passive constructions. Adjust the tone.

a. Payment will be made as soon as an itemized bill is received.
b. You will be notified.
c. Your help is appreciated.
d. Our reply to your bid will be sent next week.
e. Your request will be given our consideration.
f. My opinion of this proposal is affirmative.
g. This writer would like to be considered for your opening.

Emphasize the Positive.   Whenever you offer advice, suggestions, or recommendations, try to emphasize benefits rather than flaws.

Critical tone	Because of your division's lagging productivity, a management review may be needed.
Encouraging tone	A management review might help boost productivity in your division.

Avoid an Overly Informal Tone.   We generally do not write in the same way we would speak to friends at the local burger joint or street corner. Achieving a conversational tone does not mean lapsing into substandard usage, slang, profanity, or excessive colloquialisms. *Substandard usage* ("He ain't got none," "I seen it today," "She brang the book") fails to meet standards of educated expression. *Slang* ("hurling," "belted," "bogus," "bummed") usually has specific meaning only for members of a particular in-group. *Profanity* ("This idea sucks," "What the hell") not only displays contempt for the audience but often earns contempt for the person using it. *Colloquialisms* ("okay," "a lot," "snooze," "in the bag") are understood more widely than slang, but tend to appear more in speaking than in writing.

Slang and profanity are almost always inappropriate in school or workplace writing. The occasional colloquial expression, however, helps soften the tone of any writing—as long as the situation calls for a measure of informality. Tone is considered offensive when it violates the reader's expectations: when it seems disrespectful, tasteless, distant and aloof, too chummy, casual, or otherwise inappropriate for the topic, the reader, and the situation.

A formal, or academic, tone, is perfectly appropriate in countless writing situations: a research paper, a job application, a report for the company president. In a history essay, for example, we would not refer to Mackenzie King and Pierre Trudeau as "those dudes, Mac and Pierre."

Whenever we begin with rough drafting or brainstorming, our tone might be overly informal and is likely to require some adjustment during subsequent drafts.

Avoid Bias.   Even controversial subjects deserve unbiased treatment. Imagine you have been sent to investigate the causes of an employee-management confrontation at your company's Orillia branch. Your initial report, written for the Toronto central office, is intended simply to describe what happened. Here is how an unbiased description might read:

A factual account

> At 9:00 a.m. on Tuesday, January 21, eighty women employees set up picket lines around the executive offices of our Orillia branch, bringing business to a halt. The group issued a formal protest, claiming that their working conditions were repressive, their salary scale unfair, and their promotional opportunities limited. The women demanded affirmative action, insisting that the company's hiring and promotional policies and wage scales be revised. The demonstration ended when Garvin Tate, vice president in charge of personnel, promised to appoint a committee to investigate the group's claims and to correct any inequities.

Notice the absence of implied judgments; the facts are presented objectively. A less impartial version of the event, from a protestor's point of view, might read:

A biased version

> Last Tuesday, sisters struck another blow against male supremacy when eighty women employees paralyzed the company's repressive and sexist administration for more than six hours. The timely and articulate protest was aimed against degrading working conditions, unfair salary scales, and lack of promotional opportunities for women. Stunned executives watched helplessly as the group organized their picket lines, determined to continue their protest until their demands for equal rights were addressed. An embarrassed vice president quickly agreed to study the group's demands and to revise the company's discriminatory policies. The success of this long-overdue confrontation serves as an inspiration to oppressed women employees everywhere.

Judgmental words (*male supremacy, degrading, paralyzed, articulate, stunned, discriminatory*) inject the writer's attitude, even though it isn't called for. In contrast to this bias, the following version patronizingly defends the status quo:

A biased version

> Our Orillia branch was the scene of an amusing battle of the sexes last Tuesday, when a group of irate feminists, eighty strong, set up picket lines for six hours at the company's executive offices. The protest was lodged against supposed inequities in hiring, wages, working conditions, and promotion for women in our company. The radicals threatened to surround the building until their demands for "equal rights" were met. A bemused vice president responded to this carnival demonstration with patience and dignity, assuring the militants that their claims and demands—however inaccurate and immoderate—would receive just consideration.

Again, qualifying adjectives and superlatives slant the tone.

Being unbiased, of course, doesn't mean burying your head—and your values—in the sand. Remaining neutral about something you know to be wrong or dangerous is unethical (Kremers 59). If, for instance, you conclude that the Orillia protest was clearly justified, say so.

Avoid Sexist Usage.   When we refer to doctors, lawyers, and other professionals as *he* or *him,* while we refer to nurses, administrative assistants, and homemakers, etc., as *she* or *her,* we use sexist phrasing. In this traditional stereotype, males do the jobs that really matter and that pay higher wages, whereas females serve only as support and decoration. When females do invade traditional "male" roles, we might express our surprise at their boldness by calling them *female executives, female sportscasters, female surgeons,* or *female hockey players.* Likewise, to demean males who have settled for "female" roles, we sometimes refer to *male secretaries, male nurses, male flight attendants,* or *male models.*

The following are guidelines for non-sexist usage:

1. Use neutral expressions:

chair or chairperson	rather than	chairman
businessperson	rather than	businessman
supervisor	rather than	foreman
police officer	rather than	policeman
letter carrier	rather than	postman
homemaker	rather than	housewife
humanity or humankind	rather than	mankind
actor	rather than	actor vs. actress

2. Rephrase to eliminate the pronoun, if you can do so without altering your original meaning.

>   **Sexist**   A writer will succeed if **he** revises.

>   **Revised**   A writer who revises succeeds.

3. Use plural forms.

>   **Sexist**   A writer will succeed if **he** revises.

>   **Revised**   Writers will succeed if **they** revise (but *not* A writer will succeed if **they** revise.)

When using a plural form, avoid creating an error in pronoun-referent agreement by having the *plural* pronoun *they* or *their* refer to a *singular* referent (as in ***Each writer** should do **their** best*). Note: The Oxford English Dictionary committee now approves this "error" but it is not yet widely accepted.

4. When possible (as in direct address) use *you: **You** will succeed if **you** revise.* But use this form only when addressing someone directly. (See page 242 for discussion.)
5. Use occasional pairings (*him or her, she or he, his or hers*): *A writer will succeed if **she or he** revises.*

    But note that overuse of such pairings can be awkward: *A writer should do **his or her** best to make sure that **he or she** connects with **his or her** readers.* Most handbooks now encourage alternating use between the two pronouns, and discourage pairings and *he/she: An effective writer always focuses on **her** audience; the writer strives to connect with all **his** readers.*
6. Drop diminutive endings such as *-ess* and *-ette* used to denote females (*poetess, drum majorette, actress,* etc.).
7. Use *Ms.* instead of *Mrs.* or *Miss,* unless you know that person prefers one of the traditional titles. Or omit titles completely: *Roger Tse* and *Morag Kelly; Tse* and *Kelly.*
8. In quoting sources that have ignored present standards for non-sexist usage, consider these options: Insert [*sic*] ("thus" or "so") following the first instance of sexist terminology in a particular passage. Use ellipsis to omit sexist phrasing. Paraphrase the material, instead of quoting it directly.

Avoid Offensive Usages of All Types.   Enlightened communication respects all people in reference to their specific cultural, racial, ethnic, and national background; sexual and religious orientation; age or physical condition. References to individuals and groups should be as neutral as possible; no matter how inadvertent, any expression that seems condescending or judgmental or that violates the reader's sense of appropriateness is offensive. Detailed guidelines for reducing biased usage appear in these two works, among others:

Schwartz, Marilyn, et al. *Guidelines for Bias-Free Writing.* Bloomington: Indiana UP, 1995.

Publication Manual of the American Psychological Association, 4th ed. Washington, DC: American Psychological Association, 1994.

Below is a sampling of suggestions adapted from the previous works:

- When referring to members of a particular culture, be as specific as possible about that culture's identity: Instead of *Latin American* or *Asian* or *Hispanic,* for instance, use *Cuban American* or *Korean* or *Nicaraguan.*

  Avoid judgmental expressions: Instead of *third-world* or *undeveloped nations* or the *Far East,* use *developing* or *newly industrialized nations* or *East Asia.*

- When referring to someone who has a disability, avoid terms that could be considered pitying or overly euphemistic, such as *victims, unfortunates, challenged,* or *differently abled.* Focus on the individual instead of the disability: Instead of *blind person* or *amputee,* refer to *a person who is blind* or *a person who has lost an arm.*

  In general usage, avoid expressions that demean those who have medical conditions: *retard, mental midget, insane idea, lame excuse, the blind leading the blind, able-bodied workers,* etc.

- When referring to members of a particular age group, use *girl* or *boy* for people of age 14 or under; *young person, young adult, young man,* or *young woman* for those of high-school age; and *woman* or *man* for those of college age. (*Teenager* or *juvenile* carries certain negative connotations.) Instead of the *elderly,* use *older persons.*

Exercise 19

The following sentences contain negative emphasis, excessive informality, biased expressions, or offensive usage. Adjust the tone.

- a. If you want your workers to like you, show sensitivity to their needs.
- b. By not hesitating to act, you prevented my death.
- c. The union has won its struggle for a decent wage.
- d. The group's spokesman demanded salary increases.
- e. Each employee should submit his vacation preferences this week.
- f. While the girls played football, the men waved pom-poms.
- g. Aggressive management of this risky project will help you avoid failure.
- h. The explosion left me blind as a bat for nearly an hour.
- i. This dude would be an excellent employee if only he could learn to chill out.

CONSIDERING THE CULTURAL CONTEXT

The style guidelines in this chapter apply specifically to standard English in North America. But practices and preferences can differ widely in different cultural contexts. Certain cultures might prefer long sentences and technical language to convey an idea's full complexity. Other cultures value expressions of respect, politeness, praise, and gratitude more than clarity or directness (Hein 125–26; Mackin 349–50).

Some documents in other languages tend to be more formal than in English, and some rely heavily on passive voice (Weymouth 144). French readers, for example, may prefer an elaborate style that reflects sophisticated and complex modes of thinking. In contrast, our "plain English," conversational style might connote simple-mindedness, disrespect, or incompetence (Thrush 277).

In translation or in a different cultural context, certain words carry offensive or unfavourable connotations. A few notable disasters (Gesteland 20; Victor 44):

- The Chevrolet *Nova*—meaning "don't go" in Spanish
- The Finnish beer *Koff*—for an English-speaking market
- Colgate's *Cue* toothpaste—an obscenity in French
- A brand of bicycle named *Flying Pigeon*—imported for a North American market

Idioms ("strike out," "ground rules") hold no logical meaning for other cultures. Slang ("bogus," "fat city") and colloquialisms ("You bet," "Gotcha") can strike readers as too informal and crude.

Any form of offensive style (including profanity and inappropriate humour) can alienate audiences—toward your company *and* your culture (Sturges 32).

AVOIDING RELIANCE ON AUTOMATED TOOLS

Many of the strategies in this chapter could be executed rapidly with word-processing software. By using the *global search and replace function* in some programs, you can command the computer to search for ambiguous pronoun references. The computer will also detect overuse of passive voice, *to be* verbs, *There* and *It* sentence openers, negative constructions, clutter words, needless prefaces and qualifiers, overly technical language, jargon, sexist language, etc. With an on-line dictionary or thesaurus, you can check definitions or see a list of synonyms for a word you have used in your document.

But these editing aids can be extremely imprecise. No amount of automation is likely to eliminate the writer's burden of *choice*. None of the rules or advice offered in this chapter applies universally. Ultimately, the informed writer's sensitivity to meaning, emphasis, and tone—the human contact—determines the effectiveness of any document.

USING OBJECTIVE SELF-EVALUATION TOOLS

This chapter presents many tools for polishing writing. However, you may not know where to start, or even if much editing is needed. Moreover, you may have to evaluate the readability of your own writing, a difficult task at best. If so, you will find the following objective indexes useful.

1. *Calculate the percentage of SVO and SVC sentences.* Most clear writing features a high percentage (75 percent or better) of basic sentence patterns: **subject-verb-object** (SVO) and **subject-verb-complement** (SVC).

Hockey players need skating skills. (SVO)

Hockey players take risks on the ice. (SVO + Complementary prepositional phrase)

Hockey players are resilient. (SVC)

Pro hockey players are reluctant to play in Canada. (SVC + Complementary prepositional phrase)

In most cases, pro hockey players hire player agents. (Adverbial opening + SVO)

Consequently, they seldom participate in contract negotiations. (Adverbial opening + SVC)

Quick index: Examine a passage of 10 sentences or more. Confirm that 75 percent or more of the sentences use an SVO or SVC sentence pattern.

NOTE *Other, more complicated, patterns include compound sentences (two or more independent clauses), complex sentences (one independent clause and one or more dependent clauses), and compound-complex sentences (two or more independent clauses and at least one dependent clause).*

2. *Count the words before the subject.* Readable sentences quickly get to the point, even though many sentences start with transition words.

   Quick index: In a passage of 10 or more sentences, calculate the average number of words before the sentences' subjects. **That average should be three words or less.**

3. *Count the words between subject and verb.* Occasionally, you can afford to place words between the subject and verb (John, *a large, ungainly man with a pronounced limp,* slowly walked home). But you should **restrict the average number of intervening words to three or less.**

4. *Calculate the percentage of linking verbs.* Forms of the verb *to be* (*is, are, has been, will be, would have been*) and words like *seems, appears,* and *looks* link subjects and predicates. In most cases, linking verbs contribute to wordy phrasing and weaker phrasing than do active verbs. (Compare "he is a trainer of dogs" with "he trains dogs.")

   Quick index: Count all the verbs in a passage of at least 100 words. Then calculate the percentage of linking verbs; **these less efficient verbs should make up 30 percent or less of the total number of verbs** in the passage.

5. *Calculate a Fog Index.* During World War II, two *New York Times* editors, Gunning and Mueller, worried that the part of their newspaper's rapidly growing readership would not easily understand the paper's stories. So the editors devised a readability index based on these principles:

   ◆ the longer the sentence, the harder it is to understand
   ◆ "big" words (of three or more syllables) contribute to reading difficulty
   ◆ a document's readability can be gauged by taking random samples
   ◆ each sample passage needs to be about 100 words (or more) to give an accurate measurement of readability
   ◆ each sample passage must consist of complete sentences
   ◆ proper nouns of three or more syllables ("Saskatchewan") and words that *become* three syllables by adding "es" ("excesses") or "ed" ("exceeded") should not be counted as "big" words

A Fog Index calculation includes several steps. The following calculation is for the paragraph in point number 4, above. (A sentence is defined as ending with a period. "30 percent" is counted as two words. Numerals are not counted among the "big" words.)

1. No. of sentences	6
2. No. of words	102
3. Avg. sentence length	17
4. No. of "big" words	7
5. Percentage of "big" words	6.86
6. Subtotal (17 + 6.86)	23.86
7. Multiply by 0.4, a constant (23.86 × 0.4)	**9.54**
	Fog Index

A Fog Index of 9.54 roughly translates to a Grade 10 reading level. But the bottom line index tells only part of the story. The examined passage has a low rating primarily because only 7 of the 102 words have three or more syllables.

By contrast, the paragraph on page 247 ("Many of the strategies . . . have used in your document.") has an index of 18.4, mainly because 22.5 percent of the words have three or more syllables. Literally interpreted, that index says the passage requires a Ph.D. candidate's reading level. But is the paragraph really that difficult to follow? Probably not—words such as *strategies, executed, rapidly,* and *word-processing* should not pose a problem for the readers of this textbook.

This text tries to maintain a Fog Index of 11 to 13. Most of your business and technical writing should have an index of 10 to 12. Remember, though, to not panic if the index is higher; the high reading *may* result from a high percentage of three-syllable words, most of which your readers know very well. If a high reading results from a high average sentence length, then you should edit to correct that problem.

NOTE   *Many word-processing programs contain objective evaluation tools. Refer to your user's manual for more details.*

EXERCISES

1. Calculate Fog Index readings for various passages in this text. For each reading, consider:

    ◆ Does that reading reflect the degree of difficulty?

    ◆ Does the reading match the reading levels for anticipated readers of this book?

2. Use the five objective self-evaluation tools to assess the readability of a document important to you (application letter, progress report, class term-project report).

3. Find a passage that you find very difficult to read. Determine some of the causes of that difficulty by using the five objective indexes on pages 248–49 of this chapter.

COLLABORATIVE PROJECT (ONGOING)

Form a proofreading/editing group of three to four people. Evaluate and proofread each other's assignments. Use the editing advice in this chapter to improve conciseness, clarity, and naturalness. Compare the objective readability ratings for your writing to those of other writers in your group.

Designing Visuals

A visual is any pictorial representation used to clarify a concept, emphasize a particular meaning, illustrate a point, or analyze ideas or data. Besides saving space and words, visuals help audiences process, understand, and remember information. Because they offer powerful new ways of looking at data, visuals reveal trends, problems, and possibilities that otherwise might remain buried in lists of facts and figures.

In printed or on-line documents, in oral presentations or multi-media programs, visuals are a staple of communication today. This chapter covers four main types of visuals: tables, graphs, charts, and illustrations.

WHY VISUALS ARE ESSENTIAL

Readers expect more than just raw information; they want the information processed for their understanding. Readers want to understand the message at a glance. Visuals help us answer many of the questions asked by readers as they process information:

TYPICAL READER QUESTIONS IN PROCESSING INFORMATION

- Which information is most important?
- Where, exactly, should I focus?
- What do these numbers mean?
- What should I be thinking or doing?
- What should I remember about this?

- What does it look like?
- How is it organized?
- How is it done?
- How does it work?

More receptive to images than to words, today's readers resist pages of mere printed text. Visuals help diminish a reader's resistance in several ways:

- *Visuals enhance comprehension by displaying abstract concepts in concrete, geometric shapes.* "How does the metric system work?" (Figure 12.1).
- *Visuals make meaningful comparisons possible.* "Which industrial sectors report the largest on-site releases of environmental pollutants?" (Figure 12.2). "How does one pound compare with one kilogram?" (Figure 12.1).
- *Visuals depict relationships.* "How does seasonal change affect the rate of construction in our city?" (see Figure 12.10). "What is the relationship between Fahrenheit and Celsius temperature?" (Figure 12.1).
- *Visuals serve as a universal language.* In the global workplace, carefully designed visuals can transcend cultural and language differences and thus facilitate international communication (Figure 12.1).
- *Visuals provide emphasis.* To emphasize the change in death rates for heart disease and cancer since 1970, a table (Table 12.1) or bar graph (Figure 12.4) would be more vivid than a prose statement.
- *Visuals condense and organize information, making it easier to remember and interpret.* A simple table, for instance, can summarize a long and difficult printed passage, as in the example that follows.

All You Need to Know About Metric
(For Your Everyday Life)

10 **Metric is based on the Decimal system**

The metric system is simple to learn. For use in your everyday life you will need to know only ten units. You will also need to get used to a few new temperatures. Of course, there are other units which most persons will not need to learn. There are even some metric units with which you are already familiar; those for time and electricity are the same as you use now.

BASIC UNITS

METRE: a little longer than a yard (about 1.1 yards)
LITRE: a little larger than a quart (about 1.06 quarts)
GRAM: a little more than the weight of a paper clip

(comparative sizes are shown)

| 1 METRE |
| 1 YARD |

25 DEGREES FAHRENHEIT

COMMON PREFIXES
(to be used with basic units)

milli: one-thousandth (0.001)
centi: one-hundredth (0.01)
kilo: one-thousand times (1000)

For example
1000 millimetres = 1 metre
100 centimetres = 1 metre
1000 metres = 1 kilometre

OTHER COMMONLY USED UNITS

millimetre: 0.001 metre diameter of a paper clip wire
centimetre: 0.01 metre a little more than the width of a paper clip (about 0.4 inch)
kilometre: 1000 metres somewhat farther than 1/2 mile (about 0.6 mile)
kilogram: 1000 grams a little more than 2 pounds (about 2.2 pounds)
millilitre: 0.001 litre five of them make a teaspoon

OTHER USEFUL UNITS
hectare: about 2 1/2 acres
metric ton: about one ton

WEATHER UNITS:

FOR TEMPERATURE
degrees celsius

FOR PRESSURE
kilopascals are used
100 kilopascals = 29.5 inches of Hg (14.5 psi)

| °C | −40 | −20 | 0 | 20 | 37 | 60 | 80 | 100 |
| °F | −40 | 0 | 32 | 80 | 98.6 | | 160 | 212 |

water freezes   body temperature   water boils

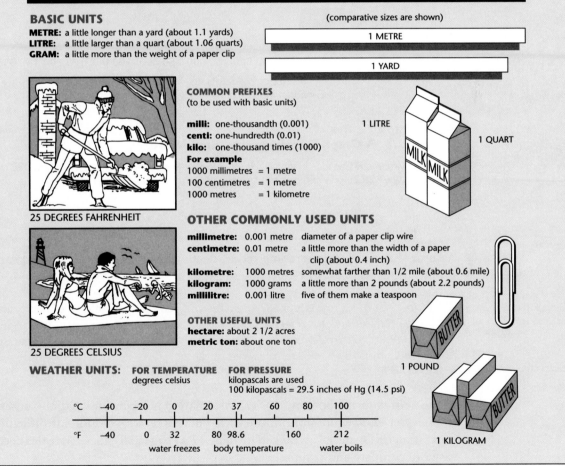

1 LITRE / 1 QUART

1 POUND / 1 KILOGRAM

Figure 12.1 Visuals that Clarify and Simplify

Source: National Institute of Standards and Technology, 1992.

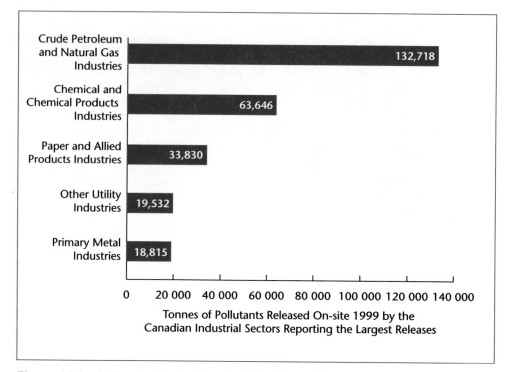

Figure 12.2 A Graph Displaying the "Big Picture"

Source: Environment Canada's National Pollutant Release Inventory—National Overview 1999.
Available: www.ec.gc.ca/pdb/npri/documents/web_1999_NPRI_Overview.pdf

Assume that you are researching recent death rates for heart disease and cancer. From various sources, you collect these data:

Technical data in printed form can be hard to interpret

1. In 1970, 471.2 males and 267.4 females per 100 000 people died of heart disease; 227.6 males and 151.6 females died of cancer.
2. In 1980, 392.1 males and 213.5 females per 100 000 people died of heart disease; 240.3 males and 148.3 females died of cancer.
3. In 1990. . . .

In the written form above, numerical information is repetitious, tedious, and hard to interpret. As the amount of numerical data increases, so does our difficulty in processing this material. Arranged in Table 12.1 (page 255), these statistics become easier to compare and comprehend.

Along with your visual, analyze or interpret the important trends or the essential message you want your readers to see:

A caption explaining the numerical relationships

As Table 12.1 indicates, both male and female death rates from heart disease decreased from 1970 to 1997, but females showed a slightly larger decrease. Cancer deaths during this period increased slightly for males but decreased slightly for females.

Table 12.1   Data Displayed in a Table

Death Rates for Heart Disease and Cancer, 1970–1997

| | Number of Deaths (per 100 000) Population | | | |
| | Heart Disease | | Cancer | |
Year	Male	Female	Male	Female
1970	471.2	267.4	227.6	151.6
1980	392.1	213.5	240.3	148.3
1990	267.5	150.6	246.6	153.2
1992	255.8	141.4	244	152.8
1997	230.8	129.7	229.7	148.5
Percent change, 1970–1997	−51.0	−51.5	+0.9	−2.0

Based on Statistics Canada, Catalogue no. 82-221-XDE and Statistics Canada's 2001 Canada Year Book, p. 14.

Besides their value as presentation devices, visuals help us analyze information. Table 12.1 is one example of how visuals enhance critical thinking by helping us identify and interpret crucial information and discover meaningful connections.

WHEN TO USE A VISUAL

Translate your writing into visuals whenever they make your point more clearly than the prose.[1] Use visuals to *clarify* and to enhance your discussion, not to *decorate* it. Use a visual display to direct the audience's focus or to help them remember something, as in the following situations (Dragga and Gong 46–48):

Use visuals in situations like these

◆ when you want to instruct or persuade
◆ when you want to draw attention to something immediately important
◆ when you expect the document to be consulted randomly or selectively (e.g., a manual or other reference work) instead of being read in its original sequence (e.g., a memo or letter)
◆ when you expect the audience to be relatively less educated, less motivated, or less familiar with the topic
◆ when you expect the audience to be in a distracting environment

An effective visual advances the writer's purpose and the reader's understanding.

---

1. One alternative approach to the writing process is to begin with one or more key visuals and then compose the text to introduce and interpret the visual.

WHAT TYPES OF VISUALS TO CONSIDER

Different types of visuals serve different functions. The following overview sorts visual displays into four categories: tables, graphs, charts, and graphic illustrations. Each type of visual offers readers a new way of seeing—a different perspective.

Tables Display Organized Lists of Data.   Tables display data (as numbers or words) in rows and columns for comparison. Use tables to present exact numerical values and to organize data so that readers can sort out relationships for themselves. Complex tables usually are reserved for more specialized readers.

Numerical tables present data for analysis, interpretation, and exact comparison.

TABLE 1 Charting the Lesson			
Lesson	Page	Page	Page
A	1	3	6
B	2	2	5
C	3	1	4

Prose tables organize verbal descriptions, explanations, or instructions.

TROUBLESHOOTING		
Problem	Cause	Solution
• power	• cord	• plug-in
• light	• bulb	• replace
• flicker	• tube	• replace

Graphs Display Numerical Relationships.   Graphs translate numbers into shapes, shades, and patterns by plotting two or more data sets on a coordinate system. Use graphs to sort out or emphasize specific numerical relationships for readers. The visual representation helps readers grasp, at a glance, the approximate values, the point being made about those values, or the relationship being emphasized.

Bar graphs often show comparisons.

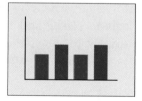

Line graphs often show changes over time.

Charts Display the Parts of a Whole.   Charts depict relationships without the use of a coordinate system by using circles, rectangles, arrows, connecting lines, and other designs.

Pie charts show the parts or percentages of a whole.

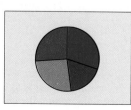

Organization charts show the links among departments, management structures, or other elements of a company.

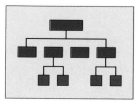

Flowcharts trace the steps (or decisions) in a procedure or stages in a process.

Tree charts show how the parts of an idea or a concept interrelate.

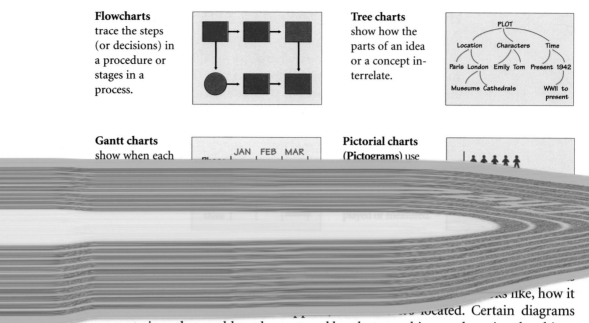

Gantt charts show when each

Pictorial charts (Pictograms) use

...s like, how it ...located. Certain diagrams present views that could not be captured by photographing or observing the object.

Representational diagrams present a realistic but simplified view, usually with essential parts labelled.

Exploded diagrams show the item pulled apart, to reveal its assembly.

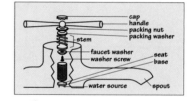

Cutaway diagrams eliminate outer layers to reveal inner parts.

Block or schematic diagrams present the conceptual elements of a principle, process, or system to depict *function* instead of appearance.

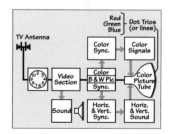

Maps enable readers to visualize a specific location or to comprehend data about a specific geographic region.

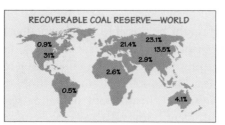

Photographs present an actual picture of the item, process, or procedure.

HOW TO SELECT VISUALS FOR YOUR PURPOSE AND AUDIENCE

You usually will have more than one way to display information in a visual format. To select the most effective display, consider carefully your specific purpose and the abilities and preferences of your audience.

QUESTIONS ABOUT A VISUAL'S PURPOSE AND INTENDED AUDIENCE

What is my purpose?

◆ What do I want the audience to do or think (know facts and figures, follow directions, make a judgment, understand how something works, perceive a relationship, identify something, see what something looks like, pay attention, other)?

◆ Do I want viewers to focus on one or more exact values, compare two or more values, or synthesize a range of approximate values?

Who is my audience?

◆ What is their technical background on this topic?

◆ What is their level of interest in this topic?

◆ Would they prefer the raw data or interpretations of the data?

◆ Are they accustomed to interpreting visuals?

Which type of visual might work best in this situation?

◆ What forms of information should this visual depict (numbers, shapes, words, pictures, symbols)?

◆ Which visual display would be most compatible with the type of judgment, action, or understanding I seek from this audience?

◆ Which visual display would this audience find most accessible?

Here are a few examples of the choices you must consider in selecting visuals:

Choices to consider in selecting visuals

◆ If you just want the audience to know facts and figures, a table might be sufficient, but if you want them to make a particular judgment about these data, a bar graph, line graph, or pie chart might be preferable.

◆ The operating parts of a mechanism might be better shown by an exploded or cutaway diagram than by a photograph.

◆ Expert audiences tend to prefer numerical tables, flowcharts, schematics, and complex graphs or diagrams they can interpret for themselves.

◆ General audiences tend to prefer basic tables, graphs, diagrams, and other visuals that direct their focus and that interpret key points extracted from the data.

Although several alternatives might be possible, one particular type of visual (or a combination) usually is superior for a given purpose and audience. None of the above examples is, of course, immutable. Your particular audience or organization may express its own preferences. Or your choices may be limited by lack of equipment (software, scanners, digitizers), insufficient personnel (graphic designers, technical illustrators), or insufficient budget. In any case, your basic task is to enable the intended audience to interpret the visual correctly.

TABLES

Tables can display exact quantities, compare sets of data, and present information systematically and economically. Numerical tables such as Table 12.1 present *quantitative information* (data that can be measured). In contrast, prose tables present *qualitative information* (brief descriptions, explanations, or instructions). Table 12.2, for example, names pollutant facilities, lists pertinent numerical data, and gives each facility's explanation for that data.

PREFERRED DISPLAYS FOR SPECIFIC VISUAL PURPOSES

Purpose	Preferred Visual
◆ Organize numerical data	Table
◆ Show comparative data	Table, bar graph, line graph
◆ Show a trend	Line graph
◆ Interpret or emphasize data	Bar graph, line graph, pie chart, map
◆ Introduce an unfamiliar object	Photo, representational diagram
◆ Display a project schedule	Gantt chart
◆ Show how parts are assembled	Photo, exploded diagram
◆ Show how something is organized	Organization chart, map
◆ Give instructions	Prose table, photo, diagrams, flowchart
◆ Explain a process	Flowchart, block diagram
◆ Clarify a concept or principle	Block or schematic diagram, tree chart
◆ Describe a mechanism	Photo, representational diagram, or cutaway diagram

---

No table should be overly complex for its intended audience. An otherwise impressive-looking table, such as Table 12.3, is difficult for non-specialists to interpret because it presents too much information at once. We can see how an unethical writer might use a complex table to bury numbers that are questionable or embarrassing (R. Williams 12). Can you discover any hidden facts in Table 12.3? (For instance, which industry has been slowest in cleaning up its act?) Readers need to understand how the table is organized, where to find what they need, and how to interpret the information they find (Hartley 90).

Tables are constructed with various tools: (1) tab markers and tab keys on a word processor, (2) row-and-column displays in a spreadsheet program, (3) the "Table" command in better word-processing programs. The "Table" command option offers a full range of table editing features: cut and paste, adjust spacing, insert text between rows, add rows or columns, adjust column width, etc.

Table Guidelines. Whichever table options you employ, use the following general guidelines:

Guidelines for using tables

- ◆ Use a table only when you are reasonably sure it will enlighten—rather than frustrate—readers. For non-specialized readers, use fewer tables and keep them simple.
- ◆ Try to limit the table to one page. Otherwise write "continued" at the bottom, and begin the second page with the full title, "continued," and the original column headings.

Table 12.2   **A Prose Table**

Reporting the Largest Increase in On-site Releases from 1998 to 1999

Facility	NPRI Pollution Reported Releases	Comments Provided by Facility
Safety-Kleen Ltd Corunna, ON	Lead (and its compounds) ◆ increase of 2 211.74 tonnes (+4 264.9%)	The increased disposal numbers are due to variation in waste management business.
Agrium Fort Saskatchewan, AB	Ammonia (total) ◆ increase in 751.3 tonnes (+86.3%)	The increase in underground injection remediation activities. The materials injected are from previous activities.
Petro-Canada Edmonton Refinery Edmonton, AB	Ammonia (total) ◆ increase of 447.62 tonnes (+28.9%)	No comments provided by facility.

Source: Environment Canada's 1999 National Pollutant Release Inventory.
Available: www.ec.gc.ca/pdb/npri/documents/National_FS_E.pdf

Table 12.3   **A Complex Table**

NPRI Pollutants Released On-site in the Largest Quantities (values in tonnes) 1999

Substance	Air	Underground Injection	Water	Land	Total
Hydrogen sulphide	7 976.5	119 871.8	18.8	0.0	127 868.7
Ammonia (total)	17 314.1	8 397.7	11 154.6	409.7	37 280.4
Methanol	20 566.7	4 238.2	1 844.7	111.4	26 775.2
Zinc (and its compounds)	709.8	0.1	206.9	15 744.6	16 669.5
T Calcium fluoride	19.5	0.0	0.0	13 035.7	13 056.2
Hydrochloric acid	11 630.6	0.0	20.8	9.6	11 665.8
Sulphuric acid	9 369.2	0.0	62.6	19.8	9 456.8
Toluene	7 191.4	72.0	2.4	10.5	7 289.5
Xylene (mixed isomers)	6 909.7	46.3	3.6	5.4	6 977.9
Nitrate ion in solution at pH >= 6.0	71.8	191.9	6 274.1	230.1	6 769.8
Methyl ethyl ketone	5 079.7	790.0	0.0	0.0	5 876.3
Carbon disulphide	4 245.1	0.0	0.0	0.0	4 246.1
Manganese (and its compounds)	143.3	0.0	790.2	3 196.3	4 141.9
T Hydrogen flouride	3 541.0	0.0	0.0	0.0	3 542.0
T Lead (and its compounds)	481.5	0.0	14.3	2 995.2	3 495.3
n-Hexane	3 405.2	15.3	0.4	1.8	3 428.9
Ethylene glycol	283.7	532.3	28.1	1 804.2	2 653.2
T Dichloromethane	2 387.5	0.0	0.0	0.0	2 388.9
Ethylen	2 167.4	0.0	0.0	0.0	2 168.5
Styrene	2 093.0	0.4	0.0	0.0	2 097.9
Isopropyl alcohol	1 953.3	8.4	0.9	0.0	1 970.4
T Asbestos (friable form)	0.0	0.0	.0	1 725.1	1 725.7
T Formaldehyde	1 610.7	4.7	36.4	2.4	1 656.3
2-Butoxyethanol	1 566.0	0.0	0.0	0.0	1 567.9
T Benzene	1 424.4	93.0	1.1	0.6	1 523.1
Largest On-site Releases	**112 141.0**	**134 262.0**	**20 459.7**	**39 302.4**	**306 292.1**
National Total	**127 311.8**	**135 562.2**	**20 789.7**	**43 833.5**	**327 694.9**
% of National Total	**88.1**	**99.0**	**98.4**	**89.7**	**93.5**

T   CEPA-toxic or Carcinogenic Pollutant

Source: Environment Canada's 1999 National Pollutant Release Inventory. National Overview.
Available: www.ec.gc.ca/pdb/npri/ documents/National_FS_E.pdf

◆ If the table is too wide for the page, turn it 90 degrees (landscape) and place its top toward the inside of the binding. Or divide the data into two tables. (Few readers may bother rotating the page to read the table broadside.)

◆ In your discussion, refer to the table by number, and explain what readers should be looking for; or include a prose caption with the table. Specifically, introduce the table, show it, and then interpret it.

For more specific information about creating tables, see the Table Construction Guidelines accompanying Table 12.4.

Table 12.4   Table Construction Guidelines

How to Construct a Table

TABLE 1 ■ Science and Engineering Graduates in 1993 and 1994: 1995 Career Status

STUB HEAD DEGREE AND FIELD	Graduates 1993 and 1994 (1000)	1995—PERCENT DISTRIBUTION				Median salary ($1,000)
		In school[a]	Employed		Not employed	
			In S&E[b]	In other		
All science fields	**580.2**	**25**	**10**	**59**	**6**	**22.8**
Computer science/math	69.2	13	32	51	4	29.0
Life sciences	121.1	37	10	47	5	21.8
Physical sciences	33.2	39	27	30	4	25.5
Social sciences	356.7	X	5	67	7	21.2
All engineering fields[c]	**118.4**	**15**	**62**	**20**	**4**	**33.5**
Civil	18.1	13	67	17	3	31.0
Electrical/electronics	38.6	11	64	21	4	35.0
Industrial	6.4	9	59	28	3	34.0
Mechanical	28.9	12	66	17	4	31.5

[a]Full-time grad. students.   [b]Science & engineering.   [c]Other fields not shown. (X) Not available.
Source: National Science Foundation/SRS, National Survey of Recent College Graduates: 1995.
Statistical Abstract of the United States: 1997 (117th edition). Washington: GPO: 611.

1. Number the table in its order of appearance and provide a title that describes exactly what is being compared or measured.

2. Label stub, row, and column heads (*Degree and Field, Median salary, Computer Science*) so readers know what they are looking at.

3. Stipulate all units of measurement using familiar symbols and abbreviations ($, hr., no.). Define specialized symbols or abbreviations (Å for *angstrom, db* for *decibel*) in a footnote.

4. Compare data vertically (in columns) instead of horizontally (in rows). Columns are easier to compare than rows. Try to include row or column averages or totals, as reference points for comparing individual values.

5. Use horizontal rules to separate headings from data. In a complex table, use vertical rules to separate columns. In a simple table, use as few rules as clarity allows.

6. List the items in a logical order (alphabetical, chronological, decreasing cost). Space listed items so they are not cramped or too far apart for easy comparison. Keep prose entries as brief as clarity allows.

7. Convert fractions to decimals, and align decimals vertically. Keep decimal places for all numbers equal. Round insignificant decimals to the nearest whole number.

8. Use *X, NA,* or a dash to signify any omitted entry, and explain the omission in a footnote ("Not available," "Not applicable").

9. Use footnotes to explain entries, abbreviations, or omissions. Label footnotes with lowercase letters so readers do not confuse the notation with the numerical data.

10. Cite data sources beneath any footnotes. When adapting or reproducing a copyrighted table for a work to be published, obtain written permission from the copyright holder.

Tables work well for displaying exact values, but for easier interpretation, readers prefer graphs or charts. Geometric shapes (bars, curves, circles) generally are easier to remember than lists of numbers (Cochran et al. 25).

Any visual other than a table usually is categorized as a *figure,* and so titled (*Figure 1 Aerial View of the Panhandle Site*). Figures covered in this chapter include graphs, charts, and illustrations.

Like all other components in the document, visuals are designed with audience and purpose in mind (Journet 3). An accountant doing an audit might need a table listing exact amounts, whereas the average public stockholder reading an annual report would prefer the "big picture" in an easily grasped bar graph or pie chart (Van Pelt 1). Similarly, an audience of scientists might find Table 12.3 perfectly appropriate, but a less specialized audience (say, environmental groups) might prefer the clarity and simplicity of Figure 12.2.

GRAPHS

Graphs translate numbers into pictures. Plotted as a set of points (a *series*) on a coordinate system, a graph shows the relationship between two variables.

Graphs have a horizontal and a vertical axis. The horizontal axis carries categories (the independent variables) to be compared, such as years within a period (1990, 1995, 2000). The vertical axis shows the range of values (the dependent variables) for comparing or measuring the categories, such as the number of people who died from heart failure in a specific year. A dependent variable changes according to activity in the independent variable (e.g., a decrease in quantity over a set time, as in Figure 12.3). In the equation $y = f(x)$, x is the independent variable and y is the dependent variable.

Graphs are especially useful for displaying comparisons, changes over time, patterns, or trends. When you decide to use a graph, choose the best type for your purpose: bar graph or line graph.

BAR GRAPHS

Easily understood by most readers, bar graphs show discrete comparisons, as on a year-by-year or month-by-month basis. Each bar represents a specific quantity. Use bar graphs to help readers focus on one value or compare values that change over equal time intervals (expenses calculated at the end of each month, sales figures totalled at yearly intervals). Use a bar graph only to compare values that are noticeably different. Otherwise, all the bars will appear almost identical.

Simple Bar Graphs.   The simple bar graph in Figure 12.3 displays one relationship taken from the data in Table 12.1, the rate of male deaths from heart disease. To aid interpretation, you can record exact values above each bar—but only if readers need exact numbers.

Multiple-bar Graphs.   A bar graph can display two or three relationships simultaneously, each relationship plotted as a separate series. Figure 12.4 displays two comparisons from Table 12.1, the rate of male deaths from both heart disease and cancer.

Whenever a graph shows more than one relationship (or series), each series of numbers is represented by a different pattern, colour, shade, or symbol, and the patterns are identified by a *legend.*

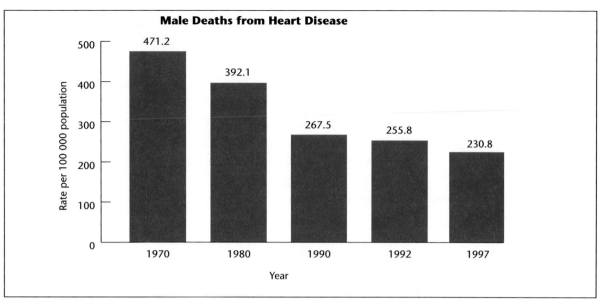

Figure 12.3 A Simple Bar Graph

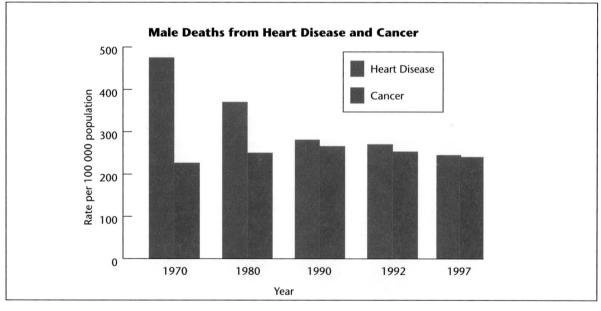

Figure 12.4 A Multiple-bar Graph

The more relationships a graph displays, the harder it is to interpret. As a rule, plot no more than three series of numbers on one graph.

Horizontal-bar Graphs. To make a horizontal-bar graph, turn a vertical-bar graph (and scales) on its side. Horizontal-bar graphs are good for displaying a large series of bars arranged in order of increasing or decreasing value, as in Figure 12.5. The horizontal format leaves room for labelling the categories horizontally *(Prof & Technical,* etc.). A vertical-bar graph leaves no room for horizontal labelling.

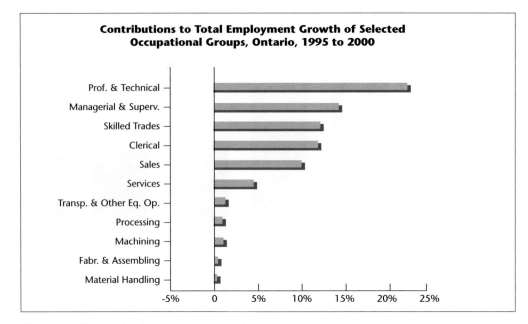

Figure 12.5   A Horizontal-bar Graph

Source: Labour Market Information and Research, Ontario Ministry of Education and Training

Stacked-bar Graphs.   Instead of side-by-side clusters of bars, you can display multiple relationships by stacking bars. Stacked-bar graphs are especially useful for showing how much each item contributes to the whole. Figure 12.6 displays other comparisons from Table 12.1.

Display no more than four or five relationships in a stacked-bar graph. Excessive sub-divisions and patterns create visual confusion.

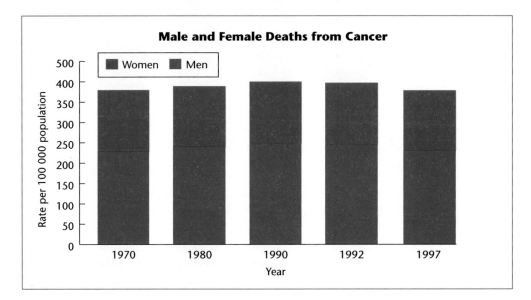

Figure 12.6   A Stacked-bar Graph

Deviation Bar Graphs.   The deviation bar graph can display both positive and negative values, as in Figure 12.7. Notice how the vertical axis extends to the negative side of the zero baseline, following the same incremental division as above the baseline.

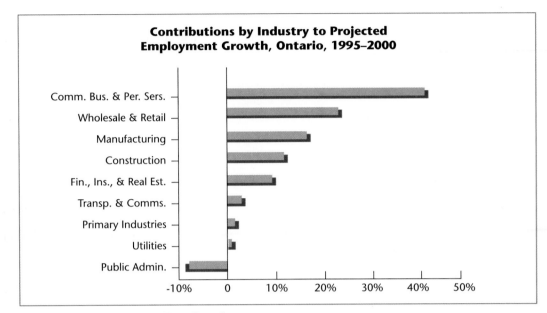

Figure 12.7   Deviation Bar Graph
Source: Adapted from forecast using FOCUS-Ontario and PRISM-Ontario models, the Institute for Policy Analysis, University of Toronto

3-D Bar Graphs.   Graphics software enables you to shade and rotate images and produce three-dimensional views. The 3-D perspectives in Figure 12.8 engage our attention and add visual emphasis to the data.

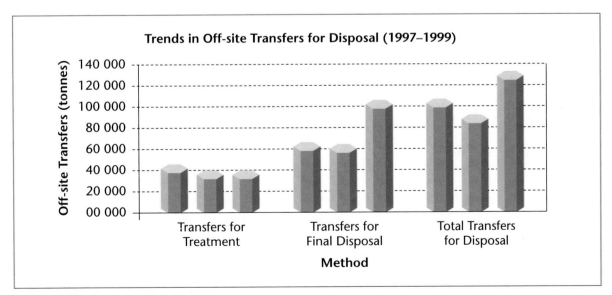

Figure 12.8   3-D Bar Graphs
Source: Environment Canada's 1999 National Pollutant Release Inventory
Available: www.ec.gc.ca/pdb/npri/documents/National_FS_E.pdf

Although 3-D graphs can enhance and dramatize a presentation, an overly complex graph can be almost impossible to interpret. Never sacrifice clarity and simplicity for the sake of visual effect.

Bar Graph Guidelines.   Once you decide on a type of bar graph, use the following suggestions for presenting the graph to your audience.

Bar graph guidelines

- Keep the graph simple and easy to read. Avoid plotting more than three types of bars in each cluster. Avoid needless visual details.
- Number your scales in units the audience will find familiar and easy to follow. Units of 1 or multiples of 2, 5, or 10 are best (Lambert 45). Space the numbers equally.
- Label both scales to show what is being measured or compared. If space allows, keep all labels horizontal for easier reading.
- Use *tick marks* to show the points of division on your scale. If the graph has many bars, extend the tick marks into *grid lines* to help readers relate bars to values.

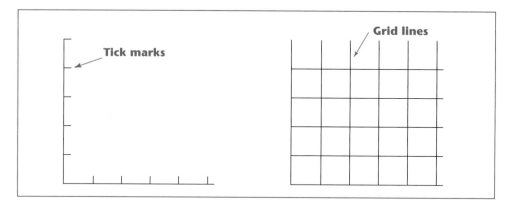

- To avoid confusion, make all bars the same width (unless you are overlapping them). If you must produce your graphs by hand, use graph paper to keep bars and increments evenly spaced.
- In a multiple-bar graph, use a different pattern, colour, or shade for each bar in a cluster. Provide a legend identifying each pattern, colour, or shade.
- If you are trying for emphasis, be aware that darker bars are seen as larger, closer, and more important than lighter bars of the same size (Lambert 93).
- Cite data sources beneath the graph. When adapting or reproducing a copyrighted graph for a work to be published, you must obtain written permission from the copyright holder.
- In your discussion, refer to the graph by number ("Figure 1"), and explain what readers should be looking for; or include a prose caption along with the graph.

Many computer graphics programs automatically employ most of the design features discussed above. Anyone producing visuals, however, should know all the conventions.

LINE GRAPHS

A line graph can accommodate many more data points than a bar graph (e.g., a 12-month trend, measured monthly). Line graphs help readers synthesize large bodies of information in which exact quantities need not be emphasized. Whereas bar graphs display quantitative differences among items (cities, regions, yearly or monthly intervals), line graphs display data whose value changes over time, as in a trend, forecast, or other change during a specified time (profits, losses, growth). Some line graphs depict cause-and-effect relationships (e.g., how seasonal patterns affect sales or profits).

Simple Line Graphs.   A simple line graph, as in Figure 12.9, uses one line to plot time intervals on the horizontal scale and values on the vertical scale. The relationship depicted here would be much harder to express in words alone.

Multiple-line Graphs.   A multiple-line graph displays several relationships simultaneously, as in Figure 12.10. For legibility, use no more than three or four curves in a single graph. Explain the relationships readers are supposed to see.

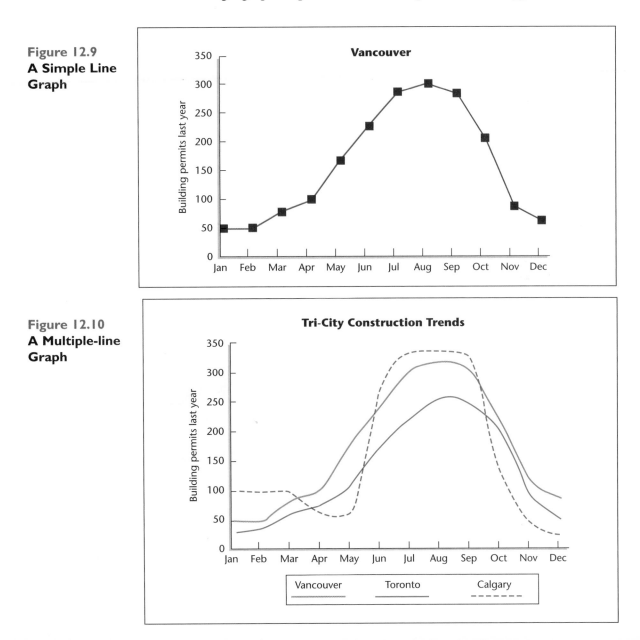

Figure 12.9
A Simple Line Graph

Figure 12.10
A Multiple-line Graph

Building permits in all three cities increased steadily as the weather warmed, but Calgary's increase was more erratic. Its permits declined for April–May, but then surpassed Vancouver's and Toronto's for June–September.

Band or Area Graphs.   For emphasis and appeal, fill the area beneath each plotted line with a pattern. Figure 12.11 is a version of the Figure 12.9 line graph.

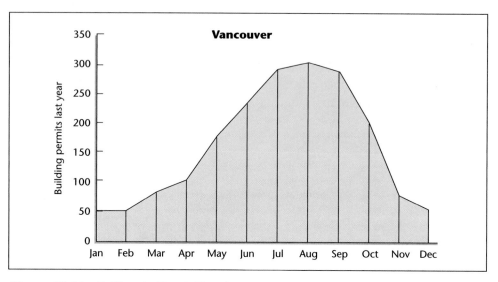

Figure 12.11   A Simple Band Graph

The multiple-bands in Figure 12.12 depict relationships among sums instead of the direct comparisons depicted in the equivalent Figure 12.10 line graph.

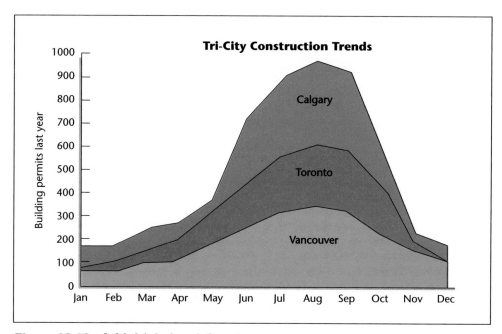

Figure 12.12   A Multiple-band Graph

Despite their visual appeal, multiple-band graphs are easy to misinterpret: In a simple-line graph, each line depicts its own distance from the zero baseline. But in a multiple-band graph, the very top line depicts the *total*, each band below it being a part of that total (like stacked-bar graph segments). Always clarify these relationships for viewers.

Line Graph Guidelines.   Follow bar graph guidelines (page 266), with these additions:

Line graph guidelines

- ◆ Display no more than three or four lines on one graph.
- ◆ Mark each individual data point used in plotting each line.
- ◆ Make each line visually distinct (using colour, symbols, etc).
- ◆ Label each line so readers will know what it represents.
- ◆ Avoid grid lines that readers could mistake for plotted lines.

CHARTS

The terms *chart* and *graph* often appear interchangeably. But a chart is more precisely a figure that displays relationships (quantitative or cause-and-effect) that are not plotted on a coordinate system. Commonly used charts include pie charts, organization charts, flowcharts, tree charts, and pictorial charts (pictograms).

PIE CHARTS

Considered easy for readers to understand, a pie chart depicts the percentages or proportions of the parts that make up a whole. In a pie chart, readers can compare the parts to each other as well as to the whole (to show how much was spent on what, how much income comes from which sources, and so on). Figure 12.13 shows a pie chart.

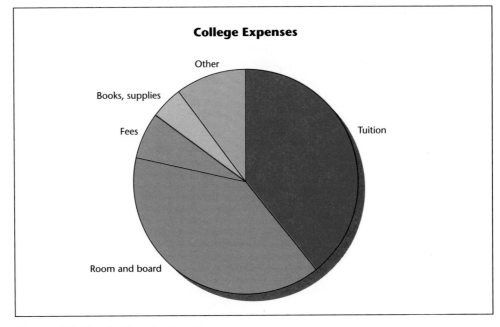

Figure 12.13 A Simple Pie Chart

Figure 12.14 shows two other versions of the pie chart in Figure 12.13. Version (a) displays dollar amounts and version (b), the percentage relationships among these dollar amounts.

**Figure 12.14
Two Other
Versions of
Figure 12.13**

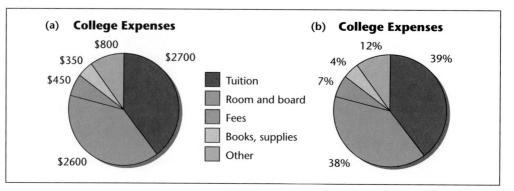

Pie Chart Guidelines.   For constructing pie charts, follow these suggestions:

Pie chart guidelines

- Be sure the parts add up to 100 percent.
- If you must produce your charts by hand, use a compass and protractor for precise segments. Each 3.6-degree segment equals 1 percent. Include any number from two to eight segments. A pie chart containing more than eight segments is difficult to interpret, especially if the segments are small (Hartley 96).
- Combine small segments under the heading "Other."
- Locate your first radial line at 12 o'clock and then move clockwise from large to small (except for "Other," usually the final segment).
- For easy reading, keep all labels horizontal.

ORGANIZATION CHARTS

An organization chart divides an organization into its administrative or managerial parts. Each part is ranked according to its authority and responsibility in relation to other parts and to the whole, as in Figure 12.15.

**Figure 12.15
An
Organization
Chart for One
Corporation**

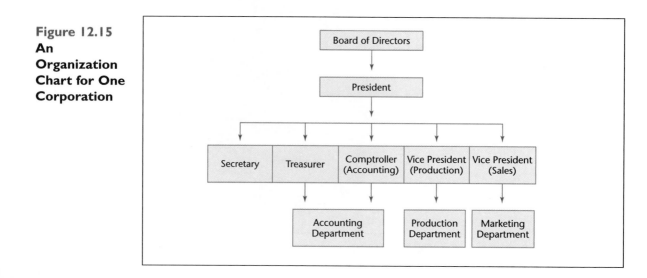

FLOWCHARTS

A flowchart traces a procedure or process from beginning to end. In displaying the steps in a manufacturing process, the flowchart would begin at the raw materials and proceed to the finished product. Figure 12.16 traces the procedure for producing a textbook. (Other flowchart examples appear on pages 97 and 98, and elsewhere throughout the text.)

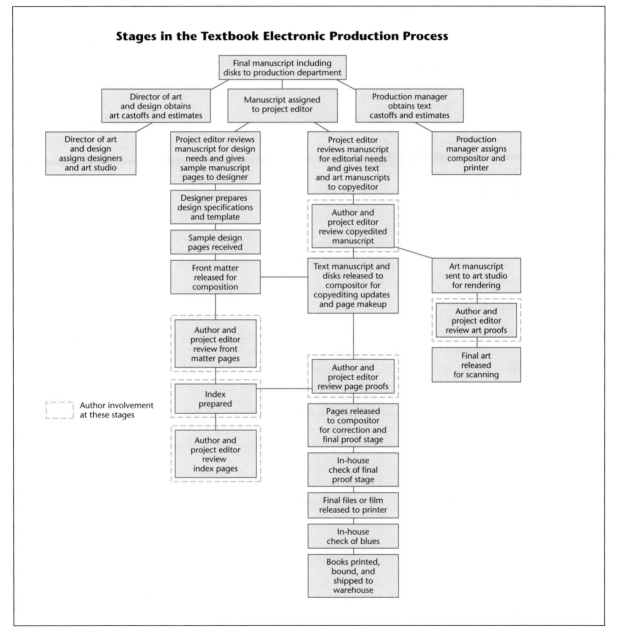

Figure 12.16 A Flowchart for Producing a Textbook

Source: Adapted from Harper & Row Author's Guide

TREE CHARTS

Whereas flowcharts display the steps in a process, tree charts show how the parts of an idea or concept relate to each other. Figure 12.17 displays the parts of an outline for this chapter so that readers can better visualize relationships. The tree version seems clearer and more interesting than the prose listing.

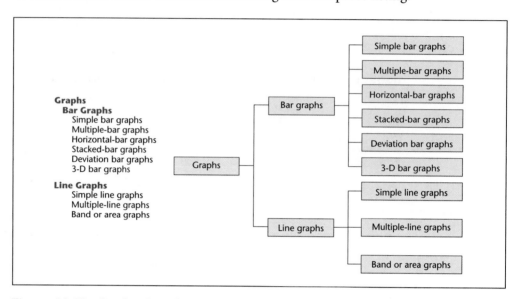

Figure 12.17 An Outline Converted to a Tree Chart

PICTOGRAMS

Pictograms depict numerical relationships with icons or symbols (cars, houses, smokestacks) of the items being measured, instead of using bars or lines. Each symbol represents a stipulated quantity, as in Figure 12.18. Many graphics programs provide an assortment of pre-drawn symbols.

2001 Canadian Population (All ages: 31,084,887) Each figure = 200,000

Age	
0–19	7,930,399
	3,764,518
	4,165,881
20–19	9,160,973
	4,231,577
	4,929,396
40–59	8,784,139
	5,046,637
	3,737,502
60 and over	5,209,376
	3,438,395
	1,770,981

Figure 12.18 A Pictogram
Source: Based on data from Statistics Canada, various sources

Use pictograms when you want to make information more interesting for non-technical audiences.

GANTT CHARTS

Named for engineer H.L. Gantt, a Gantt chart depicts progress as a function of time. A series of bars or lines (time lines) indicates start-up and completion dates for each phase or task in a project, relative to the other phases or tasks. Gantt charts are especially useful for planning a project (as in a proposal) and tracking it (as in a progress report). The Gantt chart illustrated in Figure 12.19 shows tasks whose time lines can be simultaneous, overlapping, or consecutive.

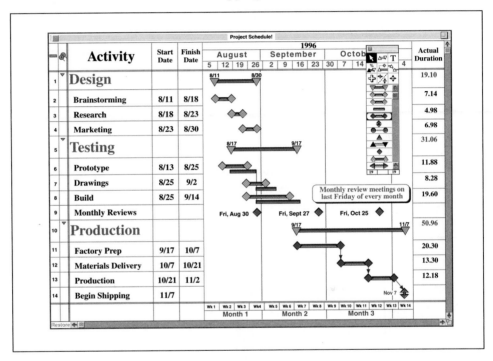

Figure 12.19   A Gantt Chart
Source: Courtesy of AEC Software, ©1996

GRAPHIC ILLUSTRATIONS

Illustrations consist of diagrams, maps, drawings, and photographs depicting relationships that are physical rather than numerical. Good illustrations help readers understand and remember the material (Hartley 82). Consider this information from a government pamphlet, explaining the operating principle of the seat belt:

> The safety-belt apparatus includes a tiny pendulum attached to a lever, or locking mechanism. Upon sudden deceleration, the pendulum swings forward, activating the locking device to keep passengers from pitching into the dashboard.

Without the illustration in Figure 12.20 the mechanism would be difficult to visualize. Clear and uncluttered, a good diagram eliminates unnecessary details

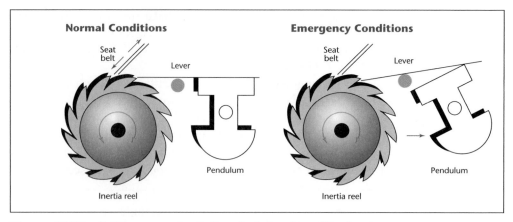

Figure 12.20   A Diagram of a Safety-belt Locking Mechanism
Source: Safety Belts. U.S. Department of Transportation

and focuses only on material useful to the reader. The following pages illustrate some commonly used diagrams.

DIAGRAMS

Exploded diagrams, like that of a brace for an adjustable basketball hoop in Figure 12.21, show how the parts of an item are assembled; they often appear in repair or maintenance manuals. Notice how all parts are numbered for the reader's easy reference in the written instructions.

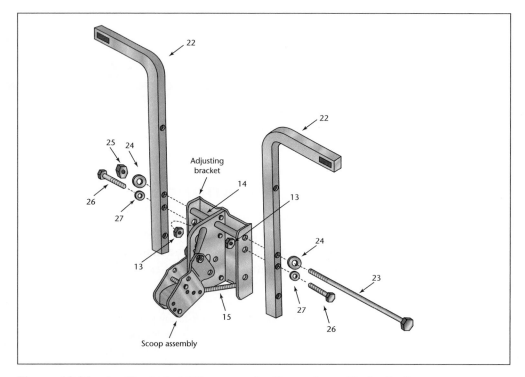

Figure 12.21   An Exploded Diagram of a Brace for a Basketball
Source: Courtesy of Spalding

Cutaway diagrams show the item with its exterior layers removed in order to reveal interior sections, as in Figure 12.22. Unless the specific viewing perspective is immediately recognizable (as in Figure 12.22), define for readers the angle of vision: "top view," "side view," etc.

Block diagrams are simplified sketches that represent the relationship between the parts of an item, principle, system, or process. Because block diagrams are designed to illustrate *concepts* (such as current flow in a circuit), the parts are represented as symbols or shapes. The block diagram in Figure 12.23 on page 276 illustrates how any process can be controlled automatically through a feedback mechanism.

Figure 12.24 on page 276 shows the feedback concept applied as the cruise-control mechanism on a motor vehicle.

Increasingly available are electronic drawing programs, clip-art programs, image banks, and other resources for creating visuals or downloading pre-drawn

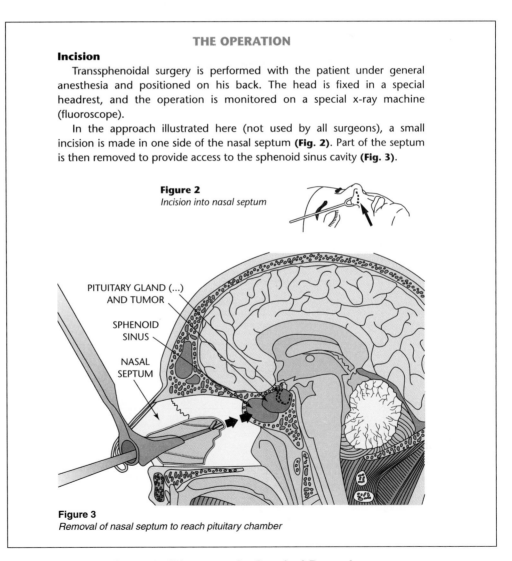

THE OPERATION

Incision

Transsphenoidal surgery is performed with the patient under general anesthesia and positioned on his back. The head is fixed in a special headrest, and the operation is monitored on a special x-ray machine (fluoroscope).

In the approach illustrated here (not used by all surgeons), a small incision is made in one side of the nasal septum **(Fig. 2)**. Part of the septum is then removed to provide access to the sphenoid sinus cavity **(Fig. 3)**.

Figure 2
Incision into nasal septum

PITUITARY GLAND (...)
AND TUMOR

SPHENOID
SINUS

NASAL
SEPTUM

Figure 3
Removal of nasal septum to reach pituitary chamber

Figure 12.22 A Cutaway Diagram of a Surgical Procedure

Source: Transsphenoidal Approach for Pituitary Tumor. ©1986 by the Ludann Co., Grand Rapids, MI

images. Specialized diagrams, however, often require the services of graphic artists or technical illustrators. The client requesting or commissioning the visual provides the art professional with an *art brief* (often prepared by writers and editors) that spells out the visual's purpose and specifications for the visual.

For example, part of the brief addressed to the medical illustrator for Figure 12.22 might read as follows:

An art brief for
Figure 12.22

Purpose: to illustrate transsphenoidal adenomectomy for laypersons

◆ View: full cutaway, axial

◆ Range: descending from cranial apex to a horizontal plane immediately below the upper jaw and second cervical vertebra

◆ Depth: medial cross-section

◆ Structures omitted: cranial nerves, vascular and lymphatic systems

◆ Structures included: gross anatomy of bone, cartilage, and soft tissue—delineated by colour, shading, and texture

◆ Structures highlighted: nasal septum, sphenoid sinus, and sella turcica, showing the pituitary embedded in a 1.5 cm tumour invading the sphenoid sinus via an area of erosion at the base of the sella

Figure 12.23
**A Block
Diagram
Illustrating the
Concept of
Feedback**

Figure 12.24
**A Block
Diagram
Illustrating a
Cruise-control
Mechanism**

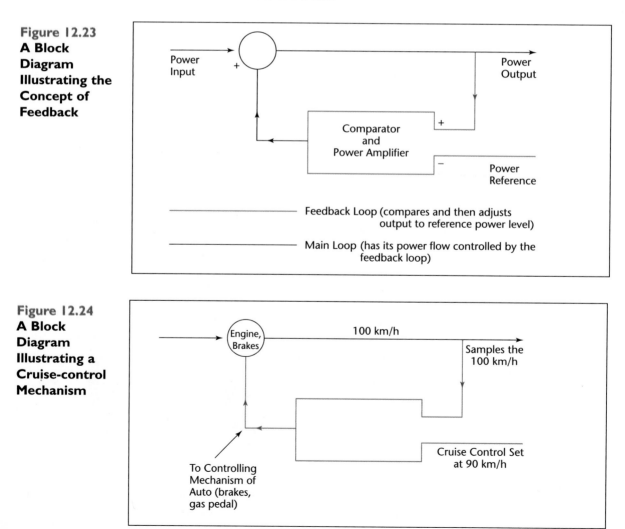

PHOTOGRAPHS

Photographs are especially useful for showing what something looks like (Figure 12.25) or how something is done (Figure 12.26).

No matter how visually engaging, a photograph is difficult to interpret if it includes needless details or fails to identify or emphasize the important material. One graphic design expert offers this practical advice for technical documents:

> *To use pictures as tools for communication, pick them for their capacity to carry meaning, not just for their prettiness as photographs, . . . [but] for their inherent significance to the [document].* (White, *Great Pages* 110, 122)

A Fixed-Platform Oil Rig
Source: SuperStock

Figure 12.25   **A Photograph that Shows the Appearance of Something**

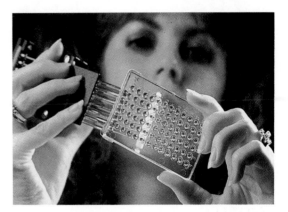

Antibody Screening Procedure
Source: SuperStock

Figure 12.26   **A Photograph that Shows How Something Is Done**

Specialized photographs often require the services of a professional who knows how to use angles, lighting, and special film to obtain the desired focus and emphasis.

Photograph Guidelines.   Whenever you plan to include photographs in a document or a presentation, observe these guidelines:

Guidelines for using photographs

- ◆ Try to simulate the approximate angle of vision readers would have in identifying or viewing the item or, for instructions, in doing the procedure (Figure 12.27).
- ◆ Trim (or crop) the photograph to eliminate needless details (Figures 12.28 and 12.29).
- ◆ For emphasizing selected features of a complex mechanism or procedure, consider using diagrams in place of photographs or as a supplement (Figures 12.30 and 12.31).
- ◆ For an image unfamiliar to readers, provide a sense of scale by including a person, a ruler, or a familiar object (such as a hand) in the photo.
- ◆ If your document will be published, obtain a signed release from any person depicted in the photograph and written permission from the copyright holder. Beneath the photograph, cite the photographer and the copyright holder.
- ◆ In your discussion, refer to the photograph by figure number and explain what readers should be looking for; or include a prose caption.

Titration in Measuring Electron-spin Resonance
Source: SuperStock

Figure 12.27 **A Photograph that
Shows a Realistic Angle
of Vision**

Replacing the Microfilter Activation Unit
Source: SuperStock

Figure 12.28 **A Photograph that
Needs to Be Cropped**

Source: SuperStock

Figure 12.29 **The Cropped
Version of
Figure 12.28**

Sapphire Tunable Laser
Source: SuperStock

Figure 12.30 **A Photograph of a
Complex Mechanism**

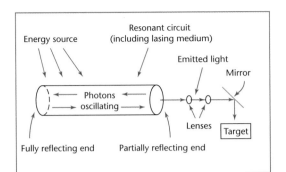

Major Parts of the Laser

Figure 12.31   A Simplified Diagram

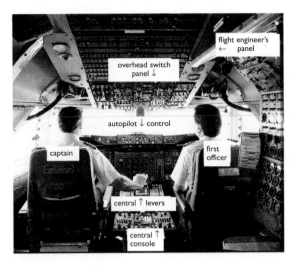

Standard Flight Deck for a Long-Range Jet
Source: SuperStock

Figure 12.32   A Photograph with Essential Features Labeled

Digital-imaging technology allows photographs to be scanned and stored electronically. These stored images then can be retrieved, edited, and altered. Such capacity for altering photographic content creates unlimited potential for distortion and raises questions about the ethics of digital manipulation (Callahan 64–65).

COMPUTER GRAPHICS

Computer technology transfers many of the tasks formerly performed by graphic designers and technical illustrators to individuals with little formal training in graphic design. In whatever career you anticipate, you probably will be expected to produce high-quality graphics for conferences, presentations, and in-house publications.

Today's computer systems create sophisticated, multi-colour graphic displays and multi-media presentations. Among the virtually countless types of computer-generated visuals are these examples:

◆ With an electronic stylus (a pen with an electronic signal), you can draw pictures on a graphics tablet to be displayed on the monitor, stored, or sent to other computers.

◆ You can create three-dimensional effects, showing an object from different angles through the use of shading, shadows, on-screen rotation, background lighting, or other techniques.

◆ You can recreate the visual effect of a mathematical model, as in writing equations to explain what happens when high winds strike a tall building. (As the wind deforms the structure, the equations change. Then you can take those new equations and represent them visually.)

◆ You can create a design, build a model, simulate the physical environment, and let the computer forecast what will happen with different variables.

- You can integrate computer-assisted design (CAD) with computer-assisted manufacturing (CAM), so that the design will direct the machinery that makes the parts themselves (CAD/CAM).
- You can create animations, to see how bodies move (as in a car crash or in athletics).
- You can practise dealing with toxic chemicals, operating sophisticated machines, or making other rapid decisions in medical or technical environments, without the cost of or danger in actual situations.
- Through various types of scientific visualization, you can do "what-if" projections and explore countless ways of conceptualizing and understanding your data. Because the computer can generate and evaluate many possibilities rapidly, it enables you to test hypotheses without doing the calculations.

SELECTING DESIGN OPTIONS

Computer graphics systems allow you to experiment with scales, formats, colours, perspectives, and patterns. By testing design options on the screen, you can revise and enhance your visual repeatedly until it achieves your exact purpose.

Here is a sampling of design options:

- Update charts and graphs whenever the data change. The software will calculate the new data and plot the relationships.
- Edit your graphics on the screen, adding, deleting, or moving material as needed.
- Create your image at one scale, and later specify a different scale for the same image.
- Overlay images in one visual.
- Adjust bar width and line thickness.
- Fill a shape with a colour or pattern.
- Scan, edit, and alter pages, photos, or other images.

Most of these options are available via simple keystrokes or pull-down menus.

USING CLIP ART

Clip art is a generic term for collections of ready-to-use images (of computer equipment, maps, machinery, medical equipment, etc.), all stored electronically. Various clip-art packages enable you to import into your document countless images like the one in Figure 12.33. By running the image through a drawing program, you can enlarge, enhance, or customize it, as in Figure 12.34.

Figure 12.33
A Clip-art Image

Source: Desktop
Art®; Business 1
© Dynamic Graphics, Inc.

Figure 12.34
A Customized Image

Source: Professor R.
Armand Dumont

Although a handy source for images, clip art often has a generic or crude look that makes a document appear unprofessional. Consider using clip art for icons only, or for in-house documents.

One form of clip art especially useful in technical writing is the *icon* (an image with all non-essential background removed). Icons convey a specific idea visually, as in Figure 12.35. Icons appear routinely in computer documentation and in other types of instructions because the image provides readers with an immediate signal of the desired action.

Whenever you use an icon, be sure it is "intuitively recognizable" to your readers ("Using Icons" 3). Otherwise, readers are likely to misinterpret its meaning—in some cases with disastrous results.

Figure 12.35
Icons

Source: Desktop Art®;
Business 1 and Health
Care 1, © Dynamic
Graphics, Inc.

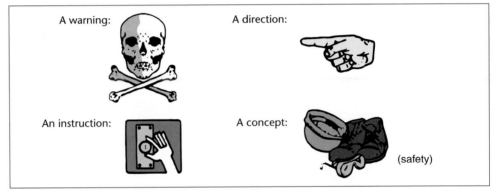

Keep in mind that certain icons have offensive connotations in certain cultures. Hand gestures are especially problematic: some Arab cultures consider the left-hand unclean. A pointing index finger—on either hand—as in Figure 12.35, is a sign of rudeness in Venezuela and Sri Lanka (Bosley 5–6).

USING COLOUR OR SHADING

Colour or shading often makes a presentation more esthetically pleasing, more interesting to look at. But colour and shading serve purposes beyond visual appeal. Used effectively in a visual, they draw and direct readers' attention and help them identify the various elements.

Colour or shading can help clarify a concept or dramatize how something works, as shown in Figure 12.36. The use of bright colours against a darker, duller background enables readers to *visualize* concepts, as does the use of dark shading against white.

Along with shape, type style, and position of elements on a printed page, colour or shading can guide readers through the material. Used effectively on a printed page, colour and shading help organize the reader's understanding, provide orientation, and emphasize important material. Following are just a few among many possible uses of colour and shading (White, *Color* 39–44; Keyes 647–49).

Use Colour and Shading to Organize.   Readers look for ways of organizing their understanding of a document (Figure 12.37). Colour or shading can reveal structure and break up material into discrete blocks that are easier to locate, process, and digest:

Figure 12.36
Colour Used as a Visualizing Tool
Source: Courtesy of National Audubon Society

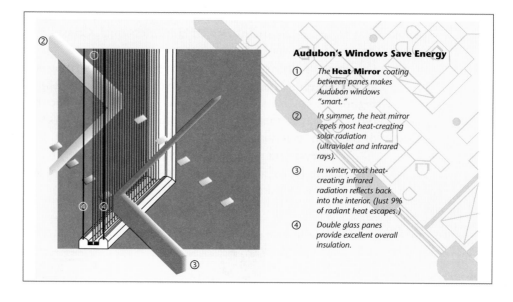

Audubon's Windows Save Energy

① The **Heat Mirror** coating between panes makes Audubon windows "smart."

② In summer, the heat mirror repels most heat-creating solar radiation (ultraviolet and infrared rays).

③ In winter, most heat-creating infrared radiation reflects back into the interior. (Just 9% of radiant heat escapes.)

④ Double glass panes provide excellent overall insulation.

- A colour or shaded background screen can set off like elements such as checklists, instructions, or examples.
- Horizontal rules can separate blocks of text, such as sections of a report or areas of a page.
- Vertical rules can set off examples, quotations, captions, and so on.

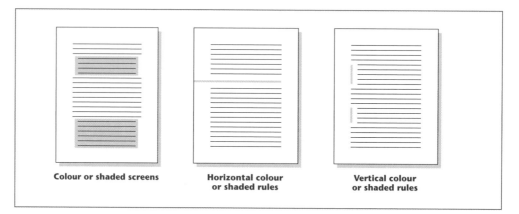

Colour or shaded screens   Horizontal colour or shaded rules   Vertical colour or shaded rules

Figure 12.37   Colour or Shading Used to Organize Page Elements

Use Colour or Shading to Orient.   Readers look for clear bearings and signposts that help them find their place and find what they need (Figure 12.38):

- Colour or shading can help headings stand out from the text and differentiate major from minor headings.
- Coloured or shaded tabs and boxes can serve as location markers.
- Coloured or shaded sidebars (for marginal comments), callouts (for labels), and leader lines (for connecting a label to its referent) can guide the eyes.

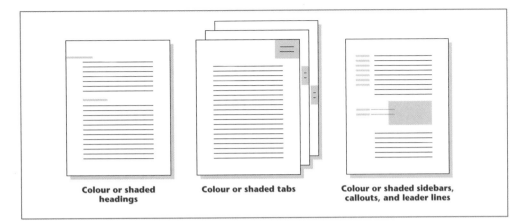

Figure 12.38   **Colour or Shading Used as an Orientation Device**

Use Colour or Shading to Emphasize.   Readers look for places to focus their attention in a document (Figure 12.39).

- Colour or shaded typefaces can highlight key words or ideas.
- Colour or shading can call attention to cross-references.
- A coloured or shaded ruled box can frame a warning, caution, note, or hint.

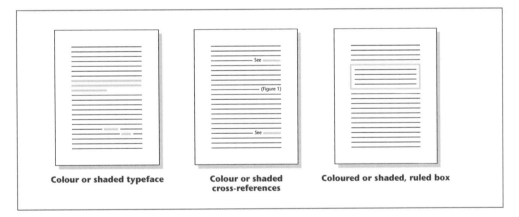

Figure 12.39   **Colour or Shading Used for Emphasis**

Colour or Shading Guidelines.   Whichever colour or shading options you use, employ the following general guidelines:

Guidelines for using colour or shading

- Colour or shading gains impact when it is used selectively. It loses impact when it is overused. (*Aldus Guide* 39). Use colour or shading sparingly, and use no more than three or four distinct colours when using colour—including black and white (White, *Great Pages* 76).
- Apply colour or shading consistently to elements throughout your document. Inconsistent use of colour or shading can distort readers' perception of the relationships (Wickens 117).

- ◆ Make colour redundant. Be sure all elements are first differentiated in black and white: by shape, location, texture, type style, or type size. Different readers perceive colours differently or, in some cases, not at all. A sizable percentage of readers have impaired colour vision (White, *Great Pages* 76).
- ◆ Use a darker colour or shade to make a stronger statement. The darker the colour or shade, the more important the material. Readers perceive differently the sizes of variously coloured or shaded objects. Darker items can seem larger and closer than lighter objects of identical size.
- ◆ Make coloured or shaded type larger than body type. Try to avoid colour or light shading for body type, or use a high-contrast colour or shade (dark against a light background). Colour is less visible on the page than black ink on a white background. The smaller the image or the thinner the rule, the stronger or brighter the colour (White, *Editing* 229, 237).
- ◆ For contrast in a colour screen, use a very dark type against a very light background, for example, a 10–20 percent screen (Gribbons 70). The bigger the area of the screen, the paler the background colour (Figure 12.40).

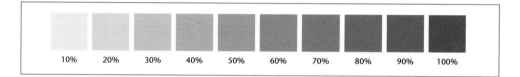

Figure 12.40   A Colour-Density Chart

- ◆ A colour's connotations can vary from culture to culture. In North America, for example, red signifies danger and green traditionally signifies safety. But in Ireland, green or orange carry political connotations in certain contexts. In Muslim cultures, green is a holy colour (Cotton 169).

USING WEBSITES FOR GRAPHICS SUPPORT

The World Wide Web offers a growing array of visual resources. Following is a sample of useful websites (Martin 135–36).

- ◆ Image banks such as Graphics Web, Photo Web, and the Internet Font Directory offer clip-art, photo, and font databases that can be browsed and from which material can be purchased on-line and downloaded to a personal computer.
- ◆ Some royalty-free images are available from the Digital Picture Archive.
- ◆ Samples of computer-generated and traditional art work can be browsed and purchased via Artists on Line.
- ◆ On-line versions of graphics magazines offer helpful design suggestions.
- ◆ Graphic design firms offer samples of work that often embody innovative design ideas.

Whether you seek design ideas, tools for creating your own visuals, or electronic catalogues of completed artwork, the Web is a good bet.

Be extremely cautious about downloading visuals (or any material, for that matter) from the Web and then using it. Review the copyright law (pages 132–33). Originators of any work on the Web own the work and the copyright.

HOW TO AVOID VISUAL DISTORTION

Although you are perfectly justified in presenting data in its best light, you are ethically responsible for avoiding misrepresentation. Any one set of data can support contradictory conclusions. Even though your numbers may be accurate, their visual display could be misleading.

PRESENT THE REAL PICTURE

Visual relationships in a graph should portray accurately the numerical relationships they represent. Begin the vertical scale at zero. Never compress the scales to reinforce your point.

Notice how visual relationships in Figure 12.41 become distorted when the value scale is compressed or fails to begin at zero.

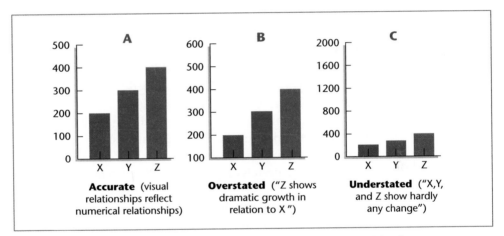

Figure 12.41   An Accurate Bar Graph and Two Distorted Versions

In version A, the bars accurately depict the numerical relationships measured from the value scale. In version B, item Z (400) is depicted as three times X (200). In version C, the scale is overly compressed, causing the shortened bars to understate the quantitative differences.

Deliberate distortions are unethical because they imply conclusions contradicted by the actual data.

PRESENT THE COMPLETE PICTURE

Without bogging down in needless detail, an accurate visual includes all essential data. Figure 12.42 shows how distortion occurs when data that would provide a complete picture are selectively omitted.

Version A accurately depicts the numerical relationships measured from the value scale. In version B, too few points are plotted.

Decide carefully what to include and what to leave out of your visual display.

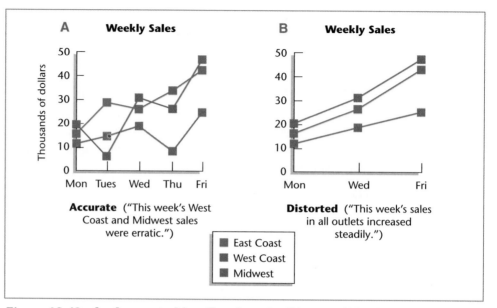

Figure 12.42   An Accurate Line Graph and a Distorted Version

NEVER MISTAKE DISTORTION FOR EMPHASIS

When you want to emphasize a point (a sales increase, a safety record, etc.), be sure your data support the conclusion implied by your visual. For instance, don't use inordinately large visuals to emphasize good news or small ones to downplay bad news (R. Williams 11). When using clip art, pictograms, or drawn images to dramatize a comparison, be sure the relative size of the images or icons reflects the quantities being compared.

A visual accurately depicting a 100 percent increase in phone sales at your company might look like version A in Figure 12.43, below. Version B overstates the good news by depicting the larger image four times the size, instead of twice the size, of the smaller. Although the larger image is twice the height, it is also twice the *width,* so the total area conveys the visual impression that sales have *quadrupled.*

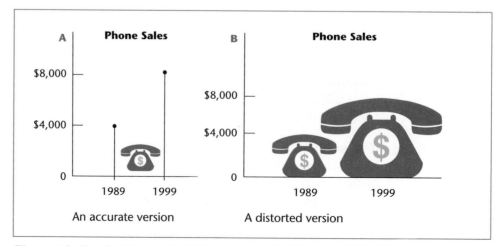

Figure 12.43   An Accurate Pictogram and a Distorted Version

Visuals have their own rhetoric and persuasive force, which we can use to advantage—for positive or negative purposes, for the reader's benefit or detriment (Van Pelt 2). Avoiding visual distortion is ultimately a matter of ethics.

HOW TO INCORPORATE VISUALS WITH THE TEXT

An effective visual enables readers to locate and extract the information they need. To simplify the reader's task, visual and verbal elements in a document should complement each other. For example, a visual should be able to stand alone in meaning—without being isolated from the verbal text.

Visual and Verbal Element Guidelines.   Following are specific guidelines for incorporating visual and verbal elements effectively.

Guidelines for fitting visuals with text

- ◆ Place the visual where it will best serve your readers. If it is central to your discussion, place the visual as close as possible to the material it clarifies. (Achieving proximity often requires that you ignore the traditional "top or bottom" design rule for placement of visuals on a page.) If the visual is peripheral to your discussion or of interest to only a few readers, place it in an appendix so that interested readers can refer to it as they wish. Tell readers when to consult the visual and where to find it.
- ◆ Never refer to a visual that readers cannot easily locate. In a long document, don't be afraid to repeat a visual if you discuss it again later.
- ◆ Never crowd a visual into a cramped space. Set your visual off by framing it with plenty of white space, and position it on the page for balance. To save space and to achieve proportion with the surrounding text, consider carefully the size of each visual and the amount of space it will occupy.
- ◆ Number the visual and give it a clear title and clear labels. Your title should tell readers what they are seeing. Label all of the important material.
- ◆ Match the visual to your audience. Don't make it too elementary for specialists or too complex for non-specialists. Be sure your intended audience will be able to interpret the visual correctly.
- ◆ Introduce and interpret the visual. In your introduction, tell readers what to expect:

Informative	As shown in Table 2, operating costs have increased 7 percent annually since 1999.
Uninformative	See Table 2.

Visuals alone make ambiguous statements (Girill, *Technical Communication and Art* 35); pictures need to be interpreted. Instead of leaving readers to struggle with a page of raw data, explain the relationships displayed. Follow the visual with a discussion of its important features:

Informative	This cost increase means that . . .

Always tell readers what to look for and what it means.

◆ Use prose captions to explain important points made by the visual. Captions help readers interpret a visual (as in Table 12.1). When possible, use a smaller type size so that captions don't compete with text type (*Aldus Guide* 35).

◆ Never include excessive information in one visual. Any visual that contains too many lines, bars, numbers, colours, or patterns will overwhelm readers, causing them to ignore the display. In place of one complicated visual, use two or more straightforward ones.

◆ Be sure the visual's meaning can stand alone. Even though it repeats or augments information already in the text, the visual should contain everything readers will need to interpret it correctly.

The Checklist for Revising Visuals and the Visual Plan Sheet, Figure 12.44, will help ensure that your visuals enhance your meaning.

Checklist FOR REVISING VISUALS

Use this checklist to revise your visuals.

Content

◆ Does the visual serve a legitimate purpose (clarification, not mere ornamentation) in the document?
◆ Is the visual titled and numbered?
◆ Is the level of complexity appropriate for the audience?
◆ Are all patterns in the visual identified by label or legend?
◆ Are all values or units of measurement specified (grams per ounce, millions of dollars)?
◆ Are the numbers accurate and exact?
◆ Do the visual relationships represent the numerical relationships accurately?
◆ Are explanatory notes added as needed?
◆ Are all data sources cited?
◆ Has written permission been obtained for reproducing or adapting a visual from a copyrighted source in any type of work to be published?
◆ Is the visual introduced, discussed, interpreted, integrated with the text, and referred to by number?

◆ Can the visual itself stand alone in meaning?

Arrangement

◆ Is the visual easy to locate?
◆ Are all design elements (title, line thickness, legends, notes, borders, white space) positioned for balance?
◆ Is the visual positioned on the page to achieve balance?
◆ Is the visual set off by adequate white space or borders?
◆ Does the top of a wide visual face the inside binding?
◆ Is the visual in the best report location?

Style

◆ Is this the best type of visual for your purpose and audience?
◆ Are all decimal points in each column vertically aligned?
◆ Is the visual uncrowded and uncluttered?
◆ Is the visual engaging (patterns, colours, shapes), without being too busy?
◆ Is the visual in good taste?
◆ Is the visual ethically acceptable?

VISUAL PLAN SHEET

Focusing on Your Purpose

- What is this visual's purpose (to instruct, persuade, create interest)? _____
- What forms of information (numbers, shapes, words, pictures, symbols) will this visual depict? _____

- What kind of relationship(s) will the visual depict (comparison, cause-effect, connected parts, sequence of steps)? _____
- What judgment, conclusion, or interpretation is being emphasized (that profits have increased, that toxic levels are rising, that X is better than Y, that time is being wasted)? _____

- Is a visual needed at all? _____

Focusing on Your Audience

- Is this audience accustomed to interpreting visuals? _____
- Is the audience interested in specific numbers or an overall view? _____
- Should the audience focus on one exact value, compare two or more values, or synthesize a range of approximate values (Wickens 121)? _____
- Which type of visual will be most accurate, representative, accessible, and compatible with the type of judgment, action, or understanding expected from the audience? _____

- In place of one complicated visual, would two or more straightforward ones be preferable? _____

Focusing on Your Presentation

- What enhancements, if any, will increase audience interest (colours, patterns, legends, labels, varied typefaces, shadowing, enlargement or reduction of some features)? _____

- Which medium—or combination of media—will be most effective for presenting this visual (slides, transparencies, handouts, large-screen monitor, data projection unit, flip chart, report text)? _____
- To achieve the greatest utility and effect, where in the presentation does this visual belong? _____

Figure 12.44   A Planning Sheet for Preparing Visuals

EXERCISES

1. The following statistics are based on data from three colleges in a large western city. They give the number of applicants to each college over six years.

   ◆ In 1999, X College received 2341 applications for admission, Y College received 3116, and Z College 1807.

   ◆ In 2000, X College received 2410 applications for admission, Y College received 3224, and Z College 1784.

   ◆ In 2001, X College received 2689 applications for admission, Y College received 2976, and Z College 1929.

   ◆ In 2002, X College received 2714 applications for admission, Y College received 2840, and Z College 1992.

   ◆ In 2003, X College received 2872 applications for admission, Y College received 2615, and Z College 2112.

   ◆ In 2004, X College received 2868 applications for admission, Y College received 2421, and Z College 2267.

   Illustrate these data in a line graph, a bar graph, and a table. Which version seems most effective for a reader who (1) wants exact figures, (2) wonders how overall enrollments are changing, or (3) wants to compare enrollments at each college in a certain year? Include a caption interpreting each of these versions.

2. Devise a flowchart for a process in your field or area of interest. Include a title and a brief discussion.

3. Devise a chart showing the lines of responsibility and authority in the organization where you hold a part-time or summer job.

4. Devise a pie chart to depict your yearly expenses. Title and discuss the chart.

5. Obtain enrollment figures at your college for the past five years by sex, age, race, or any other pertinent category. Construct a stacked-bar graph to illustrate one of these relationships over the five years.

6. Keep track of your pulse and respiration at 30-minute intervals over a four-hour period of changing activities. Record your findings in a line graph, noting the times and specific activities below your horizontal coordinate. Write a prose interpretation of your graph and give it a title.

7. In textbooks or professional journal articles, locate each of these visuals: a table, a multiple-bar graph, a multiple-line graph, a diagram, and a photograph. Evaluate each according to the revision checklist, and discuss the most effective visual in class.

8. Anywhere on campus or at work, locate at least one visual that needs revision for accuracy, clarity, appearance, or appropriateness. Look in computer manuals; lab manuals; newsletters; financial aid or admissions or placement brochures; student, faculty, or employee handbooks; newspapers; or textbooks. Use the Visual Plan Sheet and the Checklist for Revising Visuals as guides to revise and enhance the visual. Submit to your instructor a copy of the original, along with a memo explaining your improvements. Be prepared to discuss your revision in class.

COLLABORATIVE PROJECT

Compile a list of 12 websites that offer graphics support by way of advice, image banks, design ideas, artwork catalogues, and the like. Provide the address for each site, along with a description of the resources offered and the cost. Report your findings in a format stipulated by your instructor.

CHAPTER 13

Designing Pages and Documents

Page design, the layout of words and graphics, determines the look of a page. Well-designed pages invite readers in, guide them through the material, and help them understand and remember it.

Readers' *first* impression of a document tends to be a purely visual, esthetic judgment. Readers are attracted by documents that appear inviting and accessible.

PAGE DESIGN IN WORKPLACE WRITING

Page design becomes especially significant when we consider these realities about writing and reading in the workplace:

1. *Technical information generally is designed differently from material in novels, news stories, and other forms of writing.* To be accessible, a technical document requires more than just an unbroken sequence of paragraphs. To find their way through complex material, readers may need the help of charts, diagrams, lists, various type sizes and typefaces, different headings, and other page-design elements.

2. *Technical documents rarely get readers' undivided attention.* Readers may be skimming the document while they jot down ideas, talk on the phone, or drink coffee. Or they may refer to sections of the document during a meeting or presentation. Amid frequent distractions, readers must be able to leave the document and return easily.

3. *People read work-related documents only because they have to.* Novels, general newspapers, and magazines are read for relaxation, but work-related documents (trade magazines, reports, newsletters) often require the reader's *labour*. The more complex the document, the harder readers have to labour.

4. *As computers generate more and more paper, any document is forced to compete for readers' attention.* Even brilliant writing is useless unless it gets read by its intended audience. Suffering from information overload, today's readers resist any document that appears overwhelming. They want formats that will help them find the information they really need. A user-friendly document has an accessible format: At a glance, readers can see how the document is organized, where they are in the document, which items matter most, and how the items relate.

Having decided at a glance whether your document is inviting and accessible, readers will draw conclusions about the value of your information, the quality of your work, and your credibility.

Notice how the information in Figure 13.1 (page 293) resists our attention. Without design cues, we have no way of grouping this information into organized units of meaning. Now look at Figure 13.2 (page 294), which shows the same information after a design overhaul. Notice, also, that the material in Figure 13.2 appears in a slightly different order than in Figure 13.1.

DESKTOP PUBLISHING

Planning, drafting, and writing at their workstations, writers themselves are increasingly responsible for all stages in document preparation. Because they combine word processing, typesetting, and graphics, *desktop publishing* (DTP) systems mean less reliance on inputters (word processing clerical/secretarial staff, etc.), print shops, and graphic artists.

Types of Geothermal Heat Pump Systems

The energy crisis has intensified research into geothermal heating. As a result, several types of geothermal heat pump (GHP) systems have been developed. This report considers two main types of GHP systems, open loop and closed loop. Each of the types can be installed with vertical or horizontal piping.

In an open loop system, groundwater is drawn from an aquifer through a well and is passed through the pump's heat exchanger. Then, the water is pumped back to the aquifer through a second well, at some distance from the first. An open loop system can therefore present problems—local groundwater chemicals can foul the heat pump's exchanger and distributed water.

Therefore, closed loop systems have become more common; when properly installed, closed loops are economical, efficient, and reliable (Rafferty 10). Such systems circulate a water/anti-freeze solution through a continuous buried pipe. The length of piping depends on ground temperature, the ground's thermal conductivity, soil moisture, and system design.

Variations of vertical and horizontal pipe are used. Horizontal closed loop installations are cost effective for small installations, if sufficient land is available. Pipes are buried in trenches; up to six pipes are placed in each trench, with 10 to 15 feet between trenches. By contrast, vertical loops are used for large installations, or for situations where the soil is too shallow for trenching, or for installations where not enough land is available for horizontal trenches. In closed loop systems, a U-tube is installed in a drill hole 100 to 400 feet deep. Most installations require several drill holes; the pipes are joined in parallel or series-parallel patterns.

Two variations of the horizontal type are gaining favour. Pond closed loop systems place loops on the bottom of a pond or stream; others have also appeared in lakes or offshore in the ocean. A special variation is called "slinky" loops—overlapping coils of polyethylene pipe increase the heat exchange per foot of trench. Slinky coil systems can be placed in earth trenches or in water.

Figure 13.1 Ineffective Page Design

TYPES OF GEOTHERMAL HEAT PUMP SYSTEMS

The energy crisis has intensified research into geothermal heating. As a result, several types of geothermal heat pump (GHP) systems have been developed. This report considers two main types of GHP systems, open loop and closed loop. Each of the types can be installed with vertical or horizontal piping.

Open Loop Systems

In an open loop system, groundwater is drawn from an aquifer through a well and is passed through the pump's heat exchanger. Then, the water is pumped back to the aquifer through a second well, at some distance from the first. An open loop system can therefore present problems—local groundwater chemicals can foul the heat pump's exchanger and distributed water.

Closed Loop Systems

The problems with open loops mean that closed loop systems have become more common; when properly installed, closed loops are economical, efficient, and reliable (Rafferty 10). Such systems circulate a water/anti-freeze solution through a continuous buried pipe. The length of piping depends on

- ground temperature,
- the ground's thermal conductivity,
- soil moisture, and
- system design.

Horizontal closed loop

Horizontal closed loop installations are cost effective for small installations, **if** sufficient land is available. Pipes are buried in trenches; up to six pipes are placed in each trench, with 10 to 15 feet between trenches.

Two variations of the horizontal type are gaining favour.

1. Pond closed loop systems place loops on the bottom of a pond or stream; others have also appeared in lakes or offshore in the ocean.

2. "Slinky" loops (overlapping coils of polyethylene pipe) increase the heat exchange per foot of trench. Slinky coil systems can be placed in earth trenches or in water.

Vertical closed loop

Unlike the horizontal systems, vertical loops are used for large installations, or for situations where the soil is too shallow for trenching, or for installations where not enough land is available for horizontal trenches. In closed loop systems, a U-tube is installed in a drill hole 100 to 400 feet deep. Most installations require several drill holes; the pipes are joined in parallel or series-parallel patterns.

Figure 132 **Effective Page Design**

Using page-design software, scanners, and laser printers, writers at their desks can control the entire production cycle: designing, illustrating, laying out, and printing the final document:

- Text can be keyed or scanned into the program and then edited, checked for spelling and grammar, displayed in columns or other spatial arrangements, set in a variety of sizes and typefaces—or sent electronically.
- Page highlights and orienting devices can be added: headings, ruled boxes, vertical or horizontal rules, coloured background screens, marginal sidebars or labels, page-locator tabs, shadowing, shading, and so on.
- Images can be drawn directly or imported into the program via scanners; charts, graphs, and diagrams can be created with graphics programs. These visuals then can be enlarged, reduced, cropped, and pasted electronically on the text pages.
- At any point in the process, entire pages can be viewed and evaluated for visual appeal, accessibility, and emphasis, then revised as needed.
- All work at all stages can be stored in the computer for later use, adaptation, or upgrading. Documents or parts of documents used repeatedly (*boiler plate*) can be retrieved as needed, modified, or inserted in some other document.

You might work alone or in collaboration with colleagues and graphic design professionals. In either case, you will need to understand the basic principles of effective page design.

As automated design continues to improve the look of workplace writing, audiences raise their expectations for inviting and accessible documents.

PAGE-DESIGN GUIDELINES

Approach your design decisions from the top down. First, consider the overall look of your pages; next, the shape of each paragraph; and finally, the size and style of individual letters and words (Kirsh 112). Figure 13.3 depicts how design considerations follow a top-down sequence, moving from large matters to small. (All design considerations are influenced by the size of the budget for a publication. For instance, adding a single colour to major heads can double the printing cost.)

If your organization prescribes no specific guidelines, the following design principles should satisfy most readers' expectations.

SHAPING THE PAGE

In shaping a page, we consider its look, feel, and overall layout. The following suggestions will help you shape appealing and usable pages.

Use the Right Paper and Ink.   For routine documents (memos, letters, in-house reports), key or print in black ink, on 21.5 cm x 28 cm (8 1/2" x 11") low-gloss, white paper. Use rag-bond paper (20 pound or heavier) with a high fibre content (25 percent minimum). Shiny paper produces glare and tires the eyes. Flimsy or waxy paper feels inferior.

For documents that will be published (manuals, marketing literature, etc.), the grade and quality of paper are important considerations. Paper varies in weight,

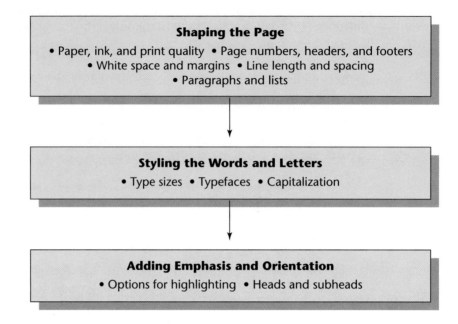

Figure 13.3   A Flowchart for Decisions in Page Design

grain, and finish—from low-cost newsprint, with noticeable wood fibre, to wood-free, specially coated paper with custom finishes. Choice of paper finally depends on the artwork to be included, the type of printing, and the intended esthetic effect; for example, specially coated, heavyweight, glossy—for an elegant effect in an annual report (Cotton 73).

Use High-quality Type or Print.   Print hard copy on an inkjet or laser printer. If your inkjet's output is blurry, consider purchasing special inkjet paper for the best output quality.

Use Consistent Page Numbers, Headers, and Footers.   For a long document, count your title page as page i, without numbering it, and number all front-matter pages, including the table of contents and abstract, with lower case Roman numerals (ii, iii, iv). Number the first text page and subsequent pages with Arabic numerals (1, 2, 3). Along with page numbers, *headers* or *footers* appear in the top or bottom page margins, respectively. These provide chapter or article titles, authors' names, dates, or other publication information. (See, for example, the headers at the top of the pages in this book.)

Use Adequate White Space.   White space is all of the space not filled by text. White space divides printed areas into small, digestible chunks. For instance, it separates sections in a document, headings and visuals from text, and paragraphs on a page (Figure 13.4). White space can be designed to enhance a document's appearance, clarity, and emphasis.

Well-designed white space imparts a shape to the whole document, a shape that orients readers and lends a distinctive visual form to the printed matter by

1. keeping related elements together
2. isolating and emphasizing important elements
3. providing breathing room between blocks of information

Use white space to orient readers

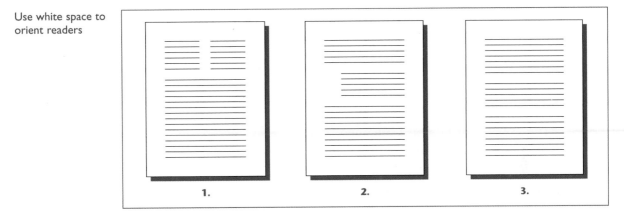

Figure 13.4   Using White Space

Pages that look uncluttered, inviting, and easy to follow convey an immediate sense of user-friendliness.

Provide Ample and Appropriate Margins.   Small margins make a page look crowded and difficult. On your 21.5 cm x 28 cm (8½" × 11") page, leave margins of at least 2.5 cm (1"). For a document that will be bound, widen the left margin an extra 1.25 cm (½").

Choose between *unjustified* text (uneven or "ragged" right margins) and *justified* text (even right margins). Each arrangement creates its own "feel."

Justified lines

In right justified text, the spaces vary between words and letters on a line, sometimes creating channels or rivers of white space. The reader's eyes are then forced to adjust continually to these space variations within a line or paragraph. Because each line ends at an identical vertical space, the eyes must work harder to differentiate one line from another (Felker 85). Moreover, to preserve the even margin, words at ends of lines are often hyphenated, and frequently hyphenated line endings can be distracting.

Unjustified lines

Unjustified text, on the other hand, has equal spacing between letters and words on a line, and an uneven right margin. For some readers, a ragged right margin makes reading easier. These differing line lengths can prompt the eye to move from one line to the next line (Pinelli 77). In contrast to justified text, an unjustified page seems to look less formal, less distant, and less official.

Justified text seems preferable for books, or annual reports, or newsletters, or other publications that use two columns. Unjustified text seems preferable for more personal forms of communication such as letters, memos, and in-house reports.

Keep Line Length Reasonable. Long lines tire the eyes. The longer the line, the harder it is for readers' eyes to return to the left margin and locate the next line (White, *Visual Design* 25).

Notice how your eye labours to follow the apparently endless message that here seems to stretch in lines that continue long after your eye was prepared to move down to the next line. After reading more than a few of these lines, you begin to feel tired and bored and annoyed, without hope of ever reaching the end.

Short lines force the eyes to move back and forth (Felker 79). "Too-short lines disrupt the normal horizontal rhythm of reading" (White, *Visual Design* 25).

> Lines that are too
> short cause your eye
> to stumble from one
> fragment to another
> at a pace that too
> soon becomes
> annoying, if not
> nauseating.

A reasonable line length is 70–80 characters (or 12 to 15 words) per line for a 21.5 cm x 28 cm ($8^1/_2$" x 11") single-column page. The number of characters will depend on print size. Longer lines call for larger type and wider spacing between them (White, *Great Pages* 70).

Line length, of course, is affected by the number of columns (vertical blocks of print) on your page. Two-column pages often appear in newsletters and brochures, but research indicates that single-column pages work best for complex, specialized information (Hartley 148).

Keep Line Spacing Consistent For any document likely to be read completely (letters, memos, instructions), single space within paragraphs and double space between. Instead of indenting the first line of single-spaced paragraphs, separate them with a line of space.

Tailor Each Paragraph to Its Purpose. Readers often skim a long document to find what they want. Most paragraphs therefore begin with a topic sentence forecasting the content.

Shape each paragraph

Use a long paragraph (no more than 15 lines) for clustering material that is closely related (such as history and background, or any body of information best understood in one block).

Use short paragraphs for making complex material more digestible, for giving step-by-step instructions, or for emphasizing vital information.

Instead of indenting a series of short paragraphs, separate them by inserting an extra line of space (as here).

Avoid *widow* and *orphan* lines: The last line of a paragraph printed at the top of a page is called a *widow.* The first line of a paragraph printed on the bottom of a page is called an *orphan.*

Make Lists for Easy Reading. Readers prefer information in list form rather than in continuous prose paragraphs (Hartley 51).

Types of items you might list include advice or examples, conclusions and recommendations, criteria for evaluation, errors to avoid, materials and equipment for a procedure, parts of a mechanism, or steps or events in a sequence. Notice how the preceding information becomes easier to grasp and remember when displayed in the list below.

Types of items you might list:

◆ advice or examples
◆ conclusions and recommendations
◆ criteria for evaluation
◆ errors to avoid
◆ materials and equipment for a procedure
◆ parts of a mechanism
◆ steps or events in a sequence

A list of brief items usually needs no punctuation at the end of each line. A list of full sentences or questions requires appropriate punctuation after each item.

Depending on the list's contents, set off each item with some kind of visual or verbal signal. If the items require a strict sequence (as in a series of steps, or parts of a mechanism), use Arabic numbers (1, 2, 3) or the words *First, Second, Third,* and so on. If the items require no strict sequence (as in the bulleted list above), use dashes, asterisks, or bullets. For a checklist use open boxes. See Figure 13.5.

Use lists to help readers organize their understanding

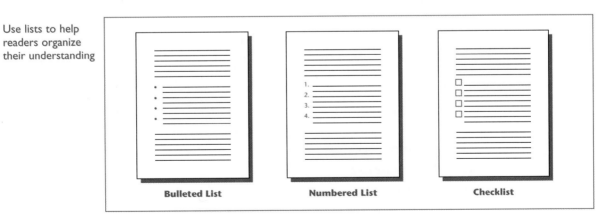

Bulleted List   **Numbered List**   **Checklist**

Figure 13.5   Types of Lists

Introduce your list with an explanation. Phrase all listed items in parallel grammatical form. If the items suggest no strict sequence, try to impose some logical ranking (most important to least important, alphabetical, or some such). Set off the list with extra white space above and below.

Keep in mind that a document with too many lists appears busy, disconnected, and splintered (Felker 55). And long lists could be used by unethical writers to camouflage bad or embarrassing news (Williams, R. 12). As with all format options, use restraint and good judgment.

STYLING THE WORDS AND LETTERS

In styling words and letters, we consider typographic choices that will make the text easy to read.

Use Standard Type Sizes.   Word-processing programs offer a wide variety of type sizes:

Select the
appropriate type
size

9 point

10 point

12 point

14 point

18 point

24 point

The standard type size for most documents is 10 to 12 point. Use larger or smaller sizes for headings, titles, captions (brief explanation of a visual), sidebars (marginal comments), or special emphasis. Use a consistent type size for similar elements throughout the document.

Select Appropriate Typefaces.   A typeface, commonly referred to, in the world of digital type, as a font, is the style of individual letters and characters. Each typeface has its own *personality:* "The typefaces you select for . . . [heads], subheads, body copy, and captions affect the way readers experience your ideas" (*Aldus Guide* 24).

Particular typefaces can influence reading speed by as much as 30 percent (Chauncey 26).

Word-processing programs offer a variety of typefaces like the examples below, listed by name.

Select a typeface for
its personality

11-point New York

11-point Courier

11-point Palatino

11-point Geneva

11-point Monaco

11-point Chicago

11-point Helvetica

11-point Times

Can you assign a personality to each of the above typefaces?

For greater visual unity, try to use different sizes and versions (**bold**, *italic*, e x p a n d e d, condensed) of the same typeface throughout your document—with the possible exception of headings, captions, sidebars, or visuals.

All typefaces divide into two broad categories: *serif* and *sans serif*. Serifs are the fine lines that extend horizontally from the main strokes of a letter.

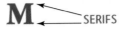

Decide between serif and sans serif type

Serif type makes body copy more readable because the horizontal lines "bind the individual letters" and thereby guide the reader's eyes from letter to letter—as in the type you are now reading (White, *Visual Design 14*).

In contrast, sans serif type is purely vertical (like this). Clean looking and "businesslike," sans serif is considered ideal for technical material (numbers, equations, etc.), marginal comments, headings, examples, tables, and captions to pictures and visuals, and any other material set off from the body copy (White 16).

European readers generally prefer sans serif type throughout their documents, and other cultures have their own preferences as well. Learn all you can about the design conventions of the culture you are addressing.

Except for special emphasis, stick to the more conservative styles and avoid ornate ones altogether.

Avoid Sentences in Full Caps.   Sentences or long passages in full capitals (uppercase letters) are difficult to read because all uppercase letters and words have the same visual outline (Felker 87). The longer the passage, the harder readers work to grasp your emphasis.

MY DOG HAS MANY FLEAS.

My dog has many fleas.

FULL CAPS are good for emphasis but hard to read

Hard   ACCORDING TO THE NATIONAL COUNCIL ON RADIATION PROTECTION, YOUR MAXIMUM ALLOWABLE DOSE OF LOW-LEVEL RADIATION IS 500 MILLIREMS PER YEAR.

Easier   According to the National Council on Radiation Protection, your MAXIMUM allowable dose of low-level radiation is 500 millirems per year.

Lowercase letters take up less space, and the distinctive shapes make each word easier to recognize and remember (Benson 37).

Use full caps as section headings (INTRODUCTION) or to highlight a word or phrase (WARNING: NEVER TEASE THE ALLIGATOR). As with other highlighting options discussed below, use full caps sparingly in your document.

HIGHLIGHTING FOR EMPHASIS

Effective highlighting helps readers distinguish the important from the less important elements. Highlighting options include typefaces, type sizes, white space, and other graphic devices that:

- emphasize key points
- make headings prominent
- separate sections of a long document
- set off examples, warnings, and notes

You can highlight with <u>underlining,</u> FULL CAPS, dashes, parentheses, and asterisks.

You can indent to set off examples, explanations, or any material that should be distinguished from other elements in your document.

Using ruled (or keyed) horizontal lines, you can separate sections in a long document:

Using ruled lines, broken lines, or ruled boxes, you can set off crucial information such as a warning or a caution:

Caution: A document with too many highlights can appear confusing, disorienting, and tasteless.

In adding emphasis and orientation, consider design elements that will direct readers' focus and help them navigate the text. See pages 281–284 for more on background screens, ruled lines, and ruled boxes.

Word-processing software offers highlighting options that might include **boldface,** *italics,* small caps, varying type sizes and typefaces, and colour. For specific highlighted items, some options are better than others:

Not all highlighting is equal

Boldface works well for emphasizing a single sentence or brief statement, and is perceived by readers as being "authoritative" (*Aldus Guide* 42).

Italics suggest a more subtle or "refined" emphasis than boldface (Aldus Guide 42). Italics can highlight words, phrases, book titles, etc. But multiple lines (like these) of italic type are difficult to read.

SMALL CAPS WORK FOR HEADINGS AND SHORT PHRASES. BUT ANY LONG STATEMENT ALL IN CAPS IS DIFFICULT TO READ.

Small type sizes (usually sans serif) work well for captions, labels for visuals, or to set off other material from the body copy.

Large type sizes and dramatic typefaces are both difficult to miss and difficult to digest. Be conservative—unless you really need to convey forcefulness.

Colour is appropriate only in some documents, and only when used sparingly. (Pages 281–284 discuss how colour or shading can influence readers' perception and interpretation of a message.)

Whichever highlights you select for a document, be consistent. Make sure that all headings at one level are highlighted identically, that all warnings and cautions are set off identically, and so on. Use the "styles" feature of your word processor to ensure this consistency.

Never combine too many highlights.

Using Headings for Access and Orientation

Readers of long documents often look back or jump ahead to sections that interest them most.

Headings announce how a document is organized, point readers to what they need, and divide the document into accessible blocks or "chunks." An informative heading can help a reader decide whether a section is worth reading (Felker 17). Besides cutting down on reading and retrieval time, headings help readers remember information (Hartley 15).

Make Headings Informative. A heading should be informative but not wordy. Informative headings orient readers, showing them what to expect. Vague or general headings can be more misleading than no headings at all (Redish et al. 144). Whether your heading takes the form of a phrase, a statement, or a question, be sure it advances thought.

Uninformative leading	Document Formatting

What should we expect here: specific instructions, an illustration, a discussion of formatting policy in general? We can't tell.

Informative versions	How to Format Your Document
	Format Your Document in This Way
	How Do I Format My Document?

When you use questions as headings, phrase the questions in the same way as readers might ask them.

Make Headings Specific as Well as Comprehensive.   Focus the heading on a specific topic. Do not preface a discussion of the effects of acid rain on lake trout with a broad heading such as "Acid Rain." Use instead "The Effects of Acid Rain on Lake Trout."

Also, provide enough headings to contain each discussion section. If chemical, bacterial, and nuclear wastes are three *separate* discussion items, provide a heading for each. Do not simply lump them under the sweeping heading "Hazardous Wastes." If you have prepared an outline for your document, adapt major and minor headings from it.

Make Headings Grammatically Consistent.   All major topics or all minor topics in a document share equal rank; to emphasize this equality, express topics at the same level in identical—or parallel—grammatical form.

Non-parallel headings	How to Avoid Damaging Your Disks:
	1. Clean Disk-Drive Heads
	2. Keep Disks Away from Magnets
	3. Refraining from Exposing the Magnetic Surface
	4. It is Crucial that Disks Be Kept Away from Heat
	5. Disks Should Be Kept Out of Direct Sunlight
	6. Keep Disks in a Dust-free Environment

In items 3, 4, and 5, the lack of parallelism (no verbs in the imperative mood) obscures the relationship between individual steps and confuses readers. This next version emphasizes the equal rank of these items.

Parallel headings	3. Refrain from Exposing the Magnetic Surface
	4. Keep Disks Away from Heat
	5. Keep Disks Out of Direct Sunlight

Parallelism helps make a document readable and accessible.

Make Headings Visually Consistent.   "Wherever heads are of equal importance, they should be given similar visual expression, because the regularity itself becomes an understandable symbol" (White, *Visual Design* 104). Use identical type size and typeface for all headings at a given rank. (Use the "styles" feature of your word-processing software.)

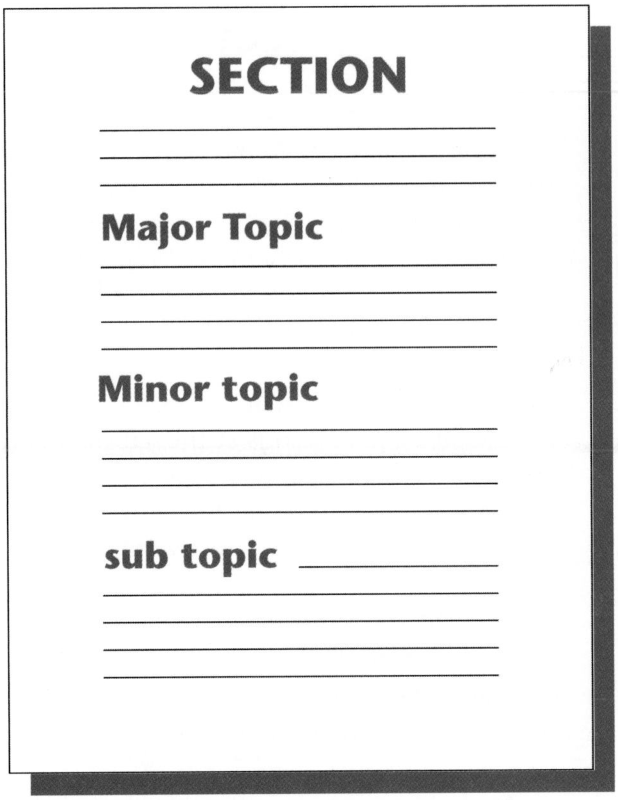

SECTION

Major Topic

Minor topic

sub topic _____

Figure 13.6 **Four Levels of Headings**

Lay Out Headings by Rank. Like a road map, your headings should announce clearly the large and small segments in your document. (Use the logical divisions from your outline as a model for heading layout.) Think of each heading at a particular rank as an "event in a sequence" (White, _Visual Design_ 95).

Figure 13.7 shows how headings vary in positioning and highlighting, depending on their rank. However, because of space considerations, Figure 13.7 does not show that each higher-level heading yields at least two lower-level headings.

SECTION HEADING

In formal reports, always centre section headings at the top of a new page. Use a type size roughly 4 points larger than body copy (say, 16-point section heads for 12-point body copy). Avoid *overly* large heads, and use no other highlights. Fully capitalize the heading. (Some documents use colour for section headings and capitalize just the first letter of each word.) Leave a full line space above the following text (as in this example).

Major Topic Heading

Place major topic headings at the left margin (flush left), and begin each word with an uppercase letter. Use a type size roughly 2 points larger than body copy, in boldface. Start the copy immediately below the heading (as shown), or leave one space below the heading.

Minor topic heading

Use boldface and the same type size as the body copy, with no other highlights. Start the copy immediately below the heading (as shown), or leave one space below the heading.

sub-topic heading. Incorporate sub-topic headings into the body copy they head. Place sub-topic heads flush left and set them off with a period. Use boldface and the same type size as in the body copy, with no other highlights.

1. *alternative sub-topic heading.* If numbering is appropriate, place the sub-topics in a list, with the numbers flush left and the body copy indented. Use italics *and* boldface if you want to draw particular attention to this fourth level of heading.

- **bulleted variation.** When the sequence of items in a list is not important, use bullets to precede the indented sub-topic headings.

Figure 13.7   **Recommended Format for Headings**

Many variations of the above format are used successfully. One such variation is shown in Figure 13.8, page 307. Another variation is demonstrated by the engineering report in Figure 19.6, starting on page 458. That report uses decimal notation, which is discussed in Chapter 10, page 200.

SECTION HEADING

Use a type size roughly 4 points larger than body copy (say, 16-point section heads for 12-point body copy). Use colour to draw attention to the main heading.

Major Topic Heading

Place major topic headings at the left margin (flush left), and begin each word with an uppercase letter. Use a type size roughly 2 points larger than body copy, in boldface. Start the text immediately below the heading.

Minor topic heading

Indent minor topic headings. Use boldface and the same type size as the body copy, with no other highlights. Start the body copy immediately below the heading (as shown).

sub-topic heading. Incorporate sub-topic headings into the body copy they head. Indent sub-topic heads two tabs and set them off with a period. Use boldface and the same type size as in the body copy, with no other highlights.

Figure 13.8   Alternate Format for Headings

The layout in Figures 13.7 and 13.8 embody the following guidelines:

♦ *Ordinarily, use no more than four levels of heading (section, major topic, minor topic, sub-topic).* Excessive heads and subheads can make a document seem cluttered or fragmented.

♦ *Make sure that the headings system has a consistent, logical progression.* Each succeeding heading level in Figure 13.7 and Figure 13.8 has less capitalization than the preceding level.

♦ *Never begin the sentence right after the heading with "this," "it," or some other pronoun referring to the heading.* Make the sentence's meaning independent of the heading.

♦ *Use boldface for all headings.*

♦ *Never leave a heading floating on the final line of a page.* Unless two lines of text can fit below the heading, carry it over to the top of the next page.

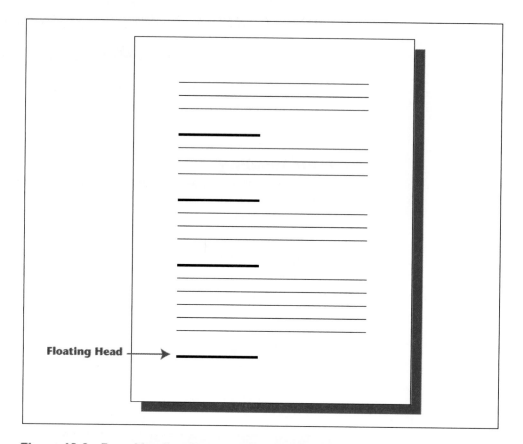

Figure 13.9   Poor Heading Format (Floating Head)

AUDIENCE CONSIDERATIONS IN PAGE DESIGN

Like any writing decisions, page design choices are by no means random. An effective writer designs a document for specific use by a specific audience.

In deciding on a format, work from a detailed audience/purpose profile (Wight 11). Know who your readers are and how they will use your information. Design a document to meet their particular needs and expectations, as in these examples:

- If readers will use your document for reference only (as in a repair manual), make sure you have plenty of headings.
- If your relationship with your readers is formal, use *topical headings* ("Advantages of Treated Pipe"); however, if that relationship is more direct and informal, consider using *talking headings* ("We Should Buy Treated Pipe"). For examples of both topical and talking headings, see the sample reports in Chapter 19 and Chapter 21.
- If readers will follow a sequence of steps, show that sequence in a numbered list.
- If readers will need to evaluate something, give them a checklist of criteria (as in this book at the end of most chapters).

- If readers need a warning, highlight the warning so that it cannot possibly be overlooked.
- If readers have asked for your one-page résumé, save space by using a 10-point type size.
- If readers will be facing complex information or difficult steps, widen the margins, increase all white space, and shorten the paragraphs.

How effectively you combine your options will depend mostly on how carefully you have analyzed your audience. But regardless of the audience, never make the document look "too intellectually intimidating" (White, *Visual Design* 4).

Consider also your audience's specific cultural expectations. For instance, Arabic and Persian text is written from right to left instead of left to right (Leki 149). In other cultures, readers move up and down the page, instead of across. Be aware that a particular culture might be offended by certain icons or by a typeface that seems too plain or too fancy (Weymouth 144). Ignoring a culture's design conventions can be interpreted as a sign of disrespect.

Finally, keep in mind that even the most brilliant page design cannot redeem a document whose content is worthless, organization chaotic, or style unreadable. The value of a document ultimately depends on elements that are more than skin deep.

DESIGNING ON-SCREEN PAGES

To be read on a computer screen, pages must accommodate small screen size, reduced resolution, and reader resistance to scrolling—among other restrictions. Some special design requirements of on-screen pages are:

Elements of
on-screen page
design

- Sentences and paragraphs are short and more concise than their hard-copy equivalents.
- The main point usually appears right up front on each screen.
- Each "page" often stands alone as a discrete "module," or unit of meaning. Instead of a traditional introduction-body-conclusion sequence of pages, material is displayed in screened-sized chunks, each linked as hypertext.
- Links, navigation bars, hot buttons, and help options are displayed on each page.

Special authoring software (Adobe FrameMaker® or RoboHelp®) automatically converts hard-copy documents to various on-screen formats, chunked and linked for easy navigation.

Provide margins for on-screen pages, so that your text won't bump against (or run off) the edge of the user's screen. Also, avoid using underlines for emphasis because these might be confused with hyperlinks (Munger).

Checklist FOR PAGE DESIGN

Use this checklist to revise your page design.

- ◆ Is the paper white, low-gloss, rag bond, with black ink?
- ◆ Is all type or print neat and legible?
- ◆ Does the white space adequately orient the readers?
- ◆ Are the margins ample?
- ◆ Is the line length reasonable?
- ◆ Is the right margin unjustified?
- ◆ Is the line spacing appropriate and consistent?
- ◆ Does each paragraph begin with a topic sentence? If not, why not?
- ◆ Does the length of each paragraph suit its subject and purpose?
- ◆ Are all paragraphs free of widow and orphan lines?
- ◆ Do parallel items in strict sequence appear in a numbered list?

- ◆ Do parallel items of any kind appear in a list whenever a list is appropriate?
- ◆ Are pages numbered consistently?
- ◆ Is the body type size 10 to 12 points?
- ◆ Do full caps highlight only words or short phrases?
- ◆ Is the highlighting consistent, tasteful, and subdued?
- ◆ Are all format patterns distinct enough so that readers will find what they need?
- ◆ Are there enough headings for readers to know where they are in the document?
- ◆ Are headings informative, comprehensive, specific, parallel, and visually consistent?
- ◆ Are headings clearly differentiated according to rank?
- ◆ Is the overall design inviting without being overwhelming?
- ◆ Does this design respect the cultural conventions of my audience?

EXERCISES

1. Find an example of effective page design, in a textbook or elsewhere. Photocopy a selection (two or three pages), and attach a memo explaining to your instructor and colleagues why this design is effective. Be specific in your evaluation. Now do the same for an example of poor formatting. Bring your examples and explanations to class, and be prepared to discuss why you chose them.

2. These are headings from a set of instructions for listening. Rewrite the headings to make them parallel.

   - ◆ You Must Focus on the Message
   - ◆ Paying Attention to Non-verbal Communication
   - ◆ Your Biases Should Be Suppressed
   - ◆ Listen Critically
   - ◆ Listening for Main Ideas
   - ◆ Distractions Should Be Avoided
   - ◆ Provide Verbal and Non-verbal Feedback
   - ◆ Making Use of Silent Periods
   - ◆ Are You Allowing the Speaker Time to Make His or Her Point?
   - ◆ Keeping an Open Mind Is Important

3. Using the checklist above, redesign an earlier assignment or a document you've prepared on the job. Submit to your instructor the revision and the original, along with a memo explaining your improvements. Be prepared to discuss your format design in class.

4. Anywhere on campus or at work, locate a document with a design that needs revision. Candidates include career counselling handbooks, financial aid handbooks, student or faculty handbooks, software or computer manuals, medical information, newsletters, or registration procedures. Redesign the document or a two-to-five page selection from it. Submit to your instructor a copy of the original, along with a memo explaining your improvements. Be prepared to discuss your revision in class.

COLLABORATIVE PROJECT

Working in small groups, redesign a document you select or your instructor provides. Prepare a detailed explanation of your group's revision. Appoint a group member to present your revision to the class, using an opaque or overhead projector, a large-screen computer monitor, data projection unit, or photocopies.

Descriptive Writing

Brian Schnitzler, Saskatoon

After a professional hockey career, Brian Schnitzler had no idea that his new career would focus on technical writing. Now, he says "writing is almost all I do." He studied mechanical engineering at Saskatoon's Kelsey Institute and went to work for Cambrian Engineering (now part of Amec Engineering and Construction Services).

After learning the ropes, Brian became a project manager. He now manages approvals processes for oil and gas pipeline construction. These complex projects require constant, clear communications at each stage. Brian gathers background data to present to the client; he helps draftspersons interpret aerial photos, topographical maps, and on-site observations; he distributes the drafted route plans; he requests approvals from, among others, municipalities; landowners; provincial hydro, water, and highways authorities; railways; and the National Energy Board.

In addition to writing long technical tenders, Brian writes a flood of correspondence, especially during an approvals process. In each description, letter, memo, or e-mail, he must "choose exactly the right level of background and plan details. Each reader, each agency, has special requirements and expectations." The secret to successful descriptive writing, he says, is to "know your readers' needs and give them what they want. . . . The approvals process can be long and onerous, so I try to get everything right the first time, to keep the project moving."

CHAPTER 14

Definitions

PURPOSE OF DEFINITIONS

ELEMENTS OF DEFINITIONS

TYPES OF DEFINITION

EXPANSION METHODS

SAMPLE SITUATIONS

PLACEMENT OF DEFINITIONS

Checklist **for Revising and Editing Definitions**

When you define a term, you explain the precise meaning you intend by using that term. Clear writing depends on definitions that both reader and writer understand. Unless you are sure readers know the exact meaning you intend, always define the term upon first use.

PURPOSE OF DEFINITIONS

Every specialty has its own technical language. Engineers, architects, or programmers talk about "torque," "tolerances," or "microprocessors"; lawyers, real estate brokers, and investment counsellors discuss "easements," "liens," "amortization," or "escrow accounts." Whenever such terms are unfamiliar to an audience, they need defining.

For colleagues, you rarely have to define specialized terms (unless the term is new), but reports often are written for the layperson—the client or some other general reader. When you write for non-specialists, clarify your meaning with definitions.

Most of the specialized terms previously mentioned are concrete and specific. Once "microprocessor" has been defined for the reader, its meaning will not differ appreciably in another context. When a term is highly technical, a writer can figure out that it should be defined for some readers. However, familiar terms like "disability," "guarantee," "tenant," "lease," or "mortgage" acquire very specialized meanings in specialized contexts. Here definition becomes crucial. What "guarantee" means in one situation is not necessarily what it means in another. Contracts are detailed (and legal) definitions of the specific terms of an agreement.

Assume you're shopping for disability insurance to protect your income in case of injury or illness. Besides comparing prices, you want each company to define "physical disability." Although Company A offers the cheapest policy, it might define physical disability as inability to work at *any* job. Therefore, if a neurological disorder prevents you from continuing work as designer of electronic devices, without disabling you for some menial job, you might not qualify as "disabled." In contrast, Company B's policy, although more expensive, might define physical disability as inability to work at your *specific* job. Both companies use the term "physical disability," but each defines it differently. Because they are legally responsible for the documents they prepare, all communicators rely on the technique of clear definition.

Growth in technology, specialization, and global communication makes definition increasingly vital. Know for whom you're writing, and why. If you are unable to pinpoint your audience, assume a general readership and define generously.

ELEMENTS OF DEFINITIONS

For all definitions, use these guidelines:

PLAIN ENGLISH

Clarify meaning by using language readers understand.

Unclear	A tumour is a neoplasm.
Better	A tumour is a growth of cells that occurs independently of surrounding tissue and serves no useful function.

Unclear	A solenoid is an inductance coil that serves as a tractive electromagnet. *(A definition appropriate for an engineering manual, but too specialized for general readers.)*
Better	A solenoid is an electrically energized coil that converts electrical energy to magnetic energy capable of performing mechanical functions.

BASIC PROPERTIES

Convey the properties of an item that differentiate it from all others. A thermometer has a singular function: it measures temperature. Without this essential information, a definition would have no real meaning for uninformed readers. Any other data about thermometers (types, special uses, materials used in construction) are secondary. A book, on the other hand, cannot be defined according to functional properties because books have multiple functions. A book can be used to write in or to display pictures, to record financial transactions, to read, and so on. Also, other items (individual sheets of paper, posters, newspapers, picture frames) serve the same functions. The basic property of a book is physical: it is a bound volume of pages. Readers would have to know this *first,* to understand what a book is.

OBJECTIVITY

Unless readers understand that your purpose is to persuade, omit your opinions from a definition. "Bomb" is defined as "an explosive weapon detonated by impact, proximity to an object, a timing mechanism, or other predetermined means." If you define a bomb as "an explosive weapon devised and perfected by hawkish idiots to blow up the world," you are editorializing, *and* ignoring a bomb's basic property.

Likewise, in defining "diesel engine," simply tell what it is and how it works. You might think that diesels are too noisy and sluggish for automobiles, but omit these judgments from your definition.

TYPES OF DEFINITION

Definitions vary greatly in length and detail: from a few words in parentheses, to one or more complete sentences, to multiple paragraphs or pages.

Your choice of definition type depends on what information readers need, and that, in turn, depends on why they need it. "Carburetor," for instance, could be defined in one sentence, briefly telling readers what it is and how it works. But this definition would be expanded for the student mechanic who needs to know the origin of the term, how the device was developed, what it looks like, how it is used, and how its parts interact. Audience needs should guide your choice.

PARENTHETICAL DEFINITION

A parenthetical definition explains the term in a word or phrase, often as a synonym in parentheses following the term:

Parenthetical
definitions

> The effervescent (bubbling) mixture is highly toxic.
> The leaching field (sievelike drainage area) requires crushed stone.

Another option is to express your definition as a clarifying phrase:

> The trees on the site are mostly deciduous; that is, they shed their foliage at season's end.

Use parenthetical definitions to convey the general meaning of specialized terms so that readers can follow the discussion where these terms are used. A parenthetical definition of "leaching field" might be adequate in a progress report to a client whose house you are building. But a public-health report titled "Groundwater Contamination from Leaching Fields" would call for an expanded definition.

SENTENCE DEFINITION

Elements of sentence definitions

A definition may require one or more sentences with this structure: (1) the item or term being defined, (2) the class (specific group) to which the term belongs, and (3) the features that differentiate the term from all others in its class.

Term	Class	Distinguishing Features
carburetor	a mixing device	in gasoline engines that blends air and fuel into a vapour for combustion within the cylinders
transit	a surveying instrument	that measures horizontal and vertical angles
diabetes	a metabolic disease	caused by a disorder of the pituitary gland or pancreas and characterized by excessive urination, persistent thirst, and decreased ability to metabolize sugar
liberalism	a political concept	based on belief in progress, the essential goodness of people, and the autonomy of the individual, and advocating protection of political and civil liberties
brief	a legal document	containing all the facts and points of law pertinent to a specific case, and filed by a lawyer before the case is argued in court
stress	an applied force	that strains or deforms a body
laser	an electronic device	that converts electrical energy to light energy, producing a bright, intensely hot, and narrow beam of light
fibre optics	a technology	that uses light energy to transmit voices, video images, and data through hair-thin glass fibres

These elements can be combined into one or more sentences.

> Diabetes is a metabolic disease caused by a disorder of the pituitary gland or pancreas. This disease is characterized by excessive urination, persistent thirst, and decreased ability to metabolize sugar.

Sentence definition is especially useful for stipulating the precise working meaning of a term that has several possible meanings. State your working definitions at the beginning of your report:

> Throughout this report, the term "disadvantaged student" means . . .

Classifying the Term. Be specific and precise in your classification. The narrower your class, the more specific your meaning. "Transit" is correctly classified as a "surveying instrument," not as a "thing" or an "instrument." "Stress" is classified as "an applied force"; to say that stress "takes place when . . ." or "is something that . . ." fails to reflect a specific classification. Be sure to select precise terms of classification: "Diabetes" is precisely classified as "a metabolic disease," not as "a medical term."

Differentiating the Term. Differentiate the term by separating the item it names from every other item in its class. Make these distinguishing features narrow enough to pinpoint the item's unique identity and meaning, yet broad enough to be inclusive. A definition of "brief" as "a legal document introduced in a courtroom" is too broad because the definition doesn't differentiate "brief" from all other legal documents (wills, written confessions, etc.). Conversely, differentiating "carburetor" as "a mixing device used in automobile engines" is too narrow because it ignores the carburetor's use in all other gasoline engines.

Also, avoid circular definitions (repeating, as part of the distinguishing features, the word you are defining). Thus, "stress" should not be defined as "an applied force that places stress on a body." The class and distinguishing features must express the item's basic property ("an applied force that strains or deforms a body").

Categorical vs. Operational Definitions. So far, our sample definitions have placed the defined term in a category; we could use the term, *categorical definition,* for this common method of defining things in sentences. Categorical definitions are static, but a second type of sentence definition, *operational definition,* defines things in active terms. Here's an example:

> **Technologists** translate engineering designs into working plans and then see that these plans are carried out.

In the example, technologists are defined in terms of *what they do,* rather than in terms of *what they are.*

Operational definitions work best in proposals, progress reports, and résumés because the active verbs contribute to the sense of an active and successful person, plan, or activity. Also, operational definitions use fewer words to convey meaning. Compare the above example to its categorical equivalent:

> A technologist **is someone who** translates engineering designs into working plans and then sees that these plans are carried out.

EXPANDED DEFINITION

An expanded definition can include parenthetical and sentence definitions, but it provides greater detail for readers who need it. The sentence definition of "solenoid" on page 314 is good for a general reader who simply needs to know what a solenoid is. But a manual for mechanics or mechanical engineers would define solenoid in detail (as on pages 322–23); these readers need to know also how a solenoid works and how to use it.

The problem with defining an abstract and general word, such as "condominium" or "loan," is different. "Condominium" is a vaguer term than "solenoid" (solenoid A is pretty much like solenoid B) because the former refers to many types of ownership agreements.

Concrete, specific terms such as "diabetes," "transit," and "solenoid" often can be defined in a sentence, and they require expanded definition only for certain audiences. But terms such as "disability" and "condominium" require an expanded definition for almost any audience. The more general or abstract the term, the more need for an expanded definition.

An expanded definition may be a single paragraph (as for a simple tool) or may extend to scores of pages (as for a digital dosimeter—a device for measuring radiation exposure); sometimes the definition itself *is* the whole report.

EXPANSION METHODS

How you expand a definition depends on the questions you think readers need answered, as shown in Figure 14.1. Begin with a sentence definition, and then use only those expansion strategies that serve your readers' needs.

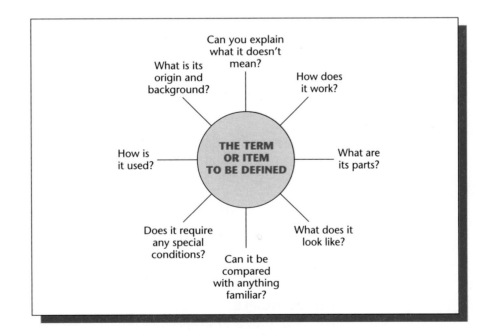

Figure 14.1   Directions in Which a Definition Can Be Expanded

ETYMOLOGY

A word's origin (its development and changing meanings) can clarify its definition. *Biological control* of insects is derived from the Greek "bio," meaning *life* or *living organism,* and the Latin "contra," meaning *against* or *opposite.* Biological control, then, is the use of living organisms against insects. College dictionaries contain etymological information, but your best bet is *The Oxford English Dictionary* and encyclopedic dictionaries of science, technology, and business.

Some technical terms are acronyms, derived from the first letters or parts of several words. *Laser* is an acronym for *light amplification by stimulated emission of radiation.*

Sometimes a term's origin can be colourful as well as informative. *Bug* (jargon for *programming error*) is said to derive from an early computer at Harvard that malfunctioned because of a dead bug blocking the contacts of an electrical relay. Because programmers, like many of us, were reluctant to acknowledge error, the term became a euphemism for *error.* Correspondingly, *debugging* is the correcting of errors in a program.

HISTORY AND BACKGROUND

The meaning of specialized terms such as "radar," "bacteriophage," "silicon chips," or "X-ray" often can be clarified through a background discussion: discovery or history of the concept, development, method of production, applications, etc. Specialized encyclopedias are a good background source.

"Where did it come from?"

> The idea of lasers . . . dates back as far as 212 B.C., when Archimedes used a [magnifying] glass to set fire to Roman ships during the siege of Syracuse. (Gartiganis 22)

"How was it perfected?"

> The early researchers in fibre optic communications were hampered by two principal difficulties—the lack of a sufficiently intense source of light and the absence of a medium which could transmit this light free from interference and with a minimum signal loss. Lasers emit a narrow beam of intense light, so their invention in 1960 solved the first problem. The development of a means to convey this signal was longer in coming, but scientists succeeded in developing the first communications-grade optical fibre of almost pure silica glass in 1970. (Stanton 28)

NEGATION

Readers can grasp some meanings by understanding clearly what the term *does not* mean. For instance, an insurance policy may define coverage for "bodily injury to others" partly by using negation:

> "We will *not* pay: 1. For injuries to guest occupants of your auto."

In this next example, negation clarifies the definition of "lasers":

> "Lasers are not merely weapons from science fiction."

OPERATING PRINCIPLE

Most items work according to an operating principle, whose explanation should be part of your definition:

"How does it work?"

A clinical thermometer works on the principle of heat expansion: As the temperature of the bulb increases, the mercury inside expands, forcing a mercury thread up into the hollow stem.

Air-to-air solar heating involves circulating cool air, from inside the home, across a collector plate (heated by sunlight) on the roof. This warmed air is then circulated back into the home.

Basically, a laser [uses electrical energy to produce] coherent light, light in which all the waves are in phase with each other, making the light hotter and more intense. (Gartiganis 23)

[A fibre optics] system works as follows: An electrical charge activates the laser ..., and the resulting light ... energy passes through the optical fibre. At the other end of the fibre, a ... receiver ... converts this light signal back into electrical impulses. (Stanton 28)

Even abstract concepts or processes can be explained on the basis of their operating principle:

Economic inflation is governed by the principle of supply and demand: If an item or service is in short supply, its price increases in proportion to its demand.

ANALYSIS OF PARTS
When your subject can be divided into parts, identify and explain them:

"What are its parts?"

The standard frame of a pitched-roof wooden dwelling consists of floor joists, wall studs, roof rafters, and collar ties.

Psychoanalysis is an analytic and therapeutic technique consisting of four parts: (1) free association, (2) dream interpretation, (3) analysis of repression and resistance, and (4) analysis of transference.

In discussing each part, of course, you would further define specialized terms such as "floor joists" and "repression."

Analysis of parts is particularly useful for helping non-technical readers understand a technical subject. This next analysis helps explain the physics of lasing by dividing the process into three discrete parts:

1. [Lasers require] a source of energy, [such as] electric currents or even other lasers.
2. A resonant circuit ... contains the lasing medium and has one fully reflecting end and one partially reflecting end. The medium—which can be a solid, liquid, or gas—absorbs the energy and releases it as a stream of photons [electromagnetic particles that emit light]. The photons ... vibrate between the fully and partially reflecting ends of the resonant circuit, constantly accumulating energy—that is, they are amplified. After attaining a prescribed level of energy, the photons can pass through the partially reflecting surface as a beam of coherent light and encounter the optical elements.
3. Optical elements—lenses, prisms, and mirrors—modify size, shape, and other characteristics of the laser beam and direct it to its target. (Gartiganis 23)

Figure 1 shows the three parts of a laser.

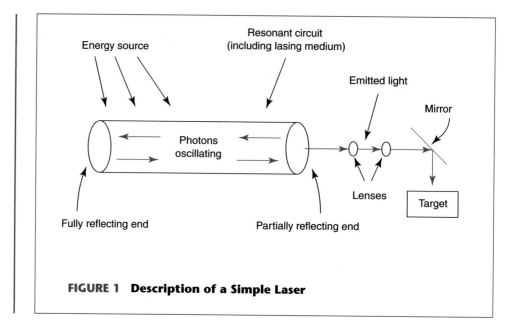

FIGURE 1   Description of a Simple Laser

VISUALS

Well-labelled visuals (such as the laser description) are excellent for clarifying definitions. Always introduce your visual and explain it. If your visual is borrowed, credit the source. Unless the visual takes up one whole page or more, do not place it on a separate page. Include the visual near its discussion.

COMPARISON AND CONTRAST

Comparisons and contrasts help readers understand. Analogies (a type of comparison) to something familiar can help explain the unfamiliar:

"Does it resemble anything familiar?"

> To visualize how a simplified earthquake starts, imagine an enormous block of gelatin with a vertical knife slit through the middle of its lower half. Gigantic hands are slowly pushing the right side forward and pulling the left side back along the slit, creating a strain on the upper half of the block that eventually splits it. When the split reaches the upper surface, the two halves of the block spring apart and jiggle back and forth before settling into a new alignment. Inhabitants on the upper surface would interpret the shaking as an earthquake. ("Earthquake Hazard Analysis" 8)

> The average diameter of an optical cable is around two-thousandths of an inch, making it about as fine as a hair on a baby's head (Stanton 29–30).

Here is a contrast between optical fibre and conventional copper cable:

"How does it differ from comparable things?"

> Beams of laser light coursing through optical fibres of the purest glass can transmit many times more information than the present communications systems. . . . A pair of optical fibres has the capacity to carry more than 10 000 times as many signals as conventional copper cable. A 1.25 cm ($^1/_2$") optical cable can carry as much information as a copper cable as thick as a person's arm. . . .

> Not only does fibre optics produce a better signal, [but] the signal travels farther as well. All communications signals experience a loss of power, or attenuation, as they move along a cable. This power loss necessitates placement of repeaters at 1.5- or 3.0-kilometre intervals of copper cable in order to regenerate the signal. With fibre, repeaters are necessary about every 50 or 65 kilometres, and this distance is increasing with every generation of fibre. (Stanton 27–28)

Here is a combined comparison and contrast:

"How is it both similar and different?"

> Fibre optics technology results from the superior capacity of lightwaves to carry a communications signal. Sound waves, radio waves, and light waves can all carry signals; their capacity increases with their frequency. Voice frequencies carried by telephone operate at 1000 cycles per second, or hertz. Television signals transmit at about 50 million hertz. Light waves, however, operate at frequencies in the hundreds of trillions of hertz. (Stanton 28)

REQUIRED MATERIALS OR CONDITIONS

Some items or processes need special materials and handling, or they may have other requirements or restrictions. An expanded definition should include this important information.

"What is needed to make it work (or occur)?"

> Besides training in engineering, physics, or chemistry, careers in laser technology require a strong background in optics (study of the generation, transmission, and manipulation of light).

Abstract concepts might also be defined in terms of special conditions:

> To be held guilty of libel, a person must have defamed someone's character through written or pictorial statements.

EXAMPLE

Familiar examples showing types or uses of an item can help clarify your definition. This example shows how laser light is used as a heat-generating device:

"How is it used or applied?"

> Lasers are increasingly used to treat health problems. Thousands of eye operations involving cataracts and detached retinas are performed every year by ophthalmologists. . . . Dermatologists treat skin problems. . . . Gynecologists treat problems of the reproductive system, and neurosurgeons even perform brain surgery—all using lasers transmitted through optical fibres. (Gartiganis 24–25)

The next example shows how laser light is used to carry information:

> The use of lasers in the calculating and memory units of computers, for example, permits storage and rapid manipulation of large amounts of data. And audiodisc players use lasers to improve the quality of the sound they reproduce. The use of optical cable to transmit data also relies on lasers. (Gartiganis 25)

And this final example shows how optical fibre can relay a video signal:

> Acting, in essence, as tiny cameras, optical fibres can be inserted into the body and relay an image to an outside screen. (Stanton 28)

Examples are a most powerful communication tool—as long as you tailor the examples to the readers' level of understanding.

Whichever expansion strategies you use, be sure to document your information sources.

SAMPLE SITUATIONS

The following definitions employ expansion strategies appropriate to their audiences' needs. Specific strategies are labelled in the margin. Each definition, like a good essay, is unified and coherent: Each paragraph is developed around one main idea and logically connected to other paragraphs. Visuals are incorporated. Transitions emphasize the connection between ideas. Each definition is at a level of technicality that connects with the intended audience.

To illustrate the importance of audience analysis in a writer's decision about "How much is enough?" this example, like many throughout the text, is preceded by an audience/purpose profile based on the worksheet on page 34.

An Expanded Definition for Semi-technical Readers

AUDIENCE/PURPOSE PROFILE. The intended readers of this material are beginning student mechanics. Before they can repair a solenoid, they will need to know where the term comes from, what a solenoid looks like, how it works, how its parts operate, and how it is used. This definition is designed as merely an *introduction,* so it offers only a general (but comprehensive) view of the mechanism.

Because the intended readers are not engineering students, they do *not* need details about electromagnetic or mechanical theory (e.g., equations or graphs illustrating voltage magnitudes, joules, lines of force). ▲

EXPANDED DEFINITION: SOLENOID

Formal sentence definition

A **solenoid** is an electrically energized coil that forms an electromagnet capable of performing mechanical functions. The term "solenoid" is derived from the word "sole,"

Etymology

which in reference to electrical equipment means "a part of," or "contained inside, or with, other electrical equipment." The Greek word *solenoides* means "channel," or "shaped like a pipe."

Description and analysis of parts

A simple plunger-type solenoid consists of a coil of wire attached to an electrical source, and an iron rod, or plunger, that passes in and out of the coil along the axis of the spiral. A return spring holds the rod outside the coil when the current is de-energized, as shown in Figure 1.

Special conditions and operating principle

When the coil receives electric current, it becomes a magnet and thus draws the iron rod inside, along the length of its cylindrical centre. With a lever attached to its end, the rod can transform electrical energy into mechanical force. The amount of mechanical force produced is the product of the number of turns in the coil, the strength of the current, and the magnetic conductivity of the rod.

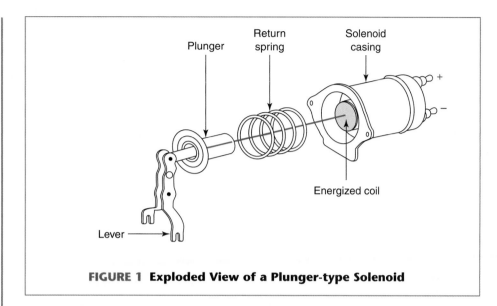

FIGURE 1 Exploded View of a Plunger-type Solenoid

Example and analysis of parts

Explanation of visual

The plunger-type solenoid in Figure 1 is commonly used in the starter motor of an automobile engine. This type is 11.5 cm (4$\frac{1}{2}$") long and 5 cm (2") in diameter, with a steel casing attached to the casing of the starter motor. A linkage (pivoting lever) is attached at one end to the iron rod of the solenoid, and at the other end to the drive gear of the starter, as shown in Figure 2.

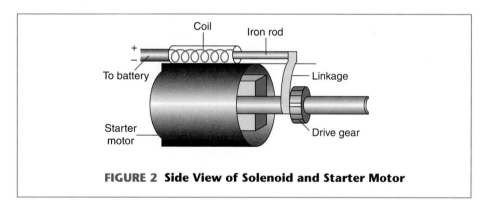

FIGURE 2 Side View of Solenoid and Starter Motor

When the ignition key is turned, current from the battery is supplied to the solenoid coil, and the iron rod is drawn inside the coil, thereby shifting the attached linkage. The linkage, in turn, engages the drive gear, activated by the starter motor, with the flywheel (the main rotating gear of the engine).

Comparison of sizes and applications

Because of the solenoid's many uses, its size varies according to the work it must do. A small solenoid will have a small wire coil, hence a weak magnetic field. The larger the coil, the stronger the magnetic field; in this case, the rod in the solenoid can do harder work. An electronic lock for a standard door would, for instance, require a much smaller solenoid than one for a bank vault.

The audience for the following definition (an entire community) is too diverse to define precisely, so the writer wisely addresses the lowest level of technicality—to ensure that all readers will understand.

An Expanded Definition for Non-technical Readers

AUDIENCE/PURPOSE PROFILE. The following definition is written for members of a community whose water supply (all obtained from wells, because the town has no reservoir) is doubly threatened: (1) by chemical seepage from a recently discovered toxic dump site, and (2) by a two-year drought that has severely depleted the water table. This definition forms part of a report that analyzes the severity of the problems and explores possible solutions.

To understand the problems, these readers first need to know what a water table is, how it is formed, what conditions affect its level and quality, and how it figures into town planning decisions. The concepts of *recharge* and *permeability* are vital to readers' understanding of the problem here, so these terms are defined parenthetically. These readers have no interest in geological or hydrological (study of water resources) theory. They simply need a broad picture. ▲

EXPANDED DEFINITION: WATER TABLE

Formal sentence
definition
Example

Operating principle

The water table is the level below the earth's surface at which the ground is saturated with water. Figure 1 shows a typical water table that might be found in the East. Wells driven into such a formation will have a water level identical to that of the water table.

The world's fresh-water supply comes almost entirely as precipitation that originates with the evaporation of sea and lake water. This precipitation falls to earth and follows one of three courses: It may fall directly onto bodies of water, such as rivers or lakes, where it is directly used by humans; it may fall onto land, and either evaporate or run over the ground to the rivers or other bodies of water; or it may fall onto land, be contained, and seep into the earth. The latter precipitation makes up the water table.

Comparison

Similar in contour to the earth's surface above, the water table generally has a level that reflects such features as hills and valleys. Where the water table intersects the ground surface, a stream or pond results.

Operating principle

A water table's level, however, will vary, depending on the rate of recharge (replacement of water). The recharge rate is affected by rainfall or soil permeability (the ease with which water flows through the soil). A water table then is never static; rather,

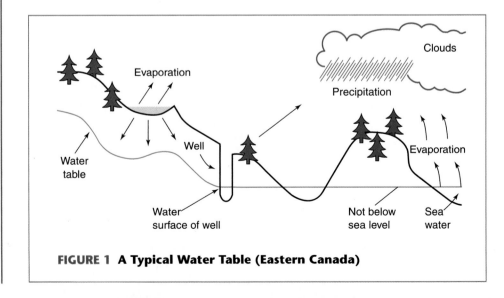

FIGURE 1 A Typical Water Table (Eastern Canada)

Example

Special conditions
and examples

Special conditions

it is the surface of a body of water striving to maintain a balance between the forces which deplete it and those which replenish it. In areas of Nova Scotia and some western provinces where the water table is depleted, the earth caves in, leaving sinkholes.

The water table's depth below ground is vital in water resources engineering and planning. It determines an area's suitability for wastewater disposal, or a building lot's ability to handle sewage. A high water table could become contaminated by a septic system. Also, bacteria and chemicals seeping into a water table can pollute an entire town's water supply. Another consideration in water-table depth is the cost of drilling wells. These conditions obviously affect an industry's or homeowner's decision on where to locate.

The rising and falling of the water table give an indication of the pumping rate's effect on a water supply (drawn from wells) and of the sufficiency of the recharge rate in meeting demand. This kind of information helps water resources planners decide when new sources of water must be made available.

PLACEMENT OF DEFINITIONS

Poorly placed definitions interrupt the information flow. If you have only a few parenthetical definitions, place them in parentheses after the terms. Any more than a few definitions per page will be disruptive. Rewrite them as sentence definitions and place them in a "Definitions" section of your introduction, or in a glossary.

If your sentence definitions are few, place them in a "Definitions" section of your introduction; otherwise, in a glossary. Definitions of terms in the report's title belong in your introduction.

Place expanded definitions in one of three locations:

1. If the definition is essential to the reader's understanding of the *entire* document, place it in the introduction. A report titled "The Effects of Aerosols on the Earth's Ozone Shield" would require expanded definitions of "aerosols" and "ozone shield" early in the discussion.
2. When the definition clarifies a major part of your discussion, place it in that section of your report. In a report titled "How Advertising Influences Consumer Habits," "operant conditioning" might be defined early in the appropriate section. Too many expanded definitions *within* a report, however, can be disruptive.
3. If the definition serves only as a reference, place it in an appendix. For example, a report on fire safety in a public building might have an expanded definition of "carbon monoxide detectors" in an appendix.

Electronic documents pose special problems for placement of definitions. In a hypertext document, for instance, each reader explores the material differently (Chapter 25). One option for making definitions available when they are needed is the "pop-up note": The term to be defined is highlighted in the text, to indicate that its definition can be called up and displayed in a small window on the actual text screen (Horton 25).

Checklist FOR EDITING AND REVISING DEFINITIONS

Use this checklist to revise your definitions.

Content

- Is the type of definition (parenthetical, sentence, expanded) suited to its purpose and readers' needs?
- Does the definition convey the basic property of the item?
- Is the definition objective?
- Is the expanded definition adequately developed?
- Are all information sources documented?
- Are visuals employed adequately and appropriately?

Arrangement

- Does the sentence definition describe features that distinguish the item from other items in the same class?

- Is the expanded definition unified and coherent (like an essay)?
- Are transitions between ideas adequate?
- Does the definition appear in the appropriate location?

Style and Page Design

- Is the definition in plain English?
- Will the level of technicality connect with the audience?
- Are sentences clear, concise, and fluent?
- Is word choice precise?
- Is the definition ethically acceptable?
- Is the page design inviting and accessible?

EXERCISES

1. Sentence definitions require precise classification and differentiation. Is each of these definitions adequate for a general reader? Rewrite those that seem inadequate. Consult dictionaries and encyclopedias as needed.

   a. A bicycle is a vehicle with two wheels.
   b. A transistor is a device used in transistorized electronic equipment.
   c. Surfing is when one rides a wave to shore while standing on a board specifically designed for buoyancy and balance.
   d. Bubonic plague is caused by an organism known as *pasteurella pestis.*
   e. Mace is a chemical aerosol spray used by the police.
   f. A Geiger counter measures radioactivity.
   g. A cactus is a succulent.
   h. In law, an indictment is a criminal charge against a defendant.
   i. A prune is a kind of plum.
   j. Friction is a force between two bodies.
   k. Luffing is what happens when one sails into the wind.
   l. A frame is an important part of a bicycle.
   m. Hypoglycemia is a medical term.
   n. An hourglass is a device used for measuring intervals of time.
   o. A computer is a machine that handles information with amazing speed.
   p. A Ferrari is the best car in the world.
   q. To meditate is to exercise mental faculties in thought.

2. Standard dictionaries define for the general reader, whereas specialized reference books define for the specialist. Choose an item in your field and copy the definition (1) from a standard dictionary and (2) from a technical reference book. For the technical definition, label each expansion strategy. Rewrite the specialized definition for a general reader.

3. Using reference books as necessary, write sentence definitions for these terms or for terms from your field.

biological insect control
generator
dewpoint
microprocessor
capitalism
local area network
marsh
artificial intelligence
economic inflation
anorexia nervosa
low-impact camping
hemodialysis

gyroscope
coronary bypass
oil shale
chemotherapy
estuary
Boolean logic
classical conditioning
hypothermia
thermistor
aquaculture
nuclear fission
modem

COLLABORATIVE PROJECT

Divide into small groups on the basis of academic majors or interests. Appoint one person as group manager. Decide on an item, concept, or process that would require an expanded definition for a layperson. For example:

◆ From computer science: an algorithm, an applications program, artificial intelligence, binary coding, top-down procedural thinking, or systems analysis

◆ From nursing: a pacemaker, coronary bypass surgery, or natural childbirth

Complete an audience/purpose profile (page 34).

Once your group has decided on the appropriate expansion strategies (etymology, negation, etc.), the group manager will assign each member to work on one or two specific strategies as part of the definition. As a group, edit, and incorporate the collected material into an expanded definition, revising as often as needed.

The group manager will assign one member to present the definition in class, using either opaque or overhead projection, a large-screen monitor, data projection unit, or photocopies.

CHAPTER 15

Descriptions and Specifications

PURPOSE OF DESCRIPTION

OBJECTIVITY IN DESCRIPTION

ELEMENTS OF DESCRIPTION

A GENERAL MODEL FOR DESCRIPTION

A SAMPLE SITUATION

SPECIFICATIONS

Checklist **for Revising and Editing Descriptions**

Description (creating a picture with words) is part of all writing. But technical descriptions convey information about a product or mechanism to someone who will use it, buy it, operate it, assemble it, manufacture it, or to someone who has to know more about it. Any item can be visualized from countless different perspectives. Therefore, *how* you describe—your perspective—depends on your purpose and the needs of your audience.

Two kinds of description are featured in this chapter, mechanism description and specifications. We start with a mechanism description (see Figure 15.1).

PURPOSE OF DESCRIPTION

Manufacturers use descriptions to sell products; banks require detailed descriptions of any business or construction venture before approving a loan; and medical personnel maintain daily or hourly descriptions of a patient's condition and treatment.

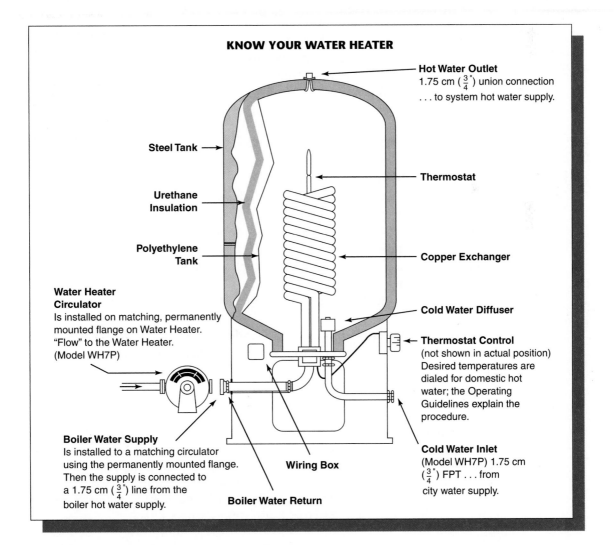

Figure 15.1   A Mechanism Description

Source: Courtesy of AMTROL Inc.

No matter what the subject of description, readers expect answers to as many of these questions as are applicable: *What is it? What does it do? What does it look like? What is it made of? How does it work? How was it put together?* The description in Figure 15.1, part of an installation and operation manual, answers applicable questions for do-it-yourself homeowners.

OBJECTIVITY IN DESCRIPTION

Each description is mainly *subjective* or *objective:* based on feelings or fact. Subjective description emphasizes the writer's attitude toward the thing, whereas objective description emphasizes the thing itself.

Essays describing "An Unforgettable Person" or "A Beautiful Moment" express opinions, a personal point of view. Subjective description aims at expressing feelings, attitudes, and moods. You create an *impression* of your subject ("The weather was miserable"), more than communicating factual information about it ("All day, we had freezing rain and gale-force winds").

Objective description shows an impartial view, filtering out personal impressions and focusing on observable details.

Except for promotional writing, descriptions on the job should be impartial, if they are to be ethical. Pure objectivity is, of course, humanly impossible. Each writer filters the facts and their meaning through her or his own perspective. Nonetheless, we are expected to communicate the facts as we know them and understand them. One writer offers this useful distinction: "All communication requires us to leave something out, but we must be sure that what is left out is not essential to our [reader's] understanding of what is put in" (Coletta 65).

An ethical writer "is obligated to express her or his opinions of products, as long as these opinions are based on objective and responsible research and observation" (MacKenzie 3). Being "objective" does not mean forsaking personal evaluation in cases in which a product may be unsafe or unsound. Even positive claims made in promotional writing (for example, "reliable," "rugged") should be based on objective and verifiable evidence.

Here are guidelines for remaining impartial.

Record the Details That Enable Readers to Visualize the Item. Ask these questions: *What could any observer recognize? What would a camera record?*

> **Subjective**   His office has an *awful* view, *terrible* furniture, and a *depressing* atmosphere.

The italicized words only *tell;* they do not *show.*

> **Objective**   His office has broken windows looking out on a brick wall, a rug with a 15-cm hole in the centre, chairs with bottoms falling out, missing floorboards, and a ceiling with plaster missing in three or four places.

Use Precise and Informative Language. Use high-information words that enable readers to *visualize*. Name specific parts without calling them "things," "gadgets," or "doohickeys." Avoid judgmental words ("impressive," "poor"), unless your judgment is requested and can be supported by facts. Instead of "large," "long," and "near," give exact measurements, weights, dimensions, and ingredients.

Use words that specify location and spatial relationships: above, oblique, behind, tangential, adjacent, interlocking, abutting, and overlapping. Use position words: horizontal, vertical, lateral, longitudinal, in cross-section, parallel.

Indefinite	Precise
a late-model car	a 2003 Ford Taurus sedan
an inside view	a cross-sectional, cutaway, or exploded view
next to the foundation	adjacent to the right side
a small red thing	a red activator button with a 2.5-cm diameter and a concave surface

Do not confuse precise language, however, with overly complicated technical terms or needless jargon. Don't say "phlebotomy specimen" instead of "blood," or "thermal attenuation" instead of "insulation," or "proactive neutralization" instead of "damage control." The clearest writing uses precise but plain language. General readers prefer non-technical language, as long as the simpler words do the job. Always think about your specific readers' needs.

ELEMENTS OF DESCRIPTION

CLEAR AND LIMITING TITLE

Promise exactly what you will deliver—no more and no less. "A Description of a Velo Ten-Speed Racing Bicycle" promises a complete description, down to the smallest part. If you intend to describe the braking mechanism only, be sure your title so indicates: "A Description of the Velo's Centre-Pull Caliper Braking Mechanism."

OVERALL APPEARANCE AND COMPONENT PARTS

Let readers see the big picture before you describe each part.

> The standard stethoscope is roughly 61 cm long and weighs about 140 grams. The instrument consists of a sensitive sound-detecting and amplifying device whose flat surface is pressed against a bodily area. This amplifying device is attached to rubber and metal tubing that transmits the body sound to a listening device inserted in the ear.
>
> Seven interlocking pieces contribute to the stethoscope's Y-shaped appearance: (1) diaphragm contact piece, (2) lower tubing, (3) Y-shaped metal piece, (4) upper tubing, (5) U-shaped metal strip, (6) curved metal tubing, and (7) hollow ear plugs. These parts form a continuous unit.

VISUALS

Use drawings, diagrams, or photographs generously. Our overall description of the stethoscope is greatly clarified by Figure 1.

FUNCTION OF EACH PART

Explain what each part does and how it relates to the whole.

> The diaphragm contact piece is caused to vibrate by body sounds. This part is the heart of the stethoscope that receives, amplifies, and transmits the sound impulse.

APPROPRIATE DETAILS

Give enough detail for a clear picture, but do not burden readers needlessly. Identify your readers and their reasons for reading your description.

Assume you set out to describe a specific bicycle model. The picture you create will depend on the details you select. How will your reader use this description? What is the reader's level of technical understanding? Is this a customer interested in the bike's appearance—its flashy looks and racy style? Is it a repair technician who needs to know how parts operate? Or is it a helper in your bicycle shop who needs to know how to assemble this bike? If you had designed the bike, you would give the manufacturer detailed specifications.

The description of the water heater in Figure 15.1 focuses on what this model looks like and what it's made of. Its intended audience of do-it-yourselfers will know already what a hot-water maker is and what it does. That audience needs no background. A description of how it was put together appears with the installation and maintenance instructions later in the manual.

Specifications for readers who will manufacture the hot-water maker would describe each part in exact detail (e.g., the steel tank's required thickness and pressure rating as well as required percentages of iron, carbon, and other constituents in the steel alloy).

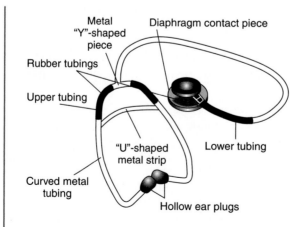

FIGURE 1   Stethoscope with Diaphragm Contact Piece (Front View)

CLEAREST DESCRIPTIVE SEQUENCE

Any item usually has its own logic of organization, based on (1) the way it appears as a static object, (2) the way its parts operate in order, or (3) the way its parts are assembled. We describe these relationships, respectively, in a spatial, functional, or chronological sequence.

Spatial Sequence.   Part of all physical descriptions, a spatial sequence answers these questions: *What is it? What does it do? What does it look like? What parts and material is it made of?* Use this sequence when you want readers to visualize the item as a static object or mechanism at rest (a house interior, a document, the CN Tower, a plot of land, a chainsaw, or a computer keyboard). Can readers best visualize this item from front to rear, left to right, top to bottom? (What logical path do the parts create?) A retractable pen would logically be viewed from outside to inside. The specifications in Figure 15.2 on page 340 proceed from the ground upward.

Functional Sequence.   The functional sequence answers: *How does it work?* It is best used in describing a mechanism in action, such as a 35-millimetre camera, a nuclear warhead, a smoke detector, or a car's cruise-control system. The logic of the item is reflected by the order in which its parts function. Like the hot-water maker in Figure 15.1, a mechanism usually has only one functional sequence. The stethoscope description on page xx follows the sequence of parts through which sound travels.

In describing a solar home-heating system, you would begin with the heat collectors on the roof, moving through the pipes, pumping system, and tanks for the heated water, to the heating vents in the floors and walls—from source to outlet. After this functional sequence of operating parts, you could describe each part in a spatial sequence.

Chronological Sequence.   A chronological sequence answers: *How has it been put together?* The chronology follows the sequence in which the parts are assembled.

Use the chronological sequence for an item that is best visualized by its assembly (such as a piece of furniture, an umbrella tent, or a pre-hung window or door unit). Architects might find a spatial sequence best for describing a proposed beach house to clients; however, they would use a chronological sequence (of blueprints) for specifying to the builder the prescribed dimensions, materials, and construction methods at each stage.

Combined Sequences.   The description of a bumper jack on pages 336–38 alternates among all three sequences: a spatial sequence (bottom to top) for describing the overall mechanism at rest, a chronological sequence for explaining the order in which the parts are assembled, and a functional sequence for describing the order in which the parts operate.

A GENERAL MODEL FOR DESCRIPTION

Description of a complex mechanism almost invariably calls for an outline. This model is adaptable to any description.

> I. Introduction: General Description[1]
> A. Definition, Function, and Background of the Item
> B. Purpose (and Audience—where applicable)
> C. Overall Description (with general visuals, if applicable)
> D. Principle of Operation (if applicable)
> E. List of Major Parts
>
> II. Description and Function of Parts
> A. Part One in Your Descriptive Sequence
> 1. Definition
> 2. Shape, dimensions, material (with specific visuals)
> 3. Sub-parts (if applicable)
> 4. Function
> 5. Relation to adjoining parts
> 6. Mode of attachment (if applicable)
> B. Part Two in Your Descriptive Sequence (and so on)
>
> III. Summary and Operating Description
> A. Summary (used only in a long, complex description)
> B. Interrelation of Parts
> C. One Complete Operating Cycle

This outline is tentative, because you might modify, delete, or combine certain parts to suit your subject, purpose, and reader.

INTRODUCTION: GENERAL DESCRIPTION

Give readers only as much background as they need to get the picture.

A Description of the Standard Stethoscope

> **INTRODUCTION**
>
> The stethoscope is a listening device that amplifies and transmits body sounds to aid in detecting physical abnormalities.
>
> This instrument has evolved from the original wooden, funnel-shaped instrument invented by a French physician, R. T. Lennaec, in 1819. Because of his female patients' modesty, he found it necessary to develop a device, other than his ear, for auscultation (listening to body sounds).

Definition and function

History and background

---

1. In most descriptions, the subdivisions in the introduction can be combined and need not appear as individual headings in the document.

Purpose and
audience

This report explains to the beginning paramedical or nursing student the structure, assembly, and operating principle of the stethoscope. [*Omit this section if you submit to your instructor an audience/purpose profile or if you write for a workplace audience.*]

Finally, give a brief, overall description of the item, discuss its principle of operation, and list its major parts, as in the overall stethoscope description on page 331.

DESCRIPTION AND FUNCTION OF PARTS

The body of your text describes each major part. After arranging the parts in sequence, follow the logic of each part. Provide only as much detail as your readers need.

Readers of this description will use a stethoscope daily, so they need to know how it works, how to take it apart for cleaning, and how to replace worn or broken parts. (Specifications for the manufacturer would require many more technical details—dimensions, alloys, curvatures, tolerances, etc.)

DIAPHRAGM CONTACT PIECE

Definition, size,
shape, and material

The diaphragm contact piece is a shallow metal bowl, about the size of a silver dollar (and twice its thickness), which is caused to vibrate by various body sounds.

Subparts

Three, separate parts make up the piece: hollow steel bowl, plastic diaphragm, and metal frame, as shown in Figure 2.

The stainless steel metal bowl has a concave inner surface, with concentric ridges that funnel sound toward an opening in the tapered base, then out through the hollow appendage. Lateral threads ring the outer circumference of the bowl to accommodate the interlocking metal frame. A fitted diaphragm covers the bowl's upper opening.

The diaphragm is a plastic disk, 2 mm thick, 10.2 cm in circumference, with a moulded lip around the edge. It fits flush over the metal bowl and vibrates sound toward the ridges. A metal frame that screws onto the bowl holds the diaphragm in place.

The stainless steel frame fits over the disk and metal bowl. A 0.75-cm ridge between the inner and outer edge accommodates threads for screwing the frame to the bowl. The frame's outside circumference is notched with equally spaced, perpendicular grooves—like those on the edge of a dime—to provide a gripping surface.

Function and
relation to adjoining
parts
Mode of attachment

The diaphragm contact piece is the heart of the stethoscope that receives, amplifies, and transmits sound through the system of attached tubing. The piece attaches to the lower tubing by an appendage on its apex (narrow end), which fits inside the tubing.

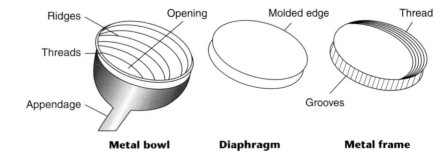

FIGURE 2   Exploded View of a Diaphragm Contact Piece

Each part of the stethoscope, in turn, is described according to its own logic of organization.

SUMMARY AND OPERATING DESCRIPTION

Conclude by explaining how the parts work together to make the whole item function.

SUMMARY AND OPERATING DESCRIPTION

How parts interrelate

One complete operating cycle

The seven major parts of the stethoscope provide support for the instrument, flexibility of movement for the operator, and ease in auscultation.

In an operating cycle, the diaphragm contact piece, placed against the skin, picks up sound impulses from the body surface. These impulses cause the plastic diaphragm to vibrate. The amplified vibrations, in turn, are carried through a tube to a dividing point. From here, the amplified sound is carried through two separate but identical series of tubes to hollow ear plugs.

A SAMPLE SITUATION

The following description of an automobile jack, aimed toward a general audience, follows our outline model.

A Mechanism Description for Non-technical Readers

AUDIENCE/PURPOSE PROFILE. Some readers of this description (written for an owner's manual) will have no mechanical background. Before they can follow instructions for using the jack safely, they will have to learn what it is, what it looks like, what its parts are, and how, generally, it works. They will not need precise dimensions (e.g., "The rectangular base is 20.25 cm long and 16.5 cm wide, sloping upward 3.75 cm from the front outer edge to form a secondary platform 2.5 cm high and 7.5 cm square"). The engineer who designed the jack might include such data in specifications for the manufacturer. Laypeople, however, need only the dimensions that will help them recognize specific parts and understand their function, for safe use and assembly.

Also, this audience will need only the broadest explanation of how the leverage mechanism operates. Although the physical principles (torque, fulcrum) would interest engineers, they would be of little use to readers who simply need to operate the jack safely. ▲

Description of a Standard Bumper Jack

INTRODUCTION—GENERAL DESCRIPTION

Definition, purpose, and function

Overall description (spatial sequence)

The standard bumper jack is a portable mechanism for raising the front or rear of a car through force applied with a lever. This jack enables even a frail person to lift one corner of a 2-ton automobile.

The jack consists of a moulded steel base supporting a free-standing, perpendicular, notched shaft (Figure 1). Attached to the shaft are a leverage mechanism, a bumper catch, and a cylinder for insertion of the jack handle. Except for the main shaft and leverage mechanism, the jack is made to be dismantled and to fit neatly in the car's trunk.

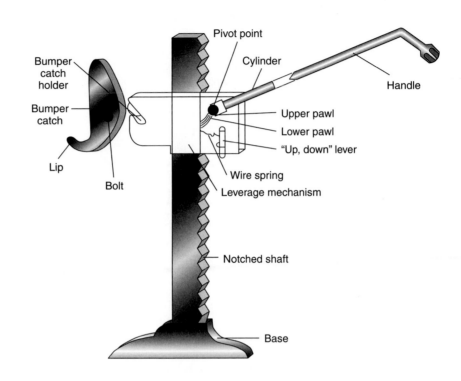

FIGURE 1   A Side View of the Standard Bumper Jack

Operating principle

The jack operates on a leverage principle, with a human hand travelling 46 cm and the car only 1 cm during a normal jacking stroke. Such a device requires many strokes to raise the car off the ground but may prove a lifesaver to a motorist on some deserted road.

List of major parts

Five main parts make up the jack: base, notched shaft, leverage mechanism, bumper catch, and handle.

DESCRIPTION OF PARTS AND THEIR FUNCTION

(chronological sequence)
First major part

Base   The rectangular base is a moulded steel plate that provides support and a point of insertion for the shaft (Figure 2). The base slopes upward to form a platform containing a 1.25-cm depression that provides a stabilizing well for the shaft. Stability is increased by a 1.25-cm cuff around the well. As the base rests on its flat surface, the bottom end of the shaft is inserted into its stabilizing well.

Definition, shape, and size
Function and mode of attachment

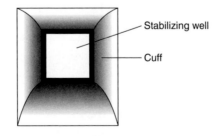

FIGURE 2   A Top View of the Jack Base

Second major part, etc.

Notched Shaft   The notched shaft is a steel bar (80 cm long) that provides a vertical track for the leverage mechanism. The notches, which hold the mechanism in its position on the shaft, face the operator.

The shaft vertically supports the raised automobile, and attached to it is the leverage mechanism, which rests on individual notches.

Leverage Mechanism   The leverage mechanism provides the mechanical advantage needed for its operator to raise the car. It is made to slide up and down the notched shaft. The main body of this pressed-steel mechanism contains two units: one for transferring the leverage and one for holding the bumper catch.

The leverage unit has four major parts: the cylinder, connecting the handle and a pivot point; a lower pawl (a device that fits into the notches to allow forward and prevent backward motion), connected directly to the cylinder; an upper pawl, connected at the pivot point; and an "up-down" lever, which applies or releases pressure on the upper pawl by means of a spring (Figure 1). Moving the cylinder up and down with the handle causes the alternate release of the pawls, and thus movement up or down the shaft—depending on the setting of the "up-down" lever. The movement is transferred by the metal body of the unit to the bumper-catch holder.

The holder consists of a downsloping groove, partially blocked by a wire spring (Figure 1). The spring is mounted in such a way as to keep the bumper catch in place during operation.

Bumper Catch   The bumper catch is a steel device that attaches the leverage mechanism to the bumper. This 23-cm moulded plate is bent to fit the shape of the bumper. Its outer 0.5 cm is bent up to form a lip (Figure 1), which hooks behind the bumper to hold the catch in place. The two sides of the plate are bent back 90 degrees to leave a 5.0-cm bumper-contact surface, and a bolt is riveted between them. This bolt slips into the groove in the leverage mechanism and provides the attachment between the leverage unit and the car.

Handle   The jack handle is a steel bar that serves both as lever and lug-bolt remover. This round bar is 56 cm long, 1.5 cm in diameter, and is bent 135 degrees roughly 13 cm from its outer end. Its outer end is a wrench made to fit the wheel's lug bolts. Its inner end is bevelled to form a bladelike point for prying the wheel covers and for insertion into the cylinder on the leverage mechanism.

CONCLUSION AND OPERATING DESCRIPTION

Assembly

One quickly assembles the jack by inserting the bottom of the notched shaft into the stabilizing well in the base, the bumper catch into the groove on the leverage mechanism, and the bevelled end of the jack handle into the cylinder. The bumper catch is then attached to the bumper, with the lever set in the "up" position.

One complete operating cycle (functional sequence)

As the operator exerts an up-down pumping motion on the jack handle, the leverage mechanism gradually climbs the vertical notched shaft until the car's wheel is raised above the ground. When the lever is in the "down" position, the same pumping motion causes the leverage mechanism to descend the shaft.

SPECIFICATIONS

Airplanes, bridges, smoke detectors, and countless other items are produced according to certain specifications. A particularly exacting type of description, specifications (or "specs") prescribe standards for performance, safety, and quality. For almost any product, specifications spell out:

◆ the methods for manufacturing, building, or installing the product
◆ the materials and equipment to be used
◆ the size, shape, and weight of the product

Because these requirements define an acceptable level of quality, specifications have ethical and legal implications. Any product "below" specifications provides grounds for a lawsuit. When injury or death results (as in a bridge collapse caused by inferior reinforcement), the contractor, sub-contractor, or supplier who cut corners is criminally liable.

Federal and provincial regulatory agencies routinely issue specifications to ensure safety. Health Canada specifies standards for a wide variety of materials and devices, from the operation of seat belts to the fire retardant qualities of cloth used for infant pyjamas. Meanwhile, the Canadian Standards Association designates safety and operating specifications for nearly every product sold in this country. Further, provincial and local agencies issue specifications in the form of building codes, electrical codes, and property development requirements, to name just a few.

Government departments (Defence, Interior, etc.) issue specifications for all types of military hardware and other equipment. A set of NASA specifications for spacecraft parts can be hundreds of pages long, prescribing the standards for even the smallest nuts and bolts, down to screw-thread depth and width in millimetres.

The private sector issues specifications for countless products or projects, to help ensure that customers get exactly what they want. Figure 15.2 on page 340 shows partial specifications drawn up by an architect for a building that will house a small medical clinic. This section of the specs covers only the structure's "shell." Other sections detail the requirements for plumbing, wiring, and interior finish work.

The detailed building specifications partially shown in Figure 15.2 provide the basis for the comprehensive agreement between the builder and the client. In addition, the specifications (along with properly drawn building plans) are important in convincing the municipal authority to issue a building permit. Subsequently, building inspectors will use the plans and specifications as part of their criteria when they inspect the clinic in various stages of the building process.

Specifications like those in Figure 15.2 must be clear enough for *identical* interpretation by the widest possible range of readers (Glidden 258–59):

◆ *the customer,* who has the big picture of what is needed and who wants the best product at the best price
◆ *the designer* (architect, engineer, computer scientist, etc.), who must translate the customer's wishes into the actual specifications
◆ *the contractor or manufacturer,* who won the job by making the lowest bid and so must preserve profit by doing only what is prescribed
◆ *the supplier,* who must provide the exact materials and equipment
◆ *the workforce,* who will do the actual assembly, construction, or installation (managers, supervisors, subcontractors, and workers—some working on only one part of the product, such as plumbing or electrical)

Ruger, Filstone, and Grant Architects

SPECIFICATIONS FOR THE POWNAL CLINIC BUILDING

Foundation

   Footings: 8" x 16" concrete (load-bearing capacity: 3000 lbs. per sq. in.)
   Frost walls: 8" x 4' @ 3000 psi
   Slab: 4" @ 3000 psi, reinforced with wire mesh over vapour barrier

Exterior Walls

   Frame: eastern pine #2 timber frame with exterior partitions set inside
       posts
   Exterior partitions: 2" x 4" kiln-dried spruce set at 16" on centre
   Sheathing: 1/4" exterior-grade plywood
   Siding: #1 red cedar with a 1/2" x 6' bevel
   Trim: finished-pine boards ranging from 1" x 4" to 1" x 10"
   Painting: 2 coats of Clear Wood Finish on siding; trim primed and finished
       with one coat of bone-white, oil base paint

Roof System

   Framing: 2" x 12" kiln-dried spruce set at 24" on centre
   Sheathing: 5/8" exterior-grade plywood
   Finish: 240 Celotex 20-year fibreglass shingles over #15 impregnated felt
       roofing paper
   Flashing: copper

Windows

   Anderson casement and fixed-over-awning models, with white exterior
       cladding, insulating glass and screens, and wood interior frames

Landscape

   Driveway: gravel base, with 3" traprock surface
   Walks: timber defined, with traprock surface
   Cleared areas: to be rough graded and covered with wood chips
   Plantings: 10 assorted lawn plants along the road side of the building

Figure 15.2   Specifications for a Building Project (Partial)

◆ *the inspectors* (such as building, plumbing, or electrical inspectors), who evaluate how well the product conforms to the specifications

Each of these parties has to understand and agree on exactly *what* is to be done and *how* it is to be done. In the case of a lawsuit over failure to meet specifications, the readership broadens to include judges, lawyers, and jury. Figure 15.3 depicts how a clear set of specifications unifies all readers (their various viewpoints, motives, and levels of expertise) in a shared understanding.

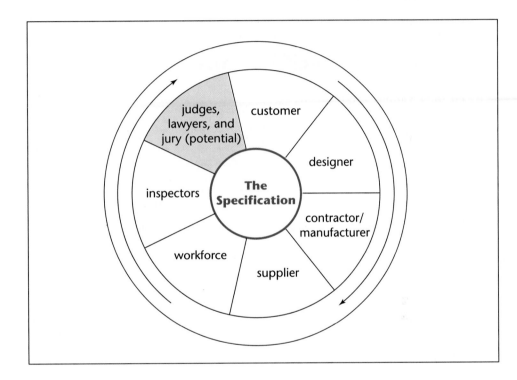

Figure 15.3 Users and Potential Users of Specifications

In addition to guiding a product's design and construction, specifications can facilitate the product's use and maintenance. For instance, specifications in a computer manual include the product's performance limits, or *ratings*: its power requirements, work or processing or storage capacity, environment requirements, the make-up of key parts, and so on. Product support literature for appliances, power tools, and other items routinely contains ratings to help readers select a good operating environment or replace worn or defective parts (Riney 186). The ratings in Figure 15.4 are taken from the owner's manual for a power supply.

Physical Dimensions

Height	4.5 inches (11.4 cm)
Width	10.5 inches (26.7 cm)
Depth	7.5 inches (19 cm)
Weight	7 lbs 6 oz (3.4 kg)
Line cord length	4 feet (1.20 metres)

Technical Specifications

Main Output Specifications		Min	Max
Voltage Range	Channel A	12.5 V	±20.00 V
	Channel B	12.5 V	±20.00 V
	Ch A and B	12.5 V	±40 V or +20/−20 V
Current Range:	Ch A/B, A and B	0.003 A	0.500 A
Load Regulation:	Ch A	1.164%	
	Ch B	1.266%	
Line Regulation:	Ch A	0.05%	
	Ch B	0.00%	
Ripple and Noise		2 mV	6–8 mV
Bandwidth			120 Hz
Transient Response:		<220us within 50 mV of set level for 90% load change	
Temperature Coefficient		<100 ppm/°C	
Output Protection:	Voltage	Outputs will withstand up to 40 V forward voltage	
	Current	Reverse protection by diode clamp = up to 1.0 A	

Figure 15.4 Specifications for the Spider Power Supply

Source: Courtesy of Curt Willis

Checklist FOR EDITING AND REVISING DESCRIPTIONS

Use this checklist to refine the content, arrangement, and style of your description.

Content

- Does the title promise exactly what the description delivers?
- Are the item's overall features described, as well as each part?
- Is each part defined before it is discussed?
- Is the function of each part explained?
- Do visuals appear whenever they can provide clarification?
- Will readers be able to visualize the item?
- Are any details missing, needless, or confusing for this audience?
- Is the description ethically acceptable?

Arrangement

- Does the description follow the clearest possible sequence?
- Are relationships among the parts clearly explained?

Style and Page Design

- Is the description sufficiently impartial?
- Is the language informative and precise?
- Will the level of technicality connect with the audience?
- Is the description written in plain English?
- Is each sentence clear, concise, and fluent?
- Is the page design inviting and accessible?

EXERCISES

1. Select an item from the following list or a device used in your major field. Using the general outline as a model, develop an objective description. Include (a) all necessary visuals; or (b) rough diagram for each visual; or (c) a "reference visual" (a copy of a visual published elsewhere) with instructions for adapting your visual from that one. (If you borrow visuals from other sources, provide full documentation.) Write for a specific use by a specified audience. Attach your written audience/purpose profile (based on the worksheet, page 34) to your document.

soda-acid fire extinguisher	algorithm
breathalyzer	Skinner box
sphygmomanometer	radio
transit	distilling apparatus
sabre saw	bodily organ
hazardous waste site	brand of woodstove
photovoltaic panel	catalytic converter

Remember, you are simply describing the item, its parts, and its function: *do not* provide directions for its assembly or operation.

As an optional assignment, describe a place you know well. You are trying to convey a visual image, not a mood; therefore, your description should be impartial, discussing only the observable details.

2. The bumper-jack description in this chapter is aimed toward a general reading audience. Evaluate it by using the revision checklist. In one or two paragraphs, discuss your evaluation, and suggest revisions.

3. Locate a description and specifications for a particular brand of automobile or some other consumer product. Evaluate this material for promotional and descriptive value and ethical appropriateness.

COLLABORATIVE PROJECTS

1. Divide into small groups. Assume that your group works in the product-development division of a large and diversified manufacturing company.

   Your division has just invented an idea for an inexpensive consumer item with a potentially vast market. (Choose one from the following list.)

   - flashlight

- nail clippers
- retractable ballpoint pen
- scissors
- stapler
- any simple mechanism—as long as it has moving parts

Your group's assignment is to prepare three descriptions of this invention:

a. for company executives who will decide whether to produce and market the item

b. for the engineers, machinists, and so on, who will design and manufacture the item

c. for the customers who might purchase and use the item

Before writing for each audience, be sure to collectively complete audience/purpose profiles (page 34).

Appoint a group manager, who will assign tasks (visuals, keying, etc.) to members. When the descriptions are fully prepared, the group manager will appoint one member to present them in class. The presentation should include explanations of how the descriptions differ for the different audiences.

2. Assume your group is an architectural firm designing buildings at your college. Develop a set of specifications for duplicating the interior of the classroom in which this course is held. Focus only on materials, dimensions, and equipment (whiteboard, desk, etc.) and use visuals as appropriate. Your audience includes teachers, school administrators, and the firm that will construct the classroom. Use the same format as in Figure 15.2, or design a better one. Appoint one member to present the completed specifications in class. Compare versions from each group for accuracy and clarity.

CHAPTER 16

Process Analyses, Instructions, and Procedures

A process is a series of actions or changes leading to a product or result. *Instructions* describe how to carry out a "process"; a *procedure* is a special kind of instructional set. This chapter discusses all three types of process-related description.

Although process descriptions and instructions both present chronological steps leading to a predicted result, that's where the similarity ends. The reasons for writing and reading a process description are quite different from the motivations for writing and reading instructions. These fundamental differences lead to the differences in content, structure, voice, mood, appearance, and style summarized in Table 16.1 on pages 348–49.

PROCESS ANALYSIS

Why readers read process descriptions

Readers of process descriptions want to know *how* and *why* the processes occur, so the writer's first order of business is to divide the process into its parts or principles. Usually, those parts occur in chronological order. Moreover, the first part or step of the process usually leads to the second step and often creates the conditions that allow that second step to occur. Then, the second step leads to the third step, and so on.

This chronological development of dependent steps is particularly noticeable to writers who analyze mechanical processes (operation of a piston-driven engine), geological processes (formation of icebergs), or chemical processes (chemical hydration within concrete). Even electronic processes, which occur at blinding speed, can be understood as a series of causative actions: It's possible to know the exact order, duration, and effect caused by each sub-process within an electronic circuit.

The electronic example raises another very interesting point about most processes. They depend on the "conditions" that cause them. The process of an electronic circuit's operation depends on the design of the circuit itself. In a less confining way, perhaps, the process by which a road is washed by heavy spring run-off depends on the physical conditions of water volumes, soil composition, and terrain.

As Table 16.1 illustrates, a process analysis must include enough detail to enable readers to follow the process step by step. That level of detail depends on the reader's needs. For example, a back-country skier who wants to avoid avalanches will be satisfied with a basic description of the forces and conditions that affect the slab-avalanche process. However, a civil engineer studying avalanches will need to know much more about snow compaction forces, changes in crystalline structures, and the forces that cause snow layers to shear apart.

Because it emphasizes the process itself, rather than the reader's role, process analysis is written in the third person. Indeed, all aspects of a process description resemble a technical essay:

◆ It uses standard paragraphs, most of which use chronological patterns.
◆ It employs serious, reflective phrasing.
◆ It employs precise, accurate vocabulary.
◆ It presents a formal appearance, usually with headings, formal illustration format, and formal documentation of sources.

To help the reader fully understand the process, the writer must carefully plan the structure of a process description. The writer's first step is to analyze the process itself, at the level of the reader's interests and needs. This analysis will help the writer produce a detailed outline.

For an idea of the components of such an outline, see the structure for process description summarized in Table 16.1, pages 348–49.

The following process description has used a structure like the one outlined in Table 16.1. The document's writer, Bill Kelly, belongs to an environmental group studying the problems of acid rain in its southern Ontario community. To gain community support, the environmentalists must educate citizens about the problem. Bill's group is publishing and mailing a series of brochures. The first brochure explains how acid rain is formed.

Here is Bill's audience/purpose profile for the document.

A Process Analysis for Non-technical Readers

> ***AUDIENCE/PURPOSE PROFILE.*** My audience will consist of general readers. Some already will be interested in the problem; others will have no awareness (or interest). Therefore, I'll keep my explanation at the lowest level of technicality (no chemical formulas, equations). But my explanation needs to be vivid enough to appeal to less aware or less interested readers. I'll use visuals to create interest and to illustrate simply. To give an explanation thorough enough for broad understanding, I'll divide the process into three chronological steps: how acid rain develops, spreads, and destroys. ▲

This is the document resulting from Bill's analysis of both his subject and his audience.

How Acid Rain Develops, Spreads, and Destroys

INTRODUCTION

Definition

Acid rain is environmentally damaging rainfall that occurs after fossil fuels burn, releasing nitrogen and sulphur oxides into the atmosphere. Acid rain, simply stated, increases the acidity level of waterways because these nitrogen and sulphur oxides combine with the air's normal moisture. The resulting rainfall is far more acidic than normal rainfall. Acid rain is a silent threat because its effects, although slow, are cumulative. This analysis explains the cause, the distribution cycle, and the effects of acid rain.

Purpose

Brief description of the process

Most research shows that power plants burning oil or coal are the primary cause of acid rain. The burnt fuel is not completely expended, and some residue enters the atmosphere. Although this residue contains several potentially toxic elements, sulphur oxide and, to a lesser extent, nitrogen oxide are the major problem, because they are transformed when they combine with moisture. This chemical reaction forms sulphur dioxide and nitric acid, which then rain down to earth.

Preview of stages

The major steps explained here are (1) how acid rain develops, (2) how acid rain spreads, and (3) how acid rain destroys.

THE PROCESS

First stage

How Acid Rain Develops   Once fossil fuels have been burned, their usefulness is over. Unfortunately, it is here that the acid rain problem begins.

Fossil fuels contain a number of elements that are released during combustion. Two of these, sulphur oxide and nitrogen oxide, combine with normal moisture to produce

Table 16.1 Process Analysis Compared to Instructions   *(continued)*

PURPOSE	Helps the reader understand how and why the process occurs
AUDIENCE	Aimed at interested persons who want to understand how something works or how it happens
CONTENT	*Explanations* are essential, in addition to straight *chronological description* of the process' stages. Description of process' *physical environment* is part of some descriptions. *Illustrations* are often very useful. Descriptions are *specific and detailed.*
STRUCTURE	*General idea* (lead-in) ◆ names and defines process and its special features ◆ where, when, why, how often the process occurs ◆ where necessary, gives background theory ◆ lists the process' main stages or actions *Individual stages* (chronological) ◆ each stage is described in detail and related to the stages that precede and follow; the importance of particularly important stages is noted ◆ each stage includes applicable measurements of time, distance, direction, density, volume, etc. *Conclusion* (lead out to practical considerations) ◆ where applicable, comments about time needed for overall process, cost, process' applications, special problems, immediate and long-term results
VOICE/MOOD	Uses indicative mood; e.g., "the next stage takes three hours . . ." Stays detached, in 3rd person; e.g., "the skier's first move . . ." Active or passive voice; e.g., "the signal travels . . ." or, "the signal is next transferred to the filtering stage . . ."
APPEARANCE AND STYLE	Usually looks formal (headings, paragraphs, standard spacing) Reads like a "serious" discussion Uses a mixture of sentence types and lengths Uses precise, accurate vocabulary

Table 16.1 Instructions Compared to Process Analysis

Helps the reader perform the process that is described	**PURPOSE**
Aimed at persons who need to complete a task or want to improve performance	**AUDIENCE**
Provides no more detail than is necessary (**Note:** Analysis and explanations <u>may</u> be necessary.) Features a very careful *chronological listing of steps* Very carefully describes *exact steps* to take Includes *frequent visual illustrations*	**CONTENT**
Introduction ◆ concisely explains the overall actions to be performed ◆ in some cases, provides background information and, where necessary, lists materials/equipment to be used or the conditions necessary for successful action ◆ in some cases, cautions reader about safety factors *Chronological list of steps* (plus necessary explanations) ◆ where appropriate, combines groups of steps together under subheadings; e.g., "Setting the Timer," "Selecting Programs" ◆ shows the interrelations and sequence of actions by using numbered steps and sequence transitions; e.g., "next," "then," "10 minutes later," "after the liquid cools" ◆ gives reasons for performing certain actions in a specific way or at a specific time ◆ uses illustrations to show the *results* of performed actions, not just the techniques for performing the actions *Brief practical conclusion* ◆ reminds the reader of expected results/performance times	**STRUCTURE**
Uses imperative mood; e.g., "Set the timer by choosing . . ." Directly addresses reader; e.g., "Your first task will be to . . ." Uses active voice; e.g., "Choose one of three settings . . ."	**VOICE/MOOD**
Uses some paragraphs, but mostly uses numbered point form Looks "user friendly" Writes in phrases or short sentences Features direct, straightforward vocabulary Employs lots of open space	**APPEARANCE AND STYLE**

sulphuric acid and nitric acid. (Figure 1 illustrates how acid rain develops.) The released gases undergo a chemical change as they combine with atmospheric ozone and water vapour. The resulting rain or snowfall is more acid than normal precipitation.

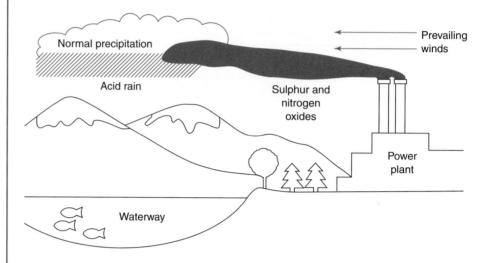

FIGURE 1 How Acid Rain Develops

Definition

Acid level is measured by pH readings. The pH scale runs from 0 through 14—a pH of 7 is considered neutral. (Distilled water has a pH of 7.) Numbers above 7 indicate increasing degrees of alkalinity. (Household ammonia has a pH of 11.) Numbers below 7 indicate increasing acidity. Movement in either direction on the pH scale, however, means multiplying by 10. Lemon juice, which has a pH value of 2, is 10 times more acidic than apples, which have a pH of 3, and is 1000 times more acidic than carrots, which have a pH of 5.

Because of carbon dioxide (an acid substance) normally present in air, unaffected rainfall has a pH of 5.6. At this time, the pH of precipitation in the northeastern United States and Canada is between 4.5 and 4. In Massachusetts, rain and snowfall have an average pH reading of 4.1. A pH reading below 5 is considered to be abnormally acidic, and therefore a threat to aquatic populations.

Second stage

How Acid Rain Spreads   Although it might seem that areas containing power plants would be most severely affected, acid rain can in fact travel thousands of miles from its source. Stack gases escape and drift with the wind currents. The sulphur and nitrogen oxides are thus able to travel great distances before they return to earth as acid rain.

For an average of two to five days after emission, the gases follow the prevailing winds far from the point of origin. Estimates show that about 50 percent of the acid rain that affects Canada originates in the United States; at the same time, 15 to 25 percent of the U.S. acid rain problem originates in Canada.

The tendency of stack gases to drift makes acid rain a widespread menace. More than 200 lakes in the Adirondacks, hundreds of miles from any industrial centre, are unable to support life because their water has become so acidic.

Third stage

How Acid Rain Destroys   Acid rain causes damage wherever it falls. It erodes various types of building rock such as limestone, marble, and mortar, which are gradually eaten away by the constant bathing in acid. Damage to buildings, houses, monuments, statues, and cars is widespread. Some priceless monuments and carvings already have been destroyed, and even trees of some varieties are dying in large numbers.

More important, however, is acid rain damage to waterways in the affected areas. (Figure 2 illustrates how a typical waterway is infiltrated.)

Because of its high acidity, acid rain dramatically lowers the pH in lakes and streams. Although its effect is not immediate, acid rain eventually can make a waterway so acidic it dies. In areas with natural acid-buffering elements such as limestone, the dilute acid has less effect. The northeastern United States and Canada, however, lack this natural protection, and so are continually vulnerable.

The pH level in an affected waterway drops so low that some species cease to reproduce. In fact, a pH level of 5.1 to 5.4 means that fisheries are threatened; once a waterway reaches a pH level of 4.5, no fish reproduction occurs. Because each creature is part of the overall food chain, loss of one element in the chain disrupts the whole cycle.

In the northeastern United States and Canada, the acidity problem is compounded by the runoff from acid snow. During the cold winter months, acid snow sits with little melting, so that by spring thaw, the acid released is greatly concentrated. Aluminum and other heavy metals normally present in soil are also released by acid rain and runoff. These toxic substances leach into waterways in heavy concentrations, affecting fish in all stages of development.

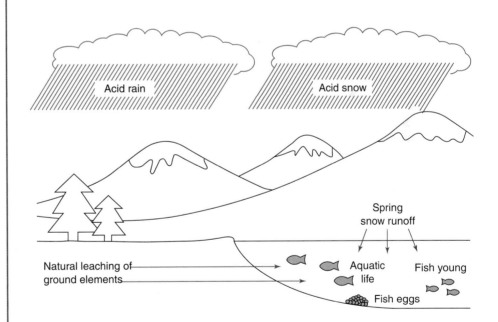

FIGURE 2 How Acid Rain Destroys

SUMMARY

Acid rain develops from nitrogen and sulphur oxides emitted by industrial and power plants burning fossil fuels. In the atmosphere, these oxides combine with ozone and water to form acid rain: precipitation with a lower-than-average pH. This acid precipitation returns to earth many miles from its source, severely damaging waterways that lack natural buffering agents. The northeastern United States and Canada are the most severely affected areas in North America.

ELEMENTS OF INSTRUCTION

Why readers need to read instructions

As consumers, we need instructions to learn how to operate everything from automobiles to VCRs. But we also seek instruction on topics that we know reasonably well. Because of our recreation activities, for instance, we often read instructional magazine articles on how to ski steep slopes or how to hit a particular golf shot or how to perform an aerobics sequence, even though we might already be able to perform the activity. Why? We want to perform the activity more skillfully!

Almost anyone with a responsible job writes and reads instructions. The new employee uses instructions for operating office equipment or industrial machinery; the employee going on vacation writes instructions for the replacement person. The person who buys a computer reads the manuals (or documentation) for instructions on connecting a printer or running a program.

Instructions carry profound ethical and legal implications. Each year, as many as 10 percent of workers are injured on the job (Clement 149). Countless injuries also result from misuse of consumer products such as power tools and car jacks—misuse often caused by defective instructions.

A reader injured because of inaccurate, incomplete, or misleading instructions can sue the writer. Courts have ruled that a writing defect in product support literature carries liability, as would a design or manufacturing defect in the product itself (Girill, *Technical Communication and Law* 37). Some legal experts argue that writing defects carry even greater liability than product defects because they are more easily demonstrated to a non-technical jury (Bedford and Stearns 128).

To ensure that your own instructions meet professional and legal requirements for accuracy, completeness, and clarity, observe the following guidelines.

CLEAR AND LIMITING TITLE

Make your title promise exactly what your instructions deliver—no more and no less. The title "Instructions for Cleaning the Drive Head of a Laptop Computer" tells readers what to expect: instructions for a specific procedure on a selected part. But the title "The Laptop Computer" gives no forecast. A reader of a document so titled might think the document contains a history of the laptop, or a description of each part, or a wide range of related information.

INFORMED CONTENT

Make sure that you know exactly what you're talking about. Ignorance on your part makes you no less liable for instructions that are faulty or inaccurate:

Never count on ignorance as an excuse

> *If the author of [a car repair] manual had no experience with cars, yet provided faulty instructions on the repair of the car's brakes, the home mechanic who was injured when the brakes failed may recover [damages] from the author.* (Walter and Marsteller 165)

Do not write instructions unless you know the procedure in detail and unless you actually have performed it.

VISUALS

In addition to showing what to do, instructional visuals attract the attention of today's graphic-oriented readers and help keep words to a minimum. Instructions also often include a persuasive dimension: to motivate interest, commitment, or action.

Types of visuals especially suited to instructions include icons, representational and schematic diagrams, flowcharts, photographs, and prose tables.

Illustrate any step that might be difficult for readers to visualize. Show the same angle of vision the reader will have when doing the activity or using the equipment—and name the angle (*side view, top view*) if you think readers will have trouble figuring it out for themselves.

The less specialized your audience, the more visuals they are likely to need. But do not illustrate any action simple enough for readers to visualize on their own, such as "Press RETURN" for any user familiar with a keyboard.

Figure 16.1 depicts an array of visuals and their specific instructional functions. Virtually each of these visuals is easily constructed and some could be further enhanced, depending on your production budget and graphics capability. Writers and editors often provide a *brief* (page 276) and a rough sketch describing the visual and its purpose for the graphic designer or art department.

APPROPRIATE LEVEL OF TECHNICALITY

Unless you know your readers have the relevant background and skills, write for general readers, and do three things:

1. Give them enough background to understand why they need your instructions.
2. Give them enough detail to understand *what* to do.
3. Give them enough examples to visualize the procedure clearly.

How you might adapt instructions titled "How to Initialize Your Blank Disk" for a general audience is shown on pages 357–58.

Background Information.   Begin by explaining the purpose of the procedure.

Tell readers why
they are doing this

> You might easily lose information stored on a floppy disk if:
>
> 1. The disk is damaged by direct sunlight, extreme temperature, or moisture.
> 2. The disk is erased by a faulty disk drive, a power surge, or a user error.
> 3. The stored information is scrambled by a nearby magnet (telephone, computer terminal, or the like).
>
> Always make a back-up copy of any important material.

Also, state your assumptions about your reader's level of technical understanding.

Tell them what they
should know
already

> To follow these instructions, you should be able to identify these parts of a Macintosh system: computer, monitor, keyboard, mouse, disk drive, and a 3.5-inch floppy disk.

how to locate something

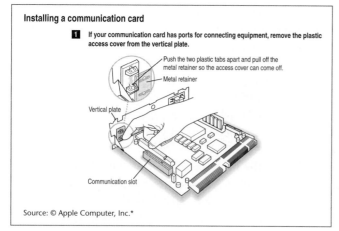

Installing a communication card

1 If your communication card has ports for connecting equipment, remove the plastic access cover from the vertical plate.

Push the two plastic tabs apart and pull off the metal retainer so the access cover can come off.

Metal retainer

Vertical plate

Communication slot

Source: © Apple Computer, Inc.*

how to operate something

Source: SuperStock

how to handle something

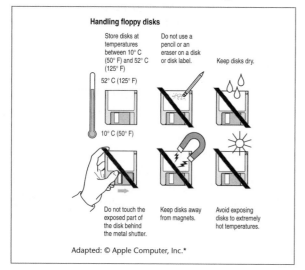

Handling floppy disks

Store disks at temperatures between 10° C (50° F) and 52° C (125° F)

Do not use a pencil or an eraser on a disk or disk label.

Keep disks dry.

52° C (125° F)

10° C (50° F)

Do not touch the exposed part of the disk behind the metal shutter.

Keep disks away from magnets.

Avoid exposing disks to extremely hot temperatures.

Adapted: © Apple Computer, Inc.*

how to assemble something

Extension Cord Retainer

1. Look into the end of the Switch Handle and you will see 2 slots. The WIDER end of the Retainer goes into the TOP slot (Figure 8).
2. Plug extension cord into Switch Handle and weave cord into Retainer, leaving a little slack (Figure 9).

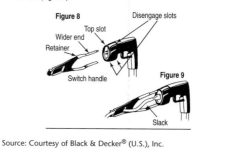

Figure 8

Disengage slots

Top slot

Wider end

Retainer

Switch handle

Figure 9

Slack

Source: Courtesy of Black & Decker® (U.S.), Inc.

how to position something

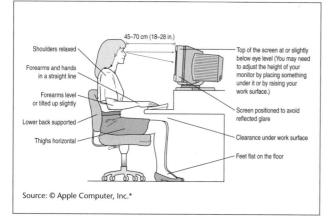

45–70 cm (18–28 in.)

Shoulders relaxed

Forearms and hands in a straight line

Forearms level or tilted up slightly

Lower back supported

Thighs horizontal

Top of the screen at or slightly below eye level (You may need to adjust the height of your monitor by placing something under it or by raising your work surface.)

Screen positioned to avoid reflected glare

Clearance under work surface

Feet flat on the floor

Source: © Apple Computer, Inc.*

how to avoid damage or injury

△ **Important:** The fixing assembly in the printer operates at very high temperatures. When you need to open the printer, be careful not to touch the fixing assembly. △

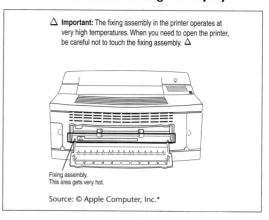

Fixing assembly.
This area gets very hot.

Source: © Apple Computer, Inc.*

Figure 16.1   Common Types of Instructional Visuals and Their Functions   *(continued)*

how to diagnose and solve problems

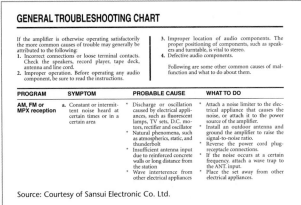

GENERAL TROUBLESHOOTING CHART

If the amplifier is otherwise operating satisfactorily the more common causes of trouble may generally be attributed to the following:
1. Incorrect connections or loose terminal contacts. Check the speakers, record player, tape deck, antenna and line cord.
2. Improper operation. Before operating any audio component, be sure to read the instructions.

3. Improper location of audio components. The proper positioning of components, such as speakers and turntable, is vital to stereo.
4. Defective audio components.

Following are some other common causes of malfunction and what to do about them.

PROGRAM	SYMPTOM	PROBABLE CAUSE	WHAT TO DO
AM, FM or MPX reception	a. Constant or intermittent noise heard at certain times or in a certain area	* Discharge or oscillation caused by electrical appliances, such as fluorescent lamps, TV sets, D.C. motors, rectifier and oscillator * Natural phenomena, such as atmospherics, static, and thunderbolt * Insufficient antenna input due to reinforced concrete walls or long distance from the station * Wave interterence from other electrical appliances	* Attach a noise limiter to the electrical appliance that causes the noise, or attach it to the power source of the amplifier. * Install an outdoor antenna and ground the amplifier to raise the signal-to-noise ratio. * Reverse the power cord plug-receptacle connections. * If the noise occurs at a certain frequency, attach a wave trap to the ANT. input. * Place the set away from other electrical appliances.

Source: Courtesy of Sansui Electronic Co. Ltd.

how to proceed systematically

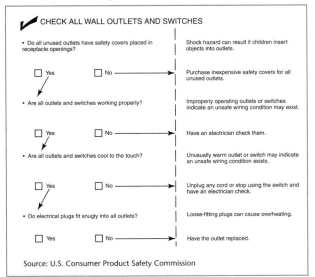

✔ CHECK ALL WALL OUTLETS AND SWITCHES

• Do all unused outlets have safety covers placed in receptacle openings?   Shock hazard can result if children insert objects into outlets.

☐ Yes   ☐ No → Purchase inexpensive safety covers for all unused outlets.

• Are all outlets and switches working properly?   Improperly operating outlets or switches indicate an unsafe wiring condition may exist.

☐ Yes   ☐ No → Have an electrician check them.

• Are all outlets and switches cool to the touch?   Unusually warm outlet or switch may indicate an unsafe wiring condition exists.

☐ Yes   ☐ No → Unplug any cord or stop using the switch and have an electrician check.

• Do electrical plugs fit snugly into all outlets?   Loose-fitting plugs can cause overheating.

☐ Yes   ☐ No → Have the outlet replaced.

Source: U.S. Consumer Product Safety Commission

how to make the right decisions

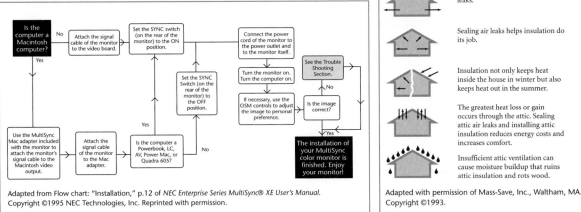

Adapted from Flow chart: "Installation," p.12 of *NEC Enterprise Series MultiSync® XE User's Manual.* Copyright ©1995 NEC Technologies, Inc. Reprinted with permission.

how to identify safe or acceptable limits

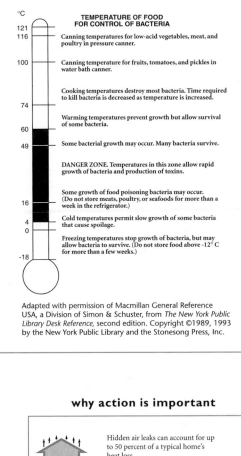

TEMPERATURE OF FOOD FOR CONTROL OF BACTERIA

°C

121
116 — Canning temperatures for low-acid vegetables, meat, and poultry in pressure canner.

100 — Canning temperature for fruits, tomatoes, and pickles in water bath canner.

Cooking temperatures destroy most bacteria. Time required to kill bacteria is decreased as temperature is increased.

74 — Warming temperatures prevent growth but allow survival of some bacteria.

60
49 — Some bacterial growth may occur. Many bacteria survive.

DANGER ZONE. Temperatures in this zone allow rapid growth of bacteria and production of toxins.

Some growth of food poisoning bacteria may occur. (Do not store meats, poultry, or seafoods for more than a week in the refrigerator.)

16
4 — Cold temperatures permit slow growth of some bacteria that cause spoilage.

0 — Freezing temperatures stop growth of bacteria, but may allow bacteria to survive. (Do not store food above -12° C for more than a few weeks.)

-18

Adapted with permission of Macmillan General Reference USA, a Division of Simon & Schuster, from *The New York Public Library Desk Reference,* second edition. Copyright ©1989, 1993 by the New York Public Library and the Stonesong Press, Inc.

why action is important

Hidden air leaks can account for up to 50 percent of a typical home's heat loss.

Total area of all air leaks can add up to 0.9 to 1.9 square metres (20 square feet); that's like leaving a door open all winter.

Insulation alone does not seal air leaks.

Sealing air leaks helps insulation do its job.

Insulation not only keeps heat inside the house in winter but also keeps heat out in the summer.

The greatest heat loss or gain occurs through the attic. Sealing attic air leaks and installing attic insulation reduces energy costs and increases comfort.

Insufficient attic ventilation can cause moisture buildup that ruins attic insulation and rots wood.

Adapted with permission of Mass-Save, Inc., Waltham, MA. Copyright ©1993.

Figure 16.1   **Common Types of Instructional Visuals and Their Functions**

* Illustrations © Apple Computer, Inc. 1993. Used with permission. Apple, the Apple logo, and Power Macintosh are registered trademarks ™ of Apple Computer, Inc. All rights reserved.

Define any specialized terms that appear in your instructions.

Tell them what each
key term means

Initialize: Before you can store or retrieve information on a disk, you must initialize the blank disk (unless you are using pre-formatted disks). Initializing creates a format the computer can understand—a directory of specific memory spaces (like post office boxes) on the disk, where you can store and retrieve information as needed.

When your reader understands *what* and *why*, you are ready to explain *how* the reader can carry out the procedure.

Adequate Detail.   Explain the procedure in enough detail for readers to know exactly what to do. Vague instructions result from the writer's failure to consider the readers' needs, as in these unclear instructions for giving first aid to an electrical shock victim:

Inadequate detail
for general readers

1. Check vital signs.
2. Establish an airway.
3. Administer external cardiac massage as needed.
4. Ventilate, if cyanosed.
5. Treat for shock.

These instructions might be clear to medical experts, but not to general readers. Not only are the details inadequate, but also terms such as "vital signs" and "cyanosed" are too technical for laypeople. Such instructions posted for workers in a high-voltage area would be useless. The instructions need illustrations and explanations, as in the following instruction for item 2, establishing an airway:

It's easy to overestimate what people already know, especially when the procedure

Adequate detail for
general readers

MOUTH-TO-MOUTH BREATHING

If there are no signs of breathing, place one hand under the victim's neck and gently lift. At the same time, push with the other hand on the victim's forehead. This will move the tongue away from the back of the throat to open the airway.

Source: Reprinted with permission of Macmillan General Reference USA, a Division of Simon & Schuster, from *The New York Public Library Desk Reference,* second edition. Copyright ©1989, 1993 by the New York Public Library and the Stonesong Press, Inc.

is almost automatic for you. (Think about when a relative or friend was teaching you to drive a car, or perhaps you tried to teach someone else.) Always assume the reader knows less than you. A colleague will know at least a little less; a layperson will know a good deal less—maybe nothing—about this procedure.

Exactly how much information is enough? These suggestions can help you find an answer:

How to provide
adequate detail

◆ Give everything readers need, so the instructions can stand alone.

◆ Give only what readers need. Don't tell them how to build a computer when they need to know only how to copy a disk.

◆ Instead of focusing on the *product* ("How does it work?"), focus on the *task* ("How do I use it?" or "How do I do it?") (Grice, "Focus" 32).

◆ Omit steps (*Seat yourself at the computer*) obvious to readers.

◆ Adjust the information rate ("the amount of information presented in a given page," Meyer 17) to readers' background and the difficulty of the task. For complex or sensitive steps, slow the information rate. Don't make readers do too much too fast.

◆ Reinforce the prose with visuals. Don't be afraid to repeat information if it saves readers from flipping pages.

◆ When writing instructions for consumer products, assume "a barely literate reader" (Clement 151). Simplify.

◆ Recognize the persuasive dimension of the instructions. You may need to persuade readers that this procedure is necessary or beneficial, or that they can complete this procedure with relative ease and competence.

Examples. Procedures require specific examples (how to load a program, how to order a part), to help readers follow the steps correctly.

Use plenty of examples

To load your program, key this command:

```
Load "Style Editor"
```

Then press RETURN.

Like visuals, examples *show* readers what to do. Examples in fact often appear as visuals.

In the sample procedure that follows, careful use of background, detailed explanation, and visual examples create a user-friendly level of technicality for readers just learning to use a computer.

FIRST STEP: HOW TO INITIALIZE YOUR BLANK DISK

Before you can copy or store information on a blank disk, you must initialize the disk (unless you are using preformatted disks). Follow this procedure:

Begin each instruction with an action verb

1. Switch the computer on.
2. Insert your application disk in the main disk drive.
3. Insert your blank disk in the external disk drive. Unable to recognize this new disk, the computer will respond with a message asking whether you wish to initialize the disk (Figure 1).

Let the visual repeat, restate, or reinforce the prose

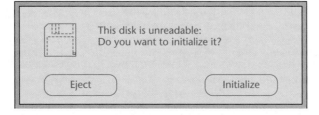

FIGURE 1 The "Initialize" Message

4. Using your mouse, place the tip of the on-screen pointer (a small arrow) inside the "Initialize" box.

5. Press and quickly release the mouse button. Within 15–20 seconds the initializing will be completed, and a message will appear, asking you to name your disk (Figure 2).

Place the visual close to the step

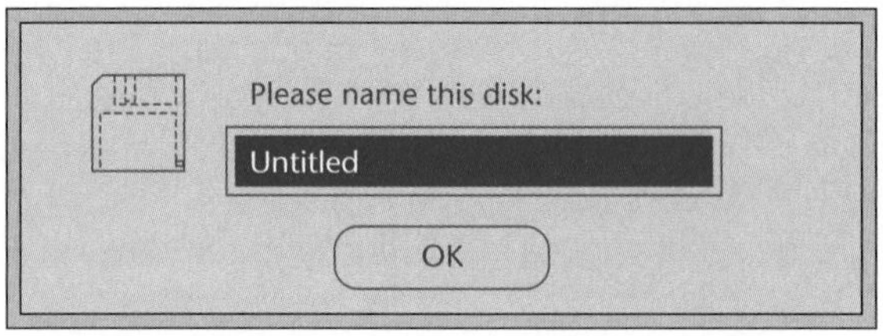

FIGURE 2   The "Disk-naming" Message

SECOND STEP: HOW TO NAME YOUR INITIALIZED DISK (AND SO ON)

Notice that in the sample, instructions, and illustrations repeat the same information—they are redundant (Weiss 100). You may recall that Chapter 11 advises writers to avoid *style redundancy* (extra words that give no extra information). For instructions, however, a writer deliberately seeks *content redundancy,* giving the same information in prose and then in visuals. Granted, this prose-visual redundancy makes a longer document, but such repetition helps prevent misinterpretation. When you can't be sure how much is enough, risk over-explaining rather than under-explaining.

LOGICALLY ORDERED STEPS

Instructions not only divide the procedure into steps; they also guide users through the steps in *chronological order.* They organize the facts and explanations in ways that make sense to readers.

Show how steps are connected

You can't splice two wires to make an electrical connection until you have removed the insulation. To remove the insulation, you will need ....

Try to keep all information for one step close together.

WARNINGS, CAUTIONS, AND NOTES

Here are the only items that should interrupt the steps in a set of instructions (Van Pelt 3):

◆ A *note* clarifies a point, emphasizes vital information, or describes options or alternatives.

NOTE: The computer will not initialize a disk that is scratched or imperfect. If your blank disk is rejected, try a new disk.

◆ A *caution* prevents possible mistakes that could result in injury or equipment damage:

CAUTION: A momentary electrical surge or power failure will erase the contents of internal memory. To avoid losing your work, every few minutes save on disk what you have just keyed into the computer.

◆ A *warning* alerts users against potential hazards to life or limb:

WARNING: To prevent electrical shock, always disconnect your printer from its power source before cleaning internal parts.

◆ A *danger* notice identifies an immediate hazard to life or limb:

DANGER: The red canister contains **DEADLY** radioactive material. **Do not break the safety seal** under any circumstances.

Use symbols to alert readers

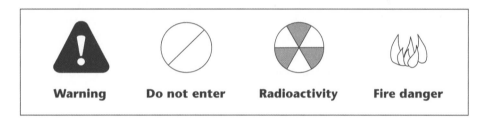

In addition to prose warnings, attract readers' attention and help them identify hazards by using symbols or icons (Bedford and Stearns 128):
Preview the warnings, cautions, and notes in your introduction, and place them, *clearly highlighted*, immediately before the respective steps.

NOTE *A recent study found that product users were six times more likely to comply with warnings included with the directions for using the product instead of on a separate warning label ("Notes" 2).*

Use notes, warnings, and cautions only when needed; overuse will dull their effect, and readers may overlook their importance.

APPROPRIATE WORDS, SENTENCES, AND PARAGRAPHS

Of all communications, instructions have the strictest requirements for clarity, because they lead to *immediate action*. Readers are impatient and often will not read the entire instructions before plunging into the first step. Poorly phrased and misleading instructions can be disastrous.

Like descriptions, instructions name parts, use location and position words, and state exact measurements, weights, and dimensions. Instructions additionally require your strict attention to phrasing, sentence structure, and paragraph structure.

Transitions to Mark Time and Sequence.   Transitional words provide a bridge between related ideas. Some transitions ("in addition," "next," "meanwhile," "finally," "10 minutes later," "the next day," "before") mark time and sequence. They help readers understand the step-by-step process:

PREPARING THE GROUND FOR A TENT

Use transitions to
mark time and
sequence

Begin by clearing and smoothing the area that will be under the tent. This step will prevent damage to the tent floor and eliminate the discomfort of sleeping on uneven ground. *First*, remove all large stones, branches, or other debris within a level 3 × 4 metre (10 × 13 foot) area. Use your camping shovel to remove half-buried rocks that cannot easily be moved by hand. *Next*, fill in any large holes with soil or leaves. *Finally*, make several light surface passes with the shovel or a large, leafy branch to smooth the area.

Carefully Shaped Paragraphs and Sentences.   Much of the introductory and explanatory material in instructions take the form of standard prose paragraphs, with enough sentence variety to keep readers interested. But the steps themselves have unique paragraph and sentence requirements. Unless the procedure consists of simple steps (as in the "Preparing the Ground for a Tent" example), separate each step by using a numbered list—one step for one activity. If the activity is especially complicated, use a new line (not indented) to begin each sentence in the step.

Instructions ordinarily employ short sentences. But brief is not always best, especially when readers have to fill in the information gaps. Never telegraph your message by omitting articles (a, an, the).

Unlike many types of documents, instructions call for very little sentence variety. Use sentences with similar structure ("Do this. Then do that.") to avoid confusing readers trying to follow the procedure.

If a single step covers two related actions, follow the sequence of required actions:

> **Logical sequence**   Insert the disk in the drive before switching on the computer.

Make your explanations easier to follow by using a familiar-to-unfamiliar sequence:

> **Difficult**   You must initialize a blank disk before you can store information on it.

This sentence is clearer if the familiar material comes first:

> **Easier**   Before you can store information on a blank disk, you must initialize the disk.

Shape every sentence and every paragraph for the reader's access.

Active Voice and Imperative Mood.   Use the active voice and imperative mood ("Insert the disk") to address the reader directly. Otherwise, your instructions can lose authority ("You should insert the disk") or become ambiguous ("The disk is inserted"). In the ambiguous example, we can't tell if the disk is to be inserted or if it already has been inserted.

> **Indirect or**   The user keys in his or her access code.
> **confusing**   You should key in your access code.
>   It is important to key in the access code.
>   The access code is keyed in.

Direct and clear	Key in your access code.

The imperative makes instructions more definite and easier to understand because the action verb—the crucial word that identifies the next action—comes first. Instead of burying your verb in mid-sentence, *begin* with an action verb (raise, connect, wash, insert, open) to give readers an immediate signal.

Affirmative Phrasing.   Research shows that readers respond more quickly and efficiently to instructions phrased affirmatively rather than negatively (Spyridakis and Wenger 205).

Weaker	Verify that your disk is not contaminated with dust.
Stronger	Examine your disk for dust contamination.

Parallel Phrasing. Like any items in a series, steps should be in identical grammatical form. Parallelism is important in all writing but in instructions especially, because repeating grammatical forms emphasizes the step-by-step organization of the instructions.

Not parallel	To log on to the VAX 950, follow these steps: 1. Switch the terminal to "on." 2. The CONTROL key and C key are pressed simultaneously. 3. Keying LOGON, and pressing the ESCAPE key. 4. Key your user number, and press the ESCAPE key.
Parallel	To log on to the VAX 950, follow these steps: 1. Switch the terminal to "on." 2. Press the CONTROL key and C key simultaneously. 3. Key LOGON, and press the ESCAPE key. 4. Key your user number, and press the ESCAPE key.

Parallelism increases readability and lends continuity to the instructions.

EFFECTIVE PAGE DESIGN

Instructions rarely get undivided attention. The reader, in fact, is doing two things more or less at once: interpreting the instructions and performing the task. Effective instructional design is *accessible* and *inviting*. An effective design conveys the sense that the task is within a qualified user's range of abilities.

Here are suggestions for designing instructions that help users find, recognize, and remember what they need:

How to design your instructions

- ◆ Use informative headings that tell readers what to expect, that emphasize what is most important, and that serve as an aid to navigation. A heading such as "How to Initialize Your Blank Disk" is more informative than "Disk Initializing." Also, the second version sounds less like a person speaking and more like a robot!
- ◆ Arrange all steps in a numbered list.
- ◆ Single-space within steps and double-space between, to separate steps visually.

(Use the Spacing Before and After Paragraph feature of your word-processing program for maximum editing flexibility.)

◆ Double-space to signal a new paragraph, instead of indenting.

◆ Use white space and highlighting to separate discussion from step.

> Set off your discussion on a separate line (like this), indented or highlighted or both. You can highlight with underlining, capitals, dashes, parentheses, and asterisks. Alternatively, you can use **boldface**, *italics*, varying type sizes, and varying `type-faces.`

◆ Set off warnings, cautions, and notes in ruled boxes or use highlighting and plenty of white space.

◆ Keep the visual and the step close together. If room allows, place the visual right beside the step; if not, right after the step. Set off the visual with plenty of white space.

◆ Strive for format variety that is appealing but not overwhelming or inconsistent. Readers can be overwhelmed by a page with excessive or inconsistent graphic patterns.

The more accessible and inviting the design, the more likely your readers will follow the instructions. Don't be afraid to experiment until you find a design that works. Then, as with all aspects of effective written instructions, test the usability of your document. For advice on usability testing, see Chapter 17.

A SAMPLE SET OF INSTRUCTIONS

The instructions for *doing something* (felling a tree) in Figure 16.2 on pages 363–66 illustrate aspects of our earlier advice. They will appear in a series of brochures for forestry students about to begin summer jobs with a Forestry Service.

Instructions for Semi-technical Readers

> **AUDIENCE/PURPOSE PROFILE.** I'm writing these instructions for partially informed readers who know how to use chainsaws, axes, and wedges but who are approaching this dangerous procedure for the first time. Therefore, I'll include no visuals of cutting equipment (chainsaws and so on) because the audience already knows what these items look like. I can omit basic information (such as what happens when a tree binds a chainsaw) because the audience already has this knowledge. Likewise, these readers need no definition of general forestry terms such as *culling* and *thinning*, but they *do* need definitions of terms that relate specifically to tree felling (*undercut, holding wood*, and so on).
>
> To ensure clarity, I will illustrate the final three steps with visuals. The conclusion, for these readers, will be short and to the point—a simple summary of major steps with emphasis on safety. ▲

INSTRUCTIONS FOR FELLING A TREE

INTRODUCTION

Forestry Service personnel fell (cut down) trees to cull or thin a forested plot, to eliminate the hazard of dead trees standing near power lines, to clear an area for construction, and the like.

These instructions explain how to remove sizable trees for personnel who know how to use a chainsaw, axe, and wedge safely.

When you set out to fell a tree, expect to spend most of your time planning the operation and preparing the area around the tree. To fell the tree, you will make two chainsaw cuts, severing the stem from the stump. Depending on the direction of the cuts, the weather, and the terrain, the severed tree will fall into a predetermined clearing.

> **WARNING:** Although these instructions cover the basic procedure, felling is very dangerous. Trees, felling equipment, and terrain vary greatly. Even professionals sometimes are killed or injured because of judgment errors or misuse of tools.
>
> Your main concern is safety. Be sure to have an expert demonstrate this procedure before you try it. Also, pay attention to warnings in steps 1 and 4.

To fell sizable trees, you will need this equipment:
   —a 3- to 5-horsepower chainsaw with a 51-cm (20-inch) blade
   —a single-blade splitting axe
   —two or more 31-cm (12-inch) steel wedges

The major steps in felling a tree are (1) choosing the lay, (2) providing an escape path, (3) making the undercut, and (4) making the backcut.

REQUIRED STEPS

1. *How to Choose the Lay*

   The "lay" is where you want the tree to fall. On level ground in an open field, which way you direct the fall makes little difference. But such ideal conditions are rare.

   Consider ground obstacles and topography, surrounding trees, and the condition of the tree to be felled. Plan your escape path and the location of your cuts depending on surrounding houses, electrical wires, and trees. Then follow these steps:

   a. Make sure the tree still is alive.

Figure 16.2 A Set of Instructions   *(continued)*

b. Determine the direction and amount of lean.

> **WARNING:** If the tree is dead and leaning substantially, do not try to fell it without professional help. Many dead, leaning trees have a tendency to split along their length, causing a massive slab to fall spontaneously.

c. Find an opening into which the tree can fall in the direction of lean or as close to the direction of lean as possible.

Because most trees lean downhill, try to direct the fall downhill. If the tree leans slightly away from your desired direction, use wedges to direct the fall.

2. *How to Provide an Escape Path*

Falling trees are unpredictable. Avoid injury by planning a definite escape path. Follow these steps:

a. Locate the path in the direction opposite of the fall (Figure 1).

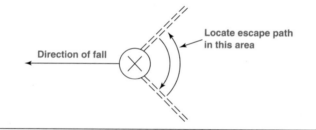

FIGURE 1 Escape-path Location

b. Clear a path 70 cm (2 feet) wide extending beyond where the top of the tree could land.

3. *How to Make the Undercut*

The undercut is a triangular slab of wood cut from the trunk on the side toward which you want the tree to fall. Follow these steps:

a. Start the chainsaw.

b. Holding the saw with blade parallel to the ground, make a first cut 70 to 92 cm (2 to 3 feet) above ground. Cut horizontally, to no more than $\frac{1}{3}$ the tree's diameter (Figure 2).

Figure 16.2   A Set of Instructions   *(continued)*

3

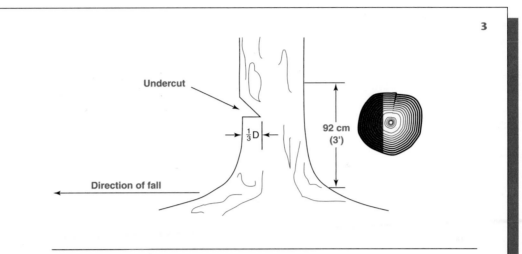

FIGURE 2 Making the Undercut

c. Make a downward-sloping cut, starting 10 to 15 cm (4 to 6 inches) above the first so that the cuts intersect at $\frac{1}{3}$ the diameter (Figure 2).

4. *How to Make the Backcut*

After completing the undercut, you make the backcut to sever the stem from the stump. This step requires good reflexes and absolute concentration.

> **WARNING:** Observe tree movement closely during the backcut. If the tree shows any sign of falling in your direction, drop everything and move out of its way. Also, do not cut completely through to the undercut; instead, leave a narrow strip of "holding wood" as a hinge, to help prevent the butt end of the falling tree from jumping back at you.

To make your backcut, follow these steps:

a. Holding the saw with the blade parallel to the ground, start your cut about 7.5 cm (3 inches) above the undercut, on the opposite side of the trunk. Leave a narrow strip of "holding wood" (Figure 3).

Figure 16.2   A Set of Instructions *(continued)*

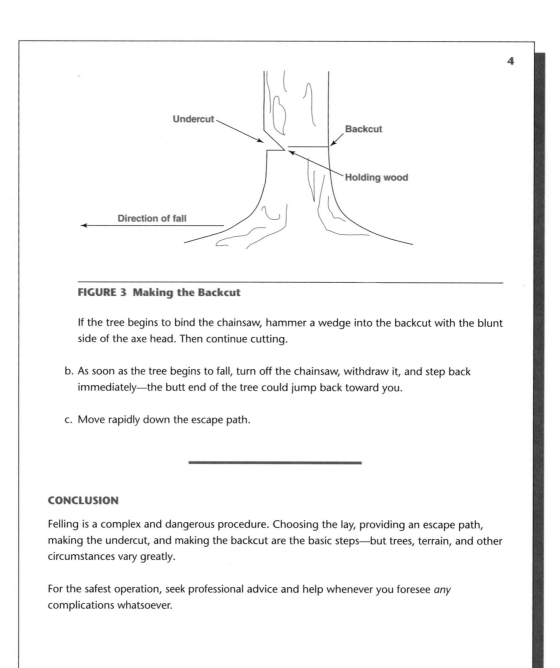

4

FIGURE 3 Making the Backcut

If the tree begins to bind the chainsaw, hammer a wedge into the backcut with the blunt side of the axe head. Then continue cutting.

b. As soon as the tree begins to fall, turn off the chainsaw, withdraw it, and step back immediately—the butt end of the tree could jump back toward you.

c. Move rapidly down the escape path.

CONCLUSION

Felling is a complex and dangerous procedure. Choosing the lay, providing an escape path, making the undercut, and making the backcut are the basic steps—but trees, terrain, and other circumstances vary greatly.

For the safest operation, seek professional advice and help whenever you foresee *any* complications whatsoever.

Figure 16.2 **A Set of Instructions**

PROCEDURES

Procedures differ from instructions in two major respects:

1. The reader already knows how to perform the tasks outlined in the procedure and thus does not need detailed instructions.
2. The reader does not necessarily know the order in which tasks are to be performed, or how and when to coordinate with other members of a work team.

Therefore, procedures are usually aimed at groups of readers; the procedures describe how the group members will coordinate their activities, or when they will perform their particular functions. Examples include evacuation procedures (in case of a fire or toxic spill), maintenance procedures, or installation procedures.

Not all procedures are aimed at groups, however. A software installation will likely be performed by a single technician who has the skills and general knowledge to do the job but who needs to know the specific variations of the procedure required by a given software program. Other procedures such as a union grievance or a procedure for reporting an on-site accident will assume that the reader needs to know whom to contact, when to contact, and what form to use. In these circumstances, the assumption is that the reader doesn't need to be told *how* to grieve the perceived infraction or *how* to phrase the accident report.

The following excerpt from a group procedure comes from a Canadian power utility. The procedure demonstrates how workers must coordinate their skills with others. In the conductor repair procedure, that coordination is essential in preventing the instant death that could result from a botched manoeuvre.

CONDUCTOR REPAIR ON 72/138 Kv H-FRAME

(INSTALL PRE-FORMED ARMOUR SPLICES USING A SECOND UNIT TO SUPPORT CONDUCTORS)

1. Operate in a crew that has a minimum of four (4) certified Journeymen Power Line Technicians.

2. Conduct a tailboard session to confirm the order in which this procedure is to be performed.

3. Position the aerial device in the best position to maintain safe working clearances while allowing two line technicians to work satisfactorily on the line. Verify safe working clearances with the measuring stick, as outlined in the General Rules.

    NOTE: Position the aerial device directly under the centre phase. Back in with the turret two (2) metres from the centre of the structure. Park the boom truck, which has a wire holder on an insulated jib, on the other side of the structure in such a way that the truck can support each conductor without having to move the truck. Install unit grounding on both vehicles.

4. *Line technicians*—Carry all tools and materials along in the buckets.

5. *Line technicians*—Test the insulators. Install a temporary jumper if it's necessary.

6. *Second unit operator*—Support the outside conductor with the jib and raise the conductor slightly.

7. *Line technician in the insulated aerial bucket*—Move into position while maintaining both phase-to-phase and phase-to-ground clearance. Disconnect the clamp from the insulators with hot sticks.

8. *Second unit operator*—Move the conductor to a position where pre-formed armour rod may be installed without reducing the limits of approach.

    The line technicians may now bond on and complete the repairs.

9. *Line technicians*—Remove the bonds and back away. Move the conductor back to the insulators and use hot line tools to attach a suspension clamp to the insulators again.

10. Repeat this operation on the other outside phase.

11. *Line technicians*—Connect the centre phase and loosen the suspension clamp with hot sticks. The clamp may be slid out on the conductor to a point where the technician in the insulated aerial device can maintain clearances and bond on to the line. After replacing the suspension clamp, you may remove the bonds and move into position to use hot line tools to apply the armour rod.

Figure 16.3   Sample Technical Procedure

Checklist FOR EDITING AND REVISING INSTRUCTIONS

Use this checklist to evaluate the usability of instructions.

Content

◆ Does the title promise exactly what the instructions deliver?
◆ Is the background adequate for the intended audience?
◆ Do explanations enable readers to understand what to do?
◆ Do examples enable readers to see how to do it correctly?
◆ Are the definition and purpose of each step given as needed?
◆ Is all needless information omitted?
◆ Are all obvious steps omitted?
◆ Do notes, cautions, or warnings appear whenever needed, before the step?
◆ Is the information rate appropriate for readers' abilities and the difficulty of this procedure?
◆ Are visuals adequate for clarifying the steps?
◆ Do visuals repeat prose information whenever necessary?

Page Design

◆ Does each heading clearly tell readers what to expect?
◆ Are steps single spaced within, and double spaced between?
◆ Do white space and highlights set off discussion from steps?

◆ Is everything accurate?
◆ Are notes, cautions, or warnings set off or highlighted?
◆ Are visuals beside or near the step, and set off by white space?

Organization

◆ Is the introduction adequate without being excessive?
◆ Do the instructions follow the exact sequence of steps?
◆ Is all the information for a particular step close together?
◆ For a complex step, does each sentence begin on a new line?
◆ Is the conclusion necessary and, if necessary, adequate?

Style

◆ Do introductory sentences have enough variety to maintain interest?
◆ Does the familiar material appear *first* in each sentence?
◆ Do steps generally have short sentences?
◆ Does each step begin with the action verb?
◆ Are all steps in the active voice and imperative mood?
◆ Do all steps have parallel phrasing?
◆ Are transitions adequate for marking time and sequence?

EXERCISES

1. Improve the readability of the following instructions by editing them for a more appropriate voice and design.

   What to Do Before Jacking Up Your Car

   Whenever the misfortune of a flat tire occurs, some basic procedures should be followed before the car is jacked up. If possible, your car should be positioned on as firm and level a surface as is available. The engine has to be turned off; the parking brake should be set; and the automatic transmission shift lever must be placed in "park," or the manual transmission lever in "reverse." The wheel diagonally opposite the one to be removed should have a piece of wood placed beneath it to prevent the wheel from rolling. The spare wheel, jack, and lug wrench should be removed from the luggage compartment.

2. Select a specialized process that you understand well and that has several distinct steps. Using the process analysis on pages 346–52 as a model,

explain this process to colleagues who are unfamiliar with it. Begin by completing an audience/purpose profile (page 34). Some possible topics: how the body metabolizes alcohol, how economic inflation occurs, how the federal deficit affects our future, how a lake or pond becomes a swamp, how a volcanic eruption occurs.

3. Choose a topic from the following list, your major, or an area of interest. Using the general outline in this chapter as a model, outline instructions that require at least three major steps. Address a general reader, and begin by completing an audience/purpose profile. Include (1) all necessary visuals; or (2) a brief (page 276) and a rough diagram for each visual; or (3) a "reference visual" (a copy of a visual published elsewhere) with instructions for adapting your visual from that one. (If you borrow visuals from other sources, provide full references.)

planting a tree	hitting a golf ball
hot-waxing skis	removing the rear
hanging wallpaper	wheel of a bicycle
filleting a fish	avoiding hypothermia

4. Select any one of the instructional visuals in Figure 16.1 and write a prose version of those instructions—without using visual illustrations or special page design. Bring your version to class and be prepared to discuss the conclusions you've derived from this exercise.

5. Locate an example of five or more visuals from the following list.
A visual that shows:

   ◆ how to locate something
   ◆ how to operate something
   ◆ how to handle something
   ◆ how to assemble something
   ◆ how to position something
   ◆ how to avoid damage or injury
   ◆ how to diagnose and solve a problem
   ◆ how to identify safe or acceptable limits
   ◆ how to proceed systematically
   ◆ how to make the right decision
   ◆ why an action or procedure is important

   Bring your examples to class for discussion, evaluation, and comparison.

COLLABORATIVE PROJECTS

Any of the exercises may be a collaborative project.

CHAPTER 17

Manuals and Usability Testing

TYPES OF MANUALS

MANUAL WRITING

PARTS OF MANUALS

FORMAT CONSIDERATIONS

USABILITY TESTING

Guidelines for Testing a Document's Usability

Checklist for Usability

This chapter describes manuals and how to write them, and then discusses aspects of testing manuals for their usability. Manuals consist primarily of instructions, often printed and bound in book form. Their use has increased, as products become increasingly complex. Even an electric wok comes with a manual. And relatively simple printers often come with 60-page operating manuals.

Why are manuals so common and so detailed? The majority of published manuals exist to help users get the most out of purchased equipment. Also, well-written manuals endear the company to consumers, instead of alienating them. Further, companies who purchase equipment and software programs do so in order to improve productivity; that goal is undercut if employees can't properly use the equipment or software.

Equipment and software suppliers have found that clearly presented, properly documented, "user-friendly" manuals reduce the number of complaints, enquiries, and warranty claims from customers. In effect, good manuals pay for themselves. Poor manuals, on the other hand, lead to overuse of the company's customer service staff and alienate users. Besides, badly written manuals require frequent revisions, an expensive process.

Even when a product supplier provides extensive employee training along with its product, user and maintenance manuals refresh employees' memories and reduce the incidence of equipment breakdowns.

TYPES OF MANUALS

There is a manual for every purpose. Table 17.1 shows a partial list:

Table 17.1   Types of Manuals and Their Purpose   *(continued)*

Type of Manual	Purpose
Software manuals	Help users navigate the intricacies of computer software
Operating manuals	Instruct users how to use and care for equipment and operating systems
Repair and service manuals	Provide detailed technical information for technicians (the complete service manual for your automobile is much more detailed and probably 20 times thicker than your operator's manual)
Maintenance and troubleshooting manuals	Provide users with detailed information on how to maintain equipment and troubleshoot common problems. They are vital for many businesses, including large-scale installations. For example, Doppelmayr's maintenance manual for its detachable chair lift system provides several hundred pages of specific maintenance instructions, covering every aspect of the system's hardware and operation

Table 17.1 Types of Manuals and Their Purpose

Type of Manual	Purpose
Test manuals	Describe techniques, equipment, and ideal results for lab and field tests of everything from soil samples to sludge samples to electronic equipment
Installation manuals	Outline how to install complicated equipment or software
Construction manuals	Describe construction procedures, pertinent building codes, and contract requirements
Reference manuals	Allow users to easily find specific information
Tutorial manuals	Take trainees through page after page of skill development exercises and explanations of how a system works
Documentation manuals	Include tutorials and descriptions (instructions) of how to do things and explanations of how certain things work (process analysis)
Business practice manuals	Describe procedures for business operation, advertising policies, customer service methods, collections and complaint handling, business travel procedures, and a multitude of business guidelines. (One of the first, and most famous, business practice manuals was written by Ray Kroc as he began to franchise the McDonald's restaurants across North America.)
User manuals	Provide detailed descriptions of how to use equipment. These are the most common types of manuals. If you want to program your VCR, you turn to the manual; if you want to re-configure your computer's hard drive, you refer to the appropriate manual.

MANUAL WRITING

A complex, collaborative process

Writing manuals is quite complex: many are book-length, with a variety of topics to cover. Full-length manuals usually require more skills and time than one person can devote to the project. Even full-time manual writers collaborate with product designers, page designers, and graphic artists, all of whom add their specialized skills and perspective to the project.

The main challenge in writing (and designing) a manual is reducing the reader's reluctance to use the document; manuals, especially long ones, can overwhelm readers. So, manual writers must employ all the aspects of "user-friendly" instructions described in Chapter 16.

PLANNING

Audience/purpose
analysis helps you
choose the content
and structure

You should begin the writing process by analyzing your manual's purpose. Each type of manual requires a primary purpose that you can clarify by analyzing who will read your manual and how they will use it. That analysis will guide your choice of what to include in the manual. The readers' needs and knowledge levels will later dictate the technical level and the amount of detail included in each section.

After your preliminary analysis of audience and purpose, you need to talk with the product developers and the project manager to learn more about the product and its uses. This consultative stage may take several meetings. Among other things, you need to determine:

- the product's features and specifications
- exactly how the product works
- the product's full range of uses and applications
- how to troubleshoot problems and malfunctions
- how to install or set up the product
- how to operate, maintain, and adjust the device
- how to obtain parts and accessories
- how to claim warranty work
- which aspects of the product's use could result in damage to the product itself or the operator or to the system in which the product's installed

The manual should be organized in the order in which readers will use various sections.

It's important to revise the manual's contents *before* completing a detailed writing outline and it's essential to examine (and possibly revise) that outline before composing the first draft. Also, you should show your completed outline to members of the manual production team before starting to write the manual. The outline should have the entire headings system in place, along with the names and locations of all illustrations and a brief description of each paragraph or block of instructions.

DRAFTING

Writing the actual instructions and technical descriptions will be relatively easy if the outline has been thoroughly prepared. Perhaps all the paragraphs and instruction blocks will be written by one person, or perhaps various sections will be written by their respective developers. In the latter case, one person or a small group of editors will revise and edit all of the sections to ensure consistent style and to avoid duplication. In some cases, the writer will create the manual's graphics; however, most large organizations employ graphic artists to create artwork and technical illustrations.

TESTING AND REVISING

Measurements of readability can be very helpful (see Chapter 11), but the best way to determine audience reaction is to have members of the target market test the manual. This is especially important for the instructions sections—ask readers to follow their literal interpretations of the instructions. You'll see which instructions are unclear or incomplete, and you can revise accordingly. Then, after revising the content and editing the phrasing, retest the manual on a new batch of readers to see if the revisions have helped.

The remaining task is to ruthlessly edit all the manual's phrasing to make it as direct, clear, and concise as possible. The advice in Chapter 11 certainly applies to the descriptive writing in manuals.

PARTS OF MANUALS

The two main parts of manuals are: (1) body sections, and (2) supplementary sections.

BODY SECTIONS

Manuals do not strictly follow a beginning/middle/closing sequence. However, an introductory section (or two) is needed to orient the reader.

Introduction. At its beginning, a manual should include:

- the manual's purpose and scope
- an overview of how to use the manual
- key definitions and principles
- introductory descriptions of equipment or procedures
- accompanying illustrations (photographs or diagrams of equipment, flow-charts that show the relationships among sets of instructions)

Early on, a manual might also provide:

- specifications for equipment and equipment operation
- background information, such as an overview of the equipment or procedures that are being replaced
- a list of accessories and/or related equipment

Use topical headings or talking headings

The heading title for an introductory section could be topical; for example, the title "Introductions and Specifications" would suit a user's manual for a sophisticated piece of electronics test equipment, especially if the manual is aimed at electronics technicians. On the other hand, the opening sections for a mass-marketed CD player might use talking headings such as:

- Key Facts About Your CD Player
- How to Install the CD Player in Your Stereo System

Remaining Sections. The number of remaining sections varies with the manual's purpose and readers. Table 17.2 illustrates this principle.

Order of Sections. The order of sections within a manual may be based on chronological development, order of importance, or spatial development, as illustrated in Table 17.3.

In organizing the components of a manual, you might consider creating two or more separate, related documents. For, example, ACCPAC International uses

Table 17.2   **Section Variations**

380-page operating and service manual for a Phillips oscilloscope:	60-page operating manual for a Sony video cassette recorder:	Onkyo Tuner/ Amplifier instruction manual:
◆ Operating Instructions ◆ Theory of Operations ◆ Maintenance Instructions ◆ Calibration Instructions	◆ Preliminaries ◆ Preparation ◆ Operation ◆ Other Information	◆ Features, Safeguards, and Precautions ◆ Before Using This Unit ◆ System Connections ◆ Controls ◆ Operations ◆ Troubleshooting Guide ◆ Specifications

Table 17.3   **Order of Manual Sections**

Order	Example/Description
Chronological development	Sets of instructions for an electronic power supply could be provided in the order in which a user would use the device: ◆ Unpacking and Installing the Power Supply ◆ Connecting the Power Supply to a Load ◆ Operating and Adjusting the Supply ◆ Maintaining and Calibrating the Supply ◆ Troubleshooting Operational Problems ◆ Ordering Parts and Accessories
Order of importance	This structure is often used for reference manuals and test manuals. The most important (and most-used) sections are placed first.
Spatial development	The physical structure of a system can form the basis for organizing that system's corresponding manual. For example, the maintenance manual for a ski lift could be divided into sections such as: ◆ Cable Maintenance ◆ Chairs and Grips ◆ Loading Ramps ◆ Drive Systems ◆ Housings, etc.

three manuals for its Simply Accounting® software: (1) Getting Started, (2) Work-book, and (3) User Guide.

Conclusion. Manuals do not usually include a separate Conclusion section at the end of the manual. If conclusion material is included, it is placed at the end of a body section. For example, the end of a set of repair instructions might indicate the standard length of time to perform the repair and might reiterate key steps to check if the repair hasn't solved the identified problem.

SUPPLEMENTARY SECTIONS

Manuals include supplementary matter to help readers use the document. Depending on the type of manual, you should include some or all of the following *front matter*.

Cover and Title Page.   Essentially, a cover's purpose is to protect the manual, but the cover also carries the manual's title and, where appropriate, the company name and logo, and equipment model number. As Figures 17.1 and 17.2 on pages 378 and 379 illustrate, a manual's corresponding title page carries more information. At a minimum, the title page names the document's intended use (in the title), the manufacturer's name and logo, the date (or year) of publication, and the name and model number of the product. Very seldom do manual writers' names appear on the title page. Well-designed covers carry design principles from the cover to the title page.

Tables of Contents and Illustrations.   The longer and more complex a manual is, the more its readers need a table of contents and a list of illustrations to find particular topics within the manual. See Chapter 20 for detailed advice on how to format a table of contents and list of figures and tables.

Warnings and Cautions.   Manuals often require warnings and cautions, especially for electrically powered equipment. Often, these warnings will appear on the inside front cover. Sometimes warranty information appears with the warnings and cautions.

Glossary.   A glossary might appear in the front matter, if a reader needs to know the meaning of certain terms before starting to use the manual. Alternatively, the introductory section could define key terms for the reader.

Depending upon the type of manual, you may need to include some back matter.

Figure 17.1 Sample Manual Cover

Source: Courtesy of Curt Willis

OPERATING AND SERVICE MANUAL

SPIDER™ SUPPLY

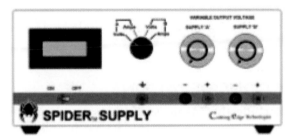

Variable Dual 20-Volt Channel Voltage Regulated Power Supply Model 001a Series 0.1.0-98a

Copyright © Cutting Edge Technologies 1998
4131 Web Cresent, Kelowna, British Columbia V1W 1V8

Figure 17.2 Sample Manual Title Page

Source: Courtesy of Curt Willis

Back Matter.   A manual's back matter might include some or all of the following:

- information on replacement parts and ordering
- schematics
- contact details for dealers and service depots
- service log
- component specifications tables
- construction codes

There are as many possible combinations of back matter as there are types of manuals. You can determine the best combination of back matter sections for your manual by examining a variety of manuals in your field.

Indexes.   Indexes are commonly used in long or complex manuals, such as computer user manuals.

FORMAT CONSIDERATIONS

In many ways, effective manuals use the same design principles that should be applied to formal reports and books. These principles are outlined in Chapters 13 and 16. However, certain special aspects of manual format should be remembered.

Section Identification

In order to constantly orient your manual's reader, use headers (or footers) that combine the page number with a diminutive section identifier. (This text uses a version of that system.) Your reader will then have three methods to easily find particular sections in the manual:

1. the table of contents
2. the index
3. the page headers/footers

Figure 17.3 shows another method of designing a section identifier. Some manuals also use sturdy card stock or plastic tabs along the right-hand edge to distinguish one section from another.

Section 6: CALIBRATION

To keep your SPIDER SUPPLY running at an optimal level, you may need to calibrate it. Calibrating the power supply is simple and will not likely need to be performed often.

Several possible circumstances could cause your power supply to drift from its specified settings: extreme temperature changes, over-drawing output current, abuse, or long-term use. The following calibration instructions will help you bring your supply back up to specifications, should the need arise.

6.1 Non-powered Tests

If you plan to calibrate your supply, you should do these quick no-power tests as well. They take little time and require minimal equipment.

To ensure that all connections are indeed *connected*, perform the following checks:

✓ Check every component solder joint and header connection to confirm that they are sound and properly affixed.

✓ Check the following resistances with a DMM:

6.1 Non-powered Tests continued . . .

✓ Confirm the following resistances with a DMM:

Table 6-5 Variable voltage resistor (dials) checks

Point	Front Panel Voltage Dial Adjustment	Resistance Measurement
Between J1 & J2	SUPPLY 'A' fully CCW	150 Ω
Between J1 & J2	SUPPLY 'A' fully CW	2.60 Ω
Between J3 & J4	SUPPLY 'B' fully CCW	150 Ω
Between J3 & J4	SUPPLY 'B' fully CW	2.60 Ω

Figure 17.3   Variations in Section Identifiers Used for the First and Second Pages of a Section

Source: Courtesy of Curt Willis

HEADINGS SYSTEM

Headings and subheadings are particularly valuable in lengthy, complex manuals, so make headings stand out by using boldface or colour. Clearly distinguish among levels of headings by using capitalization, type size, and heading placement. You might also use one kind of typeface for the main headings and another kind for the other levels of headings. Figure 17.4 shows colour and subtle shadowing being used in an electronics student's manual for the power supply he has built.

Section 1: INTRODUCTION

We hope you're as excited as we are about the new 1998 version of the industrious SPIDER SUPPLY. The SPIDER SUPPLY Dual 20VDC Voltage Regulated Power Supply is a reliable, user-friendly DC output device.

1.1 Features

The Basics (batteries are included)

The SPIDER SUPPLY comes with many easy-to-use features for your convenience. The new liquid crystal digital display allows you to view the output with more clarity and accuracy than our past analogue meters.

Our new voltage regulated 20 volt DC supplies either a single or dual channel output, ideal for use with OP-Amp circuits. Your power supply also includes a circuit common ground, as well as a true earth ground. And new cutting-edge regulators in accordance with our vented case result in improved heat dissipation, and thus a more durable output.

Figure 17.4   Colour and Subtle Shadowing Used to Distinguish Headings
Source: Courtesy of Curt Willis

Flashy graphics attract attention, but the headings must also be clearly phrased to be usable. Notice how effectively the following heading (from Hewlett-Packard's *HP Desk Jet 720 Series Printer Users' Guide*) signals the content of its section:

How to Print on Different Paper Sizes

Compare that heading to much less useful headings such as:

Printing and Paper Size Variations

or

Paper Differentiation Techniques

PAGE LAYOUT

Chapter 16 discusses effective page design for instructions. Follow that advice, but also remember to use a format that can comfortably accommodate both point-form instructions and paragraphed explanations.

Hints for Effective Page Layout. The following useful hints for effective page layout are illustrated in Figures 17.1 to 17.4:

- Use lots of open (white) space.
- Use pictures, diagrams, or graphs to show the *results* of the manual user's actions as well as how to achieve those results.
- Place lengthy explanations in boxes.
- Use typefaces wisely:
  a. bold, italicize, or change type size before changing typeface
  b. use strong typefaces (Times New Roman, **Frutiger**, **Impact**, SABON, Universe)

SYMBOLS AND DESIGN GRAPHICS

Hints, warnings, explanations, and special instructions can be highlighted by placing them in boxes and by using symbols to capture the reader's attention. Think of how you might use some of the following symbols that are found in MS Word's "Insert" pull-down menu. Many more symbols are available from various

locations on the Internet.

Design Graphics Benefits. Design graphics, such as those illustrated in Figures 17.3 and 17.4, provide three main benefits:

1. They contribute to a professional appearance.
2. They help set up a balanced page layout.
3. They help create an inviting, accessible look that encourages readers to use the manual.

Design Techniques. The simplest design technique is to use italics, bolding, underlining, and quotation marks for special effects, and to use each of those techniques for one effect. For example, you could use

- italics for emphasizing key words or phrases
- bolding for headings and definitions
- underlining for names of printed publications or for emphasizing important words, warnings, etc.
- quotation marks for quoted words or for words used in a special way

Remember to use such techniques consistently within a manual. For example, don't use italics to emphasize a phrase in one place, and bolding to emphasize a phrase in another place.

Of all the kinds of documents produced by technical writers, a manual is the most likely to be *used,* not just read. Therefore, the document's usability needs to be tested before publication and distribution. A *usable* document is safe, dependable, and easy to use.

Most of the following advice for testing a manual's usability can be applied to other types of documents, such as product brochures, catalogues, stand-alone instruction sheets, and technical reports.

USABILITY TESTING

Companies routinely measure the usability of their products—including the **documentation** that accompanies the product (warnings, explanations, instructions for assembling or operating, and so on). To keep their customers—and to avoid lawsuits—companies go to great lengths to identify flaws or to anticipate all the ways a product might fail or be misused.

The purpose of usability testing is to keep what works in a product or a document and to fix what doesn't.

HOW USABILITY TESTING IS DONE

Usability testing usually occurs at two levels (Petroski 90): (1) *alpha testing,* by the product's designers or the document's authors, and (2) *beta testing,* by the actual users of the product or document.

For a document to achieve its objectives—to be *usable*—users must be able to do at least three things (Coe, *Human Factors* 193; Spencer 74):

What a usable document enables readers to do

- ◆ easily locate the information they need,
- ◆ understand the information immediately, and
- ◆ use the information successfully.

To measure usability in a set of instructions, for instance, we ask this question: *How effectively do these instructions enable all users to carry out the task safely, efficiently, and accurately?*

Ideally, usability tests for documents occur in a setting that simulates the actual situation, with people who will actually use the document (Redish and Schell 67; Ruhs 8). But even in a classroom setting we can reasonably assess a document's effectiveness on the basis of usability criteria discussed below.

HOW USABILITY CRITERIA ARE DETERMINED

Usability criteria for workplace documents are based on answers to these questions about the task, the user, and the setting:

Questions for assessing usability

- ◆ What specific task is being documented?
- ◆ What do we know about the specific users' abilities and limitations?
- ◆ In what setting will the document be read/used?

Only after these questions are answered can we decide on criteria for evaluating the effectiveness of a specific document.

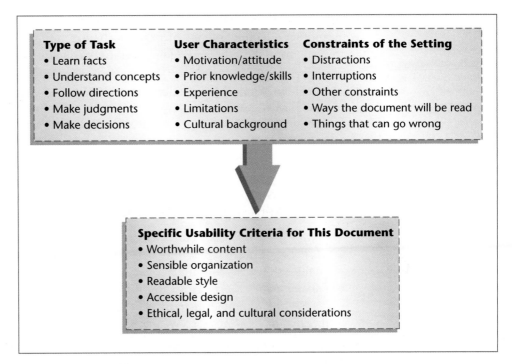

Figure 17.5   Defining a Document's Usability Criteria

The Checklist for Usability (page 389) identifies criteria shared by many technical documents. In addition, specific elements (visuals, page design) and document types (proposals, memos, instructions) have their own usability criteria.

USABILITY ISSUES IN ON-LINE OR MULTI-MEDIA DOCUMENTS

In contrast to printed documents, on-line or multi-media documents have unique usability issues (Holler 25; Humphreys 754–55):

- ◆ On-line documents tend to focus more on "doing" than on detailed explanations. Workplace readers typically use on-line documents for reference or training rather than for study or memorizing.
- ◆ Users of on-line instructions rarely need persuading (say, to pay attention or follow instructions) because they are guided interactively through each step of the procedure.
- ◆ Visuals play an increasingly essential role in on-line instruction.
- ◆ On-line documents are typically organized to be read interactively and selectively rather than in linear sequence. Users move from place to place, depending on their immediate needs. Organization is therefore flexible and modular, with small bits of easily accessible information that can be combined to suit a particular user's needs and interests.
- ◆ On-line readers can easily lose their bearings. Unable to shuffle or flip through a stack of printed pages in linear order, readers need constant orientation (to

retrieve some earlier bit of information or to compare something on one page with another). In the absence of page or chapter numbers, index, or table of contents, "Find," "Search," and "Help" options need to be plentiful and complete.

See Chapter 20 for usability considerations in Web pages and other electronic documents.

GUIDELINES

for Testing a
Document's
Usability

1. *Identify the document's purpose.* Determine how much *learning* versus *performing* is required in performing the task—and assess the level of difficulty (Mirel et al. 79; Wickens 232, 243, 250):

   ◆ *Simply learning facts.* "How rapidly is this virus spreading?"
   ◆ *Mastering concepts or theories.* "What biochemical mechanism enables this virus to mutate?"
   ◆ *Following directions.* "How do I inject the vaccine?"
   ◆ *Navigating a complex activity that requires decisions or judgments.* "Is this diagnosis accurate?" "Which treatment option should I select?"

2. *Identify the human factors.* Determine which characteristics of the user and the work setting (**human factors**) enhance or limit performance (Wickens 3):

   ◆ *Users' abilities/limitations.* What do these users know already? How experienced are they in this task area? How educated? What cultural differences could create misunderstanding?
   ◆ *Users' attitudes.* How motivated or attentive are they? How anxious or defensive? Do users need persuading to pay attention or be careful?
   ◆ *Users' reading styles.* Will users be scanning the document, studying it, or memorizing it? Will they read it sequentially (page by page) or consult the document periodically and randomly? (For example, expert users often look only for key words and key concepts in titles, abstracts, or purpose lines, then determine what sections of the document they will read.)
   ◆ *Workplace constraints.* Under what conditions will the document be read? What distractions or interruptions does the work setting pose? (For example, procedures for treating a choking victim, posted in a busy restaurant kitchen.) Will users always have the document in front of them while performing the task?
   ◆ *Possible failures.* How might the document be misinterpreted or misunderstood (Boiarsky 100)? Any potential "trouble spots" (material too complex for these users, too hard to follow or read, too loaded with information)?

GUIDELINES

for Testing a
Document's
Usability

(continued)

For a closer look at human factors, review the Audience/Purpose Profile Sheet (page 34).

3. *Design the usability test.* Ask respondents to focus on specific problems (Hart 53–57; Daugherty 17–18):

   ◆ *Content:* "Where is there too much or too little information?" "Does anything seem inaccurate?"

   ◆ *Organization:* "Does anything seem out of order, or hard to find or follow?"

   ◆ *Style:* "Is anything hard to understand?" "Are any words inexact or too complex?" "Do any expressions seem wordy?"

   ◆ *Design:* "Are there any confusing headings, or too many or too few?" "Are any paragraphs, lists, or steps too long?" "Are there any misleading or overly complex visuals?" "Could any material be clarified by a visual?" "Is anything cramped and hard to read?"

   ◆ *Ethical, legal, and cultural considerations:* "Does anything mislead?" "Could anything create potential legal liability or cross-cultural misunderstanding?"

   Figure 17.6 shows a basic usability survey (Carliner 258). Notice how the phrasing encourages users to respond with examples instead of just "yes" or "no."

4. *Administer the test.* Have respondents perform the task under controlled conditions. To ensure dependable responses, look for these qualities (Daugherty 19–20):

   ◆ *A range of responses.* Select evaluators at various levels of expertise with this task.

   ◆ *Independent responses.* Ask each evaluator to work alone when testing the document and recording the findings.

   ◆ *Reliability.* Have evaluators perform the test twice.

   ◆ *Group consensus.* After individual testing, arrange a meeting for evaluators to compare notes and to agree on needed revisions.

   ◆ *Thoroughness.* Once the document is revised/corrected, repeat this entire testing procedure.

Basic Usability Survey

1. Briefly describe why this document is used. _____
_____

2. Evaluate the *content:*
   • Identify any irrelevant information. _____
   _____

   • Indicate any gaps in the information. _____
   _____

   • Identify any information that seems inaccurate. _____
   _____

   • List other problems with the content. _____
   _____

3. Evaluate the *organization:*
   • Identify anything that is out of order or hard to locate or follow. ____
   _____

   • List other problems with the organization. _____
   _____

4. Evaluate the *style:*
   • Identify anything you misunderstood on first reading. _____
   _____

   • Identify anything you couldn't understand at all. _____
   _____

   • Identify expressions that seem wordy, inexact, or too complex. _____
   _____

   • List other problems with the style. _____
   _____

5. Evaluate the *design:*
   • Indicate any headings that are missing, confusing, or excessive. _____
   _____

   • Indicate any material that should be designed as a list. _____
   _____

   • Give examples of material that might be clarified by a visual. _____
   _____

   • Give examples of misleading or overly complex visuals. _____
   _____

   • List other problems with design. _____
   _____

6. Identify anything that seems misleading or that could create legal problems or cross-cultural misunderstanding. _____
   _____

7. Please suggest other ways of making this document easier to use. _____
   _____

Figure 17.6   A Basic Usability Survey

Checklist FOR USABILITY

Content

◆ Is all material relevant to this user for this task?

◆ Is all material technically accurate?

◆ Is the level of technicality appropriate for this audience?

◆ Are warnings and cautions inserted where needed?

◆ Are claims, conclusions, and recommendations supported by evidence?

◆ Is the material free of gaps, foggy areas, or needless details?

◆ Are all key terms clearly defined?

◆ Are all data sources documented?

Organization

◆ Is the structure of the document visible at a glance?

◆ Is there a clear line of reasoning that emphasizes what is most important?

◆ Is material organized in the sequence that users are expected to follow?

◆ Is everything easy to locate?

◆ Is the material "chunked" into easily digestable parts?

Style

◆ Is each sentence understandable the first time it is read?

◆ Is rich information expressed in the fewest words possible?

◆ Are sentences put together with enough variety?

◆ Are words chosen for exactness, and not for camouflage?

◆ Is the tone appropriate for the situation and audience?

Design

◆ Is page design inviting, accessible, and appropriate for the user's needs?

◆ Are there adequate aids to navigation (heads, lists, typestyles)?

◆ Are adequate visuals used to clarify, emphasize, or summarize?

◆ Do supplements accommodate the needs of a diverse audience?

Ethical, Legal, and Cultural Considerations

◆ Does the document indicate sound ethical judgment?

◆ Does the document comply with copyright law and other legal standards?

◆ Does the document respect users' cultural diversity?

EXERCISES

1. Select part of a technical manual in your field, or instructions for a general reader, and make a copy of the material. Use the revision/editing checklist (page 389) to evaluate the sample's usability. In a memo to your instructor, discuss the strong and weak points of the instructions; or be prepared to explain to the class why the sample is effective or ineffective.

2. Create an operating or user manual for a product or system you've developed in one of your classes. Alternatively, write a procedures manual for a series of related techniques and procedures that you've learned in your program of studies. (Examples could include CAD methods, test procedures, fabrication techniques, design principles, or procedures for complying with government regulations.)

COLLABORATIVE PROJECT

Test the usability of a document prepared for this course:

1. As a basis for your "alpha" test, adapt the guidelines on page 386, the Audience/Purpose Profile on page 34, the Checklist for Usability on page 389 and whichever specific checklist applies to this particular document (as on page xx, for example).

2. As a basis for your "beta" test, adapt the Basic Usability Survey on page 388.

3. Revise the document based on your findings.

4. Appoint a group member to explain the usability testing procedure and the results to the class.

Applications

Judith Whitehead, Ottawa

Judith Whitehead has edited over 800 reports in the past 20 years, mostly for the federal government. During that time, she has researched and written about 200 more reports. The subjects have ranged from health care administration to the impact of the spruce budworm on Canadian forests.

The challenge in these reports, says Whitehead, is to "meet the reader's needs and priorities." Everything flows from audience analysis: "That analysis guides the writer's research and, then, the report's content and structure."

Whitehead has ghost-written dozens of proposals for clients; also, she writes her own proposals as she bids for editing and writing contracts. Typically, these proposals range from 15 to 40 pages, but she has produced proposals that exceed 100 pages. According to Whitehead, the secret of writing successful proposals is to "clearly show how the reader will benefit from the proposed product or service."

As an experienced and trusted editor and writer, Judith Whitehead is on the standing offer lists of several government departments. Still, she must write clear, comprehensive proposals to win contracts. One of her many successful techniques is a "compliance matrix," which matches client requirements with components of her proposal.

Proposals

A proposal offers to do something or recommends that something be done. A proposal's general purpose is to improve conditions, authorize work on a project, present a product or service (for payment), or otherwise support a plan for solving a problem or doing a job.

Your proposal may be a letter to your college's dean of Engineering to suggest an interdisciplinary program of studies linked with the Computer Science and Arts programs; it may be a memo to your firm's general manager to request funding for computer training; or it may be a 100-page document to the provincial highways ministry to bid for a contract to design a series of highway overpasses. You may write the proposal alone or as part of a team. It may take hours or months.

TYPES OF PROPOSALS

Proposals are as varied as the situations that generate them, but we can identify three main types:

1. *sales proposals*, which promote business
2. *research proposals*, which are used in academic institutions and in business
3. *improvement proposals* (or *planning proposals*), which suggest how to improve situations or ways of doing things, or which present solutions to problems

Some proposals are sent internally; others are sent to external readers. Either way, the proposal writer has reasons for sending the proposal. But, internal or external, the success or failure of the proposal depends almost entirely on the writer's ability to look at the situation from the *reader's* point of view. Table 18.1 shows a variety of proposals whose writers have tried to use supporting arguments that appeal to their readers.

Some of the proposals in Table 18.1 have been solicited by their readers. These may have to be presented differently from proposals that have not been solicited and so will likely not be anticipated by their receivers.

A *solicited proposal* is received by a client who is not surprised or annoyed to receive that proposal. However, the client usually states definite requirements and expects to see those requirements met. Also, frequently, the client indicates the conditions and/or the format for the proposal. In these cases, the proposal writer has to pay strict attention to the reader's expectations.

The writer of an *unsolicited proposal* has a different challenge: Often, the reader initially feels reluctant to accept the proposal. Perhaps the reader doesn't see a problem or is happy with the current way of doing things. Perhaps the reader is not aware of the product you have to sell, or has other priorities. However, if you can convince the reader of the need for your proposed solution, you have a chance of persuading the reader. If you can't establish that need in the reader's mind, there's no point in proposing how to meet the need!

Table 18.1   **Types of Proposals**

Type	Internal	External
Sales	The design unit in a company that designs and manufactures auto parts solicits proposals for testing new magnesium-alloy brake parts. In keeping with the company's budgetary policies, the company's design unit can choose the company's own research lab's proposals or an external lab's proposal. Either way, the design unit will be billed for the testing. If the company lab wins the testing contract, it gains credits for salaries and new equipment.	An engineering firm that specializes in environmental studies proposes the assessment methods, timeline, staffing, and budget for assessing the potential environmental impact of a proposed golf course. The proposal responds to a "Request for Proposal" (RFP) issued by the provincial Ministry of the Environment. The RFP lists several pages of requirements that a successful proposal must meet.
Research	A truck manufacturing company's engine design team proposes a project to determine methods of reducing engine vibration. In arguing its case, the design team points to negative customer and dealer feedback about excessive engine vibration and noise. The design team also cites examples of how engine vibration has created warranty problems for the manufacturer.	A university physicist proposes a four-year computer modelling study of the properties of surface molecules. The physicist knows that his readers at the Natural Sciences and Engineering Research Council will understand his highly technical proposal, but for possible public consumption he attaches an executive summary that explains the potential applications of his research to the upper ozone layer.
Improvement	A research chemist in a petroleum firm's product development branch proposes a new way of blending methanol with gasoline. This new blending method would reduce production costs. In response to her supervisor's request, a fisheries biologist proposes a software package that tracks costs of extended research studies.	A software development firm, after reading newspaper reports about sailing delays and operating cost overruns for an East coast ferry fleet, proposes a software program that fully coordinates the ferries' sailing schedules, maintenance schedules, staff training, supply loading, and fueling. Before submitting the proposal, the software firm's partners learn all they can about who screens proposals at the ferry corporation.

THE PROPOSAL PROCESS

The basic proposal process for a solicited proposal can be summarized simply: Someone offers a plan for something that needs to be done. In business and government, this process has five stages:

1. Client X recognizes a need for a service or product.
2. Client X draws up detailed requirements and advertises its need in a Request for Proposal (RFP).
3. Firms A, B, and C research Client X and its needs. (Likely they will request a more detailed version of the RFP.)
4. Firms A, B, and C propose a plan for meeting the need.
5. Client X awards the job to the firm offering the best proposal.

The complexity of each phase will, of course, depend on the situation. In practice, the process could look like the following.

> John Sand, a computer network designer in Vernon, British Columbia, spots the following RFP in a local newspaper.

 Province of British Columbia
Ministry of Transportation

**Request for Proposal
20021209-EIS**

The Ministry of Transportation requires the computers in its Vernon engineering field office to be networked. Proposals for the design and installation of such a network are being accepted.

Proposals will be evaluated primarily on the following criteria:
- ability to integrate existing computer equipment
- understanding of the engineering office's needs
- providing the least amount of disruption to current operations
- cost and value of the proposed products and services

For details of the Vernon office's needs, contact Sandra Mazzini, 549-3099.

Submit proposals no later than December 20, 2002, to:

**Mr. Richard Janvier
Engineering Information Systems Coordinator
Ministry of Transportation
784 Jasper Avenue
Kamloops, BC V2C 1L7
Phone: (250) 371-3935
Fax: (250) 371-3928**

After phoning his partner, Tom Glavine, who is working on a project in Cranbrook, John phones Sandra Mazzini to learn more about the Vernon office's needs, current computer equipment, type of work performed, and methods of filing data and designs. He takes detailed notes and asks Sandra to confirm key points at the end of the 20-minute conversation.

The conversation with the local office confirms what John had suspected:

1. The Ministry of Transportation has a limited capital budget and therefore wants to reduce costs by using the current group of computers. The client does not wish to buy any new computers unless absolutely necessary.

2. The engineers at the office are currently very busy preparing for upcoming spring and summer projects, so they cannot be interrupted for long periods by equipment installation or staff network training.

John Sand again consults with his partner. They agree that the best way to meet the client's needs is to design a peer-to-peer network that uses the engineering office's existing computers. The office's networking and file management needs do not require a more sophisticated (and more expensive) system. Such a network will not make much money for Sand and Glavine Systems because they will not earn their normal mark-up on equipment, so the partners agree to try to make money on the contract by offering system support, which the client can purchase as an add-on. (System support does not require money from the capital budget. John's conversation with Sandra Mazzini revealed that her office has more leeway with its annual services budget.)

Other firms will compete for this project. The client will award it to the firm submitting the best proposal, based on the criteria listed in the RFP, and the following criteria, which are generally used to assess proposals:

◆ understanding of the client's needs
◆ soundness of the firm's technical approach
◆ quality of the proposed project's organization and management
◆ ability to complete the job by the deadline
◆ ability to control costs
◆ specialized experience of the firm in this type of work
◆ qualifications of staff assigned to the project
◆ the firm's record for similar projects

NOTE   *Sometimes an RFP will list the client's specific criteria in a point scale, as in the excerpt on the following page, from an RFP for harvesting timber on Crown land:*

Variation:
assessment criteria
may be weighted

Criteria	Weighting
Employment	30
Proximity	10
Existing plant	10
New capital investment	10
Labour value-added	10
Change in value-added	20
Revenue	10
Total Weighting	**100**

All applicants must submit a proposal that contains a business case for lumber manufacturing or specialty wood products manufacturing and addresses the development objectives of the Crown.

Source: Adapted from a format used by the Ministry of Forests, Province of British Columbia.

Variation:
pre-proposal
conference

Some clients hold a pre-proposal conference for the competing firms. During this briefing, the firms are informed of the client's needs, expectations, specific start-up and completion dates, criteria for evaluation, and other details to guide proposal development. Such a conference is not necessary for a relatively small project such as networking a small engineering office.

Variation: restricted
list of selected firms

Nor does the Vernon engineering office do what some municipal and government agencies do: restrict a given project to a list of pre-selected firms. Usually, this is done on a rotation basis—if, for example, a city department has 15 firms on its contractor list, it might send invitations to bid to the first five firms on the list, and then to the next five firms for the next project, and so on. Or, an agency might require firms to present their qualifications for a certain kind of project before actually soliciting proposals for it. Only those firms who clear the qualifications hurdle will be allowed to submit a proposal.

In the networking case, John Sand knows that the merits of his firm's proposed plan will be evaluated *solely* on the basis of what he puts *on paper*. Still, it helps to know who will make the final decision about the competing proposals. In the conversation with Sandra Mazzini, John had learned that Richard Janvier will make the decision at the regional office. So, being a local vendor will not be a major advantage for Sand and Glavine Systems. Also, Mazzini hints that Janvier is a no-nonsense person who dislikes high-pressure sales. With that in mind, John takes pains to design a simple, functional system and write a proposal that is businesslike and free from unnecessary jargon.

The Sand and Glavine Systems proposal appears in this chapter, starting on page 415. ▲

PROPOSAL GUIDELINES

Readers will evaluate your proposal according to how clearly, informatively, and realistically you answer these questions:

- What are you proposing?
- What problem will you solve?
- Why is your plan worthwhile?
- What is unique about your plan?
- What are your (or your firm's) credentials?
- How will the plan be implemented?
- How long will the project take to complete?
- How much will it cost?
- How will we benefit if we accept your plan?

In addition to answering the questions above, successful proposal writers adhere to the following guidelines.

Design an Accessible and Appealing Format. Format is the *look* of a document, including such features as:

- the layout of words and graphics
- typeface, type size, and white space
- highlights and lists
- headings

A poorly designed proposal suggests to readers the writer's careless attitude toward the project.

Signal Your Intent with a Clear Title. Decision makers are busy people who have no time for guessing-games. The title should clearly signal the proposal's purpose and content.

> **Unclear**   Proposed Office Procedures for Vista Freight, Inc.

What kinds of office procedures are being proposed? This title is too broad.

> **Revised**   A Proposal for Automating Vista's Freight Billing System

Don't write "Recommended Improvements" when you mean "Recommended Wastewater Treatment." A specific and comprehensive title signals the proposal's intent.

Include Supporting Material and Appropriate Supplements. Both short and long proposals may include supporting materials (maps, blueprints, specifications, calculations, and so forth). Place supporting material in an appendix to avoid interrupting the discussion.

Depending on your readers, appropriate supplements for a long proposal might include a title page, cover letter, table of contents, summary, abstract, and appendices. Readers with various responsibilities will be interested in different parts of your proposal. Some know about the problem and will read only your plan; some will look only at the summary; others will study recommendations or costs; still others will need all the details. If you're unsure as to which supplements to include in an internal proposal, ask the intended reader or study other proposals. For a solicited proposal (one written for an outside agency) follow the agency's instructions *exactly*.

Focus on the Problem and the Objective.   Readers want specific suggestions for filling specific needs. Their biggest question is "What's in this for me?" Show them that you understand their problem and offer a plan for improving their products, sales, or services.

Notice in the following example how proposal writer Gerald Beaulieu focuses on Vista's inefficient office procedures and then outlines specific solutions.

STATEMENT OF THE PROBLEM

Gives background

Vista provides two services: (1) It locates freight carriers for its clients. The carriers, in turn, pay Vista a 6 percent commission for each referral; (2) Vista handles all shipping paperwork for its clients. For this auditing service, clients pay Vista a monthly retainer.

Describes problem and its effects

Although Vista's business has increased steadily for the past three years, recordkeeping, accounting, and other paperwork still are done *manually*. These inefficient procedures have caused a number of problems, including late billings, lost commissions, and poor account maintenance. Unless its office procedures are updated, Vista stands to lose clients.

OBJECTIVE

Enables readers to visualize results

This proposal offers a realistic and efficient plan for Vista to streamline office procedures. We first identify the burden imposed on your staff by the current system, and then we show how to reduce inefficiency, eliminate client complaints, and improve your cash flow by automating most office procedures.

Treat Contingencies and Limitations Realistically.   Do not underestimate the project's complexity. Identify contingencies (occurrences subject to chance) readers might not anticipate, and propose realistic methods for dealing with the unexpected. Here is how the Vista proposal qualifies its promises:

Assesses contingencies realistically

As outlined below, Vista can realize tangible benefits by automating office procedures. But, as countless firms have learned, *imposing* automated procedures on a staff can create severe morale problems—particularly among senior staff who feel coerced. To diminish employee resistance, invite your staff to comment on this proposal. To help avoid hardware and software problems once the system is operational, we have included recommendations and a budget for staff training. (Firms have learned that inadequate training is counter-productive to the automation process.)

If the best available solutions have limitations, let readers know. Otherwise, you and your firm could be liable in the case of project failure. Avoid overstatement. Notice how the above solutions are qualified ("diminish" and "help avoid" instead of "eliminate") so as not to promise more than the writer can deliver. Ethical communication is essential.

Explain the Benefits of Implementing Your Plan.   A persuasive proposal shows readers how they (or their organization) will benefit by adopting your plan.

The Vista proposal specifies the following benefits. (Each benefit will be described at length in the body section.)

Relates benefits directly to reader's needs

Once your automated system is operational, you will be able to:

◆ identify cost-effective carriers
◆ coordinate shipments (which will ensure substantial discounts for clients)
◆ print commission bills
◆ track shipments by weight, distance, fuel costs, and destination
◆ send clients weekly audit reports on their shipments
◆ bill clients on a 25-day cycle
◆ produce weekly or monthly reports

Additional benefits include reducing repetitive tasks, improving cash flow, and increasing productivity.

Provide Concrete and Specific Information.   Vagueness is a fatal flaw in a proposal. Before you can persuade readers, you must inform them; therefore, you need to *show* as well as *tell*. Instead of writing, "We will install state-of-the-art equipment," write,

Spells out details

To meet your automation requirements, we will install twelve Power Macintosh computers with 6-Gigabyte hard drives. The system will be networked for rapid file transfer between offices. The plan also includes interconnection with four HP printers, and one HP Desk Jet colour printer.

To avoid any misunderstanding and to reflect your ethical commitment, a proposal must elicit *one* interpretation only.

Include Effective Visuals.   If they enhance your proposal, use visuals, properly introduced and discussed. Page 401 shows one visual from the Vista proposal.

Gives a framework for interpreting the visual

As the flowchart (Figure 1) illustrates, your routing and billing system creates a good deal of redundant work for your staff. The routing sheet alone is handled at least six times. Such extensive handling leads to errors, misplaced sheets, and late billing.

Use a Tone That Connects with Readers.   Your tone is the voice and personality that appear between the lines. Make the tone confident, encouraging, and diplomatic. Show readers you believe in your plan; urge them to act, and anticipate how they will react to your suggestions. But do not convey a bossy or insulting tone, as in this example:

Had Vista's managers been better trained, they could have streamlined office procedures by eliminating some of the more obvious paper-shuffling problems. Moreover, the clerical staff could have avoided errors had they been more vigilant.

Here is a more diplomatic version:

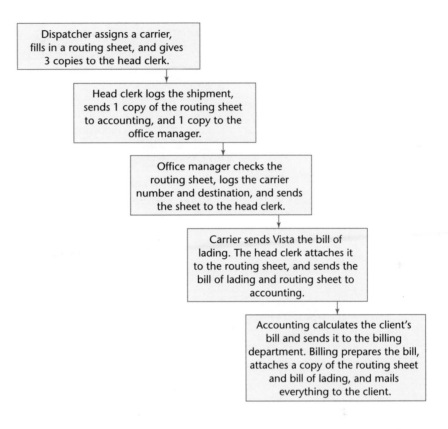

FIGURE 1   Flowchart of Vista's Manual Routing and Billing System

Avoids blaming
anyone

Vista's manual office procedures have resulted in an inefficient flow of information. The problems include repetitive tasks, late billings, lost documents, and poor account maintenance.

Analyze Audience Needs.   Proposals address diverse audiences. A research proposal might be read by experts, who would then advise the granting agency whether to accept or reject it. Planning and sales proposals might be read by colleagues, superiors, and clients (often laypersons). Informed and expert readers will be most interested in the technical details. Non-technical readers will be interested in the expected results but will need an explanation of technical details as well. Learn all you can about the needs, interests, and biases of your audience.

If the primary audience is expert or informed, keep the proposal itself technical. For uninformed secondary readers (if any), provide an informative abstract, a glossary, and appendices explaining specialized information. If the primary audience has no expertise and the secondary audience does, follow this pattern: Write the proposal itself for laypersons, and provide appendices with the technical details (formulas, specifications, calculations) that the informed readers will use to evaluate your plan.

Table 18.2   Typical Sections of Proposals   *(continued)*

Section	What to Include
Introduction	◆ Connect with your reader by referring to the reader's request or to a problem that has led to this proposal. ◆ Briefly summarize the proposed plan, service, or product. Perhaps highlight your (or your team's) qualifications. ◆ "Hook" the reader by previewing the proposal's main benefit. ◆ Describe the scope of the project and list the topics covered in the proposal. ◆ Match the length of the introduction to the length of the proposal. For example, use a single paragraph for a short letter proposal or one or more pages for a report-length document. Alternatively, you may put the introduction in the transmittal letter that accompanies the proposal.
Background	◆ Discuss the problem or need that has led to the proposal. This discussion will be more extensive in an unsolicited proposal. In a solicited proposal, show that you understand the problem completely. ◆ Discuss the general situation that has led to the current problem (when appropriate) and discuss the significance of solving that problem. ◆ Talk about the requirements for a solution.
Project Description	◆ Describe your solution to the problem: – What will be done – When – Where – By which methods ◆ Include headings such as: – Plan of the work – Schedule – Task breakdown – Projected results ◆ Incorporate the project components specified in the client's detailed guidelines, if you are responding to an RFP.

Table 18.2   Typical Sections of Proposals

Section	What to Include

Supporting Material:
 Facilities/equipment
 Personnel
 Past experience
	◆ Provide evidence of ability to complete all aspects of the project:
	   – Show that you have access to required facilities, equipment, and other resources.
	   – Show that the project leaders have the required qualifications and experience to complete the project.
	 ◆ Provide examples of similar work performed by your team.
	 ◆ Include résumés (where appropriate).

Supporting Arguments	◆ Show that the proposed plan is feasible (For example, that it can indeed be completed in the projected time frame, or that a similar solution has been successfully implemented in a similar situation). ◆ Discuss the benefits of the proposed action and, if necessary, address potential reader concerns.
Budget	◆ Complete this section carefully, to avoid problems due to potential increases in your costs—you're legally bound during the time period you specify. Some proposals need detailed cost breakdowns; others need only a bottom-line figure.
Authorization	◆ Use this closing action statement to request your reader to authorize your proposed plan, or request a meeting to present your proposal. ◆ Include the authorization request in the closing paragraph of informal proposals (along with a reminder of the proposal's main benefits). For formal proposals, place the authorization request in the transmittal letter.

A GENERAL STRUCTURE FOR PROPOSALS

To be successful, a proposal must clearly explain the proposed actions and it must present supporting arguments that recognize the reader's needs and priorities. In practice, a proposal also needs to open with an overview of its purpose, method, and benefits. Finally, it needs to close with a request for action—either the reader's authorization or a meeting to discuss the proposal in more detail.

TYPICAL SECTIONS OF PROPOSALS

Although a proposal's complexity determines exactly which sections will be included, the sections listed in Table 18.2 are usually present.

Table 18.3   Additional Sections in Formal Proposals

Section	What to Include
Copy of RFP	◆ Include a copy of the RFP when you know or suspect that the receiving organization has recently issued more than one RFP.
Transmittal Document (Letter)	◆ Use this persuasive letter to address the person who receives the proposal or is responsible for the final decision. This is your only chance to directly address the gatekeeper or decision maker who will decide the fate of the proposal, so word this letter very carefully. ◆ Refer to the RFP. ◆ Describe (briefly) the main points and benefits of the proposal. ◆ Indicate the time limit of your bid. ◆ Ask for the action you desire.
Summary	◆ Summarize the proposal's highlights in one page or less. ◆ Use the headings "Summary" or "Abstract" for technical readers. ◆ Use the heading "Executive Summary" for less technical summaries designed for managers and others. ◆ Include both types of summaries if more than one type of reader may read the proposal.
Title Page	◆ Include, in order: – The title of the proposal – The name of the client organization – The RFP number or other identifier – The author's name (if appropriate) and that of the organization – The date of submission
Table of Contents	◆ Include all section headings and the pages on which they appear. ◆ Do not list parts that appear before the table of contents page: – RFP – Transmittal letter – Summary – Title page ◆ List the names of appendices.
List of Illustrations	◆ Include a list of figures and tables if the document contains a total of three or more figures and tables.
Supporting Documents	◆ Include such appendices as: – Testimonials from satisfied clients – Résumés – Technical evidence – Relevant news or magazine articles, brochures, or photos of previous projects

ADDITIONAL SECTIONS IN FORMAL PROPOSALS

In addition to the standard seven sections found in informal proposals, longer formal proposals should include the sections shown in Table 18.3, page 404.

SAMPLE PROPOSALS

Now, let's see how the identified sections are implemented in three different proposals:
1. An informal improvement proposal
2. An informal research proposal
3. A formal sales proposal

AN INFORMAL IMPROVEMENT PROPOSAL

Here's a planning proposal titled "A Proposal for Solving the Noise Problem in the University Library."

> Jill Sanders, a library student on a co-op work placement, addresses her proposal to the chief librarian. Because this proposal is unsolicited, it must first make the problem vivid through details that arouse concern and interest. This introduction is longer than it would be in a solicited proposal, whose readers would agree on the severity of the problem. Notice that Jill has chosen to use a three-level heading system, based on the traditional *introduction/body/conclusion* pattern. ▲

Concise descriptions of problem and objective immediately alert the readers

INTRODUCTION

Statement of Problem

During the October 2003 Convocation at Margate University, students and faculty members complained about noise in the library. Soon afterward, areas were designated for "quiet study," but complaints about noise continue. To create a scholarly atmosphere, the library should take immediate action to decrease noise.

Objective

This proposal examines the noise problem from the viewpoint of students, faculty, and library staff. It then offers a plan to make areas of the library quiet enough for serious study and research.

This section comes early because it is referred to in next section

Sources

My data come from a university-wide questionnaire; interviews with students, faculty, and library staff; enquiry letters to other college libraries; and my own observations for three years on the library staff.

Figure 18.1   An Informal Improvement Proposal *(continued)*

Details enable
readers to
understand the
problem

Details of the Problem

This subsection examines the severity and causes of the noise.

Severity. Since the 2003 Convocation, the library's fourth and fifth floors have been reserved for quiet study, but students hold group-study sessions at the large tables and disturb others working alone. The constant use of computer terminals on both floors adds to the noise, especially when students converse. Moreover, people often chat as they enter or leave study areas.

On the second and third floors, designed for reference, staff help patrons locate materials, causing constant shuffling of people and books, as well as loud conversation. At the computer service desk on the third floor, conferences between students and instructors create more noise.

Shows how campus
feels about problem

The most frequently voiced complaint from the faculty members interviewed was about the second floor, where people using the Reference and Government Documents services converse loudly. Students complain about the lack of a quiet spot to study, especially in the evening, when even the "quiet" floors are as noisy as the dorms.

Shows concern is
widespread and
pervasive

More than 80 percent of respondents (530 undergraduates, 30 faculty, 22 graduate students) to a university-wide questionnaire (Appendix A) insisted that excessive noise discourages them from using the library as often as they would prefer. Of the student respondents, 430 cited quiet study as their primary reason for wishing to use the library.

The library staff recognizes the problem but has insufficient personnel. Because all staff members have assigned tasks, they have no time to monitor noise in their sections.

Causes. Respondents complained specifically about these causes of noise (in descending order of frequency):

Identifies specific
causes

1. Loud study groups that often lapse into social discussions.
2. General disrespect for the library, with some students' attitudes characterized as "rude," "inconsiderate," or "immature."
3. The constant clicking of computer terminals on all five floors.
4. Vacuuming by the evening custodians.

All complaints converged on lack of enforcement by library staff. Because the day staff works on the first three floors, quiet-study rules are not enforced on the fourth and fifth floors. Work-study students on these floors have no authority to enforce rules not enforced by the regular staff. Small, black-and-white "Quiet Please" signs posted on all floors go unnoticed, and the evening security guard provides no deterrent.

Needs

This statement of
need evolves
logically and
persuasively from
earlier evidence

Excessive noise in the library is keeping patrons away. By addressing this problem immediately, we can help restore the library's credibility and utility

Figure 18.1 An Informal Improvement Proposal   *(continued)*

as a campus resource. We must reduce noise on the lower floors and eliminate it from the quiet-study floors.

Scope

Previews the plan

The proposed plan includes a detailed assessment of methods, costs and materials, personnel requirements, feasibility, and expected results.

PROPOSED PLAN

This plan takes into account the needs and wishes of our campus community, as well as the available facilities in our library.

Methods

Tells how the plan will be implemented

Noise in the library can be reduced in three complementary phases: (1) improving publicity, (2) shutting down and modifying our facilities, and (3) enforcing the quiet rules.

Describes first phase

Improving Publicity. First, the library must publicize the noise problem. This assertive move will demonstrate the staff's interest. Publicity could include articles by staff members in the campus newspaper, leaflets distributed on campus, and a first-year student library orientation acknowledging the noise problem and asking cooperation from new students. All forms of publicity should detail the steps being taken by the library to solve the problem.

Shutting Down and Modifying Facilities. After notifying campus and local newspapers, you should close the library for one week. To minimize disruption, the shutdown should occur between the end of summer school and the beginning of the fall term.

During this period, you can convert the fixed tables on the fourth and fifth floors to cubicles with temporary partitions (six cubicles per table). You could later convert the cubicles to shelves as the need increases.

Then you can take all unfixed tables from the upper floors to the first floor, and set up a space for group study. Plans already are underway for removing the computer terminals from the fourth and fifth floors.

Describes third phase

Enforcing the Quiet Rules. Enforcement is the essential, long-term element in this plan. No-one of any age is likely to follow all the rules all the time—unless the rules are enforced.

First, you can make new "Quiet" posters to replace the present, innocuous notices. A visual-design student can be hired to draw up large, colourful posters that attract attention. Either the design student or the university print shop can take charge of poster production.

Next, through publicity, library patrons can be encouraged to demand quiet from noisy people. To support such patron demands, the library staff can begin monitoring the fourth and fifth floors, asking study groups to move to the first floor, and revoking library privileges of those who

Figure 18.1 An Informal Improvement Proposal   *(continued)*

refuse. Patrons on the second and third floors can be asked to speak in whispers. Staff members should set an example by regulating their own voices.

Costs and Materials

Estimates costs and materials needed

◆ The major cost would be for salaries of new staff members who would help monitor. Next year's library budget, however, will include an allocation for four new staff members.

◆ A design student has offered to make up four different posters for $200. The university printing office can reproduce as many posters as needed at no additional cost.

◆ Prefabricated cubicles for 26 tables sell for $150 apiece, for a total cost of $3,900.

◆ Rearrangement on various floors can be handled by the library's custodians.

The Student Fee Allocations Committee and the Student Senate routinely reserve funds for improving student facilities. A request to these organizations presumably would yield at least partial funding for the plan.

Personnel

The success of this plan ultimately depends on the willingness of the library administration to implement it. You can run the program itself by committees made up of students, staff, and faculty. This is yet another area where publicity is essential to persuade people that the problem is severe and that you need their help. To recruit committee members from among students, you can offer Contract Learning credits.

The proposed committees include an Anti-noise Committee overseeing the program, a Public Relations Committee, a Poster Committee, and an Enforcement Committee.

Feasibility

Assesses probability of success

On March 15, 2004, I mailed survey letters to 25 Canadian colleges, enquiring about their methods for coping with noise in the library. Among the respondents, 16 stated that publicity and the administration's attitude toward enforcement were main elements in their success.

Improved publicity and enforcement could work for us as well. And slight modifications in our facilities, to concentrate group study on the busiest floors, would automatically lighten the burden of enforcement.

Benefits

Offers a realistic and persuasive forecast of benefits

Publicity will improve communication between the library and the campus. An assertive approach will show that the library is aware of its patrons' needs and is willing to meet those needs. Offering the program for public inspection will draw the entire community into improvement efforts. Publicity, begun now, will pave the way for the formation of committees.

Figure 18.1   An Informal Improvement Proposal   *(continued)*

The library shutdown will have a dual effect: It will dramatize the problem to the community, and also provide time for the physical changes. (An anti-noise program begun with carpentry noise in the quiet areas would hardly be effective.) The shutdown will be both a symbolic and a concrete measure, leading to reopening of the library with a new philosophy and a new image.

Continued strict enforcement will be the backbone of the program. It will prove that staff members care enough about the atmosphere to jeopardize their friendly image in the eyes of some users, and that the library is not afraid to enforce its rules.

CONCLUSION AND RECOMMENDATION

Re-emphasizes need and feasibility and encourages actions

The noise in Margate University library has become embarrassing and annoying to the whole campus. Forceful steps are needed to restore the academic atmosphere.

Aside from the intangible question of image, close inspection of the proposed plan will show that it will work if you take the recommended steps and—most important—if daily enforcement of quiet rules becomes a part of the library's services.

Figure 18.1   An Informal Improvement Proposal

AN INFORMAL RESEARCH PROPOSAL

An informal student research proposal is illustrated on page 411. This proposal memo describes a computer student's idea for a research project required by the English 116 class that she's taking. That project will culminate in an analytical report for a "real-world" reader.

The student, Amy Suen, chooses to examine firewall systems that might provide network security for the networked systems at 2020 Design, an electronics design and manufacturing company. Amy does some preliminary research to confirm that she can find information on firewalls. Also, through a contact at 2020 Design she learns that her report would interest Mark Reuters, 2020's computer networking specialist. Mark has been too busy designing and installing his company's new array of computers to pay much attention to the external security issue. Still, 2020 is committed to marketing its products on the Internet and therefore risks incursions into its internal communications, including its proprietary designs.

Amy knows that 2020 Design will not make decisions based on a first-year computer student's report, but Mark Reuters assures her that her report could provide a springboard for his own investigation of firewall software. For that reason, he is willing to help Amy with her preliminary research.

In preparing her proposal for her instructor, Devon Koenig, Amy starts by analyzing her reader and her purpose for writing the proposal, as follows:

Audience: Devon Koenig, project supervisor. He will use my proposal to decide whether to approve my proposed approach. If my proposal is rejected, I'll have to prepare another.

His knowledge: severely limited (so I'll have to provide background)

His questions: (in the assignment memo dated January 17)

1. What's your topic? Who's the reader for your proposed report?
2. What main question does your reader want answered? (Identify the report's specific analytical purpose.)
3. How will the proposed report achieve that purpose? (Here, an attached detailed outline would help establish that you do indeed have a plan.)
4. Do you have the time, resources, and commitment to complete the project? (You should include a time budget and a resources budget.)
5. Will you find sufficient information to answer the report's overall question? (Attach a tentative bibliography.)

Purpose: Convince Prof. Koenig that the topic has merit and that the project is feasible. (Show that I can research, that I can manage the project, that I can afford it, that I can think, that I can write well enough to handle a report of this complexity.)

Audience attitude and temperament: Has high standards and expectations. Therefore, I'll have to answer all his questions thoroughly. He said he'll be skeptical, so I'll have to support all my statements. This won't be an easy sell! I don't think he cares much about how much the project will cost me, as long as I can prove I can afford it. He's a tough marker, so I'll need to phrase the memo carefully and get it proofread.

Audience expectations: He doesn't want more than three pages, not including attachments. His comments in class emphasized our need to provide proof of all positive statements. That's the only way I'll get his support. I'll need to use a direct, businesslike tone—he doesn't like extra words or pompous language. (I'd better stay away from words like "utilize.") And I'd better stick pretty close to the proposal structure he recommended: Connect with reader/provide "hook" in the proposal pre-summary/give background/show need for report/describe planned approach/propose schedule/prove that it's feasible—sources/my qualifications/budget/request authorization. ▲

Now, here's the memo that Amy Suen presents to her project supervisor.

English 116 Memorandum

Date: February 4, 2002
To: Professor Devon Koenig, English Department
From: Amy Suen, Computer Systems 1st year
Re: **Proposed Research Project for English 116**

In response to the proposal assignment that you announced in our January 25 class, this memo outlines my proposal to research firewall products. The research will lead to a recommendation of which product will best suit the needs of 2020 Design, a local electronics company. Its networking specialist, Mark Reuters, has indicated his interest in my findings.

Background

Mark Reuters has recently revamped 2020 Design's total network of computers. He is now in the process of installing new networking software, and will continue fine tuning the network and training 2020 Design personnel for the next three months. At that time he will turn his attention to the challenges of partially linking 2020's internal network to the Internet.

One of the major problems to solve will be the issue of internal network security. 2020 Designs stays competitive by developing new electronic designs, so it doesn't want outsiders tapping into its research and development work. Still, the company wants to market its products on the Internet and engage in e-commerce.

Recently, "firewall" has come to mean software products for blocking unwanted access to protected information, but its original meaning included all aspects of network security strategy—software, hardware, and personnel. Mark Reuters has asked me to focus on software products.

Developing appropriate software is very expensive, so most companies purchase rather than develop. Firewall software ranges from a few thousand dollars to about $100,000, depending on performance and user requirements. Each product has advantages and disadvantages which depend on network configurations and the method of implementing the software. Therefore, it's not easy to choose an appropriate firewall product.

Proposed Plan

My initial research into the topic indicates that the following process would be best.

1. **Determine the client's needs.** Subject to your approval of my proposal, Mark Reuters will provide me with details of the network he administers. Those details will include the special features and challenges built into that network, (which apparently is quite unique). With Mark Reuters's help, I'll be able to choose the criteria that I can use to evaluate and compare firewall software products.

2. **Research available software.** As the attached tentative bibliography shows, there seems to be plenty of information available. In addition, I have arranged to possibly interview Marsha Campbell, one of my computer instructors, and Guy Lariviere, the network administrator for our college. I'll proceed with those interviews if I receive your authorization to proceed.

3. **Evaluate information gathered about firewall products.** Please see the attached outline, which describes the general approach I plan to take. So far, I have identified three products (Alta Vista, CheckPoint, and CyberGuard), but I'll continue to look for others. I expect to discuss three to five products in the final report; that means that I may have to do a preliminary assessment to weed out inappropriate firewall products.

4. **Rigorously apply the assessment criteria and choose the best product for 2020 Designs.** Part of this assessment can come from the specifications and product information provided by the manufacturers on their websites. The full assessment will come from actually testing the software.

Figure 18.2   A Student Research Proposal   *(continued)*

Feasibility of Project

I've looked at this project quite carefully and I think it's feasible because
- Information is available and I have access to expert opinion here at the college.
- My strong interest in this subject has already prompted me to read all the sources listed in the attached bibliography and I've made notes on three of the articles.
- Professor Campbell has agreed to help me assess software on the network in computer lab 218, and she is accepting this project for credit in her Networking class.
- I can get access to firewall products through Mark Reuters, who will request demo. software from manufacturers.

Schedule

According to your January comments about task requirements, I have overestimated the time required for the following tasks, but my semester time budget can still accommodate the following:

Activity	Time Required	Dates	Document Produced
Research re: firewalls (types and leading products) and 2020 system configuration	10–15 hrs	Feb. 2–14	Refine planning outline
Interview Campbell and Lariviere	2–2 1/2 hrs	Feb. 15 Feb. 16	Refine planning outline
Evaluate data	5–8 hrs	Feb. 22	Adjusted research plan (?)
Analyze/organize data	6–10 hrs	Feb. 27–28	Working outline (due Mar. 8)
Plan/write progress report	4 hrs	Mar. 5–6	Progress report (due Mar. 8)
Write report from outline	6–10 hrs	Mar. 14–16	Analytical report—draft
Edit report and polish format	5–8 hrs	Mar. 24–27	Analytical report—due Mar. 31

Budget

Because I won't have to buy firewall products, and because I use my home Internet connection for many purposes, my budget for this project is minimal:

Photocopy articles	$20
Bus travel for interviews	$15
Report printing and binding	$14
Total:	$49

Authorization

I hope that you agree that my proposal topic and approach is appropriate for the English 116 project, because I'm committed to doing an excellent job. For one thing, I'm also doing the project for my Networking class, and I can afford to put more effort into a project that fulfills two sets of requirements. Also, I think I may be able to get a student co-op job at 2020 Design this May if I do well on this project. May I have your authorization to proceed? If you wish to contact me outside of class, please e-mail me at amysuen@silk.net

AS

Attachments: Tentative Bibliography
              Planning Outline

Figure 18.2   A Student Research Proposal   *(Continued)*

PLANNING OUTLINE: REPORT PROJECT
English 116

Topic: Firewalls

Reader (needs/reason for reading/knowledge level):

> Mark Reuters, 2020 Design he's interested in firewall products for his firm's network; he's very knowledgeable, but not about current products

Purpose: Assess firewall products suitable for 2020 Design (maybe recommend best one)

Tentative Structure/Topics:

> Reason for report
> Background re: current firewall technology
>
> > ◆ list of products (CyberGuard, Alta Vista, Check Point, and others)
> > ◆ manufacturers
> > ◆ concepts behind the products
>
> 2020 Design's network structure
> Client's general and special requirements
> Product evaluations:
>
> > ◆ supporting operating system and services
> > ◆ performance
> > ◆ price
> > ◆ installation and training
> > ◆ does it meet client's particular requirements?
>
> Conclusion re: which products satisfy the criteria/which best satisfies the criteria

Tentative Sources:

> Interviews: Prof. Campbell, Guy Lariviere, and Mark Reuters

Check Point Technologies Ltd. (Updated 2001, March 23). *Firewall-1 products and solutions.* Retrieved February 1, 2002, from http://www.checkpoint.com.

Computer Security Institute. (n.d.). *CSI firewall matrix.* Retrieved January 31, 2002, from http://www.gosci.com

Digital Equipment Corp. (copyright 2001). *Alta Vista firewall 98.* Retrieved January 29, 2002, from http://www.altavista.software.digital.com

Firewall Guide. (2002). Retrieved January 27, 2002, from http//:www.firewallguide.com

Hallogram reviews. (copyright 2001). Retrieved January 22, 2002), from http://hallogram.com/avfirewall

Markus, H.S. (Last updated 2002, February 3), *Home PC firewall guide.* Retrieved February 4, 2002, from http://www.firewallguide.com

Novell Corp. (n.d.) *Border Manager Enterprise Edition.* Retrieved February 2, 2002, through the *Firewall product overview,* at http://www.thegild.com/firewall

Rubin, A., Geer, D., and Ranum, M.J. (2001). *Web security source book.* New York: Wiley Computer Publishing.

Stein, L.S. (2000). *Web security, a step-by-step reference guide.* Reading: Addison-Wesley.

Figure 18.2   A Student Research Proposal   (attachment)

A FORMAL SALES PROPOSAL

The formal proposal that begins on page 415 presents a commercial sales pitch. Its writer, John Sand, uses a formal report format to impress his reader with a professional-looking document. John does not include a copy of the RFP; the receiver doesn't have any other RFPs circulating at the moment. Also, the proposal does not include a Conclusion section or a Summary, because the transmittal letter contains a summary of the proposal and states the author's conclusions and request for authorization.

Other aspects of the writer's audience and purpose analysis have been discussed in "The Proposal Process" section of this chapter (page 395).

Sand and Glavine Systems

Suite 106-204 Kalamalka Lake Road, Vernon, BC   V1T 7M3 Phone (250) 558-5704 Fax: (250) 558-1554

December 19, 2002

Mr. Richard Janvier
Engineering Information Systems Coordinator
Ministry of Transportation
784 Jasper Avenue
Kamloops BC   V2C 1L7

Dear Mr. Janvier:

Response to RFP 20021209-EIS, Office Computer Network

The enclosed proposal details the methodology that Sand and Glavine Systems would use to complete the computer network system outlined in your recent request for proposal.

Sand and Glavine Systems is well qualified to supply the network design and implementation outlined in RFP 20021209-EIS, "Office Computer Network." We are an established business in the Okanagan Valley with a reputation for providing outstanding quality and customer service. We have designed and installed many computer network systems including several for engineering firms.

Based on the information supplied by Sandra Mazzini and our extensive experience with computer networks, we recommend the use of a small peer-to-peer network. Networks like these are cost-effective, efficient, and easy to administer. We could supply this system by upgrading your existing hardware and software. By using this strategy, we will keep costs to a minimum and shorten the time line for getting a system in place and operating. Our proposal calls for upgrading your equipment, installing a network, and orientating the staff, all to be completed within one week of being awarded the contract.

We will design and supply this system, including all necessary hardware, software, cables, and taxes for under $5,000. This price also includes a brief orientation for staff who will be using the system.

If you have any questions about the proposal after you have had the opportunity to review and evaluate it, please call me at 558-5704.

Sincerely,

John Sand

Figure 18.3   A Formal Sales Proposal

(continued)

**PROPOSAL FOR
OFFICE COMPUTER NETWORK
RFP 20021209-EIS**

Prepared for
Richard Janvier
Engineering Information Systems Coordinator

Ministry of Transportation
Engineering Services
784 Jasper Avenue
Kamloops, BC
V2C 1L7

Prepared by

Sand and Glavine Systems
Suite 106-204 Kalamalka Lake Road
Vernon, BC
V1T 7M3

December 19, 2002

Figure 18.3   **A Formal Sales Proposal**   *(continued)*

TABLE OF CONTENTS

Figure 18.3   A Formal Sales Proposal   *(continued)*

Sand and Glavine Systems

INTRODUCTION

Sand and Glavine Systems is well qualified to supply the network design and implementation outlined in RFP 20021209-EIS, "Office Computer Network." We are an established business in the Okanagan Valley with a reputation for providing outstanding quality and customer service. We have extensive experience in networking computer systems, including several engineering offices. This proposal details our plan to create an efficient, low-maintenance computer network.

Statement of Need

Background

The Ministry of Transportation currently operates a field engineering office in Vernon. Working as a team, the office staff produces engineering plans ranging from preliminary drawings and estimates to detailed design packages. This team relies heavily on the use of computerized engineering and drafting tools as well as traditional office software. To handle the high volume and variety of documents produced, the office is equipped with seven personal computers (PCs), two plotters, and one laser printer. The PCs have varying configurations, processors, and attached peripherals. The objective of Sand and Glavine Systems is to maximize the use of the existing equipment into a seamless cost-effective network, making the job of the staff easier and the office more productive.

Impacts

With such a high emphasis on a team concept, many or all of the staff can be involved in a project. Having so many people involved in a single project using separate personal computers can lead to complications in file management and resource sharing.

Unless a stringent file management system is followed, duplicate copies of files will exist. The possibility of having files in more than one location can lead to serious complications. Modifications performed on any one copy will result in different versions of the same document. Updating one of the duplicate documents to contain all modifications wastes time, if the error is even detected. Worse, the probability of having the latest version of a file overwritten with an earlier version is high, leaving disastrous results.

To remedy this, work units sometimes rely on a catalogue of floppy disks containing the latest version of a file. Updating a file requires transferring it from the floppy disk to the hard disk of a PC, making the modifications, saving the file, and transferring it back to the floppy. A fail-safe system will have a back-up floppy of the original. Using this method will ensure data integrity but staff will spend much of their time transferring files and making back-up copies. Also, it works only if strict guidelines are followed. The longer the project, the greater the chance of a breakdown in file management occur.

– 1 –

Figure 18.3   A Formal Sales Proposal   *(continued)*

Sand and Glavine Systems

Sharing hardware resources, such as plotters, and printers, between individual PCs is difficult and will lead to productivity losses. Options include dedicating one computer to printing and plotting or connecting the plotters and printer to different computers and sharing time with the operators. Neither choice is efficient.

Unless a computer is used full time, dedicating it to printing and plotting is a waste of a costly resource. Sharing a workstation is complicated because the operator must handle printing and plotting chores as well as his or her own duties. In both cases the method for getting the information to the printer or plotter involves transferring files to and from floppy disks: a time consuming task.

An alternative used by some companies is to purchase additional equipment to spread the workload. The problem, though, is not that the printer or plotter cannot handle the existing tasks, but that people cannot freely access them when needed. Adding equipment is not a cost-effective approach for pieces of hardware that see only heavy use in spurts and sit idle for great lengths of time.

Needs

The Ministry of Transportation needs to cut productivity losses. This must be achieved at a minimal cost. Staff of the engineering field office need an effective and hassle-free way of performing their duties. Methods considered for achieving this cannot introduce new problems or unreasonable demands for the staff.

Any system used will require a safe and practical method for the engineering office staff to share and store electronic documents. It must also provide a way to produce hard copies of these documents, on demand, from any computer on the system.

To successfully meet these needs, Sand and Glavine Systems proposes a local area network be established using the ministry's existing computers.

Meeting the Need

Local area networks

Since the arrival of the personal computer there has been a drive to share information and computing resources. Local area networks (LANs) were created to fulfill that desire. A LAN is a data communications system that allows any connected device to send and/or receive information. This connectivity allows users to share information and computer resources including storage devices, software applications, data files, printers, and plotters.

LANs cover relatively short distances usually within an office, a department, or perhaps a building or campus. LANs consist of network interface cards attached to the connected computers, cables to connect the computers, and software to manage both data flow and the user interface to the network.

– 2 –

Figure 18.3   A Formal Sales Proposal *(continued)*

Sand and Glavine Systems

The main goal of a LAN is to bring connectivity and savings to a company. Once installed, a network distributes a company's informational resources to the people who need it. People can then do their jobs more quickly, efficiently, and with less trouble than with individual PCs. The LAN will also provide savings through sharing of expensive computer resources—printers, plotters, software, and information.

LAN solutions

Once a decision to network has been made, the next step is to decide how to distribute the workload. There are three ways to do this—through a dedicated server, a non-dedicated server, or a peer-to-peer network.

1. A network with a **dedicated server** has a computer set aside to handle the network traffic and tasks for the LAN. The server can be optimized with large storage disks, lots of memory, and a fast processor. A network with a dedicated server provides the greatest amount of security and is the simplest to administer. Although the most powerful arrangement of the three, the dedicated server is the most expensive in terms of equipment and administration. If the network being designed is large, having a dedicated server is the only option to consider.

2. With smaller networks a **non-dedicated server** can be considered. The server acts as both a network host and a workstation. The system software manages requests from the network as well as from the user sitting in front of the system. A priority ratio determines how the requests are handled. The setting of the priority ratio depends on whether the server will be performing more network requests or more local user requests. This set-up gets more use from the server, but if network traffic is heavy the local user will notice a degradation in performance. Also, if the applications of the local user require a great amount of processing, this processing demand will have a detrimental effect on the network.

3. The alternative to using either type of server is for each individual computer to act as **both a server and workstation**. Any computer connected to the network can share resources, such as disk space and printers, with any other computer on the network. These **peer-to-peer networks** provide a very cost-effective way to share information and limited resources. On peer-to-peer networks, users have easy access to each other's computer. Each user is also responsible for backing up the data on his or her computer. So, if security or data administration is a concern, a peer-to-peer network is not the system to choose.

Recommended solution

The needs and current practices of the Ministry of Transportation field engineering office suggest the most suitable system would be a peer-to-peer network (Figure 1).

– 3 –

Figure 18.3   A Formal Sales Proposal

(continued)

Sand and Glavine Systems

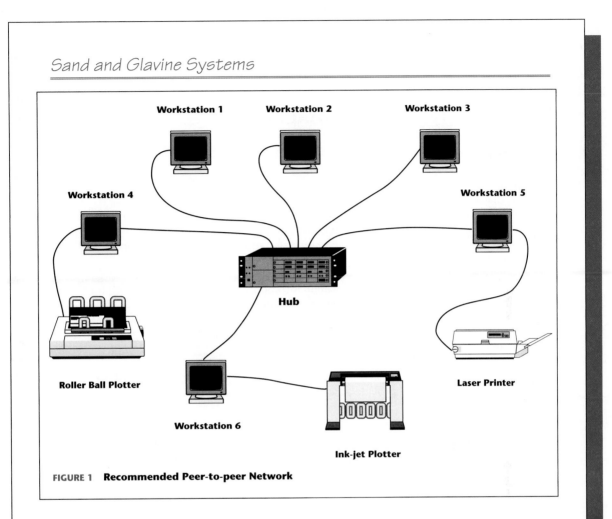

FIGURE 1   **Recommended Peer-to-peer Network**

The main needs include the ability to easily store and retrieve data, to occasionally share information, and to readily access the plotters or laser printer. This type of activity does not place a heavy burden on a network, so a network server is unjustified. A peer-to-peer network can smoothly handle these tasks without the costs and administration involved in using a system with a server.

As noted above, the drawbacks to a peer-to-peer network are data security and data administration. Since data security is not an issue at this time, a peer-to-peer network poses no threat. Data administration is already performed on an individual basis with each workstation having its own tape back-up system. A peer-to-peer network would retain the same data security and administration.

It should be noted that installing a network does not negate the need for good file management. The possibility of having duplicate files or newer files being overwritten with older versions still exists. This is especially true for a peer-to-peer network, where central storage does not exist.

– 4 –

Figure 18.3   A Formal Sales Proposal   *(continued)*

Sand and Glavine Systems

PROPOSAL PLAN

Installation

1. Cabling

Sand and Glavine Systems will install high-quality, category 5 unshielded twisted pair (UTP) cable. With category 5 UTP cable, data transmission rates of 10 megabits per second (Mbs) are possible on today's standards, and the system will comply with the future 100 Mbs transmission rates. UTP cable is quickly becoming the standard for smaller LANs because of reliability and ease of installation.

The most common practice for cable installation is through a suspended ceiling, with the cable dropped to each computer location either through the wall or a drop-post. Without seeing the installation site first, we cannot specify exact cable locations, but we suggest that all cables be hidden, eliminating the potential for network interruptions and damage that exposed cables might cause.

Our proposal calls for the plotters and printer to be linked directly to one or more of the workstations. The network will provide access to these devices as if they were connected directly to each workstation.

2. Topology and hardware

The topology for the network will be a bus-star configuration using a 12-port Ethernet switching hub. Each computer will be fitted with an Ethernet 10/100 network interface card (NIC) following the 10Base-T standard. These reliable cards supply data transmission speeds of up to 10 Mbs and will accommodate future 100 Mbs transmission rates as well.

We will not install the office's 586 Wang computer on the network, thus saving the cost of one NIC. While this computer's architecture would run the software we plan to install, the performance would be sluggish. Using this computer as a printer server for the plotter will not be required once the network is operating.

3. Software

Since each computer already has a registered copy of Windows 2000 installed, an upgrade to Windows NT 4.0 will provide the most cost-effective network software. While there is a nominal cost for this upgrade, it is less expensive than purchasing separate networking software.

Using this strategy will lower training costs and time. Most computer users are familiar with the Windows environment and they will have little to learn about the network portion of this software.

– 5 –

Figure 18.3   A Formal Sales Proposal   *(continued)*

Sand and Glavine Systems

4. Testing

Once the network is installed, extensive testing will guarantee that it is operating properly. Testing the system will include the following:

1) Executing all software applications to check for memory conflicts with the network software drivers. All conflicts will be resolved.

2) Successfully transferring files from each computer to every other connected computer to ensure that all computers are "talking" to each other.

3) Printing from every computer to each plotter and the printer to ensure they are recognized by the network and configured properly.

Training

A good computer network is invisible to the operator. The network to be installed is designed for ease of use. All computers will operate as they did before, only with added functionality. Because of this simplicity, extensive training is not required.

However, an introduction to the network's capabilities is advisable. We recommend a three-hour orientation for all employees who will use the system. We will show the staff the added functionality and the benefits of using the network. The orientation can be done in a group setting at the office and is provided at no additional cost.

Schedule

Once notice of approval has been received and a contract signed, Sand and Glavine Systems is prepared to have the network operational within one week. A break down of approximate time requirements is listed below.

1) Two days to acquire the necessary hardware and the software licences.

2) One day for an on-site visit. During this visit we will interview staff to establish a design criterion for the network. Cable requirements will also be determined and finalized.

3) One day to design the system and cable layout.

4) One day for the physical installation, including cabling. To minimize disruption to personnel and ease the process, we recommend the installation be scheduled for a weekend.

5) A three-hour network orientation for the staff. It would be best to schedule this for the Monday morning following the physical installation.

Figure 18.3   A Formal Sales Proposal *(continued)*

Sand and Glavine Systems

Support

Our standard support packages are available to purchase on a contract basis. The monthly and yearly packages offer unlimited phone support for the length of the contract. Our knowledge base is exceptional and past experience shows the majority of problems are resolved over the phone. An outline of all available packages, including prices, is in the Costs section.

Staffing

Sand and Glavine Systems is a well-established computer network business located in Vernon. We have a solid reputation for superior technical expertise and product support. Computer network design and installation is our specialty and we have successfully completed large and small projects for many businesses located in the Okanagan Valley. Appendix A contains a selective list of satisfied clients.

We will assign our most qualified and senior staff to this project.

Mr. John Sand—B.Sc., CNE
Mr. Sand is co-founder of Sand and Glavine Systems. He has a degree in project management from the University of Alberta as well as network certification from Novell. Mr. Sand will manage the project and oversee installation.

Mr. Tom Glavine—B.Sc.
Mr. Glavine is co-founder of Sand and Glavine Systems. He has a degree in computer science from the University of Toronto and specializes in computer network design. Mr. Glavine will design the network system and oversee installation.

Mr. Roger Clemenceau—CNCI
Mr. Clemenceau has been with Sand and Glavine Systems since its inception. He has certification in network cable layout and installation. Mr. Clemenceau will be responsible for the cable layout and will oversee the cable installation.

Costs

Determining the total cable needed for this project requires an on-site visit. We have approximated the cable required from the floor plan that Sandra Mazzini faxed to us. We anticipate our estimate to be very close. Should we need more cable, we will bill the Ministry of Transportation separately for this item. If we have overestimated the cable requirements, we will show the credit on the invoice.

The total cost for this project is shown in the table on the following page:

– 7 –

Figure 18.3 A Formal Sales Proposal *(continued)*

Sand and Glavine Systems

Software	Windows NT 4.0 upgrade	6 × $99	$594.00
Hardware	3Com Fast Etherlink XL 10/100	6 × $109	$654.00
	3Com 12 Port Switching Hub	1 × $780	$780.00
Cable	Category 5 Unshielded Twisted Pair	300' @ $0.30/ft.	$ 90.00
Services	System design and installation labour		$2,850.00
Total			**$4,968.00**

System support can be purchased separately on a monthly or yearly contract basis. This base entitles you to unlimited phone support with on-site service charged at an additional hourly rate. Support can also be purchased on an on-site, as-needed basis. Support costs are as follows:

Monthly—$60 per month—on site service $60/hr., minimum 1/2 hour charge
Yearly—$480 per year—on-site service $50/hr., minimum 1/2 hour charge
On-site, as-needed—$75/hr., minimum 1-hour charge

Scope of services

The prices in this proposal are valid until April 15, 2003. The scope of the network services offered is only for the computer systems located within Suite 102-2380, 35th Avenue, Vernon, British Columbia.

Figure 18.3   A Formal Sales Proposal   *(continued)*

Sand and Glavine Systems

APPENDIX A

SELECTED CLIENT LIST

Blade Architectural Engineering, Penticton

CivilTech Consulting and Engineering, Vernon

Douglas Benton & Associates, Vernon

GLP Engineering, Vernon

McAfee and Sons, Kelowna

Ministry of Forests, Kamloops

Osoyoos Growers Coop, Osoyoos

Premium Foods, Winfield

Robinson Agencies, Vernon

School District 103, Kelowna

The Video Shop, Kelowna

References are available upon request.

– 9 –

Figure 18.3   A Formal Sales Proposal

AN INTERPERSONAL PERSPECTIVE

Some of the examples on the preceding pages show external proposals, most of which attempt to win business contracts. Often, however, as employees we feel the need to improve workplace conditions or ways of doing things. If we try to ignore our feelings, we can end up feeling bitter and helpless. But, if we have methods of convincing others to make necessary changes, we can feel better about ourselves and the place where we work.

The following action gradient, which becomes more positive as it moves to the right, illustrates the values of suggesting improvements.

negative	neutral	positive
Emotional response	*Complaint re: specific incident*	*Rational analysis*
Childish reaction		Adult reaction
Emphasis on momentary relief	Emphasis on specific compensation	Emphasis on improving the overall situation
Employee feels powerless and bitter	Employee feels good if specific problem is handled	Employee feels empowered

Figure 18.4   The Action Gradient

Here's an example of the action gradient at work.

> Let's assume that you work as an engineering technologist for a civil engineering firm that specializes in developing city subdivisions. Because you're frequently placed in charge of projects, you essentially act as an assistant manager. Your office manager is technically competent and works hard, but she uses negative feedback as her primary motivational tool; the only time she comments on a person's work is when that person has made a mistake. Usually, she presents her criticisms in a hostile, aggressive manner. Employee morale and productivity are starting to suffer.
>
> By contrast, your co-workers prefer your positive motivational techniques. You give credit for work well done. When you have to comment on incomplete or shoddy work, you take care to focus on the work itself, not on the worker. Lately, though, the manager's attitude and behaviour have been particularly hard to stomach because everyone has been working extra hard to meet a contract deadline. ▲

AVOIDING ACTION

If you operate at the left side of the action gradient, you don't confront the problem directly; instead, you use passive aggression by talking behind your manager's back. This approach allows you and the other employees to take care of your resentment and frustration for a moment, but you don't feel good about yourself, and meanwhile, the interpersonal climate steadily worsens.

TAKING LIMITED ACTION

If you operate in the middle ground, you may comment about a specific incident and ask for an apology. However, even if you receive that apology, the basic situation hasn't changed because you haven't confronted the underlying problem.

TAKING POSITIVE ACTION

The advantage of operating on the right side of the gradient is that you try to improve the overall situation. This response pattern requires a rational analysis of the problem and the formulation of a workable solution. Once you believe you have a valid solution, you have a choice of presenting it as a proposal or as recommendations. In the proposal, you single-mindedly argue in favour of a definite course of action, while the recommendations report assesses two or more possible solutions and then chooses one of those solutions.

Whether you chose the proposal route or the (apparently) more objective recommendations report, you have set in motion a series of productive possibilities. For example, your proposal of an incentive plan for managers and staff may get adopted. But even if it doesn't gain approval, you will feel good about yourself for positively confronting the situation. Moreover, you could be perceived as a positive influence within the organization, which will not hurt your subsequent chances for promotion. And, in the long run, positive actions improve the working climate, which was your goal in the first place.

GRAPHICS IN PROPOSALS

Proposals tend to feature text, not graphics. Still, as Table 18.4 suggests, visuals can help make proposals more persuasive.

Table 18.4 Graphics in Proposals[1]

The Message	The Graphic
We offer high performance at low costs	◆ Line, bar, and pie charts ◆ Tables
Our plan is logical	◆ Flow chart
Our system or equipment does the job	◆ Schematic diagram ◆ Hybrid graphic (such as drawings, photos, and tabular data pasted onto a flow diagram)
The parts are easy to assemble	◆ Exploded view drawing
We can meet the schedule	◆ Timeline with milestones • Critical path diagram
We have the resources and experience	◆ Data charts ◆ Résumés with experience timelines ◆ Photos (people, facilities, and equipment)

1. Adapted from G. Edward Quimby, "Make Text and Graphics Work Together," *Intercom*, January 1996, p. 34.

Checklist FOR EDITING AND REVISING PROPOSALS

Use this checklist as a guide to revising and refining your proposals.

Format

◆ Have you chosen the best format (letter, memo, report) for your purpose and audience?

◆ Does the long proposal include adequate appendices?

◆ Does the title forecast the proposal's subject and purpose?

Content

◆ Is the problem clearly identified?

◆ Is the objective clearly identified?

◆ Does everything in the proposal support its objective?

◆ Does the proposal *show* as well as *tell*?

◆ Does the proposed plan, service, or product benefit the reader's personal or organizational needs?

◆ Are the proposed methods practical and realistic?

◆ Are all foreseeable limitations and contingencies identified?

◆ Is the proposal free of overstatement?

◆ Is the proposal's length appropriate to the subject?

Arrangement

◆ Does the proposal include all *relevant* sections of the recommended structure?

◆ Does the introduction provide sufficient orientation to the problem and the plan?

◆ Does the plan explain *how*, *where*, and *how much*?

◆ Are there clear transitions between related ideas?

Style

◆ Is the writing style clear, concise, and fluent?

◆ Is the level of technicality appropriate for the primary reader?

◆ Do supplements follow the appropriate style guidelines?

◆ Does the tone connect with the readers?

◆ Is the language convincing and precise?

◆ Is the proposal grammatical?

◆ Is the proposal ethically acceptable?

EXERCISES

1. Assume the head of your high school English department has asked you, as a recent graduate, for suggestions about revising the English curriculum to prepare students for writing. Write a proposal, based on your experience since high school. (Primary audience: the English department head and faculty; secondary audience: school committee.) In your external, solicited proposal, identify problems, needs, and benefits, and spell out a realistic plan. Review the outline on pages 402–03 before selecting specific headings.

2. After identifying your primary and secondary audience, compose a short planning proposal for improving an unsatisfactory situation in the classroom, on the job, in your dorm, or in your apartment (e.g., poor lighting, drab atmosphere, health hazards, poor seating arrangements). Choose a problem or situation whose resolution is more a matter of common sense and lucid observation than of intensive research. Be sure to (a) identify the problem clearly, give brief background, and stimulate the readers' interest; (b) state clearly the methods proposed to solve the problem; and (c) conclude with a statement designed to gain readers' support for your proposal.

3. Write a research proposal to your instructor (or an interested third party) requesting approval for the final term project (an analytical report or formal proposal). Identify the subject, background, purpose, and benefits of your planned enquiry, as well as the intended audience, scope of enquiry, data sources, methods of enquiry, and a task timetable. Be certain that adequate primary and secondary sources are available. Convince your reader of the soundness and usefulness of the project.

4. As an alternative term project to the formal analytical report (Chapter 19), develop a long proposal for solving a problem, improving a situation, or satisfying a need in your college, community, or job. Choose a subject sufficiently complex to justify a formal proposal, a topic requiring research (mostly primary). Identify an audience (other than your instructor) who will use your proposal for a specific purpose. Compose an audience/purpose profile, using the sample on pages 409–10 as a model. Here are possible subjects for your proposal:

- improving living conditions in your dorm or fraternity/sorority
- creating a daycare centre on campus
- creating a new business or expanding a business
- saving labour, materials, or money on the job
- improving working conditions
- improving campus facilities for the disabled
- supplying a product or service to clients or customers
- eliminating traffic hazards in your neighbourhood
- reducing energy expenditures on the job
- improving in-house training or job-orientation programs
- improving tutoring in the learning centre
- making the course content in your major more relevant to student needs
- changing the grading system at your school
- establishing more equitable computer use

COLLABORATIVE PROJECT

Exercises 1, 2, or 4 may be used for a collaborative project.

19

Formal Analytical Reports

A formal analytical report is used when an analysis requires lengthy discussion (usually 10 pages or more) and/or when the topic is important enough to warrant formal presentation (title page, table of contents, formal heading system, formal documentation, and so on).

Less formal reports can use memo formats, letter formats, or a relaxed, semi-formal version of the formal report format. Chapter 21 illustrates short, informal formats.

Analytical reports answer the questions:

1. *What data, observations, ideas, and background information can we gather about the topic discussed in this report? (What do we know?)*
2. *What conclusions can we draw from this material? (What does it all mean?)*
3. *What recommendations stem from our conclusions? (What should we do about it?)*

All readers of analytical reports want the first two questions answered. Most readers want the third question answered as well.

"Real-world" analytical reports answer questions for decision makers. Often, such reports provide the main basis for a reader's practical business decision. By contrast, "academic" analytical reports usually employ a more theoretical model although, increasingly, academic researchers are being approached by business firms for answers to difficult "real-world" questions.

FOUR MAIN TYPES OF ANALYSIS

As an employee you may be asked to *evaluate* a new assembly technique on the production line, or to locate and purchase (*recommend*) the best equipment at the best price. You might have to *identify the cause* of a monthly drop in sales, the reasons for low morale among employees, the causes of an accident, or the reasons for equipment failure. You might need to assess the *feasibility* of a proposal for a company's expansion or investment. There are many varieties of these four main types of analysis, but the procedure remains the same: (1) ask the right questions, (2) search for information, (3) evaluate and interpret your findings, and (4) draw conclusions and possibly recommend actions.

In general, then, your prime responsibility is to answer your reader's questions. In providing those answers, you will need to fulfill several attendant responsibilities:

- ◆ make the report's purpose clear
- ◆ use an appropriate structure for that purpose
- ◆ examine the topic at an appropriate level, and use appropriate language
- ◆ ensure that the report is readable, by evaluating it objectively
- ◆ write ethically: admit data limitations, and do not suppress contrary evidence
- ◆ forcefully make points
- ◆ make the report professional and error-free so that it gets the attention it deserves

Now, let's examine the nature of productive analysis, which allows readers to make informed decisions.

TYPICAL ANALYTICAL PROBLEMS

Far more than an encyclopedia presentation of information, the analytical report shows how you arrived at your conclusions and recommendations. Here are some typical analytical problems.

Will X Work for a Specific Purpose? Analysis can answer practical questions. For example, imagine that your employer is concerned about the effects of stress on employees. She asks you to investigate the claim that low-impact aerobics has therapeutic benefits—with an eye toward such a program for employees. You design your analysis to answer this question: *Do low-impact aerobics programs significantly reduce stress?* The analysis follows a *questions-answers-conclusions* sequence. Because the report could lead to action, you include recommendations based on your conclusions.

The questions posed in such a *feasibility report* are also termed *assessment criteria*. In order to answer the main question of whether low-impact aerobics reduce stress, supporting questions have to be asked (assessment criteria have to be applied):

- What causes stress?
- How is stress revealed physiologically?
- Can the physical manifestations of stress be measured?
- What kinds of activities reduce stress? How strenuous do they have to be?
- How long do these activities have to be followed before measurable effects are detected?

- Do the stress-reducing activities work equally well for all subjects?

Has X Worked As Well As Expected? *Evaluation (assessment) reports,* like feasibility studies, use a series of evaluation criteria to assess the performance or value of equipment, facilities, or programs. Unlike feasibility reports, however, assessment reports apply those criteria after a decision has been made.

Let's imagine, for example, that your engineering firm decided last year to network all of the firm's computers. Now, in assessing that network's performance, your report might use the following criteria to determine if the predicted gains have actually happened:

- performance gains, if any (the amount and quality of design work)
- communication within the firm (savings in meeting time)
- compatibility with the firm's design and communication software
- network reliability and down time
- the firm's ability to accept more complex projects

Is X or Y Better for a Specific Purpose? Analysis is essential in comparing machines, processes, business locations, computer systems, or the like. Assume that you manage a ski lodge and need to answer this question: *Which of the two most popular ski bindings is best for our rental skis?* In a comparative analysis of the Rossignol FTX and the Tyrolia Cyber D8SX bindings, you would assess the strengths and weaknesses of each binding on the basis of specific criteria (safety, cost, ease of repair, ease of adjustment, dependability, and so on), which you would rank in order of importance.

The comparative analysis follows a *questions-answers-conclusions* sequence and is designed to help the reader make a choice. Examples appear in magazines such as *Consumer Reports* and *Consumer's Digest.*

Why Does X Happen? The causal analysis is designed to answer questions like this: *Why do small businesses have a high failure rate?* This kind of analysis follows a variation of the questions-answers-conclusions structure: namely, *problem-causes-solution.* Such an analysis follows this sequence:

1. identify the problem
2. examine possible and probable causes, and isolate definite ones
3. recommend solutions

An analysis of low morale among employees would investigate causal relationships.

How Can X Be Improved or Avoided? Another form of problem solving focuses on desired results and recommends methods of achieving these results. This type of analysis answers questions like these: *How can we operate our division more efficiently? How can we improve campus security?*

Usually, a *recommendations report* first identifies causes of a problem or components of a desired result. Then, the report presents possible solutions and uses a consistent set of criteria to evaluate each solution in turn. Finally, the report recommends which solution or combination of solutions to implement.

Many readers of solicited recommendations prefer to see the final recommendations first, *before* the full analysis that leads to those recommendations. Chapter 21 describes when a direct recommendations pattern would be more suitable than an indirect pattern.

What Are the Effects of X? An analysis of the consequences of an event or action would answer questions like these: How has air quality been affected by the local power plant's change from burning oil to coal? Does electromagnetic radiation pose a significant health risk?

Another kind of problem-solving analysis is done to predict an effect: What are the consequences of my changing majors? Here, the sequence is *proposed action—probable effects—conclusions and recommendations.*

Is X Practical in This Situation? The feasibility analysis assesses the practicality of an idea or plan: *Will the consumer interests of Hicksville support a computer store?* In a variation of the questions-answers-conclusions structure, a feasibility analysis presents *reasons for—reasons against,* with both sides supported by evidence. Business owners often use this type of analysis.

Combining Types of Analysis. Types of analytical problems overlap considerably. Any one study may in fact require answers to two or more of the previous questions. The sample report on pages 444–451 is both a feasibility analysis and a comparative analysis. It is designed to answer these questions: *Is technical marketing the right career for me? If so, how do I enter the field?*

ELEMENTS OF ANALYSIS

Successful analytical reports feature the following elements.

CLEARLY IDENTIFIED PROBLEM OR QUESTION

Know what you're looking for. If your car's engine fails to turn over when you switch on the ignition, you would wisely check battery and electrical connections before dismantling parts of the engine. Apply a similar focus to your report.

On page 434, a hypothetical employer posed this question: *Will a low-impact aerobics program significantly reduce stress among my employees?* The aerobics question obviously requires answers to three other questions: *What are the therapeutic claims for aerobics? Are they valid? Will aerobics work in this situation?* How aerobic exercise got established, how widespread it is, who practices, and other such questions are not relevant to this problem (although some questions about background might be useful in the report's introduction). Always begin by defining the main questions and thinking through any subordinate questions they may imply. Only then can you determine the data or evidence you need.

With the main questions identified, the writer of the aerobics report can formulate her statement of purpose:

Define

> The purpose of this report is to examine and evaluate claims about the therapeutic benefits of low-impact aerobic exercise.

The writer might have mistakenly begun instead with this statement:

Vague

> This report examines low-impact aerobic exercise.

Words such as *examine* and *evaluate* (or *compare, identify, determine, measure, describe,* and so on) enable readers to understand the specific analytical activity that is the subject of the report.

Notice how the first version sharpens the focus by expressing the precise subject of the analysis: not aerobics (a huge topic), but the alleged *therapeutic benefits* of aerobics.

Define your purpose by condensing your approach to a basic question: *Does low-impact aerobic exercise have therapeutic benefits?* or *Why have our sales dropped steadily for three months?* Then restate the question as a declarative sentence in your statement of purpose.

SUBORDINATION OF PERSONAL BIAS

Interpret evidence impartially. Throughout your analysis, stick to your evidence. Do not force viewpoints on your material that are not substantiated by dependable evidence.

ACCURATE AND ADEQUATE DATA

Do not distort the original data by excluding vital points. Say you are asked to recommend the best chainsaw for a logging company. Reviewing test reports, you come across this information.

> Of all six brands tested, the Bomarc chainsaw proved easiest to operate. It also had the fewest safety features, however.

If you cite these data, present *both* findings, not simply the first—even though you may prefer the Bomarc brand. *Then* argue for the feature you think should receive priority.

As space permits, include the full text of interviews or questionnaires in appendices.

FULLY INTERPRETED DATA

Explain the significance of your data. Interpretation is the heart of the analytical report. You might interpret the chainsaw data in this way:

> Our cutting crews often work suspended by harness, high above the ground. Also, much work is in remote areas. Safety features therefore should be our first requirement in a chainsaw. Despite its ease of operation, the Bomarc saw does not meet our safety needs.

By saying "therefore" you engage in analysis—not mere information sharing. *Merely listing your findings is not enough.* Tell readers what your findings mean.

CLEAR AND CAREFUL REASONING

Each stage of your analysis requires decisions about what to record, what to exclude, and where to go next. As you evaluate your data *(Is this reliable and important?)*, interpret your evidence *(What does it mean?)*, and make recommendations based on your conclusions *(What action is needed?)*, you might have to alter your original plan. Remain flexible enough to revise your thinking if contradictory new evidence appears.

APPROPRIATE VISUALS

Use visuals generously (Chapter 12). Graphs are especially useful in an analysis of trends (rising or falling sales, radiation levels). Tables, charts, photographs, and diagrams work well in comparative analyses.

VALID CONCLUSIONS AND RECOMMENDATIONS

Along with the informative abstract, conclusions and recommendations are the sections of a long report that receive most attention from readers. The goal of analysis is to reach a valid *conclusion*—an overall judgment about what all the material means (that X is better than Y, that B failed because of C, that A is a good plan of action). Here is the conclusion of a report on the feasibility of installing an active solar heating system in a large building:

Offer a final judgment

1. Active solar space heating for our new research building is technically feasible because the site orientation will allow for a sloping roof facing due south, with plenty of unshaded space.
2. It is legally feasible because we are able to obtain an access easement on the adjoining property, to ensure that no buildings or trees will be permitted to shade the solar collectors once they are installed.
3. It is economically feasible because our sunny, cold climate means high fuel savings and faster payback (15 years maximum) with solar heating. The long-term fuel savings justify our short-term installation costs (already minimal because the solar system can be incorporated during the building's construction—without renovations).

Conclusions are valid when they are logically derived from accurate interpretation.

Having explained *what it all means,* you then recommend *what should be done.* Taking into account all possible alternatives, your recommendations urge specific action (to invest in A instead of B, to replace C immediately, to follow plan A, or the like). Here are the recommendations based on the previous interpretations:

Tell what should
be done

1. I recommend we install an active solar heating system in our new research building.
2. We should arrange an immediate meeting with our architect, building contractor, and solar-heating contractor. In this way, we can make all necessary design changes before construction begins in two weeks.
3. We should instruct our legal department to obtain the appropriate permits and easements immediately.

Recommendations are valid when they propose an appropriate response to the problem or question.

Because they culminate your research and analysis, recommendations challenge your imagination, your creativity, and—above all—your critical-thinking skills. Having reached a valid conclusion about *what is,* you now must decide *what ought to be done.* But what strikes one person as a brilliant idea might be seen by others as idiotic or offensive. Depending on whether recommendations are carefully thought out or off the wall, writers earn an audience's respect or its scorn. Figure 19.1 (page 439) depicts the types of decisions writers encounter in formulating, evaluating, and refining their recommendations.

When you do achieve definite conclusions and recommendations, express them with assurance and authority. Unless you have reason to be unsure, avoid non-committal statements ("It would seem that" or "It looks as if"). Be direct and assertive ("The earthquake danger at the reactor site is acute," or "I recommend an immediate investment"). Let readers know where you stand.

If, however, your analysis yields nothing definite, do not force a simplistic conclusion on your material. Instead, explain your position ("The contradictory responses to our consumer survey prevent me from reaching a definite conclusion. Before we make any decision about this product, I recommend a full-scale market analysis"). The wrong recommendation is far worse than no recommendation at all.

A GENERAL MODEL FOR ANALYTICAL REPORTS

Every analytical report identifies an issue to be examined or an overall question to be answered for the intended audience. That issue or question is eventually settled in the report's conclusion section. In between, the report employs a series of supporting questions (analytical criteria) to lead to the bottom-line answer. Figure 19.2 illustrates the line of reasoning one might use for a feasibility report about a proposed downtown location for a multi-purpose arena. The city council likes the idea of a downtown location but wants to be certain that the location is practical, so it instructs the consultant to question traffic flow in the area, potential parking problems, and the ability of existing city services to handle increased demands. The city also asks the consultant to predict the impact on surrounding businesses.

Figure 19.2 (page 440) shows the relationships among the introduction, the central sections, and the conclusion in one particular type of analytical report. Though other reports will use the same basic *introduction-analysis-conclusion* pattern, no general model can cover all formal analytical reports. Some reports will have one central section; others will have several. Much depends on the scope of the report and on the complexity of the analysis.

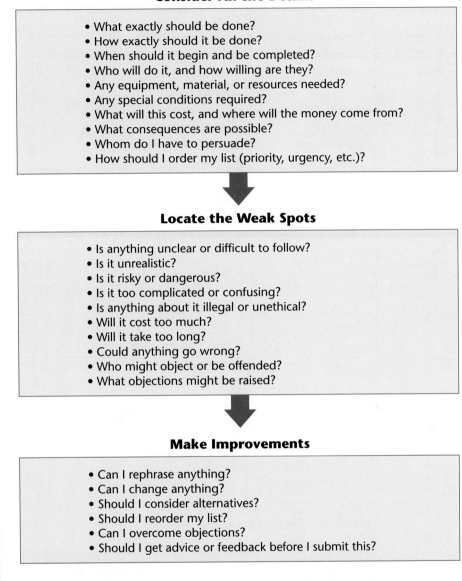

Consider All the Details

- What exactly should be done?
- How exactly should it be done?
- When should it begin and be completed?
- Who will do it, and how willing are they?
- Any equipment, material, or resources needed?
- Any special conditions required?
- What will this cost, and where will the money come from?
- What consequences are possible?
- Whom do I have to persuade?
- How should I order my list (priority, urgency, etc.)?

Locate the Weak Spots

- Is anything unclear or difficult to follow?
- Is it unrealistic?
- Is it risky or dangerous?
- Is it too complicated or confusing?
- Is anything about it illegal or unethical?
- Will it cost too much?
- Will it take too long?
- Could anything go wrong?
- Who might object or be offended?
- What objections might be raised?

Make Improvements

- Can I rephrase anything?
- Can I change anything?
- Should I consider alternatives?
- Should I reorder my list?
- Can I overcome objections?
- Should I get advice or feedback before I submit this?

Figure 19.1 How to Think Critically About Your Recommendations

Source: Questions adapted from Ruggiero, Vincent R. *The Art of Thinking*, 3rd ed. New York: Harper, 1991: 162–65.

PARTS OF A FORMAL REPORT

Table 19.1 (page 440) presents the sections that usually appear in formal reports, in the order they most often follow. (The numbers in brackets refer to the suggested order for preparing the sections; following that order of preparation will help you write the report efficiently.) Your supervisor will tell you which sections are required for the specific report you're writing.

INTRODUCTION	BACKGROUND	ASSESSMENT		CONCLUSION
Question		→		**Answer**
Report's purpose: *Assess feasibility of proposed arena location* To be evaluated on basis of: ① *Cost of land* ② *Impact on traffic* ③ *City services infrastructure (water, sewer, gas)* ④ *Parking* ⑤ *Impact on business*	Explain process by which this site was identified Explain why other sites will be assessed later: *priority given to downtown development* Explain information sources	**Cost of Land** • *City property—reserved for public* • *Equivalent value* **Impact on Traffic** • *Concerts and conventions* • *Hockey games* **Availability of Parking** • *Concerts and conventions* • *Hockey games* **City Infrastructure** • *Water* • *Sewer* • *Gas* **Parking** • *Requirements for various events* • *On-site parking* • *Parking within 4-block radius* **Impact on Business** • *Restaurants* • *Shopping*		Land cost makes location desirable Feasible in terms of traffic flow, parking, existing infrastructure Little positive impact on local business Recommendation: *Retain this site as feasible option, but look at other sites also*

Figure 19.2   A Question-to-Answer Development Pattern for Analytical Reports

Table 19.1   Parts of a Formal Report

Front Matter	Body	Back Matter
Transmittal Document (13) Cover (14) Title Page (8)	Introduction (3)	Sources Cited (6)
Summary (4)	Central Section(s) (1)	Additional Sources Consulted (7)
Table of Contents (12)	Conclusion (2)	
List of Illustrations (11)	Recommendation (2)	
Glossary (9) List of Symbols (9) Acknowledgements (10)		Appendices (5)

Strictly speaking, the transmittal document (letter or memo) does not belong in the front matter (i.e., between the title page and introduction). Transmittal documents *accompany* reports. For information and advice about transmittal documents and other parts of the front matter and back matter, see Chapter 20.

INTRODUCTION

The most important function of an analytical report's introduction is to identify the report's analytical purpose and preview how that purpose will be achieved. Usually that preview includes a list of supporting questions that will be used to answer the report's major question. In other words, the introduction lists the criteria used in the report's assessment or outlines the process used to determine causes, or previews the logical path to be used in arriving at a recommended action. In some cases, you will need to justify your choice of criteria or explain your analytical method.

An introduction may also require some or all of the following elements:

1. The context, situation, or problem prompting this report (background)
2. Type of data on which the report is based and the type of source
3. Other pertinent theoretical or background information
4. Useful illustrations

An introduction also indirectly sets the tone of the report. The phrasing reflects whether the report takes an aggressive stance or uses a more cautious or conciliatory approach. For example, a causal report's direct, no-nonsense approach is signalled as follows:

> The Forest Ministry assembled an investigation team to determine if:
>
> 1. forestry activities in the area contributed to the large destructive debris flow that killed three people, and if
> 2. additional investigation is required to assess the future risk of mass mudslides in the area.

BODY SECTIONS

Some reports need just one central section. For example, a 10-page causal analysis report might use a central section called "Contributory Causes," with a subsection for each of the factors that may have helped cause the problem or situation.

Other reports may need several central sections. For example, a 35-page assessment of three submitted proposals for a truck-leasing contract might have <u>four</u> central sections, one for each proposal and one entitled "Comparison of Alternatives." Each of four central sections would use the same set of assessment criteria to organize the analysis within each section. The report's conclusion section would identify the best of the three proposals and recommend whether to accept that proposal in its entirety, or to negotiate a modified version.

As you read the body sections of the two sample reports in this chapter, notice how the analytical criteria (supporting or exploratory questions) are presented in the introduction and then used to form logical structures in the report's main body. Notice also that the sample reports use clear, informative headings that identify exactly where you are at any point in the discussion.

CONCLUSION

The conclusion of an analytical report will interest readers because it answers the questions that sparked the analysis in the first place. Some workplace reports, therefore, place the conclusion *before* the introduction and body sections.

In the conclusion you summarize, interpret, and (perhaps) recommend. Although you have interpreted evidence at each step in the analysis, your conclusion pulls the strands together in a broader interpretation. In other words, you lead out from the material, to put it in perspective. This final section must be consistent in three ways:

1. The summary must reflect accurately the body of the report, and the bottom-line conclusions must be firmly based on information, ideas, and analysis already presented in the report. Do not introduce new material in the conclusion section.
2. Your overall interpretation must be consistent with the findings in your summary and must present an honest and objective appraisal of the material.
3. If you include recommendations, they must be consistent with the purpose of the report, the evidence presented, and the interpretations given.

Often recommendations form the last part of the conclusion section, especially if the report's primary purpose is to assess or to identify causes, but if the report's main purpose is to advise the reader what action to take, create a separate recommendations section. Remember also that not all reports require a set of recommended actions.

A SAMPLE SITUATION

The report in Figure 19.3 combines a feasibility analysis with a comparative analysis.

Richard Larkin, author of the following report, has a work-study job 15 hours weekly in his school's placement office. His supervisor, Mimi Lim (placement director), likes to keep abreast of trends in various fields. Larkin, an engineering major, has become interested in technical marketing and sales. In need of a report topic for his writing course, Larkin offers to analyze the feasibility of a technical marketing and sales career, both for himself and for technical and science graduates in general. Lim accepts Larkin's offer, looking forward to having the final report in her reference file for use by students choosing careers. Larkin wants his report to be useful in three ways: to satisfy a course requirement, to help him in choosing his own career, and to help other students with their career choices.

With his topic approved, Larkin begins gathering his primary data, using interviews, letters of enquiry, telephone enquiries, and lecture notes. He supplements these primary sources with articles in recent publications. He will document his findings in APA (author-date) style.

As a guide for designing his final report, Larkin completes the following audience/purpose profile. ▲

Audience/Purpose Profile for a Formal Report

AUDIENCE IDENTITY AND NEEDS

My primary audience consists of Mimi Lim, placement director, and the students who will be referring to my report as they choose careers. The secondary audience is my writing instructor. The data I've uncovered will help me make my own career choice.

Lim is highly interested in this project, and she has promised to study my document carefully and to make copies available to interested students. Because she already knows something about the technical marketing field, Lim will need very little background to understand my report. Many student readers, however, may know little or nothing about technical marketing, and so will need background, definitions, and detailed explanations. Here are the questions I can anticipate from my collective audience:

◆ *What, exactly, is technical marketing and sales?*
◆ *What are the requirements for this career?*
◆ *What are the pros and cons of this career?*
◆ *Could this be the right career for me?*
◆ *How do I enter the field?*
◆ *Is there more than one option for entering the field? If so, which option would be best for me?*

ATTITUDE AND PERSONALITY

Readers likely to be most affected by my document are students who will be making career choices. I would expect my readers' attitudes to vary widely.

To connect with this array of readers, I will need to persuade them that my conclusions are based on dependable data and careful reasoning.

EXPECTATIONS ABOUT THE DOCUMENT

I know that my readers are busy and impatient, so I'll want to make this report concise enough to be read in no more than 15 or 20 minutes.

Essential information will include an expanded definition of technical marketing and sales, the skills and attitudes needed for success, the career's advantages and drawbacks, and a description of various paths for entering the career. Throughout, I'll relate my material to many technical and science majors, not just engineers.

The body of this report combines a feasibility analysis with a comparative analysis. Therefore, I'll use a reasons-for and reasons-against structure in the feasibility section. In the comparison section, I'll use a block structure followed by a table that presents a point-by-point comparison of the four entry paths. Because I want this report to lead to informed decisions, I will include concrete recommendations that are based solidly on my conclusions.

To address various readers who may not want to read the entire report, I will include an informative abstract.

My tone throughout should be conversational. Because I am writing for a mixed audience (placement director, students, and writing instructor), I will use a third-person point of view. ▲

This report's front matter (title page and so on) appear in Chapter 20.

Feasibility Analysis

INTRODUCTION

The escalating cutbacks in aerospace, defense-related, and other goods-producing industries have narrowed career opportunities for many of today's science and engineering graduates. A study by the Massachusetts Institute of Technology, for example, found that leading industries hired 80 to 90 percent fewer engineers in the mid-1990s than in the mid-1980s (Solomon, 1996). This trend is expected to continue—if not worsen—in the foreseeable future.

With the notable exception of computer engineering, employment in all engineering specialties will grow at rates ranging from average to far below average to near static through the year 2006. In some specialties (e.g., mining and petroleum engineering), employment will actually decline ("Occupational Employment," 1998, p. 16).

Given such bleak employment prospects, recent graduates might consider alternative careers in which they could apply their technical training. One especially attractive alternative is in marketing and selling of technology products or services.

Customer orientation is an ever-growing part of today's business and manufacturing climate. Beginning in the 1980s, U.S. industry ceased to be "manufacturing driven" (where customers would buy whatever products were available). Technology companies became "service driven" by the demand for customized, efficiently serviced products (Basta, 1984, p. 84).

In the product-oriented industries of the 1970s, technical marketing accounted for only 39 percent of top management backgrounds, but that number nearly doubled by 1995. Also, 1998 employment listings for recent graduates showed countless major companies offering positions in technical marketing and sales (College Placement Council, 1998, p. 402). Undergraduates interested in this career need answers to these basic questions:

- *Is this the right career for me?*
- *If so, how do I enter the field?*

To help answer these questions, this report analyzes information gathered from professionals as well as from the literature.

After defining *technical marketing,* the following analysis examines the field's employment outlook, required skills and personal qualities, career benefits and drawbacks, and various entry options.

1.

Figure 19.3　An Analytical Report　　　　　　　　　　　　　　　　　**(continued)**

COLLECTED DATA

Key Factors in a Technical Marketing Career

Anyone considering technical marketing needs to assess whether this career fits his or her interests, abilities, and aspirations.

The technical marketing process. Although the terms *marketing* and *sales* are often used interchangeably, technical marketing traditionally has involved far more than sales work. The process itself (identifying, reaching, and selling to customers) entails six major activities (Cornelius & Lewis, 1983, p. 44):

1. *Market research:* gathering information about the size and character of the target market for a product or service.
2. *Product development and management:* producing the goods to fill a specific market need.
3. *Cost determination and pricing:* measuring every expense in the production, distribution, advertising, and sales of the product, to determine its price.
4. *Advertising and promotion:* developing and implementing all strategies for reaching customers.
5. *Product distribution:* coordinating all elements of a technical product or service, from its conception through its final delivery to the customer.
6. *Sales and technical support:* creating and maintaining customer accounts, and servicing and upgrading products.

Fully engaged in all these activities, the technical marketing professional gains a detailed understanding of the industry, the product, and the customer's needs (Figure 1).

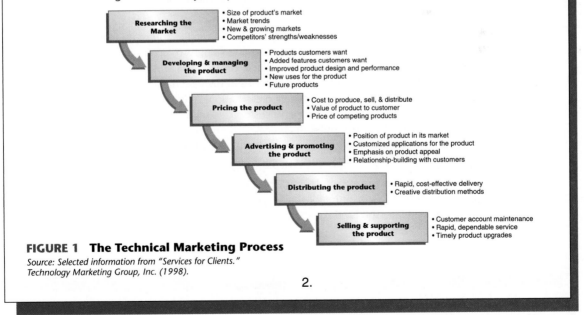

FIGURE 1   The Technical Marketing Process

Source: Selected information from "Services for Clients."
Technology Marketing Group, Inc. (1998).

2.

Figure 19.3   An Analytical Report   **(continued)**

Employment outlook. For graduates with the right combination of technical and personal qualifications, the employment outlook for technical marketing appears excellent. While engineering jobs will increase at barely one half the average growth rate for jobs requiring a bachelor's degree, marketing jobs will exceed the average growth rate (Figure 2).

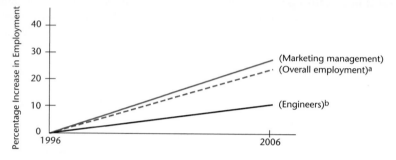

FIGURE 2   The Employment Outlook for Technical Marketing
[a]Jobs requiring a bachelor's degree.
[b]Excluding outlying rates for engineering specialties at extreme ends of the spectrum (computer engineers: +109%; mining and petroleum engineers: –14%).
Source: Data from U.S. Department of Labor. (1998, Winter). Occupational Outlook Quarterly, *11–16.*

Although highly competitive, these marketing positions call for the very kinds of technical, analytical, and problem-solving skills that engineers can offer (Solomon, 1996)—especially in an automated environment.

Technical skills required. Computer networks, interactive media, and multi-media will increasingly influence the way products are advertised and sold. Also, marketing representatives increasingly work from a "virtual" office. Using laptop computers, fax networks, and personal digital assistants, representatives in the field have real-time access to electronic catalogues of product lines, multi-media presentations, pricing for customized products, inventory data, product distribution, and customized sales contacts (Tolland, 1999).

With their rich background in computer, technical, and problem-solving skills, engineering graduates are ideally suited for (a) working in automated environments, and (b) implementing and troubleshooting these complex and often sensitive electronic systems.

Other skills and qualities required. In marketing and sales, not even the most sophisticated information can substitute for the "human factor": the ability to connect with customers on a person-to-person level (Young, 1995, p. 95). One senior sales engineer

Figure 19.3   An Analytical Report   *(continued)*

praises the efficiency of her automated sales system, but thinks that automation will "get in the way" of direct customer contact. Other technical marketing professionals express similar views about the continued importance of human interaction (94).

Besides a strong technical background, marketing requires a generous blend of those traits summarized in Figure 3.

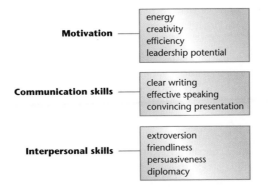

Motivation ——— energy
creativity
efficiency
leadership potential

Communication skills ——— clear writing
effective speaking
convincing presentation

Interpersonal skills ——— extroversion
friendliness
persuasiveness
diplomacy

FIGURE 3 Requirements for a Technical Marketing Career

Motivation is essential for marketing work. Professionals must be energetic and able to function with minimal supervision. Career counsellor Phil Hawkins describes the ideal candidates as people who can plan and program their own tasks, who can manage their time, and who have no fear of hard work (personal interview, February 11, 1999). Leadership potential, as demonstrated by extracurricular activities, is an asset.

Motivation alone provides no guarantee of success. Marketing professionals are paid to communicate the virtues of their products or services. This career therefore requires skill in communication, both written and oral. Documents for readers outside the organization include advertising copy, product descriptions, sales proposals, sales letters, and user manuals and on-line help. In-house writing includes recommendation reports, feasibility studies, progress reports, memos, and e-mail correspondence (U.S. Department of Labor, 1997, p. 8).

Skilled oral presentation is vital to any sales effort, as Phil Hawkins points out. Technical marketing professionals need to speak confidently and persuasively—to represent their products and services in the best possible light (personal interview, February 11, 1999). Sales presentations often involve public speaking at conventions, trade shows, and other similar forums.

Figure 19.3   **An Analytical Report**   **(continued)**

Beyond motivation and communication skills, interpersonal skills are the ultimate requirement for success in marketing (Solomon, 1996). Consumers are more likely to buy a product or service when they like the person selling it. Marketing professionals are extroverted, friendly, and diplomatic; they can motivate people without alienating them.

Advantages of the career. As shown in Figure 1, technical marketing offers diverse experience in every phase of a company's operation, from a product's design to its sales and service. Such broad exposure provides excellent preparation for countless upper-management positions.

In fact, sales engineers with solid experience often open their own businesses as "manufacturers' agents" representing a variety of companies. These agents represent products for companies that have no marketing staff of their own. In effect their own bosses, manufacturers' agents are free to choose, from among many offers, the products they wish to represent (Tolland, 1999).

Another career benefit is the attractive salary. Marketing professionals typically receive a base pay plus commissions. According to John Turnbow, managing recruiter of National Electric's technical marketing program, new marketing engineers for NE average over $60,000 in first-year wages. Moreover, salaries often reach six figures—sometimes higher than executive salaries (personal communication, April 5, 1999).

Technical marketing is especially attractive for its geographic and job mobility. Companies nationwide seek recent graduates, but especially in the Southeast and on the east and west coasts ("Electronic sales," 1996, pp. 1134–37). In addition, the types of interpersonal and communication skills that marketing professionals develop are highly portable. This is especially important in our current, rapidly shifting economy, in which job security is disappearing in the face of more and more temporary positions (Jones, 1997, p. 51).

Drawbacks of the career. Technical marketing is by no means a career for every engineer. Sales engineer Roger Cayer cautions that personnel might spend most of their time travelling to meet potential customers. Success requires hard work over long hours, evenings, and occasional weekends. Above all, the job is stressful because of constant pressure to meet sales quotas (phone interview, February 8, 1999). Anyone considering this career should be able to work and thrive in a highly competitive environment.

A Comparison of Entry Options
Engineers and other technical graduates enter technical marketing through one of four options. Some join small companies and learn their trade directly on the job. Others join companies that offer formal training programs. Some begin by getting experience in their

Figure 19.3   An Analytical Report   *(continued)*

technical specialty. Others earn a graduate degree beforehand. These options are compared below.

Option 1: Entry-level marketing with on-the-job training. Smaller manufacturers offer marketing positions in which people learn on the job. Elaine Carto, president of ABCO Electronics, believes small companies offer a unique opportunity; entry-level salespersons learn about all facets of an organization and have a good possibility for rapid advancement (personal interview, February 10, 1999). Career counsellor Phil Hawkins says, "It's all a matter of whether you prefer to be a big fish in a small pond or a small fish in a big pond" (personal interview, February 11, 1999).

Entry-level marketing offers immediate income and a chance for early promotion. A disadvantage, however, might be the loss of any technical edge one might have acquired in college.

Option 2: A marketing and sales training program. Formal training programs offer the most popular entry into sales and marketing. Mid-size to large companies typically offer two formats: (a) a product-specific program, focused on a particular product or product line, or (b) a rotational program, in which trainees learn about an array of products and develop the various skills outlined in Figure 1. Programs last from weeks to months.

Former trainees Roger Cayer, of Allied Products, and Bill Collins, of Intrex, speak of the diversity and satisfaction such programs offer: specifically, solid preparation in all phases of marketing, diverse interaction with company personnel, and broad knowledge of various product lines (phone interviews, February 8, 1999).

Like direct entry, this option offers the advantage of immediate income and early promotion. With no chance to practise in their technical specialty, however, trainees might eventually find their technical expertise compromised.

Option 3: Prior experience in one's technical specialty. Instead of directly entering marketing, some candidates first gain experience in their specialty. This option combines direct exposure to the workplace with the chance to sharpen technical skills in practical applications. In addition, some companies, such as Roger Cayer's, will offer marketing and sales positions to outstanding staff engineers, as a step toward upper management (phone interview, February 8, 1999).

Although this option delays a candidate's entry into technical marketing, industry experts consider direct workplace and technical experience key assets for career growth in any field. Also, work experience becomes an asset for applicants to top MBA programs (Shelley, 1997, pp. 30–31).

Figure 19.3   An Analytical Report   *(continued)*

Feasibility Analysis 7

Option 4: Graduate program. Instead of direct entry, some people choose to pursue an MS degree in their specialty or an MBA. According to engineering professor Mary McClane, MS degrees are usually unnecessary for technical marketing unless the particular products are highly complex (personal interview, April 2, 1999).

In general, jobseekers with an MBA have a distinct competitive advantage. More significantly, new MBAs with a technical bachelor's degree and one to two years of experience command salaries from 10 to 30 percent higher than MBAs who lack work experience and a technical bachelor's degree. In fact, no more than 3 percent of job candidates offer a "techno-MBA" specialty, making this unique group highly desirable to employers (Shelley, 1997, p. 30).

A motivated student might combine graduate degrees. Dora Anson, president of Susimo Cosmic Systems, sees the MS/MBA combination as ideal preparation for technical marketing (1999).

One disadvantage of a full-time graduate program is lost salary, compounded by school expenses. These costs must be weighed against the prospect of promotion and monetary rewards later in one's career.

An overall comparison by relative advantage. Table 1 compares the four entry options on the basis of three criteria: immediate income, rate of advancement, and long-term potential.

TABLE 1   Relative Advantages Among Four Technical-Marketing Entry Options

Option	Relative Advantages		
	Early, immediate income	Greatest advancement in marketing	Long-term potential
Entry level, no experience	yes	yes	no
Training program	yes	yes	no
Practical experience	yes	no	yes
Graduate program	no	no	yes

Figure 19.3   An Analytical Report   *(continued)*

CONCLUSION Feasibility Analysis

Summary of Findings

Technical marketing and sales involves identifying, reaching, and selling the customer a product or service. Besides a solid technical background, the field requires motivation, communication skills, and interpersonal skills. This career offers job diversity and excellent income potential, balanced against hard work and relentless pressure to perform.

College graduates interested in this field confront four entry options: (1) direct entry with on-the-job training, (2) a formal training program, (3) prior experience in a technical specialty, and (4) graduate programs. Each option has benefits and drawbacks based on immediacy of income, rate of advancement, and long-term potential.

Interpretation of Findings

For graduates with a strong technical background and the right skills and motivation, technical marketing offers attractive career prospects. Anyone contemplating this field, however, needs to be able to enjoy customer contact and thrive in a highly competitive environment.

Those who decide that technical marketing is for them can choose among the various entry options:

- For hands-on experience, direct entry is the logical option.
- For sophisticated sales training, a formal program with a large company is best.
- For sharpening technical skills, prior work in one's specialty is invaluable.
- If immediate income is not vital, graduate school is an attractive option.

Recommendations

If your interests and abilities match the requirements, consider these suggestions:

1. To get a firsthand view, seek the advice and opinions of people in the field.
2. Before settling on an entry option, consider all its advantages and disadvantages and decide whether this option best coincides with your career goals. (Of course, you can always combine options during your professional life.)
3. When making any career decision, consider career counsellor Phil Hawkins' advice: "Listen to your brain and your heart" (personal interview, February 11, 1999). Choose an option or options that offer not only professional advancement but also personal satisfaction.

8

REFERENCES

[The complete list of references is shown and discussed in Chapter 8, page 176.]

[**Note:** References normally start on a new page. In this report, The REFERENCES page would be page 9.]

Figure 19.3   An Analytical Report

THE PROCESS OF WRITING REPORTS

The efficient writing process described in Chapter 3 certainly applies to writing lengthy reports. Also, the research advice in Chapters 6, 7, and 8 applies to gathering, recording, and documenting information for formal reports. To avoid unnecessary effort and to save time in writing a lengthy report, follow the advice in Chapters 3 and 6.

USING OUTLINES

You can use three types of outlines to write top-quality reports efficiently:

1. Use a *planning outline* to guide your research and initial planning. That outline will change as you gather material, but such an outline will continually remind you of the report's purpose and the analytical criteria to achieve that purpose. Figure 19.4, on page 453, shows a planning outline for a comparative assessment report.

2. When you have chosen, evaluated, and analyzed the material for your report, write a detailed formal *working outline*, including:

   ◆ the report's working title
   ◆ a purpose/audience statement to remind yourself of the reason for the report
   ◆ a thesis statement to further remind yourself that everything in the report contributes to a "bottom-line" answer
   ◆ all headings and subheadings, named and formatted as they will be in the finished report's body
   ◆ a brief description of every paragraph in the finished report
   ◆ the name and number of each illustration, placed where it will appear in the report

   This working outline will take some time to write because it forms a complete blueprint for the first draft, but a thorough, well-conceived outline will dramatically decrease the time required to compose, revise, and edit your first draft. Also, *each keystroke that goes into the working outline will appear in that first draft;* the headings and illustration labels will all be in place, and even the paragraph description phrases will likely end up in their respective paragraphs. Compare the headings and paragraph descriptions in Figure 19.5 on pages 454–456 with the corresponding report in Figure 19.6 on pages 463–74.

3. As you use the working outline to compose the first draft, you can refer to your note cards to establish the exact content of each paragraph, or you can write a brief, informal *paragraph outline* for each one. See page 456 for an example. Such "quickie" outlines don't have to be neat; they merely help you write coherent, unified paragraphs quickly. You may prefer to create a paragraph outline for each new paragraph as you come to it, or you may prefer to write outlines for several paragraphs in succession.

Now, let's see this sequence of outlines at work. First we'll look at the outline for writing the report that was referred to in the proposal in Chapter 3. In Chapter 21, we'll look at the outline for writing the progress report.

The writer is Art Basran, Kelowna branch manager for MMT Consulting. The primary reader is Jessica Proctor, director of Mining Development for Jackson Mining Ltd. of Grand Forks, British Columbia. First, Art's planning outline forms the basis for his project plan and the report that will present the project's results. ▲

Project: Determine best method to haul coal from the proposed Othello mine.

Jackson Mining has suggested four haul methods, based on its previous experience:
- Diesel trucks
- Diesel trains
- Electric trains
- Electric fleets

Method:
To help Jackson Mining select the most cost-effective method of hauling coal from the Othello mine, we'll need to calculate:

1. Annual Costs (fuel consumption, labour, maintenance): Will survey rough-cleared roadway; will calculate all grades and energy needed to haul loads on those grades; will calculate tractive resistances for each of four haul methods; will project haul times; will investigate potential for electric fleet energy recycling
2. Construction Costs: For roads and for rail roads (road bed and track); for electric installations; for maintenance building
3. Capital Costs: For each of the four haul methods—Note: Start by calculating number of units required to haul daily mine output (client does not plan to store coal on site, for environmental reasons). Then, use Merrill Lynch Guide for unit costs.
4. Communication System Costs: GPS, radio system and back-ups, on-board computers

Special Problem:
Need a method to uniformly compare all components of all four types of operations—Is there such a method? (maybe ask Courtland Petersen at KPMG)

Also:
- Should we consider expansion of haul fleet, in case production exceeds projections?
- What about interest rates?
- Who's going to be the primary reader of the report? How much civil engineering and transport engineering background does he/she have?

Information Sources:
Merrill Lynch Industrial Guide
Chartwell's *Power Consumption Calculations*
Find specs books for diesel and electric trains
Internet websites for Western Star Trucks, Kenworth, Volvo Industrial Trucks
Jackson Mining's Minesite report for Environment Ministry

Figure 19.4   A Planning Outline

After weeks of hard work by Art Basran's team, and after consulting an accountant (see Chapter 3, page 51), Art analyzes all the data his team has generated and organizes it into the following working outline. Notice that the detailed calculations are placed in 38 pages of appendices to ensure that the calculations don't clutter the main body of the report.

Title: Cost Analysis of Four Alternatives for Hauling Coal from the Proposed Othello Mine

Purpose/Audience
The report will help the client, Jackson Mining Company, select the most cost-effective method of hauling coal from its proposed Othello mine. <u>Primary reader</u>: Jessica Proctor, Director of Mining Development.

Thesis Statement
If the investors know the interest rate and the amortization period for capitalizing the Othello mine, and if other factors in choosing a coal hauling method are less important than cost, the investors can select the haul method from the EUAC tables.

1.0   INTRODUCTION

    1.1   The Situation
- Jackson Mining's new find: Where, when developed, name, MMT's involvement
- Four transport alternatives to be analyzed and main criterion of analysis
- Guidelines for analysis

    1.2   Limitations
- Basic assumption

    1.3   Background
- Sales and expansion overview (transition paragraph)

    1.3.1   Diesel trucks
- Why?

    1.3.2   Diesel trains
- Haul requirements

    1.3.3   Electric trains
- Haul requirements

    1.3.4   Electric fleets
- Description of planned operation
- Sidings
- Transition: Refer to specifications in Appendix G

2.0   COST ANALYSIS

    2.1   Annual Costs
- Annual cost components

    2.1.1   Fuel consumption
- Why important, and how calculated (refer to Appendix A)
- Table 1: Tractive and Grade Resistances for Diesel Trucks
- Explain Table 1
- Refer to Appendix E—lead into next table

The conclusion section of this report is more tentative than in most analytical reports, so the thesis reflects that tentative "bottom line."

Figure 19.5   A Working Outline   *(continued)*

- Table 2: Speed, Power, and Fuel Consumption for Diesel Trucks
- Explain Table 2

2.1.2   Labour
- Method of calculating (refer to Appendix B)

2.1.3   Maintenance
- Components

2.2   Capital Costs
- Three components (list)

2.2.1   Transportation units
- Calculation method

2.2.2   Construction costs
- Three main factors (list)
- Calculation method
- Maintenance buildings

2.2.3   Communication system
- Name components

2.3   Cost Comparisons
- Refer to Table 3
- Table 3: Capital and Operating Costs
- Explain Table 3 and lead into Equivalent Uniform Annual Costs

2.4   Equivalent Uniform Annual Costs
- Definition and purpose of EUAC comparison
- Refer to Figure 1: What to see there
- Why the sharp decline in EUAC
- Refer to Appendix D (transition paragraph)
- Figure 1: EUAC for Four Alternatives at 6%

3.0   POTENTIAL FOR EXPANSION
Why examine potential for expansion (lead-in)

3.1   Diesel Trucks
- Explain how an immediate increase of 23.3% could be achieved

3.2   Diesel and Electric Trains
- Show room for a 53.5% increase in production and refer to increases past this amount

3.3   Electric Fleets
- Show room for a 130% increase in production and refer to increases past this amount

4.0   CONCLUSION
- Tables in Appendix D show that the diesel and electric train options give much better results than the diesel trucks and the electric train fleets
- Which is better, diesel train option or electric train option?
- Table 4: Diesel Trains as Best Alternative
- Table 5: Electric Trains as Best Alternative
- Comment on Tables 4 and 5
- Factors beyond the scope of this report which must be decided by the investors

Figure 19.5   A Working Outline   *(continued)*

APPENDIX A: FUEL CONSUMPTION CALCULATIONS
- Tables 6, 8, 10, and 12: Tractive and Grade Resistances
- Tables 7, 9, 11, and 13: Speed, Power, and Fuel Consumption
- Figures 2, 3, 4, and 5: Fuel Consumption by Section

APPENDIX B: ANNUAL OPERATING COSTS
- Tables 14, 15, 16, and 17: Annual Operating Costs

APPENDIX C: CAPITAL COSTS
- Tables 18, 19, 20, and 21: Capital Costs

APPENDIX D: EQUIVALENT UNIFORM ANNUAL COSTS
- Tables 22, 23, 24, 25, 26, and 27: Capital and Operating Cost Comparisons
- Figures 6, 7, 8, 9, and 10: EUAC for Alternatives at 6%, 7%, 8%, 9%, and 10%

APPENDIX E: SAMPLE CALCULATIONS RE: ENERGY REQUIREMENTS

APPENDIX F: MINESITE TOPOGRAPHICAL MAP

APPENDIX G: UNIT SPECIFICATIONS

Figure 19.5   A Working Outline

As an example of how Art Basran develops paragraph outlines to efficiently write paragraphs, consider his working outline note for section 2.1.1:

2.1.1   Fuel consumption
Why important, and how calculated (refer to Appendix A)

The third-level heading is already in place, and the basic idea of the paragraph has been identified. He develops this idea in the following paragraph outline:

2.1.1   Fuel consumption
Major factor because calculations are complicated
Appendix A—calculations based on example: Table 1

Now, with the key points in place, Art writes sentences in a *statement* + *explanation* paragraph sequence:

2.1.1   Fuel consumption
Fuel consumption is a major factor because of the large amounts of energy required to transport 5 million tons of coal annually. Calculating fuel requirements is complicated—

Appendix A contains the full breakdown of fuel calculations for each of the alternative hauling methods. Such calculations are based on the total resistance faced by each particular method of hauling the coal. For an example of how total resistance is calculated, see the following table, which calculates the resistance for diesel trucks on the proposed road.

Clearly, Art writes in a deliberate, painstaking way. He writes this way to avoid duplicating effort—nearly everything he keys into his working and paragraph outlines appear in the first draft of his report. This disciplined approach suits Art's working style. You may prefer to write your paragraph outlines in pen or pencil, perhaps in whole sequences of paragraphs, and then key the paragraphs as you compose sentences. Find the method that best suits your working style, but use outlines!

A FORMAL ANALYTICAL REPORT

The following pages illustrate excerpts from the report Art Basran produced from his working outline. For the purposes of this textbook, only a few of the report's 38 appendix pages have been included.

As you review the report, notice the following aspects of the format:

- *Margins, headers, and page numbering:* They are used to distinguish first pages of a section from the other pages of that section.
- *Spacing and indentation:* Paragraphs are not indented; rather, double spacing is used between paragraphs. Note: Only main headings are followed by a line space.
- *Headings system:* Multiple-decimal:
    - *First order* (1.0): 16 point, bolded, fully capitalized, at left margin, 4 cm (1.5") from top edge of page, followed by one line space
    - *Second order* (1.1): 14 point, bolded, first letter of each word capitalized, indented one tab from left margin, followed by a line of type
    - *Third order* (1.3.1): 12 point, bolded, indented two tabs from left margin, first letter of entire heading capitalized, followed by a line of type
- *Documentation system:* APA format
- *Letter of transmittal:* Accompanies report

MMT Consulting

#12-1542 Dickson Avenue, Kelowna, British Columbia V1V 2W6   Ph. (250) 868-4040   Fax (250) 868-5060

April 14, 2005

Ms. Jessica Proctor
Director of Mining Development
Jackson Mining Ltd.
174 Central Avenue
Grand Forks, British Columbia
V2P 6W9

Dear Ms. Proctor:

Re: Othello Mine Transportation Alternatives

Enclosed is a copy of the report, "Cost Analysis of Four Alternatives for Hauling Coal from the Proposed Othello Mine." As you requested, MMT Consulting has considered the annual costs and the capital costs for each of diesel trucks, diesel trains, electric trains, and electric fleets.

We found that the best method of comparing the four transportation alternatives relies on the accounting vehicle known as Equivalent Uniform Annual Costs (EUAC). Using this vehicle, we have calculated costs for all alternatives in a consistent manner. After all costs are compared, your choice will depend on the cost of borrowing and on the length of time you operate the Othello mine. For example, at a rate of 6%, diesel trains provide the best alternative for the first 12 years of operation, while at the same rate, electric trains give the best value for operations lasting beyond 12 years.

I would be pleased to discuss this report's findings with your Board of Directors. Please phone me at (250) 868-4040 during business hours, or call my cell number, (250) 868-3445 at other times.

This project has been very interesting, in all respects. If you would like to discuss how MMT Consulting could further assist with the Othello project, you will find me eager to do so.

Sincerely,

Art Basran, P.Eng.
Kelowna Branch Manager

Figure 19.6   A Formal Analytical Report—Transmittal Document   *(continued)*

COST ANALYSIS OF FOUR ALTERNATIVES FOR HAULING COAL FROM THE PROPOSED OTHELLO MINE

For the Jackson Mining Company
Grand Forks, British Columbia

Prepared by MMT Consulting
Kelowna, British Columbia

Submitted:
April 14, 2005

Figure 19.6   A Formal Analytical Report   *(continued)*

MMT Consulting   ii.

EXECUTIVE SUMMARY

The proposed Othello mine has an expected annual production of 5 million tons. The mine requires a cost-effective method of transporting the coal from the minesite to the mainline railroad.

Four suggested alternatives were analyzed, considering various interest rates and study periods:
1. Diesel trucks
2. Diesel trains
3. Electric trains
4. Electric fleets

Capital costs and annual operating costs were determined for each alternative, and combined to form an Equivalent Uniform Annual Cost. Figure 1 shows the results for the preferred interest rate of 6%.

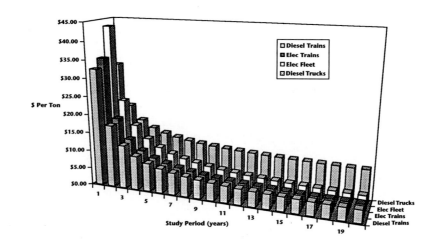

FIGURE 1: EUAC for Alternatives at 6%

Diesel trains and electric trains were the two preferred alternatives. The EUAC calculations revealed that the cost curves for diesel trains and electric trains cross at different points, depending on the interest rate. At an interest rate of 6%, the curves cross at 12 years. Therefore, from 1 to 12 years, diesel trains are preferred, and after 12 years, electric trains are preferred. This crossover year increases with increasing interest rates.

Therefore, if both the study period and the interest rate are known to the investors, the preferred alternative can be chosen from the calculations provided in this report. Other factors may affect the investors' choice of transportation for Othello mine coal, but these factors are outside the scope of this report.

Figure 19.6 **A Formal Analytical Report**   *(continued)*

TABLE OF CONTENTS

Figure 19.6　A Formal Analytical Report　　　　　*(continued)*

MMT Consulting   iv.

LIST OF ILLUSTRATIONS

Figure 19.6   **A Formal Analytical Report**   *(continued)*

MMT Consulting

1.0 INTRODUCTION

1.1 The Situation

The Jackson Mining Company has discovered a high-grade coal deposit in the mountains about 55 miles due north of Grand Forks, British Columbia. The company expects the market for this grade of coal to increase dramatically in the near future, and thus anticipates developing the deposit if development and transportation costs are acceptable. Jackson Mining has tentatively named its proposed minesite as the Othello mine. MMT Consulting has been authorized to determine the most cost-effective method(s) of transporting the coal from that minesite to the mainline railroad, a distance of approximately 50 miles (Jackson Mining, 2003, p.16).

Four transport alternatives will be considered in the following analysis:

1. Diesel trucks
2. Diesel trains
3. Electric trains
4. Electric fleets

The criterion used to evaluate each option is the Equivalent Uniform Annual Cost (the EUAC). EUAC is a cost measure that considers start-up (capital) costs as well as annual operating and maintenance costs. The EUAC discussed in this report is in dollars per ton, or how much it costs to haul one ton of coal from the minesite to the railhead.

The alternatives will be studied over study periods ranging from one to twenty years, and at interest rates varying from 6% to 10%. A short evaluation of each alternative's capacity to handle an increase in production is also provided.

1.2 Limitations

All calculations and recommendations are based on the assumption that fuel and electricity costs will remain constant or comparable throughout the study period.

1.3 Background

The Othello mine under study has anticipated coal sales of 5 million tons annually (Jackson Mining, 1999, p. 6). Each of the following transportation alternatives would have to handle this volume. Also, the ability to expand in the future would be an asset.

1.3.1 Diesel trucks

The CAT 777B truck was chosen for this study because of its load capacity of 95 tons and its output of 920 horsepower. These trucks would be running 24 hours a day.

1.3.2 Diesel trains

Each train would be running with six diesel engines and 98 coal cars. It takes 11.2 hours to complete a round trip, and the time it takes to produce the amount of coal the train hauls is 17.2 hours. Thus, the train would haul a complete load and then idle for about 6 hours until the next load was ready.

Figure 19.6   A Formal Analytical Report   *(continued)*

1.3.3 Electric trains

Each train would run four electric engines and 98 coal cars. The haul times are the same as quoted for the diesel train.

1.3.4 Electric fleets

Each fleet would consist of three of the above electric trains. The first train would be loaded and sent off. While it was enroute, the second train would be loading, and this load-and-go process would be repeated for the third train. The departures would be timed so that the second train leaves just as the first train crests the highest point of the pass through which the rail line runs. This would also be repeated for the third train. In this way, the electricity generated by the first and second trains on their downgrades could be conserved and used to assist in powering the second and third trains on their upgrades.

Four sidings would need to be constructed in order to allow the outgoing and returning trains to pass each other. However, these sidings are proposed for options 2 and 3 as well.

Complete specifications for all four alternatives can be found in Appendix G.

Figure 19.6   A Formal Analytical Report   *(continued)*

MMT Consulting   3.

2.0 COST ANALYSIS

Data has been collected for the capital costs and for the annual operating costs for all alternatives. Much of this data is available in Merrill Lynch's *Industrial Guides* (Jacobs, 1994, pp.140–175 and Ringness, 2001, pp. 242–263).

2.1 Annual Costs

The annual costs involve three main components:

1. fuel
2. labour, and
3. maintenance.

2.1.1 Fuel consumption

Fuel consumption is a major factor because of the large amounts of energy required to transport 5 million tons of coal annually. Calculating fuel requirements is complicated—Appendix A contains the full breakdown of fuel calculations for the alternative hauling methods. Such calculations are based on the total resistance faced by a particular method of hauling the coal. For an example of how total resistance is calculated, see Table 1, which calculates the resistance for diesel trucks on the proposed road.

Table 1: Tractive and Grade Resistances for Diesel Trucks

Section	Grade (%)	Truck Wt (ton)	Load Wt	Total Wt	Rt (lb)	Rg	Tr
G	2.0	66.3	95.0	161.3	16,125.0	6,450.0	22,575.0
F	1.8	66.3	95.0	161.3	16,125.0	5,805.0	21,930.0
E	2.0	66.3	95.0	161.3	16,125.0	6,450.0	22,575.0
D	2.1	66.3	95.0	161.3	16,125.0	6,772.5	22,897.5

The first column in Table 1 lists the sections into which the road has been divided for analytical purposes. These sections of road have varying grades. Grade resistance (Rg) and tractive resistance (Tr) are found for each section. (Sample calculations for these are given in Appendix E.) The calculated total resistance (Tr) is then used in the next table (Table 2).

Table 2: Speed, Power, and Fuel Consumption for Diesel Trucks

Section	Grade (%)	Tr (lb)	Power (hp)	Speed (mph)	Dist (mi)	Time (hr)	BTU (x1000)	Fuel Burned	Total Fuel
G	2.0	22,575.0	920.0	12.55	8.4	0.67	1,572.5	11.34	42.62
F	1.8	21,930.0	920.0	12.92	7.2	0.56	1,309.3	9.44	35.49
E	2.0	22,575.0	920.0	12.55	5.4	0.43	1,010.9	7.29	27.40
D	2.1	22,897.5	920.0	12.38	4.8	0.39	911.4	6.57	24.70

As in Table 1, the speed for each section is calculated, given the maximum power output of the alternative, to a maximum of 30 mph. The distance and time are then calculated and used to find total BTU per section of road. Finally, the total fuel consumed per section is determined; this depends on the fuel efficiency of that transportation alternative. All the calculated values for the various sections of road are added together to find the fuel consumed per trip. Then, given the cost of fuel, a total cost per trip is found.

Figure 19.6   A Formal Analytical Report   *(continued)*

MMT Consulting   4.

2.1.2 Labour

The next component of annual cost is labour. The method used to calculate labour differs from the trucks to the trains. Since the trucks are assumed to run 24 hours a day, the total hours per year is multiplied by the number of drivers and further by their hourly wage rate. In the other approach, the time per trip for the trains is calculated. That figure is multiplied by the number of engine staff and train persons needed, and further by their hourly rate. The results of these calculations can be found in Appendix B.

2.1.3 Maintenance

This third component of annual operating costs can be further divided into unit maintenance and road maintenance. Road or rail maintenance is given as a cost per mile, and then multiplied by the total distance covered. The results of these calculations appear in Appendix B.

2.2 Capital Costs

Capital costs for each of the four alternatives consist of:

1. cost of the transportation units
2. cost of construction, and
3. cost of communication systems.

2.2.1 Transportation units

The number of required units was found by taking the annual output of the mine and calculating the number of units that would haul the coal with the maximum efficiency. It was found that 30 diesel trucks and trains of 98 cars were needed. Seven extra trucks would be purchased, to allow for expansion and also to allow trucks to be taken off the road for maintenance.

2.2.2 Construction costs

Construction costs can be determined by examining three main factors:

1. road and rail construction
2. the maintenance building, and
3. electrical installations (where applicable).

Road and rail construction were found by multiplying the cost per mile by the total distance covered. For the rail alternatives, the costs of switches, loops, and sidings must be added. Overall, then, the railroad construction costs outweigh road construction costs by more than $45 million. In the case of the maintenance building costs, the truck maintenance facility would be $1.6 million more than the building required for the train options.

The two electric train options involve an additional expense—they require an electric installation, at a cost of $15.7 million. This figure was calculated by multiplying the cost per mile by the total distance and adding the cost of tapping into existing power sources. The full calculations are given in Appendix C.

Figure 19.6   A Formal Analytical Report   **(continued)**

2.2.3 Communication system

This is the third main component of the capital costs for this project. Such a system is necessary to control the movements of any of the four alternative transportation methods, so the same figure has been used for all four alternatives. A Global Positioning System, on-board computers, and two-way radios make up the complete system.

2.3 Cost Comparisons

The costs for the four alternatives are provided in Table 3 below.

Table 3: Capital and Operating Costs

Option	Capital Costs	Annual Operating Costs
Diesel Trucks	$115,539,000	$38,027,700
Diesel Trains	$149,083,000	$4,574,400
Electric Trains	$161,783,000	$3,061,000
Electric Fleets	$201,463,000	$2,766,600

Table 3 makes it clear that there is a direct correlation between the two cost parameters: The more expensive the system is to purchase and install, the less it will cost to operate. However, the investors must also consider the cost of borrowing the required capital funds. This expense, which can be considerable, is factored into a comprehensive comparative tool called the Equivalent Uniform Annual Costs; this measure is discussed next.

2.4 Equivalent Uniform Annual Costs

Once the capital and annual costs are found, they can be combined with varying interest rates and study periods to calculate tables of Equivalent Uniform Annual Costs (EUAC). The purpose of converting all costs into an EUAC is to provide a common means of comparing all the alternatives. The capital costs are spread over the length of the study period, and added to the annual costs. Thus, a comparison of EUACs will qualitatively assess the alternatives to go with the quantitative assessment suggested by Table 3 (B. Winters, letter, October 7, 2004).

Figure 1 (page 6) illustrates EUAC for the four alternatives, with study periods ranging from one to twenty years, at an interest rate of 6%. The graph's format makes it easy to compare the four alternatives at this interest rate. For example, at an interest rate of 6% diesel trains provide the most cost-effective method of transporting coal from the mine for study periods from two to twelve years.

The sharp decline in EUAC after the early study periods reflects the fact that the capital costs are being spread over longer and longer study periods. For shorter study periods, the capital costs play a larger role in determining EUAC, while for longer periods the annual operating cost is the more significant determinant. A full summary of EUAC for interest rates of 6%, 7%, 8%, 9%, and 10% appears in Appendix D.

Figure 19.6   A Formal Analytical Report   *(continued)*

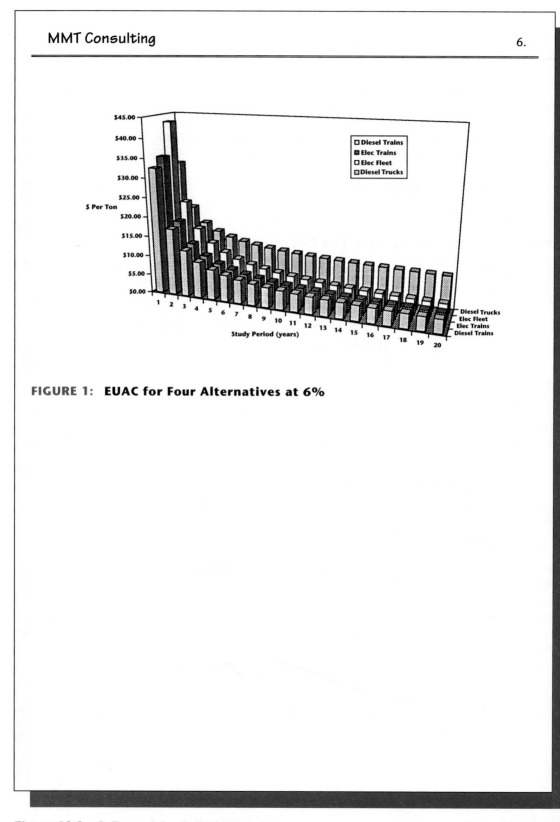

FIGURE 1:   **EUAC for Four Alternatives at 6%**

Figure 19.6   **A Formal Analytical Report**   *(continued)*

3.0 POTENTIAL FOR EXPANSION

Market conditions or additional discoveries of coal seams may result in mine expansion. If this proves to be the case, it would require an increase in the transportation of coal. Thus, it is prudent to examine each transport alternative's potential for expansion.

3.1 Diesel Trucks

To protect against breakdowns, 37 diesel trucks would be purchased, although only 30 are needed to effectively haul the coal from the minesite at a time. This leaves seven trucks for immediate expansion. In other words, an immediate increase of 23.3% could be achieved. Also, it is very easy for a trucking system to grow with the mine, because one or two additional units can readily be purchased as needed.

3.2 Diesel and Electric Trains

It takes both the diesel and electric trains 11.2 hours to complete a round trip, while it takes the mine 17.2 hours to produce enough coal for a full load. Therefore, six hours are spent idle waiting for the next load. This leaves room for a 53.5% increase in production. Increases past this amount would result in either adding more cars and engines to the trains, or switching to fleets of trains. Both of these are expensive capital options.

3.3 Electric Fleets

It takes the electric fleets 22.4 hours to complete a round trip, while it takes the mine 51.6 hours to produce enough coal for a full load. Therefore, 29.2 hours are spent idle waiting for the next load. This idle time leaves room for a 130% increase in production. Increases past this amount would result in either adding more cars and engines to the trains, or adding more fleets. Again, this expansion involves expensive capital options.

Figure 19.6 **A Formal Analytical Report**   **(continued)**

MMT Consulting   8.

4.0 CONCLUSION

The tables in Appendix D (pages 30–41) contain the Equivalent Uniform Annual Cost values for the four options at varying interest rates. Using the EUAC method clearly shows that the diesel and electric train options give much better results than the diesel trucks and the electric train fleets.

However, it is not easy to choose between the diesel train and electric train options, as the following tables show. The diesel train option is the better alternative for the first 12 or so years, after which the electric train option becomes slightly better.

Table 4: Diesel Trains as Best Alternative

% Interest Rate	Study Period (yrs)
6	1 to 12
7	1 to 13
8	1 to 14.5
9	1 to 16
10	1 to 19

Table 5: Electric Trains as Best Alternative

% Interest Rate	Study Period (yrs)
6	1 to 12
7	over 13
8	over 14.5
9	over 16
10	over 17

Although Tables 4 and 5 show that the diesel train option has a slight overall edge in EUAC, the figures throughout the study periods are quite close. For more details of just how close these figures are, see the following appendices.

In any event, the choice between diesel and electric trains may depend on additional factors, not just transportation cost analysis. One prime consideration will be how long the investors plan to operate the mine. This in turn will depend on factors such as the continuing quality of the coal deposits, the market demand, and the market price of coal. Other criteria might include the availability of diesel fuel or electricity, expected fuel or wage increases, or environmental concerns.

These additional factors are beyond the scope of this report and must be decided by the investors.

Figure 19.6   A Formal Analytical Report   **(continued)**

MMT Consulting 9.

REFERENCES CITED

Jackson Mining. (2003). *Othello minesite proposal.* Presented to the British Columbia Ministry of the Environment and the Ministry of Small Business, Tourism, and Culture.

Jacobs, G. (1994). *Merrill Lynch industrial calculation guide.* New York: Wiseman Press.

Ringness, A. (2001). 2001 *Update: Merrill Lynch industrial calculation guide.* New York: Wiseman Press.

Figure 19.6 A Formal Analytical Report ***(continued)***

MMT Consulting 10.

ADDITIONAL SOURCES CONSULTED

Chartwell, Y. K. (January 2000). Power consumption calculations: A planners' guide. *Transportation Today*, pp. 123–182.

Connors, C. S. (May 24, 2001). *Mining feasibility of the proposed Othello mine.* Confidential report for Jackson Mining, prepared by Connors Bowles, Mining Engineer Consultants, Calgary.

Siemens Electrical Corporation. (2003). *Siemens electric engines: Specifications and fleet rates.* Seattle: Siemens International Sales Department.

Figure 19.6   A Formal Analytical Report

(continued)

MMT Consulting 11.

APPENDIX A: FUEL CONSUMPTION CALCULATIONS

Note: For purposes of this text, only page 12 has been included.

Figure 19.6 A Formal Analytical Report[1] ***(continued)***

1. The remainder of the 37 pages has been omitted to save space.

MMT Consulting   12.

Table 6: Tractive and Grade Resistances for Diesel Trucks

Section		Grade (%)	Truck Wt (ton)	Load Wt	Total Wt	Rt (lb)	Rg	Tr
Trip from mine	G	2.0	66.3	95.0	161.3	16,125.0	6,450.0	22,575.0
	F	1.8	66.3	95.0	161.3	16,125.0	5,805.0	21,930.0
	E	2.0	66.3	95.0	161.3	16,125.0	6,450.0	22,575.0
	D	2.1	66.3	95.0	161.3	16,125.0	6,772.5	22,897.0
	C	-1.7	66.3	95.0	161.3	16,125.0	-5,482.0	10,642.0
	B	-2.2	66.3	95.0	161.3	16,125.0	-7,095.0	9,030.0
	A	-2.0	66.3	95.0	161.3	16,125.0	-6,450.0	9,675.0
Trip to mine	A	2.0	66.3	0.0	66.3	6,625.0	2,650.0	9,275.0
	B	2.2	66.3	0.0	66.3	6,625.0	2,915.0	9,540.0
	C	1.7	66.3	0.0	66.3	6,625.0	2,252.5	8,877.5
	D	2.1	66.3	0.0	66.3	6,625.0	2,782.5	9,407.5
	E	2.0	66.3	0.0	66.3	6,625.0	2,650.0	9,275.0
	F	-1.8	66.3	0.0	66.3	6,625.0	-2,385.0	4,240.0
	G	-2.0	66.3	0.0	66.3	6,625.0	-2,650.0	3,975.0

Figure 19.6   A Formal Analytical Report

Checklist FOR EDITING AND REVISING ANALYTICAL REPORTS

Use this checklist to refine the content, arrangement, and style of your report.

Content

◆ Does the report grow from a clear statement of purpose?

◆ Is the report's length adequate and appropriate for the subject?

◆ Are all limitations of the analysis clearly acknowledged?

◆ Are visuals used whenever possible to aid communication?

◆ Are all data accurate?

◆ Are all data unbiased?

◆ Are all data complete?

◆ Are all data fully interpreted?

◆ Is the documentation adequate, correct, and consistent?

◆ Are the conclusions logically derived from accurate interpretation?

◆ Do the recommendations constitute an appropriate response to the question or problem?

Arrangement

◆ Is there a distinct introduction, body, and conclusion?

◆ Are headings appropriate and adequate?

◆ Are there enough transitions between related ideas?

◆ Is the report accompanied by all needed front matter?

◆ Is the report accompanied by all needed end matter?

Style and Page Design

◆ Is the level of technicality appropriate for the stated audience?

◆ Is the writing style throughout clear, concise, and fluent?

◆ Is the language convincing and precise?

◆ Is the writing in the report grammatical?

◆ Is the page design inviting and accessible?

EXERCISE

Prepare an analytical report, using some sequence of these guidelines:

a. Choose a subject for analysis from the list your instructor provides, from your major, or from a subject of interest.

b. Identify the problem or question so that you will know exactly what you are looking for.

c. Restate the main question as a declarative sentence in your statement of purpose.

d. Identify an audience—other than your instructor—who will use your information for a specific purpose.

e. Hold a private brainstorming session to generate major topics and sub topics.

f. Use the topics to make an outline based on the model outline in this chapter. Divide as far as necessary to identify all points of discussion.

g. Make a tentative list of all sources (primary and secondary) that you will investigate. Verify that adequate sources are available.

h. Write your instructor a proposal memo, describing the problem or question and your plan for analysis. Attach a tentative bibliography.

i. Use your planning outline as a guide to research and observation. Evaluate sources and evidence, and interpret all evidence fully. Modify your outline as needed.

j. Read Chapter 21 and then submit a progress report to your instructor describing work completed, problems encountered, and work remaining. Attach a detailed working outline.

k. Compose an audience/purpose profile. (Use the sample on page 443 as models, along with the profile worksheet on page 34.)

l. Write the report for your stated audience. Work from a clear statement of purpose, and be sure that your reasoning is shown clearly. Verify that your evidence, conclusions, and recommendations are consistent. Be especially careful that your recommendations observe the critical-thinking guidelines in Figure 19.1.

m. After writing your first draft, make any needed changes in the outline and revise your report according to the revision checklist. Include all necessary supplements.

n. Exchange reports with a colleague for further suggestions for revision.

o. Prepare an oral report of your findings for the class as a whole.

COLLABORATIVE PROJECTS

1. Divide into small groups. Choose a subject for group analysis—preferably, a campus issue—and partition the topic by group brainstorming. Next, select major topics from your list and classify as many items as possible under each major topic. Finally, draw up a working outline that could be used for an analytical report on this subject.

2. Prepare a questionnaire based on your work above, and administer it to members of your campus community. List the findings of your questionnaire and your conclusions in clear and logical form. (Review pages 117–123, on questionnaires and surveys.)

CHAPTER 20

Adding Document Supplements

Supplements are reference items generally added to a long report or proposal to make the document more accessible. A document's supplements accommodate readers with various interests: The title page, letter of transmittal, table of contents, and abstract give summary information about the content of the document. The glossary, appendices, and list of works cited can either provide supporting data or help readers follow technical sections. According to their needs, readers can refer to one or more of these supplements or skip them altogether. All supplements, of course, are written only after the document itself has been completed.

Some companies and organizations require that a full range of supplements routinely accompany any long document; others do not. For situations in which your audience has not stipulated its requirements, select only those supplements that enhance the informative value of your particular document. Avoid using supplements merely as decoration.

PURPOSE OF SUPPLEMENTS

Documents must be accessible to varied readers for many purposes. Supplements address these workplace realities:

- ◆ *Confronted by a long document, many readers will try to avoid reading the whole thing.* Instead they look for the least information they need to complete the task, make the decision, or take some other action.
- ◆ *Different readers often use the same document for different purposes.* Some look for an overview; others want details; others want only conclusions and recommendations, or the "bottom line." Technical personnel might focus on the body of a highly specialized report and on the appendices for supporting data (maps, formulas, calculations). Executives and managers might read only the transmittal letter and the executive summary. If the latter audience reads any parts of the report proper, they are likely to focus on conclusions and recommendations.

A document with carefully planned supplements accommodates the needs of diverse audiences.

Report supplements can be classified into two groups:

1. *supplements that precede your report* (front matter): transmittal document, cover, title page, summary, table of contents, and list of illustrations.

2. *supplements that follow your report* (back matter): glossary, appendices, footnotes, endnote pages.

COVER

Use a sturdy, plain cover with page fasteners. With the cover on, the open pages should lie flat. Use covers only for long documents.

Centre the report title and your name 10–13 cm (4–5") below the upper edge of your page (many workplace reports include a company name and logo instead of the report author's name).

THE FEASIBILITY OF A TECHNICAL MARKETING CAREER

by

Richard B. Larkin

TRANSMITTAL DOCUMENT

What to include in
a transmittal
document

For college reports, the transmittal document (memo or letter) sometimes follows the title page and is bound as part of the report. For workplace reports, the transmittal document usually is not bound in the report but is presented separately. Include a transmittal document with any formal report or proposal addressed to a specific reader. Your letter or memo adds a note of courtesy and gives you a place for personal remarks. For instance, your letter or memo might:

- acknowledge those who helped with the report
- refer to sections of special interest: unexpected findings, key visuals, major conclusions, special recommendations, and the like
- discuss the limitations of your study, or any problems gathering data
- discuss the need and approaches for follow-up investigations
- describe any personal (or off-the-record) observations
- suggest some special uses for the information
- urge the reader to immediate action

The transmittal document can be tailored to a particular reader, as is Richard Larkin's in Figure 20.1 on page 480. If a report is being sent to a number of people who are variously qualified and bear various relationships to the writer, individual transmittal documents may vary within the following basic structure:

Introduction.   Open with reference to the reader's original request. Briefly review the reasons for your report or include a brief descriptive abstract. Maintain a confident and positive tone throughout. Indicate pride and satisfaction in your work. Avoid implied apologies, such as "I hope this report meets your expectations."

Body.   In the body, include items from your prior list of possibilities (acknowledgments, special problems). Although your informative abstract will summarize major findings, conclusions, and recommendations, your letter or memo gives a brief and personal overview of the *entire project.*

Conclusion.   State your willingness to answer questions or discuss findings. End positively with something like, "I believe that the data in this report are accurate, that they have been analyzed rigorously and impartially, and that the recommendations are sound."

TITLE PAGE

How to prepare a
title page

The title page (see page 481) lists the report title, author's name, name of person(s) or organization to whom the report is addressed, and date of submission.

7409 Trinity Court
Niagara Falls, ON   L2H 3A6
April 29, 2000

Ms. Mimi Lim
Placement Director
Seneca College
1750 Finch Avenue East
North York, ON   M2J 2X5

Dear Ms. Lim:

Here is my analysis to determine the feasibility of a career in technical marketing. In preparing my report, I've learned a great deal about the requirements and modes of access to this career, and I believe my information will help other students as well.

Although committed to their specialities, some technical and science graduates seem interested in careers in which they can apply their technical knowledge to customer and business problems. Technical marketing may be an attractive choice of career for those who know their field, who can relate to different personalities, and who are good communicators.

Technical marketing is competitive and demanding, but highly rewarding. In fact, it is an excellent route to upper-management and executive positions. Specifically, marketing work enables one to develop a sound technical knowledge of a company's products, to understand how these products fit into the marketplace, and to perfect sales techniques and interpersonal skills. This is precisely the kind of background that paves the way to top-level jobs.

I've enjoyed my work on this project, and would be happy to answer any questions.

Sincerely,

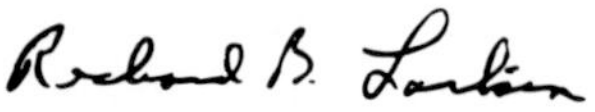

Richard B. Larkin

Richard B. Larkin

Figure 20.1   A Letter of Transmittal for a Formal Report

FEASIBILITY OF A CAREER
IN TECHNICAL MARKETING

for

Mimi Lim

Placement Director

Seneca College

North York, Ontario

by

Richard B. Larkin,

English 266 Student

April 29, 2000

Figure 20.2 A Title Page for a Formal Report

Title. Your title announces the report's purpose and subject. The previous title (given as an example for the cover) is clear, accurate, comprehensive, and specific. But even slight changes can distort this title's signal.

An unclear title   A TECHNICAL MARKETING CAREER

The version above is unclear about the report's purpose. Is the report *describing* the career, *proposing* the career, *giving instructions* for career preparation, or *telling one person's career story*? Insert descriptive words ("analysis," "instructions," "proposal," "feasibility," "description," "progress") that accurately state your purpose.

To be sure that your title forecasts what the report delivers, write its final version *after* completing the report.

Placement of Title Page Items.   Do not number your title page but count it as page i of your preliminary pages. Centre the title horizontally, 8–10 cm (3–4") below the upper edge. Place other items in the spacing and order shown in Figure 20.2 on page 481, the title page to a report. Or devise your own system, as long as your page is balanced.

SUMMARY

Chapter 9 defines varieties of summary writing, including the summary (or executive summary) that *accompanies* a formal report or proposal (as shown in Figure 20.3). Many readers who don't have the time or willingness to read your entire report will consider the summary to be the most useful part of the material you present.

Chapter 9 also recommends a step-by-step process that will help you find and condense the elements of a good summary:

◆ the issue or need that led to the report
◆ the report's key facts, statistics, findings, and in some cases, illustrations—this is the material your reader *must* know
◆ the report's conclusions and recommendations

When you write and edit the summary, make clear connections between the report's data and interpretations. If you find yourself unable to do this in the summary, you'll probably need to revise the original report.

Follow these guidelines for the report summary:

◆ Make the summary about one-tenth the length of the original, but remember that summaries rarely are shorter than three-quarters of a page or longer than five pages.
◆ Where appropriate, use a table to summarize key facts and findings.
◆ Add no new information. Simply give the report's highlights.
◆ Use the same order of topics in the summary as in the original.
◆ Adjust the vocabulary to suit the intended reader. An executive summary, for example, includes little technical jargon. When you send report copies to readers with varying levels of expertise, write a different summary for each type of reader.
◆ Include a graph or other figure in the summary *only* if the illustration is absolutely necessary in understanding the report.

SUMMARY

The feasibility of technical marketing as a career is based on a college graduate's interests, abilities, and expectations, as well as on possible entry options.

Technical marketing is a feasible career for anyone who is motivated, who can communicate well, and who knows how to get along with people. Although this career offers job diversity and excellent potential for income, it entails almost constant travel, competition, and stress.

College graduates enter technical marketing through one of four options: entry-level positions that offer hands-on experience, formal training programs in large companies, prior experience in one's specialty, or graduate programs. The relative advantages and disadvantages of each option can be measured in resulting immediacy of income, rapidity of advancement, and long-term potential.

Anyone considering a technical marketing career should follow these recommendations:

- Speak with people who work in the field.
- Weigh carefully the implications of each entry option.
- Consider combining two or more options.
- Choose options for personal as well as professional benefits.

Figure 20.3   A Summary

TABLE OF CONTENTS

Your table of contents serves as a road map for readers and a checklist for you. If you are using a high-end word-processing program, you can generate a table of contents automatically provided that you have assigned styles or codes to all of the headings in your report. If your word-processing program does not have this feature, compose the table of contents by assigning page numbers to headings from your outline. Keep in mind, however, that not all levels of outline headings appear in your table of contents or your report. Excessive headings can fragment the discussion.

Follow these guidelines:

How to prepare a table of contents

- ◆ List front matter (summary, list of illustrations), numbering the pages with small Roman numerals. (The title page, though not listed, is counted as page i.) List back

TABLE OF CONTENTS

Figure 20.4 **A Table of Contents for a Formal Report**

matter such as glossary, appendices, and endnotes. Number these pages with Arabic numerals, continuing the page sequence of your main report.

- Include in the table of contents only headings or subheadings that are in the report; the report may, however, contain subheadings not listed in the table of contents.
- Phrase headings in the table of contents exactly as in the report.
- List various levels of headings with varying typefaces and indentations.
- Use *leader lines* (. . . . .) to connect the heading text to the page number. Align rows of dots vertically, each above the other.

LIST OF FIGURES AND TABLES

Following the table of contents is a list of figures and tables, if needed. When a report has three or more visuals, place this table on a separate page. List the figures first, then the tables. Figure 20.5 on page 486 shows the list of figures and tables for Larkin's report. Also see the extensive list of illustrations in Figure 19.6 on page 462.

DOCUMENTATION

As Chapter 8 demonstrates, reference pages are organized according to the format required by your workplace or academic discipline. MLA and APA formats require references to be listed in alphabetical order on a Works Cited page or References page. The CBE format lists sources in the same numerical order as they are cited in the report.

Whichever documentation format you use, consider including a separate list of sources you consulted but did not cite. For these, use a heading like, "Additional Sources Consulted." Thus, you will

- show your reader that you have indeed covered all the bases,
- acknowledge that certain sources influenced your thinking, even if you didn't have reason to cite them specifically, and
- serve the reader who wishes to explore the topic further.

GLOSSARY

A glossary alphabetically lists specialized terms and their definitions, following or preceding your report. Specialized reports often contain glossaries, especially when written for both technical and non-technical readers. A glossary makes key definitions available to non-technical readers without interrupting the flow of the report for technical readers. If fewer than five terms need defining, place them in the report introduction as working definitions, or use footnote definitions. If you use a separate glossary, inform readers of its location: "(see the glossary at the end of this report)." Note, though, that some readers prefer the glossary in the front matter, just before the introduction.

Follow these guidelines for a glossary:

How to prepare a glossary

- Define all terms unfamiliar to a general reader (an intelligent layperson).
- Define all terms that have a special meaning in your report (e.g., "In this report, a small business is defined as").
- Define all terms by giving their class and distinguishing features, unless some terms need expanded definitions.
- List your glossary and its first page number in your table of contents.

FIGURES AND TABLES

Figure 20.5   A List of Figures and Tables for a Formal Report

◆ List all terms in alphabetical order. Boldface each term. You may use a colon to separate the term from its single-spaced definition. This is not necessary if you use a tabular format.

◆ Define only terms that need explanation. In doubtful cases, over-defining is safer than under-defining.

Figure 20.6 shows part of a non-tabular glossary for a comparative analysis of two techniques of natural childbirth, written by a nurse practitioner for expectant mothers and student nurses. Figure 20.7 illustrates a partial tabular glossary.

GLOSSARY

Analgesic: a medication given to relieve pain during the first stage of labour.

Cervix: the neck-shaped anatomical structure that forms the mouth of the uterus.

Dilation: cervical expansion occurring during the first stage of labour.

Episiotomy: an incision of the outer vaginal tissue, made by the obstetrician just before the delivery, to enlarge the vaginal opening.

First stage of labour: the stage in which the cervix dilates and the baby remains in the uterus.

Induction: the stimulating of labour by puncturing the membranes around the baby or by giving an oxytoxic drug (uterine contractant), or both.

Figure 20.6   A Non-tabular Glossary (Partial)

GLOSSARY

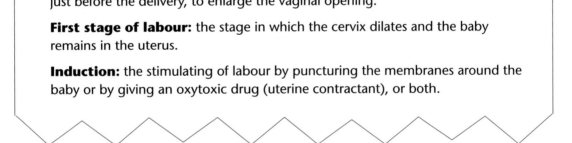

| **Analgesic:** | A medication given to relieve pain during the first stage of labour. |
| **Cervix:** | The neck-shaped anatomical structure that forms the mouth of the uterus. |

Figure 20.7   A Tabular Glossary (Partial)

APPENDICES

An appendix follows the text of your report. It expands items discussed in the report without cluttering the report text. Typical items in an appendix include:

What an appendix might include

- ◆ complex formulas
- ◆ details of an experiment
- ◆ interview questions and responses
- ◆ long quotations (one or more pages)
- ◆ maps or photographs
- ◆ material more essential to secondary readers than to primary readers
- ◆ related correspondence (letters of enquiry, and so on)
- ◆ sample questionnaires and tabulated responses
- ◆ sample tests and tabulated results
- ◆ some visuals occupying more than one full page
- ◆ statistical or other measurements
- ◆ texts of laws and regulations

The appendix is a catch-all for items that are important but difficult to integrate within your text. Figure 20.8 on page 489 shows an appendix to a budget proposal. Also see appendix pages in the sample report in Figure 19.6 on pages 473–74.

Do not stuff appendices with needless information. Do not use them unethically for burying bad or embarrassing news that belongs in the report proper. Follow these guidelines:

How to prepare an appendix

- ◆ Include only material that is relevant.
- ◆ Use a separate appendix for each major item.
- ◆ Title each appendix clearly: "Appendix A: Projected Costs."
- ◆ Limit an appendix to a few pages, unless more length is essential.
- ◆ Mention your appendix early in your introduction, and refer readers to it at appropriate points in the report: "(see Appendix A)."

Use an appendix for any material that is essential but might harm the unity and coherence of your report. Remember that readers should be able to understand your report without having to turn to the appendix. Distill the essential facts from your appendix and place them in your report text.

APPENDIX A

Table 1   Allocations and Performance of Five College Newspapers

	Prairie College	Drake College	Kelsey College	Hollander College	Northern College
Enrollment	1,600	1,400	3,000	3,000	5,000
Fee paid (per year)	$65.00	$85.00	$35.00	$50.00	$65.00
Total fee budget	$88,000.00	$119,000.00	$105,000.00	$150,000.00	$334,429.28
Newspaper budget	$10,000.00	$6,000.00	$25,300.00	$37,000.00	$21,500.00
					$25,337.14[a]
Yearly cost per student	$6.25	$4.29	$8.43	$12.33	$4.06
					$5.20[a]
Format of paper	Weekly	Every third week	Weekly	Weekly	Weekly
Average no. of pages	8	12	18	12	20
Average total pages	224	120	504	336	560
					672[a]
Yearly cost per page	$44.50	$50.00	$50.50	$110.11	$38.25
					$38.69[a]

[a]These figures are next year's costs for the Northern *Torch*.

Source: Figures were quoted by newspaper business managers in April 2002.

Figure 20.8   An Appendix

EXERCISES

1. These titles are intended for investigative, research, or analytical reports. Revise each inadequate title to make it clear and accurate.

    a. The Effectiveness of the Prison Furlough Program in Our Province
    b. Drug Testing on the Job
    c. The Effects of Nuclear Power Plants
    d. Woodburning Stoves
    e. Interviewing
    f. An Analysis of Vegetables (for a report assessing the physiological effects of a vegetarian diet)
    g. Wood as a Fuel Source
    h. Oral Contraceptives
    i. Lie Detectors and Employees

2. Prepare a title page, transmittal document (for a definite reader who can use your information in a definite way), table of contents, and informative abstract for a report you have written earlier.

3. Find a short but effective appendix in one of your textbooks, in a journal article in your field, or in a report from your workplace. In a memo to your instructor and colleagues, explain how the appendix is used, how it relates to the main text, and why it is effective. Attach a copy of the appendix to your memo. Be prepared to discuss your evaluation in class.

COLLABORATIVE PROJECT

Collect samples of various document supplements. As a group, critique the format and content of the supplements.

Short Reports

FORMATS

A STRUCTURE FOR ALL PURPOSES

PROGRESS REPORTS

PROJECT COMPLETION REPORTS

PERIODIC ACTIVITY REPORTS

INCIDENT REPORTS

INSPECTION REPORTS

COMPLIANCE REPORTS

FIELD TRIP REPORTS

MEETING MINUTES

FEASIBILITY REPORTS

CAUSAL ANALYSES

ASSESSMENT REPORTS

RECOMMENDATIONS REPORTS

LAB REPORTS

FORM REPORTS

WRITING A SHORT REPORT EFFICIENTLY

Checklist for Editing and Revising
Short Reports

Short reports form the bulk of the writing done by technologists, technicians, research scientists, and engineers. Such reports provide information and analysis that readers use to stay informed and to make practical decisions.

Some reports emphasize *information*:

◆ progress reports
◆ field trip reports
◆ field observations
◆ project completion reports
◆ periodic activity reports
◆ meeting minutes

Other reports focus on *analysis*:

◆ feasibility reports
◆ causal analyses
◆ assessment reports
◆ yardstick (comparison) reports
◆ justification reports
◆ recommendations reports

FORMATS

How to choose formal or semi-formal report format

When a document is quite short (1 to 3 pages), or when its subject matter suits a direct, informal address to the reader, use a *correspondence format* such as a letter, memo, or e-mail. For details of letter, memo, and e-mail formats, see Chapter 22.

When a document is somewhat longer (4 to 10 pages), and when its subject matter is serious enough to warrant a more formal approach, use a *semi-formal report format*. However, use a *formal report format*, as illustrated in Chapter 19, when you want to influence policy within your organization: The more formal the document, the better chance of its being heeded. So, consider "dressing up" even 6- to 10-page reports in formal report clothing if you really want to impress the reader. Table 21.1 summarizes the differences between semi-formal and formal report formats.

Table 21.1   **Semi-formal and Formal Report Formats**   *(continued)*

Item	Semi-formal	Formal
Length	4–10 pages	6 pages +
Required Sections		
Transmittal document	optional	yes
Cover	no	optional
Title page	optional—title could be placed on first page of report body	yes
Summary	no	yes
Table of contents	optional	yes
List of illustrations	optional	yes
Glossary	incorporate into text	optional
Introduction	yes	yes
Background	optional	optional
Central analysis	one or more sections	one or more sections
Conclusion	yes	yes

Table 21.1   Semi-formal and Formal Report Formats

Item	Semi-formal	Formal
Recommendations	optional	optional
Sources cited	optional	yes (if sources were cited)
Sources consulted	optional	optional
Format and Appearance		
Headings system	more relaxed; seldom more than two levels of headings; main headings placed where they come on the page	formal; usually three or four levels; each main placed at the top of a page
Page numbering	all page numbers placed at same location on page	different for first page of a section than for subsequent pages in that section
Margins	all pages use same margins layout	top and bottom margins for first page of a section are larger than for subsequent pages
Indentation	paragraphs not indented (double-space between paragraphs); bulleted and numbered lists may be indented	paragraphs not indented; (double-space between paragraphs); bulleted and numbered lists may be indented
Headers	seldom used	optional, but headers used in most professional reports

A STRUCTURE FOR ALL PURPOSES

The action structure gives readers what they want

Readers of business reports are generally busy people. They want reports to be as brief as possible. They usually want a report's main point in the first or second paragraph and they want clear, logical idea patterns. The *action structure* illustrated in Figure 21.1, on page 494, satisfies those readers' desires.

ACTION OPENING

Using the action structure, you immediately "connect" with your reader by:

- referring to an issue that concerns the reader (and you); or
- referring to comments made in a recent meeting; or
- responding to the reader's previous communication (memo, e-mail, letter, telephone call) on the subject

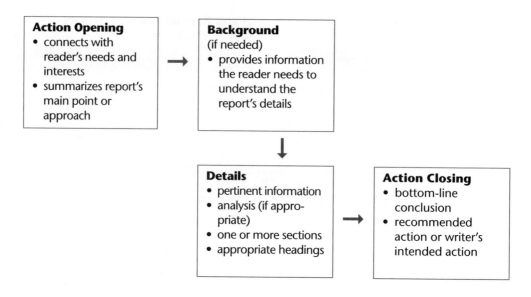

Figure 21.1   An Action Structure for Informal and Semi-formal Reports

By making such a connection, you gain the reader's attention. Then, in the same paragraph you summarize the report's main point *or* you preview the approach taken in the report.

BACKGROUND

Next, you provide any background needed by your reader to understand the report's detailed information and analysis. You may have to:

◆ review the circumstances leading to the issue discussed in the report
◆ define terms
◆ provide technical background
◆ review a problem or a proposed solution

Not all reports require background information. Your audience/purpose analysis will help you determine whether your reader needs to be briefed. You may be able to go straight from the opening paragraph to the Details section of the report.

Most readers appreciate subject headings for both the Background and the Details sections, even for a one-page report. Clearly, the number of headings you use depends on the depth and complexity of the report's topic. See this chapter's sample reports for examples of headings usage. As you read the reports, distinguish between standard topical headings ("Project Costs") and talking headings, which speak directly to the reader ("How Much We've Spent").

DETAILS

Essentially, a report's Details section answers most, if not all, of the questions posed by any good reporter: *What? Where? When? Who? How? How much? Why?* The order in which you answer such questions depends on the subject matter and your reader's priorities.

ACTION CLOSING

Finally, the report provides bottom-line conclusions. Most reports also discuss what should be done next. Some reports recommend action to be taken by the reader or the reader's organization; others state what action the writer intends or proposes. Still other reports simply list possible actions without indicating who might take responsibility.

The action structure can be adapted to develop any type of correspondence or short report. The remainder of this chapter demonstrates the action structure's valuable adaptability to various reports:

- ◆ progress reports
- ◆ project completion reports
- ◆ periodic activity reports
- ◆ incident reports
- ◆ inspection reports
- ◆ field trip reports
- ◆ meeting minutes
- ◆ feasibility reports
- ◆ causal analyses
- ◆ assessment reports
- ◆ recommendations reports
- ◆ lab reports
- ◆ form reports

Let's look at each of these reports.

PROGRESS REPORTS

Progress reports serve many purposes

Large organizations depend on progress (or status) reports to keep track of activities, problems, and progress on various projects. Daily progress reports are vital in a business that assigns crews to many projects. Managers use progress reports to evaluate a project and its supervisor, and to decide how to allocate funds. Managers also need to know about delays that could dramatically affect outcomes and project costs.

Also, managers need information to coordinate the efforts of the work groups. For example, a hydro manager responsible for restoring power transmission lines after a severe ice storm will have to coordinate clean-up crews, construction crews, and line crews. A large project such as a major power line restoration would require several written *periodic progress reports.*

When work is performed for an external client, the reports explain to the client how time and money have been spent and how difficulties have been overcome. The reports can therefore be used to assure the clients that the work will be completed on schedule and on budget. Many contracts stipulate when progress will be reported. Failure to report on time may invoke contractual penalties.

To meet managers' and clients' needs, progress reports must answer these questions:

1. *How much has been accomplished since the last report?*
2. *Is the project on schedule?*

3. *If not, what went wrong?*
   (a) *How was the problem corrected?*
   (b) *How long will it take to get back on schedule?*
4. *What else needs to be done?*
5. *What is the next step?*
6. *Have you encountered any unexpected developments?*
7. *When do you anticipate completion?* Or (on a long project): *When do you anticipate completion of the next phase?*

The following general structures adapt the action structure to periodic progress reporting. Exactly how you structure the report depends on the nature of the project and on the aspect you want to stress. You could organize the report by:

- the different tasks or subcontracts involved
- the amount of work completed versus the amount remaining
- phases or time frames identified by the contact or the project manager

Figures 21.2 to 21.4, below and on page 497, illustrate outlines for the three possible structures. These outlines relate to the third in a series of reports for a six-month highway reconstruction project. Which structure would you use?

Any of the three variations could incorporate a special category that lists "Problems and Setbacks," where appropriate. This category could also describe methods for overcoming any problems and setbacks. Alternatively, the report might discuss problems and setbacks under other headings, if the difficulties do

---

VARIATION 1: Organized by Task

- Intro: connect with reader/refer to project
- Project description or summary

ROAD BASE PREPARATION
 Progress to date
- in previous reporting period(s)
- work done in period just closing
 Work to be completed
- work planned for next work period
- work planned for periods thereafter

PAVING
 Progress to date
- in previous reporting period(s)
- work done in period just closing
 Work to be completed
- work planned for next work period
- work planned for periods thereafter

COST ANALYSIS
- costs to date
- costs in period just closing

CONCLUSION
- overall appraisal of work to date
- cost appraisal
- conclusions and recommendations re: work

Figure 21.2   A Progress Report Organized by Project Task

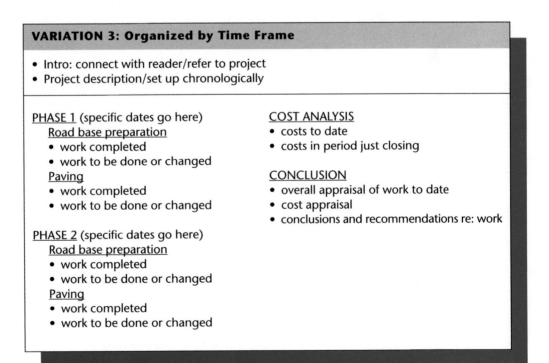

VARIATION 2: Organized by Work Completed

- Intro: connect with reader/refer to project
- Project description or summary

WORK COMPLETED
    Road base preparation
- in previous reporting period(s)
- work done in period just closing
    Paving
- work planned for next work period
- work planned for periods thereafter

WORK REMAINING
    Road base preparation
- work planned for next work period
- work planned for periods thereafter
    Paving
- work planned for next work period
- work planned for periods thereafter

COST ANALYSIS
- costs to date
- costs in period just closing

CONCLUSION
- overall appraisal of work to date
- cost appraisal
- conclusions and recommendations re: work

Figure 21.3 **A Progress Report Organized by Degree of Project Completed**

VARIATION 3: Organized by Time Frame

- Intro: connect with reader/refer to project
- Project description/set up chronologically

PHASE 1 (specific dates go here)
    Road base preparation
- work completed
- work to be done or changed
    Paving
- work completed
- work to be done or changed

PHASE 2 (specific dates go here)
    Road base preparation
- work completed
- work to be done or changed
    Paving
- work completed
- work to be done or changed

COST ANALYSIS
- costs to date
- costs in period just closing

CONCLUSION
- overall appraisal of work to date
- cost appraisal
- conclusions and recommendations re: work

Figure 21.4 **A Progress Report Organized by Project Time Frame**

not require special treatment. Also, notice that the three sample structures each show only two main sections for reporting progress, but three or more sections may be required.

TWO SAMPLE PROGRESS REPORTS

A workplace progress report

Figure 21.5, on pages 499–504, illustrates a periodic progress report with its information organized in a manner that distinguishes between "work completed" and "work in progress."

> Art Basran is the writer of the report shown in Figure 21.5. As you may recall, he manages the Kelowna, British Columbia, office of MMT Consulting. Art's report is the fourth in a series of oral and written progress reports for his regional manager, Brenda Backstrom, regarding a project that is very important to MMT Consulting. (Chapter 19 features the formal analytical report that Art later presents to the client, Jackson Mining.)
>
> For this progress report, Art has chosen a semi-formal report format instead of a memo. He has selected this format because of the six-page length and because he is trying to persuade his supervisor to support a new method of analyzing combined engineering and accounting data. ▲

A student progress report

The second sample progress report, illustrated in Figure 21.6, on pages 505–506, is written by an engineering technology student, Tim Anders.

> Tim is reporting progress on a research project to his project supervisor. Tim has had some setbacks in the project, so he discusses those setbacks before discussing his recent progress. He modifies the "Work Completed/Work to be Completed" structure to suit his particular situation. The attachments ("Bibliography," "Working Outline," and "Work Schedule") listed at the end of Tim's report have not been included in this book, to save space. ▲

Occasional progress reports are written for short-term projects that do not have scheduled reporting dates. Either the writer responds to a supervisor's request (or a client's request), or the writer reports progress in order to elicit reader support.

PROGRESS REGARDING MMT's COST ANALYSIS OF FOUR ALTERNATIVES FOR HAULING COAL FROM JACKSON MINING'S PROPOSED OTHELLO MINE

For Brenda Backstrom, Regional Manager
MMT Consulting, Calgary Regional Office

Prepared by Art Basran
Kelowna Branch Office Manager
MMT Consulting

Submitted:
March 5, 2005

Figure 21.5 Semi-formal Progress Report Organized by Degree of Completion

(continued)

INTRODUCTION

The Jackson Mining Company has discovered a high-grade coal deposit in the mountains about 55 miles due north of Grand Forks, British Columbia. The company expects the market for this grade of coal to increase dramatically in the near future, and thus anticipates developing the deposit if development and transportation costs are acceptable. Jackson Mining has tentatively named its proposed minesite as the Othello mine. MMT Consulting has been authorized to determine the most cost-effective method(s) of transporting the coal from that minesite to the mainline railroad, a distance of approximately 50 miles.

Four transport alternatives are being considered:

1. Diesel trucks
2. Diesel trains
3. Electric trains
4. Electric fleets

The criterion used to evaluate each alternative is the Equivalent Uniform Annual Cost (EUAC). The EUAC discussed in this report is in dollars per ton for the haul to the railhead. The four alternatives will be studied over periods ranging from one to twenty years, and at interest rates varying from 6% to 10%.

   The EUAC formula provides clear comparisons of the transportation alternatives. Using this method, and capitalizing on hard work by our staff and good surveying weather, the project seems certain to be completed under budget and ahead of schedule. Details are provided in the following sections.

BACKGROUND

The Othello mine under study has anticipated coal sales of 5 million tons annually. Each of the following transportation alternatives would have to handle this volume. Also, the ability to expand in the future would be an asset.

1. **Diesel trucks**
   The CAT 777B truck was chosen for this study because of its load capacity of 95 tons and its output of 920 horsepower. These trucks would be running 24 hours a day.
2. **Diesel trains**
   Each train would be running with six diesel engines and 98 coal cars. It would take 11.2 hours to complete a round trip, and 17.2 hours to produce the amount of coal the train would haul. Thus, the train would haul a complete load and then idle for about 6 hours until the next load was ready.
3. **Electric trains**
   Each train would run four electric engines and 98 coal cars, with the same haul times as for the diesel train.
4. **Electric fleets**
   Each fleet would consist of three of the above electric trains. The first train would be loaded and then depart. While enroute, the second train would be loaded, and this load-and-go process would be repeated for the third train. The departures would be timed so that the second train would leave just as the first train crested the highest point of the rail line pass. In this way, the electricity generated by the downgrade trains would be conserved and used to assist in powering the upgrade trains.

1.

Figure 21.5   **Semi-formal Progress Report Organized by Degree of Completion**

(continued)

MMT'S INVOLVEMENT

After the initial negotiations between the respective head offices of Jackson Mining and MMT, our Kelowna office prepared a detailed proposal which Jackson Mining approved on January 6. Since then, two engineers and three technologists from our Kelowna office have researched this project with advice from Brendan Winters, a Penticton chartered accountant who has special expertise in mining projects.

Work Completed

Since January 6, our team has completed the following tasks:

- completed a preliminary survey of the proposed rail line and haul road; the elevations and other data collected from this survey are being used to help calculate tractive and grade resistances for the diesel trucks and for the two types of trains
- researched construction costs for both a rail line and a haul road, and for associated construction costs (such as maintenance buildings)
- researched maintenance costs for both a rail line and a haul road
- researched labour costs for operating the trucks and the trains
- researched the costs of purchasing trucks, diesel trains, and electric trains
- designed and researched the full costs of a complete communications system, which is required to control the proposed mine's transportation system

Work in Progress

We now have all the data required to complete our calculations and analyses. This work will progress in three stages:

1. Using the data collected from our surveys, we will calculate tractive and grade resistances for the diesel trucks and for the two types of trains. A sample of those calculations is provided below in Table 1. Next, to supplement the information regarding resistances, we will calculate the speed, power, and fuel consumption of each transportation type. A sample of those calculations appears in Table 2 on page 3.

Table 1: Tractive and Grade Resistances for Diesel Trucks

Section	Grade (%)	Truck Wt (ton)	Load Wt	Total Wt	Rt (lb)	Rg	Tr
G	2.0	66.3	95.0	161.3	16,125.0	6,450.0	22,575.0
F	1.8	66.3	95.0	161.3	16,125.0	5,805.0	21,930.0
E	2.0	66.3	95.0	161.3	16,125.0	6,450.0	22,575.0
D	2.1	66.3	95.0	161.3	16,125.0	6,772.5	22,897.5

The first column in Table 1 lists the sections into which the road has been divided for analytical purposes. These sections of road have varying grades. Grade resistance (Rg) and tractive resistance (Rt) are found for each section.

2.

Figure 21.5   Semi-formal Progress Report Organized by Degree of Completion

(continued)

Table 2:   Speed, Power, and Fuel Consumption for Diesel Trucks

Section	Grade (%)	Tr (lb)	Power (hp)	Speed (mph)	Dist (mi)	Time (hr)	BTU (x1000)	Fuel (gal)	Total Fuel
G	2.0	22,575.0	920.0	12.55	8.4	0.67	1,572.5	11.34	42.62
F	1.8	21,930.0	920.0	12.92	7.2	0.56	1,309.3	9.44	35.49
E	2.0	22,575.0	920.0	12.55	5.4	0.43	1,010.9	7.29	27.40
D	2.1	22,897.5	920.0	12.38	4.8	0.39	911.4	6.57	24.70

The final calculations for resistances and fuel consumption will be added to labour and maintenance costs to determine annual operating costs for each of the alternatives under study. Thus, we will be able to complete the Cost Comparison table, a draft of which is shown in Draft Table 3 below.

Draft Table 3:   Cost Comparisons

Option	Capital Costs	Annual Operating Costs
Diesel Trucks	$115,539,000	$
Diesel Trains	$149,083,000	$
Electric Trains	$161,783,000	$
Electric Fleets	$201,463,000	$

2. Once the capital and annual costs are found, they can be combined with varying interest rates and study periods to calculate tables of Equivalent Uniform Annual Costs (EUAC). The purpose of converting all costs into an EUAC is to provide a common means of comparing all the alternatives. The capital costs are spread over the length of the study period and added to the annual costs. Thus, a comparison of EUACs will yield a qualitative assessment of the alternatives to go with the quantitative assessment suggested by Draft Table 3.

To provide an idea of how these EUAC tables (and corresponding graphs) will look in the final analysis, we have estimated figures for the four transportation alternatives and placed them in Draft Table 4 and Draft Figure 1 (Appendix).

The final phase of our work for this project will be to produce a comprehensive report of our findings. This report has been outlined, and some of the preliminary material has been drafted. We expect the final version of this report to total about 40 pages, with the bulk of those pages presenting our detailed calculations. We're using the format that we learned from Mykon Communications last March and which proved successful in our final report to the City of Kelowna regarding the proposed multi-purpose arena.

3.

Figure 21.5   Semi-formal Progress Report Organized by Degree of Completion

(continued)

CONCLUSION

Appraisal of Work to Date

We're satisfied with both the quantity and the quality of the data gathered for the required analysis. Part of this is due to the hard work performed by our Kelowna project team, part is due to our recent connection to the Internet, and part of the credit should go to Brendan Winters whose experience in gathering and analyzing financial data has been invaluable.

Also, we had some good fortune with the weather during the survey work. Unseasonably warm weather and light snowfalls allowed the surveys to finish one week ahead of our anticipated February 2 completion date, and under budget by 15%. (We would have been even more under budget, but we had to rent a helicopter for three days more than anticipated because of very difficult terrain in the Kelso Pass area.)

Overall, it seems that we will be about $14,000 under budget for the project, partly because of the savings in the surveys and partly because Brendan Winters' research expertise has cut our estimated research time. Full figures will be provided in our final progress report.

Anticipated Completion Date

We expect to have all calculations completed by March 7. After that, it should take two of our staff (one engineer and one technologist) two working days to produce a final draft of the report for examination at Regional Office. Allowing for examination time and for final editing at the Kelowna office, we should be able to have the finished copy of our report produced by March 14, which is 14 days ahead of the date we had scheduled for delivering our analysis to Jackson Mining.

Recommendations

We have forged a positive working relationship with Jessica Proctor, director of Mining Development at Jackson Mining. Jessica has indicated that her company would like us to bid on the construction engineering contract for the anticipated road/rail line construction project, which may begin as early as this July. In connection with this possible contract, we recommend two courses of action:

1. Have the Jackson Mining report produced professionally by a graphics firm. We have money to spend because we're under budget for this project. Also, a graphics firm such as Apex Graphics (which does annual reports and similar documents for major businesses in the Okanagan) could produce a high-quality colour report in one day if we provide a disk prepared in Microsoft Office, the software we use. My main reason for suggesting this extra expenditure of under $1,000 is that Jackson Mining's directors place a high value on professional work. I think it would help project a positive image which will benefit any future proposals we make to this company.
2. Do some preliminary investigation of the parameters and requirements of road construction and rail line construction for Othello Mine's proposed transportation routes. If Jackson Mining presents a Request for Proposal on this construction project in the near future, MMT Consulting will be prepared.

4.

Figure 21.5 Semi-formal Progress Report Organized by Degree of Completion

(continued)

APPENDIX

Draft Table 4: EUAC for Alternatives at 6% (Based on preliminary estimates)

Study Period	Diesel Trains	Electric Trains	Electric Fleets	Trucks
1	$32.52	$34.91	$43.26	$32.10
2	17.18	18.26	22.53	20.21
3	12.07	12.72	15.63	16.25
4	9.52	9.95	12.18	14.27
5	7.99	8.29	10.12	13.09
6	6.98	7.19	8.75	12.30
7	6.26	6.41	7.77	11.74
8	5.72	5.82	7.04	11.33
9	5.30	5.37	6.48	11.00
10	4.97	5.01	6.03	10.75
11	4.70	4.71	5.66	10.54
12	4.47	4.47	5.36	10.36
13	4.28	4.27	5.10	10.22
14	4.12	4.09	4.89	10.09
15	3.98	3.94	4.70	9.98
16	3.87	3.81	4.54	9.89
17	3.76	3.70	4.40	9.81
18	3.67	3.60	4.27	9.74
19	3.59	3.51	4.16	9.68
20	3.51	3.43	4.07	9.62

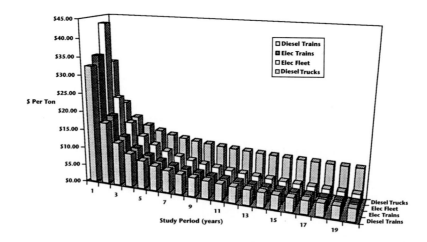

Draft Figure 1: EUAC for Four Alternatives at 6% (based on preliminary estimates)

5.

Figure 21.5   Semi-formal Progress Report Organized by Degree of Completion

Okanagan University College   *Memo*

To: Don Klepp, English 124 Instructor   **Date:** February 19, 2003

From: Tim Anders, Civil Engineering Technology 1

Re: **Progress of My Research Project**

As I mentioned during our conversation yesterday, my research project is progressing despite the delay caused by choosing a dead-end topic. This memo presents the details of my progress and outlines the work remaining.

Background

You'll recall that I started to research the process used to choose the site for the city's new multi-purpose arena. I found information about the site itself (city maps and infrastructure services), but the site selection process seems to have been cut and dried. The arena is being built on city-owned land; no other site came close to matching the current site's cost or other advantages, as I learned from Jim McElroy's report submitted for last year's English 124 class.

So, following your suggestion, I'm exploring the feasibility of using the Internet to develop technical writing skills.

Progress: Planning and Research

I started on February 10 by identifying TerraTech Engineering as the reader of my report for two reasons:

1. According to a friend who works in TerraTech's Cranbrook office, TerraTech needs to upgrade the report writing skills of employees in several of its locations because the firm has been restructured and almost everybody in the firm is now responsible for producing reports for clients. TerraTech is not sure how to improve its employees' writing skills in a cost-effective manner. Perhaps the Internet could form the basis for guided self-improvement.
2. I hope to get a co-op position with TerraTech this summer and fall. Maybe my report will show my interest in the firm.

As instructed, I wrote a planning outline, which is attached. I've used this outline to guide my research in two main areas:

1. Locating and examining websites for several topic areas—grammar and English language basics; letters, memos, and e-mail; job reports (progress reports, project completion reports, test reports; analytical reports; proposals; graphics and illustrations; and technical talks.

2. Locating and reading articles in education journals. The attached bibliography lists two articles that I'm finding very useful. The material I've gleaned from those articles, along with an interview I did with Billie Martin in the Distance Education department, has led to the section called "Relevant Learning Theory," which appears in the attached working outline.

The bulk of that outline sets up a report that will describe and assess the various websites on the basis of the

1. usefulness of the site's material compared to *Technical Communication* and your lectures
2. clarity of the explanations in each site
3. site's ease of use

See my attached working outline for more details.

Figure 21.6   A Student Research Progress Report   *(continued)*

Professor Klepp February 19, 2003 Page 2

Analysis and Organization of Gathered Data

As the working outline shows, I have completed my analysis of all the website data, except for the job reports. This week, my friend at TerraTech is sending me a list of all the short job reports written by TerraTech employees. I'll see if the reports on this list are discussed on any websites.

My assessments are leading to definite conclusions about the feasibility of using Internet sites to teach oneself to be a better business communicator. Also, I'm pleased with the report's structure. The order of assessed sites follows the order of topics in English 124. This order presents a logical progression for a self-directed study program.

Work to Be Completed

When I receive your feedback on my working outline, I'll compose the report, probably on the March 6–7 weekend. Then I'll edit the report on the following weekend and exchange it with Jan Bath's and Phil Strain's reports—we have formed a proofreading group. I look forward to your March 12 lecture on editing and finishing techniques.

The attached Work Schedule shows the hours required and deadline dates for all remaining tasks.

Conclusion

I'm now confident that this project will be completed before the submission deadline of March 22. In case I have any last-minute questions, may I book 10 minutes of your office time on March 19?

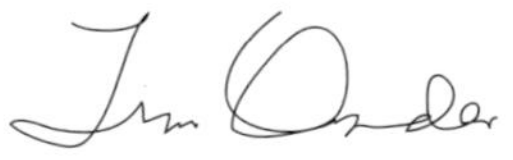

Tim Anders

Attachments: Bibliography
Planning Outline
Work Schedule
Working Outline

Figure 21.6 A Student Research Progress Report

PROJECT COMPLETION REPORTS

A project completion report is presented as the concluding progress report for a lengthy project, or the only report arising from a short project. Either way, the action structure can form the report's backbone. (Note, though, that in Table 21.2 the Details section is expanded into Project Highlights and Exceptions.)

Table 21.2 Project Completion Report Structure

Section	Contents
Action Opening	The reader connection depends on the type of reader: client? writer's supervisor? (What are the reader's main concerns?) State that the project is complete; briefly describe the outcome.
Background	Review the job's features: purpose, schedule, budget, location, people involved.
Project Highlights	Describe the project's main accomplishments (work completed, targets met, results obtained). Discuss problems encountered, the impact of these problems on the project, and how the problems were handled.
Exceptions	Describe the deviations from the contract or project plan (if any)—the work not completed or done differently than planned. Give the reason for each deviation and explain the effect of each on the final project result.
Action Closing	Analyze the reader's main concerns. What types of follow-ups are needed?

PERIODIC ACTIVITY REPORTS

The periodic activity report is similar to a progress report in that it summarizes activities over a specified period. But unlike a progress report, which describes activity on a given *project*, a periodic reports summarizes the general activities during a given *period*. Manufacturers requiring periodic reports often have prepared forms, because most of the tasks are quantifiable; i.e., units produced, etc. Still, not all jobs lend themselves to prepared-form reports, and you will probably have to develop your own format. If so, the action structure can be readily adapted to your purpose.

INCIDENT REPORTS

An incident report resembles news accounts of events. Most of the description is provided as past tense narrative. Note, however, that some incident reports also use present tense to refer to the current situation and future tense to discuss what needs to be done.

INSPECTION REPORTS

Building inspectors, park wardens, gas inspectors, quality-control technicians, and others sometimes use forms to report the results of their inspections. However, often a form is not available or does not suit a particular inspection. In these cases, the following adaptation of the action structure works very well.

Table 21.3   **Inspection Report Structure**

Section	Reader Questions to Answer
Action Opening	◆ Why should I read this report? ◆ What is the main result of the inspection?
Background	◆ Why was this inspection conducted? ◆ What was inspected? ◆ Who did the inspection? ◆ When and where did the inspection occur?
Details	◆ What did the inspection reveal? 1. Conditions found: What did the inspectors observe re: the quality of work performed or items provided at the site? In what condition were equipment, facilities, or materials? 2. Deficiencies: What conditions, if any, need to be corrected? Does any work need to be done or re-done?
Action Closing	◆ Overall, what is the state of the site (facilities, equipment, etc.)? ◆ Does the writer suggest specific actions?

The inspection report on page 509 deals with a troubling incident: a family, against its will, had been evacuated from its home because of dangerous carbon monoxide levels in the home. The writer, who is relatively new in her position, phrases her observations and opinions very cautiously. In particular, notice that she records details very clearly and that she uses passive voice wherever possible to emphasize the *results* of the inspection, *not her part in the inspection and subsequent actions.*

COMPLIANCE REPORTS

At every level of government, regulatory agencies require organizations to report the degree to which the organizations have complied with the agencies' regulations. Also, industrial and professional associations require their members to report how the members have complied with codes of business conduct. Some compliance reports can be completed on forms provided by the watchdog agency, but more often the reporting organization has to create its own format and structure for the report. In such cases, the structure outlined in Table 21.4 could form the basis for any particular compliance report.

Prairie POWER Corporation

INSPECTION MEMORANDUM

DATE: January 7, 2005

TO: Randall Johnson, Gas & Electrical Inspections Coordinator

FROM: Miranda Ocala, Gas Inspector

RE: **Clogged Masonry Chimney – 322 Montcalm Crescent, Saskatoon, SK**

On the evening of January 3, an elderly member of the Smith family resident at 322 Montcalm Crescent was rushed to the U of S Hospital Emergency Department. An alert resident suspected CO poisoning and alerted SaskEnergy. Later that day, Melvin Trask of SaskEnergy advised the occupants of the two-storey, single family residence to vacate because:

- the chimney was blocked with ice, and
- CO concentrations of .02% were present apparently due to spillage of gas combustion products.

On January 4, Keith McLeod and I inspected the gas equipment and found:

- the masonry chimney was blocked with ice. (We noted a white, lime-like substance on the exterior portion of the chimney that is exposed in the garage.)
- the furnace and the water heater were spilling.
- the home had evidence of excessive moisture – the windows were frozen shut. A serviceperson from Prairie Heating was present; she opened a small passageway at the top of the chimney's interior. Soon after a draft was established, (i.e., in 30 minutes), the ice began to melt.
- the gas equipment was in good condition. That equipment consists of:
  1. a 137,000 BTUH standard Lennox furnace with a 6" vent draft hood
  2. a 36,000 John Wood water heater with draft hood (3" vent).

The furnace and the heater operated satisfactorily as soon as the chimney passage was reasonably clear.

- The 1", two-outlet supply pipe was in good condition
- The masonry chimney, constructed of bricks and concrete and lined with tile throughout, seemed in good condition, although our initial inspection was unable to confirm the chimney's interior condition because of the ice buildup.

On January 6, after the ice had thawed, our subsequent inspection revealed damaged tile liner around the vent connector's entry point. This defect, coupled with the exposure of all four sides of the chimney and the recent cold weather, seems to have led to the icing condition. There is evidence that severe icing has occurred before: There is a white substance on the chimney exterior in the garage.

Because the tile liner measures 6 1/4" by 6 1/4", and a flexible liner measures 6 3/8" OD, I have approved the use of a traditional (shop made) sectional 6" aluminum liner. This is the most economical method of any acceptable corrections.

Still, the owner, Rod Smith, is annoyed that corrections are necessary to a house built just 14 years ago. He is also very angry that an owner's defect has been issued precluding occupancy until satisfactory corrections have been made. He has threatened to sue for the costs of housing his family in a hotel until the residence is cleared for occupancy. I suggest that our customer service people speak with Mr. Smith to explain all the ramifications of allowing a family to occupy a home with potential for CO-induced deaths.

Miranda Ocala

Miranda Ocala

Figure 21.7 A Sample Inspection Report

Table 21.4   Compliance Report Structure

Section	Content	Comments
Action Opening	Name the report's purpose. Refer to the act or to the body of the regulations prompting report. State the degree to which the requirements have been met; list the time frame covered by the report.	The reader connection might be placed in a transmittal letter or memo if a semi-formal format is used. In a letter or memo report, use a subject line that names the specific act and its sections. Often, a compliance report has legal implications, so do not connect with the reader in an informal, friendly manner.
Background	Supply necessary details about regulations for a reader who is not familiar with them. For a report to the regulatory agency, possibly review your organization's compliance record.	Background about the regulations will not be necessary in a report to the regulatory agency.
Details of Compliance	Describe specific actions taken to comply with specific regulations, in the order the regulations appear. The report could follow a time pattern—the order in which you've met the reader's expectations. Or, organize in the order of importance. Or, use an actions-results pattern.	Use bolded headings, bullets, numbered lists, italics, white space, and indents to help make the document easy to read.
Action Closing	Summarize key aspects of the reported activities. State the bottom line, including results, where appropriate. List future methods of complying with regulations or expectations.	In some cases, the report might describe costs of problems of compliance; the conclusion might argue for a relaxed interpretation of the regulations.
Attachments (optional)		Photos, affidavits, test data, shift reports, or other detailed backup data might be appropriate.

FIELD TRIP REPORTS

Usually, a field assignment is complete only after you have reported on what you observed and what you did. The field trip may have involved a four-hour hike along an abandoned forestry road, or required two weeks of testing pollution levels in salmon spawning rivers. Regardless of the trip's duration and complexity, you need to perform two main tasks in advance of the inevitable report:

1. Make careful, detailed observations and record them in a notebook or on an audio cassette recorder.

2. Organize those notes to answer your reader's questions in the order your reader would prefer.

Table 21.5 illustrates a structure that could be used for most field trip reports.

Table 21.5   Field Trip Report Structure

Section	Reader Questions to Answer
Action Opening	◆ Why are you reporting this? (optional) ◆ In brief, what have you been doing? What did you accomplish?
Background	◆ Who went where? When? Why? ◆ On whose authority? (optional) ◆ How did the writer travel? (optional) ◆ What was the project? (optional)
Details of Work Accomplished	◆ What did you do? What routine work? Which work specifications were followed? ◆ What work did you perform beyond the routine requirements? ◆ What did you observe? ◆ What meetings, if any, did you have? With whom? What were the results?
Problems Encountered	◆ What were the specific problems, if any? Did you identify the causes of these problems? ◆ What specific actions did you take to solve the problems? ◆ Were you successful? If not, why not?
Action Closing	◆ What remains to be done? What resources are necessary? Who should perform the work? Have you assigned the work? ◆ Are you requesting support or authorization from me, your reader?

MEETING MINUTES

Many team or project meetings require someone to record the proceedings. Minutes are the records of such meetings. Copies of minutes are distributed to all members and interested parties, to track the proceedings and to remind members of their designated responsibilities. The appointed secretary records the minutes. When you record minutes, answer these questions:

◆ *What group held the meeting? When, where, and why?*
◆ *Who chaired the meeting? Who else was present?*
◆ *Were the minutes of the last meeting approved (or disapproved)?*
◆ *Who said what? Was anything resolved?*
◆ *Who made which motions and what was the vote? What discussion preceded the vote?*
◆ *Who was given responsibility for which actions?*

See Figure 21.8 on page 512 for a sample set of minutes.

MEETING OF THE CAMPUS RECYCLING COMMITTEE
Room 125, Student Services Building, March 28, 2006

Chair: T. Maguire, Jordan College Student
Association President

Present: R.W. Siggia, V.P., Administration   T. Singh, 3rd year Arts
J. Klym, Campus Services   J. Cormier-Bauer, 2nd year Engineering
M. O'Connor, Print Services   H. Calvin, 4th year Business
                                          P. Masinkowski, 4th year Phys. Ed.

Guest: John Maravich, Canadian Waste Disposal, Ltd.

1. Approval of Agenda

T. Maguire asked to add a presentation by John Maravich and suggested that discussion of bottle recycling as a student association fundraiser be tabled to the next meeting, in order to accommodate the address. T. Maguire called for approval of the amended agenda.

Passed Unanimously.

2. Other Business

John Maravich proposed a business arrangement wherein Canadian Waste Disposal would have exclusive rights to recycle paper products at Jordan College in return for an annual $5,000 scholarship to a Jordan College Business student, and a commitment to hire Jordan College students on a part-time basis. The projected total volume of business was discussed, along with other details provided in Canadian's written proposal (see attached). The committee agreed to hold a special meeting in two weeks to discuss the proposal.

Action by: T. Singh and J. Cormier-Bauer will press class presidents to poll students re: their on-campus paper usage.

M. O'Connor and R.W. Siggia will review administrative and academic paper usage.

J. Klym will determine recycling potential for calendars, phone books, and all other campus service publications.

3. Approval of Previous Meeting

After revision of the numbers relating to recycling of library culls, the previous meeting minutes were accepted.

MOVED: R.W. Siggia   SECONDED: J. Klym   **Passed Unanimously.**

4. Recycling Ink Products

H. Calvin and J. Klym presented a report on the types and volume of ink used by on-campus photocopiers and computer printers. Their main findings were that in the last fiscal year:
1. $212,700 was spent on ink cartridges for laser printers, inkjet printers, and photocopiers
2. of that amount, $192,654 was used to purchase new cartridges and the remainder was spent on recycled cartridges. On average, recycled cartridges cost 62% as much as new ones.
3. the latest editions of Consumer Journal and Computer Equipment Monthly report 93% reliability with recycled cartridges.

MOVED: J. Klym   SECONDED: H. Calvin

That Jordan College adopt a one-year trial period of using recycled cartridges. Discussion centred on the issue of whether this committee has a legitimate right to take this action. R.W. Siggia contended it is Administration's prerogative, but agreed to approach President Monroe with the committee's decision.

Passed Unanimously.

Call for adjournment at 5:05 p.m.

Carried.

Figure 21.8   A Sample Set of Meeting Minutes

The reports discussed so far in this chapter focus on factual information. Now the focus shifts to short analytical reports, all of which logically arrive at a conclusion. Readers of analytical reports want more than the facts; they want to know what the facts mean.

The four major varieties of formal analytical reports (feasibility reports, causal analyses, assessment reports, and recommendations reports) have all been discussed in Chapter 19. Therefore, this chapter discusses only how to modify them for letter, memo, and semi-formal report versions. Like short informational reports, short analytical reports can profitably use the four-part action structure.

As you examine the structures suggested for short analytical reports, notice a key difference between formal and informal analytical reports. A formal report places its main point, its "bottom line," at the end of the report and satisfies a reader's immediate "need to know" by placing a Summary in the front matter. An informal report, on the other hand, places the report's main finding in the first or second paragraph, to satisfy the reader's curiosity. While this "front-loading" technique suits most informal reports, it doesn't suit every situation, as you'll see in the discussion of indirect recommendations reports, later in this chapter.

FEASIBILITY REPORTS

Your reader for a feasibility analysis will likely want your answer at the beginning, so the structure in Table 21.6 should work well. See Figure 21.9 on page 514 for a sample report.

Table 21.6   Short Feasibility Report Structure

Section	Content	Comments
Action Opening	Refer to reader's request or the situation requiring analysis. State whether the examined project or equipment is feasible.	The reader connection might be placed in a transmittal document if a semi-formal format is used. A letter or memo does not need a heading for the opening paragraph.
Background	Describe the situation leading to this feasibility study. Explain exactly what kind of feasibility is studied and list the assessment criteria.	The amount of background depends on the reader's familiarity with the subject. The criteria may have to be justified.
Details of Assessment	Apply each assessment criterion, step by step, to the data. Choose suitable criteria—a proposed equipment purchase, for example, could look at cost, warranty, equipment reliability, performance, and compatibility with current equipment.	The title of this section will depend on the kind of feasibility being discussed, and on the reader's priorities.
Action Closing	Summarize the results of applying all criteria and state the bottom-line conclusion. If appropriate, recommend approval.	A summary table could be effective. Brochures, test data, financial projections, or other detailed supporting data could be attached.

Ministry of Transportation   **Internal Memo**

DATE: January 10, 2004
TO: Richard Janvier, Engineering Information Systems Coordinator
FROM: Sheri Prasso, Science and Technical Officer
RE: **Sand and Glavine Proposal for Office Networking
             (RFP 20021209-EIS)**

As you requested, this report assesses the proposal submitted by Sand and Glavine Systems for an office network for our Vernon Engineering Services office. The proposal has merit, but requires changes before the Ministry of Transportation can accept it.

The RFP placed its focus on making better use of the computer information systems by creating a local area network (LAN). Therefore, I used the following criteria to assess the Sand and Glavine computer network proposal:

1. Technical considerations
2. Cost
3. Training and support
4. Efficiency gains

Information for the assessment was collected from current books on the subject, staff at the Vernon Engineering office, and local businesses.

Technical Considerations and Cost
The proposed network will meet Engineering Information Systems' requirements, with minor changes. (In particular, the Wang 486 needs to be retained as part of the network (see the attached technical analysis for more detail). These changes put the cost of the network slightly over budget. However, anticipated reductions in cable requirements and installation time should lower the cost. The overall cost of the modified network will be close to the proposal's quoted price of $4,000.

Training and Support
Technically, the Sand and Glavine Systems proposed network is simple. Because of this simplicity, and the competence of the staff at the Vernon Engineering Office, the proposed training will be sufficient. Unlike training, support was not included in the proposal's quoted price. It was offered at additional cost through monthly service contracts. The Ministry of Transportation has qualified computer support personnel on staff. Purchasing support from Sand and Glavine Systems would duplicate service and add to the direct cost of this network.

Efficiency Gains
Sand and Glavine's proposed computer network will meet Engineering Service's Information Systems' objective of increasing the efficiency of its Vernon office. The network will save time and allow staff to focus their efforts on engineering rather than on file management. Also, Sand and Glavine will install the system on a weekend, saving two days of down time.

Recommendation
If Sand and Glavine Systems re-submits the proposal with the requested changes, it should be adopted.

Sheri Prasso

Attachments: Technical analysis (cable, topology, hardware, and software)
             Cost analysis
             Task time comparisons

Figure 21.9   A Sample Feasibility Report

CAUSAL ANALYSES

Your reader's immediate interest in what caused a problem, a failure, or an incident should direct you to state the report's main point in the lead paragraph as in Table 21.7. Incidentally, not all causal analyses deal with problems, breakdowns, or failures. You might be asked to identify the causes of an unexpected rise in factory productivity, a steady increase in fish breeding stock, or an improvement in the employee turnover rate.

The causal analysis illustrated in Figure 21.10 on pages 516–518 was commissioned by a homeowner who suspected that a gaping hole under her driveway resulted from her neighbour's faulty driveway design. TerraTech, the consulting firm engaged by the homeowner, presents its findings in detached, carefully measured language. The report writer uses the classic causal analysis technique of identifying and eliminating possible causes until the most likely, primary cause has been found and proved.

A letter format is used for this report because of its length and relatively straightforward content.

Table 21.7 Informal or Semi-formal Causal Analysis Structure

Section	Content	Comments
Action Opening	Refer to the reader's request or to the writer's role in analyzing the identified situation. State whether the cause(s) can be identified and, if so, name the main cause.	The reader connection might be placed in a transmittal letter or memo if a semi-formal format is used. A letter or memo report does not need a heading for the opening paragraph or two.
Background	Describe the situation(or environment) in which the event occurred or in which the problem developed. Provide back-ground about similar problems or situations.	This section should not exceed two paragraphs. If more detail is necessary, it can be placed as attachments.
Details of Analysis	Describe the step-by-step analytical process and give the results of that process. and prior conditions. See Chapter 7	Causal analysis usually names possible causes identified from previous experience and based on the relation between an event re: correlation and causation.
Action Closing	Summarize the report's main findings. State the bottom line. If appropriate, recommend remedial or preventative action.	A summary table could be effective. Attached brochures, performance tests, financial projections, or other detailed supporting data could be appropriate.

TerraTech ENGINEERING

1714 Kalamalka Lake Road, Vernon, British Columbia V1G 2N6   Ph. (250) 545-0919
Fax (250) 545-2020   terra@junction.net

June 21, 2003

Ms. L. P. Garcia
306 Melville Court
Vernon, BC   V1B 2W9

Dear Ms. Garcia:

Re: Causes of the Undercut Driveway at 306 Melville Court

At your request, we have examined the extent of the undercutting of your 17-month old driveway and have identified the primary cause of that problem to be water directed from a neighbouring driveway.

Background

In similar situations that we have analyzed, we've determined that open space can develop under a concrete pad if the base soil has not been properly packed, or if flowing water has eroded soil away, or if water has pooled under the pad and thus caused the soil to settle.

Much depends on the soil's composition. In the case of your driveway, the base soil primarily consists of glacial till, a combination of sand, gravel, larger rocks, and small amounts of organic material. Typically, when glacial till gets saturated with water, it turns into a slurry that either flows downhill or collapses into itself, as all loose spaces among the particles are filled.

Investigation and Results

We conducted the investigation in four stages:

1. We determined the nature and extent of damage by examining the concrete deck and by probing the empty space beneath it. Our visual inspection of the concrete revealed no cracks or sagging of the deck; the driveway has stood up very well. We then interviewed your builder, Mark Lambton, who showed us his construction notes. The notes say that the concrete subcontractor used twice the normal amount of reinforcing bar within the concrete slab.

   The driveway has not pulled away from the rebar fitted into the garage footings, despite 9" of open space under the concrete along the intersection of driveway and footings. The drawing on page 2 shows the extent of open space under the driveway, and other features of the existing situation. (Numbers show the depth of settling or erosion at various points.) The undercutting is quite extensive, as the drawing shows.

2. We ruled out inadequate packing as a cause of the settling. Mark Lambton's notes reveal that the driveway base was packed uniformly, that it was watered between packings, and that it was packed five times over a four-day period. This exceeds normal practice.

Figure 21.10   A Causal Analysis   *(continued)*

Ms. L. P. Garcia
June 21, 2003
Page 2

3. We ruled out erosion as a major causative factor. There may have been some minor initial erosion along the north edge of the driveway but erosion could not have caused the irregular pattern of open space beneath the slab. That irregular pattern suggests that pooled water has led to irregular soil settling.

4. We determined the source of the amounts of water required to cause the degree of observed settling. We do not believe that the water has come from your driveway, say in last fall's heavy rains. Your driveway has been designed to channel water down toward a collection drain located two metres from the centre of the garage door. To test our belief, we ran water onto your driveway from two hoses simultaneously. All the water was easily channelled down to the drain.

Then, we investigated water flow from the north-neighbouring driveway that slopes away from the garage, toward the street. That driveway is higher than yours at every point of its length, and the entire driveway slopes down toward the rock-covered depression between the two drives. Moreover, a 1-metre-wide diagonal depression in the neighbouring driveway channels water toward the catchment area immediately adjacent to the spot where your driveway begins to be undercut (see the diagram).

With your neighbour's permission, we ran water onto his driveway for 45 minutes. All of the released water ran down to the catchment area A, from which the water seeped under your driveway and formed pools of settling water. When we re-inspected in three hours, all that water had soaked in.

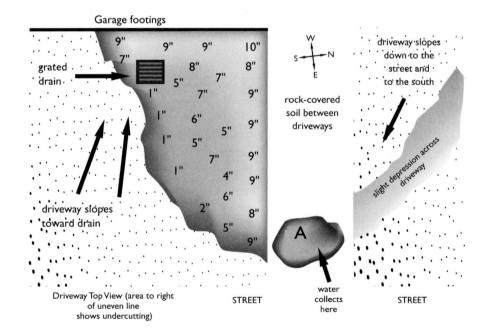

Conclusion
We're confident that the driveway undercutting has resulted from water-induced settling and that the majority of the water has come from your north neighbour's driveway.

Figure 21.10   A Causal Analysis *(continued)*

Ms. L. P. Garcia
June 21, 2003
Page 3

Corrective action will require shoring up the space under the driveway with an impermeable base and diverting water away from possible entry points along the north edge of your driveway. We can recommend Majestic Mudjacking for the former task—this company will pump a rapidly hardening slurry of cement, sand, water, clay, and loam under the exposed slab. If you wish, we'll undertake the water diversion.

We believe that you have grounds for requiring your neighbour to pay the costs of all repair work; local bylaws require each homeowner to control the passage of water off their property, so that the water doesn't flow onto a neighbour's property. Please let us know if we can be of further service. Our invoice is enclosed.

Sincerely,
TerraTech Engineering

Dimitri J. Jones, C.E.T.

Figure 21.10   A Causal Analysis

ASSESSMENT REPORTS

Assessment reports essentially use the same approach as feasibility reports, except that assessments (also known as *evaluation reports or investigation reports*) are conducted *after* a project has been conducted, or *after* changes have been made.

One special type of assessment report is the *yardstick* or *benchmark report*, which assesses and compares two or more alternative solutions or equipment proposals. In this type of analysis, appropriate criteria act as yardsticks by which the competing alternatives can be compared. Thus, the analysis is handled consistently and fairly. See Table 21.8 for the structure of these types of reports.

Table 21.8 Yardstick Assessment Report Structure

Section	Content	Comments
Action Opening	Define the situation requiring assessment—the problem or challenge. Report the main conclusion.	Only one or two paragraphs are necessary for a letter or memo. Use the heading, "Introduction," for a semi-formal report.
Background	Briefly explain possible alternative solutions. Name the assessment criteria and explain how these were developed and selected.	The section could be called "Background" for "Assessment Method." Extensive technical data should be attached to the end of the report, not placed in the Background section.
Details of Assessment	Evaluate each alternative, according to the assessment criteria. The assessment could be organized by criteria such as cost, durability, product support). Or the alternatives could be examined, one by one: All the criteria would be applied to the first alternative before moving on to the next alternative.	Use a title like "Data and Assessment." Organize by criteria to allow the alternatives to be ranked by each criterion in turn (e.g., the least to the most expensive). Focus on each alternative in turn.
Action Closing	Conclude which alternative, or combination of alternatives, best meets the assessment criteria.	An implementation plan could be included, but it shouldn't be the focus of an assessment report.

RECOMMENDATIONS REPORTS

Many recommendations reports respond to reader requests for a solution to a problem; others originate with the writer, who has recognized a problem and developed a solution. This latter type is often called a *justification report*.

Both kinds of recommendations reports could use either a direct pattern or an indirect pattern, depending on the reader's needs and attitudes.

The *direct pattern* suits situations where you can anticipate reader support. Perhaps the reader has accepted similar recommendations in the past. Perhaps your recommended choice is so obvious and clear-cut that there is no other course of action. (See Figure 21.11, on page 521, for such a situation.) Perhaps your recommended action matches company policy and practice. Perhaps the recommendation reflects the reader's own preferred approach or administrative bias.

Table 21.9 on page 520 shows the direct pattern for the kind of recommendations that would be received favourably.

Table 21.9   Direct Recommendations Report Structure

Section	Content	Comments
Action Opening	Refer to the need or problem in a way that your reader will recognize and accept. Use active verbs to recommend action.	Use a heading like "Recommendation" or "Problem and Solution."
Background	Name the alternative solutions you considered and explain the criteria used to assess the alternatives. Briefly explain why you discarded alternatives other than the one you selected.	In some cases, it may not be wise to quickly dismiss potential solutions; your approach may seem arbitrary. In such cases, fully apply the assessment criteria to all alternatives.
Details of Assessment	Discuss the benefits, comparative advantages, and drawbacks of your recommended solution. Detail the required resources and costs of your recommended solution.	Use a title like "Features of the Solution," or "Advantages and Requirements," or "How Our Firm Will Benefit."
Action Closing	Summarize the main reason for choosing the recommended solution and provide a plan for implementing it. Request authorization for your actions or specify the actions you're asking of the reader.	You could attach detailed supporting information such as performance tests, brochures, quotes, and financial projections.

Direct POWER Corporation   MEMORANDUM

DATE: April 17, 2004
TO: Martin Scherre, Head Engineer, Correl Park Power Station
FROM: Judy Shohat, Plant Engineer, Correl Park Power Station *JS*

RE: Ash Line Replacement

As I mentioned in our plant meeting two weeks ago, some of the ash lines will need replacing. Shimon Barak and I have since examined Production Units 1 through 6 and found potential line failures for the lines leading from Unit 5 and Unit 6.

Recommendation
Correl Park should purchase 915 m of 300 mm (319 mm OD, 9.5 mm WT) commercial-grade black steel pipe for replacement ash line, at an estimated cost of $60,000.

Background
Approximately 2/3 of all ash line failures are detected and repaired with only some welding time required. However, if an ash line fails in the early evening and is not detected until morning, the line downstream of the failure will plug due to reduced flow velocity. When this has happened, we have had to contract a high-pressure washing truck to clean the line at a cost of from $5,000 to $10,000. In order to maximize line life, we rotate the lines 1/3 turn every three or four years, to distribute the wear around the inner circumference of the pipe.

Findings
Correl Park Unit 6 has three ash lines. 6A and 6B each have one rotation to go, so they will be fine for at least three more years. However, 6C ash line received its third and final turn in May 2004. Last week this line was examined and rotated in whatever direction exposed the thickest remaining wall to the area of highest wear. A random sampling of thickness readings along this pipe showed an average thickness of 4.47 mm, which is less than the original thickness of 9.5 mm. Experience has shown that an average thickness of equal to or less than one-half the original significantly increases the failure frequency. About 765 metres needs replacing.

Currently, No. 5 ash pit is being cleaned out, with the ash being used for road construction. Once the cleaning is complete, it will be necessary to install about 150 metres of ash line to make this pit functional again. (We could then restore Unit 5 to service.) We considered installing the required line with used pipe that we have in stock, but that used pipe is no better than the 6C line that needs to be replaced.

Authorization
I request authorization for the recommended replacement so that bidding for the pipe supply contract can proceed. A Purchase Recommendation and a Technical Specification for the required pipe are attached.

Att.

Figure 21.11 A Direct Recommendations Report

The direct recommendations report in Figure 21.11, on page 521, responds to the following identified need.

At a coal-fuelled electrical power plant, steel outflow pipes carry ash from the coal furnaces to storage lagoons (known as "ash pits" or "lagoons"). Some of these pipes (known as "ash lines") have been eroded and are in danger of failing. The writer knows, as does the reader, that the plant must replace the lines when they are in danger of failing; no other alternative action is available. Still, the recommended action needs to be justified. Also, the report needs to include some background information for the reader, who has recently transferred from a hydroelectric plant. ▲

An *indirect approach* works better in situations where your reader may be unreceptive to your recommendations, or where the recommendations deal with a sensitive issue such as workplace harassment or strained employer-employee relations. Table 21.10 suggests the sequence for an indirect recommendations report.

Table 21.10   Indirect Recommendations Report Structure

Section	Content	Comments
Action Opening	Refer to the situation in such a way that your reader realizes there's a problem or a need to be addressed. Briefly describe the approach used in this report.	A semi-formal report could have an "Introduction"; a letter or memo would not need a heading for this section. In a letter or memo, "connect" with the reader in the first sentence.
Background	Show the extent of the need or problem by presenting quotes, examples, or supporting statistics. List alternative solutions and explain the criteria used to assess the alternatives.	Either semi-formal or informal reports could use a heading such as "Background" or "Problem and Solutions."
Details of Assessment	Evaluate the alternatives with the identified criteria, starting with the least applicable solution. Present the best alternative last; apply the criteria vigorously.	Use titles like "Assessment of Alternative Solutions" or "Possible Solutions."
Action Closing	Summarize your recommendation. Show how it can be implemented. Ask for authorization or specify the actions you're asking your reader to take.	You might attach detailed supporting information such as performance tests, brochures, quotes, and financial projections.

MEMORANDUM

TO: Professor Stroud, English Department   **DATE:**   October 16, 2002

FROM: Kim Briere, Applied English 140

RE: **Improving Performance in English 140**

As you have stated in class, we have a severe problem in English 140: The overwhelming majority of the 38 students have failed at least two of the first three assignments so far. Students like me, who are committed to success, are very concerned. So, we've consulted a College Counsellor and the Engineering Dean and we've identified the solution that is presented in this memo.

Thirty-one of us have met three times to discuss the issue. We need to solve the problem of low grades and we need to pass this class to move on to English 150 next semester. Many of us have also discussed the issue with our Department Chair, who warns us that we need a solid grounding in English to do well in our engineering program.

We see a problem for you, too. It must be difficult to have to correct so many things in our memos, letters, and reports. Also, we ask so many questions about grammar and sentence basics in class that you don't have time to present your full lecture.

The cause of the problem seems to be our "inadequate grasp of the English language" (your comment in the October 12 class). We agree with your assessment. In our meeting yesterday, 29 of us found that we lost an average 21 marks for mechanical errors and poor paragraphing on the last assignment!

So, what can we do? We must satisfy three criteria. We need to:

1. improve our English grades to succeed in our program of study
2. build our writing skills for future careers
3. find a practical, <u>immediate</u> solution

Following the advice of an Academic Counsellor, who showed us the Harvard Case Study model, we've applied the above criteria to four options:

1. Work harder and spend more time on our assignments
2. Lobby the college to reinstate the drop-in Writing Centre, that disappeared after last year's budget cuts
3. Drop the English 140 course now
4. Arrange for special tutorial sessions

The table on page 2 summarizes our thoughts about the four options.

Figure 21.12 An Indirect Recommendations Report   *(continued)*

Options	Improve Grades	Build for Future?	Practical and Immediate?
Work harder and longer	Not likely—we still need basic skills	No—we're held back by lack of basic skills	No—we carry seven classes each and spend too much time on English now!
Lobby for writing centre	Yes—individual examples and instruction could help us develop basics	Yes—could build the base we need	No—the college is still in a deficit situation and the bureaucracy moves too slowly
Drop the course	Perhaps, if we take the course later, we'll succeed	We don't know how to build writer skills on our own	No—we need a solution this semester
Arrange for tutor(s)	Yes—we can develop the skills, individually and collectively	Yes—we need to build skills to get to the next level	This is the only practical possibility of the four options, **if** we can get the needed assistance

As you can see, we have only one viable option. In order to make that option work, two things have to happen:

1. We need times and a place to meet. Dean Cartwright has arranged a classroom for 4–7 p.m. on Mondays, Wednesdays, and Fridays. She has also found $1,000 to pay a tutor or tutors.

2. We need your help to direct a tutor (or tutors); you have the best idea of our needs. Also, can you help us find one or two tutors? Perhaps you know capable retired professors or graduate students.

Please support our recommendation action. We'd like to start no later than next Monday, so may we have your response in Thursday's class?

Kim Briere

Kim Briere

Figure 21.12   An Indirect Recommendations Report

To illustrate the indirect pattern at work, consider the challenge faced by Kim Briere and her colleagues in an Engineering Technology Diploma Program.

> Like many of her colleagues, Kim's writing skills are not adequate for the Applied English 140 course offered in the first semester of her five-semester program. After lengthy consultations with her colleagues, with a college academic counsellor, and with the chairperson of the Engineering Technology Program, Kim sends the recommendation illustrated in Figure 21.12 (see page 523) to Jake Stroud, the Applied English 140 instructor.
>
> As a result of her audience/purpose analysis, Kim realizes that her needs and those of her colleagues are different from those of Professor Stroud. She's also aware that he is opposed to teaching what he calls "remedial English." He has said that the English 140 class "applies university-level writing skills to real-world situations; there's no time to develop basic skills that university students should bring with them."
>
> Kim suspects that Professor Stroud will philosophically oppose her recommendations and that he'll resist her recommended action because it means more work for him. Therefore, she places her recommendations at the end of the report, after leading her reader through an analysis that shows that the majority of the Applied English 140 students really have only one viable option. ▲

LAB REPORTS

The academic lab reports you write during your studies are different from the reports you'll write in industry. First, the names are different; at college, you submit *lab reports*; at work, you'll produce *laboratory reports* or *test reports*. The purpose also differs: Your college lab reports help you learn material or prove a theory, while "real-world" laboratory tests have practical applications such as the following:

- ◆ Water samples are tested to determine if a water treatment plant is working properly.
- ◆ Car seat child-restraint systems are tested to see if they are safe and effective.
- ◆ Soil core samples are tested to determine if a PCB-contaminated site has been decontaminated.

Various formats and requirements exist for academic and industrial settings. You will have to adapt to the specific requirements at your workplace. However, all laboratory and test reports use the general pattern shown on page 526.

Table 21.11   Lab Report Structure

Section	Content	Comments
Introduction	Name and define the subject; review the subject's significance. Or, indicate how this test fits into a project or routine procedure.	Could also include the scope of the research. Might discuss the rationale for the research or the objective of the research. Sometimes this is expressed as a question to be answered, sometimes as a hypothesis to be proved or disproved.
Equipment and Procedure	Where appropriate, describe the design of the investigation. List materials, instruments, and equipment.	Might also be called "Materials and Methods."
	Describe, step by step, how the test, experiment, or study was done.	Use passive voice, third person narrative, in the past tense.
	Describe methods for observing, recording, and interpreting results.	Use passive voice, third person narrative, in the past tense.
Results	List recorded observations. Provide detailed relevant calculations.	Relate these results to the methods used to achieve them.
Conclusion	Analyze the percentage of error and the possible causes of error. Answer: ♦ Do the results answer the questions? ♦ Was the research objective met? ♦ Do you have doubts about the results? Why?	Often called "Discussion." Might also answer: ♦ Was the hypothesis proved? ♦ Are these results consistent with other research? ♦ Are there implications for further research?

FORM REPORTS

Many reporting situations can be handled with pre-printed forms (or electronic templates). Daily and weekly progress reports, for example, often use forms to keep clients and supervisors informed about a project. In many jobs, the best time to learn how to use job-specific forms is during the orientation period—the first two to three weeks on the job. Ask questions about the purpose of each form and the expected standard of completion. Also ask to see completed sample forms. (Some supervisors will prefer to show you how to use a given form only when that form is needed, and not before.)

Employers have passed on the following hints for successfully completing form reports:

♦ Read headings or questions on the form carefully. If necessary, ask directions, or look for models (precedents). Do not assume you've guessed correctly.

- Before writing or keying the form, read the entire form and make some quick notes of what to include.
- Choose *exact* words and phrases, not approximate descriptive language.
- Use jargon only if necessary; perhaps a non-technical person will read your report.
- Analyze the reader and the purpose for the report, and provide *all* necessary detail.
- Write or print neatly. On multiple-copy forms, press firmly!
- Check for errors in facts and figures, spelling, or logic.
- Know deadlines and stick to them. Remember that form reports are not designed for the writer's convenience; they're used to help you provide information quickly, while the information is still useful to the reader.

WRITING A SHORT REPORT EFFICIENTLY

Most short reports are written in the middle of busy days, with other tasks waiting to be completed. Thus, it's important to use an efficient writing process, such as the following, which assumes a report of two to three pages.

1. Analyze the report's purpose and audience. (*2 to 3 minutes*)
2. Quickly list the report's content; then evaluate and revise that content. (*up to 5 minutes*)
3. Construct a rough outline, perhaps with headings in place. (*up to 5 minutes*)
4. Then, **and only then,** compose the first draft. (*40 to 60 minutes*)
5. Revise the content where necessary. (*5 minutes*)
6. Edit phrasing to ensure it's clear and readable, and that the tone suits the proposal and the reader's likely reaction. (*10 minutes*)
7. Proofread for errors and correct them. (*5 minutes*)

Checklist FOR EDITING AND REVISING SHORT REPORTS

Use this checklist as a guide to revising and refining your short reports.

- Have you chosen the best report format for your purpose and audience?
- Does the letter or memo use proper format?
- Does the subject line forecast the letter or memo's contents?
- Does the semi-formal report format contain the appropriate elements?
- Are readers given enough information for an informed decision?
- Are the conclusions and recommendations clear?
- Did you make the right choice between the direct and indirect patterns of presenting the report's bottom line?

- Are paragraphs single spaced within and double spaced between?
- Do headings, charts, or tables appear whenever needed?
- If more than one reader is receiving copies, does the letter or memo include a distribution notation to identify other readers?
- Does the semi-formal report's title page name other readers?
- Is the writing style clear, concise, exact, fluent, appropriate, and direct?
- Does the document's appearance create a favourable impression?
- Have you included useful details such as supplementary attachments, enclosures, or appendices?

EXERCISES

1. Identify a dangerous or inconvenient area or situation on campus or in your community (endless cafeteria lines, a poorly lit intersection, slippery stairs, a poorly adjusted traffic light). Observe the problem for several hours during a peak-use period. Write a justification report to a *specifically identified* decision maker, describing the problem, listing your observations, making recommendations, and encouraging reader support or action.

2. Assume you have received a $10,000 scholarship, $2,500 yearly. The only stipulation for receiving installments is that you send the scholarship committee a yearly progress report on your education, including courses, grades, school activities, and cumulative average. Write the report.

3. In a memo to your instructor, outline your progress on your term project. Describe your accomplishments, plans for further work, and any problems or setbacks. Conclude your memo with a specific completion date.

4. Keep accurate minutes for one class session (preferably one with debate or discussion). Submit the minutes in memo form to your instructor.

5. Conduct a brief survey (e.g., of comparative interest rates from various banks on a car loan, comparative tax and property evaluation rates in three local towns, or comparative prices among local retailers for an item). Arrange your data and report your findings to your instructor in a memo.

6. Recommendations Report (choose one)

   a. You are a consulting engineer to an island community of 200 families suffering a severe shortage of fresh water. Some islanders have raised the possibility of producing drinking water from salt water (desalination). Write a report for the Island Trust, summarizing the process and describing instances in which desalination has been used successfully or unsuccessfully. Would desalination be economically feasible for a community of this size? Recommend a course of action.

   b. You are a health officer in a town less than one kilometre from a massive radar installation. Citizens are disturbed about the effects of microwave radiation. Do they need to worry? Should any precautions be taken? Find the facts and write your report.

   c. You are an investment broker for a major firm. A long-time client calls to ask your opinion. She is thinking of investing in a company that is fast becoming a leader in fibre optics communication links. "Should I invest in this technology?" your client wants to know. Find out, and give her your recommendations in a short report.

   d. The "coffee generation" wants to know about the properties of caffeine and the chemicals used on coffee beans. What are the effects of these substances on the body? Write your report, making specific recommendations about precautions that coffee drinkers can take.

   e. As a consulting dietitian to the school cafeteria in Blandville, you've been asked by the school board to report on the most dangerous chemical additives in foods. Parents want to be sure that foods containing these additives are eliminated from school menus, insofar as possible. Write your report, making general recommendations about modifying school menus.

   f. Dream up a scenario of your own in which information and recommendations would make a real difference. (Perhaps the question could be one you've always wanted answered.)

COLLABORATIVE PROJECT

Organize into groups of four or five and choose a topic upon which all group members can take the same position. Here are some possibilities:

- Should your college abolish core requirements?
- Should every student in your school pass a writing proficiency exam before graduating?
- Should courses outside one's major be graded pass/fail at the student's request?
- Should your school drop or institute student evaluation of teachers?
- Should all students be required to be computer literate before graduating?
- Should campus police carry guns?
- Should dorm security be improved?
- Should students with meal tickets be charged according to the type and amount of food they eat, instead of paying a flat fee?

As a group, decide your position on the issue. Brainstorm collectively to justify your recommendation to a stipulated primary audience in addition to your colleagues and instructor. Complete an audience/purpose profile (page 34), and compose a justification report. Appoint one member to present the report in class.

Workplace Correspondence: Letters, Memos, and E-mail

A report may be completed by a team of writers for multiple readers, but letters, memos, and e-mail (electronic mail) are usually written by a single writer for one or more definite readers. Also, workplace correspondence is more direct and personal than reports and often has a persuasive purpose, so proper tone is essential. You want your reader to be on your side. Because successful business depends on a two-way transaction in which both participants meet their needs, you must use a "you" attitude. You must look at the situation from the reader's viewpoint.

This chapter looks at four correspondence media: letters, memos, e-mail, and faxes. The chapter presents current formats, introduces considerations common to all four media, and discusses issues that concern each different medium.

Letters appeared on the scene first, followed in the mid-twentieth century by memos. Then, as that century drew to a close, new technologies brought facsimile transmission (faxes) and e-mail messaging. As we begin a new century, we can be sure of three things:

1. Rapid technological change will bring new varieties of electronic correspondence, **but**
2. We will want relatively permanent records of that correspondence, **and**
3. The basic principles of successful correspondence will work in any correspondence medium, paper-based or electronic.

LETTER FORMAT

Letter formats have evolved

North American business correspondence blends ordered, elaborate formatting with streamlined, direct phrasing. (By contrast, European letters often *look* less formal but are phrased more formally and elaborately than Canadian and American letters.)

As the twenty-first century begins, two traditional formats are out of fashion and a third may soon lose favour. The *semi-block* and *block formats,* shown in Figures 22.1 and 22.2 on page 531, were both popular in the 1960s and early 1970s in Canada, but neither format is widely used now.

In the 1970s and early 1980s, IBM and others championed the *full-block format* because its left-justified set-up saved keying time. The full-block format (Figure 22.3 on page 531) led the way through the 1980s and 1990s, but as the 1990s drew to a close, it started to give way to an even more streamlined layout, which had been introduced as the "simplified letter" by the U.S. National Office Management Association in the early 1960s. Figure 22.4 on page 531 shows a contemporary version of the *simplified format.*

Today's version of the simplified format responds to a growing discomfort with the formal greeting ("Dear . . .") and complimentary close ("Yours ..."), which seem characteristic of an earlier era. Notice that a subject line replaces the salutation and that the complimentary close is eliminated. In other respects, a simplified letter copies a full-block letter, although the simplified format's overall appearance resembles contemporary memos. In other words, letter and memo formats are becoming similar.

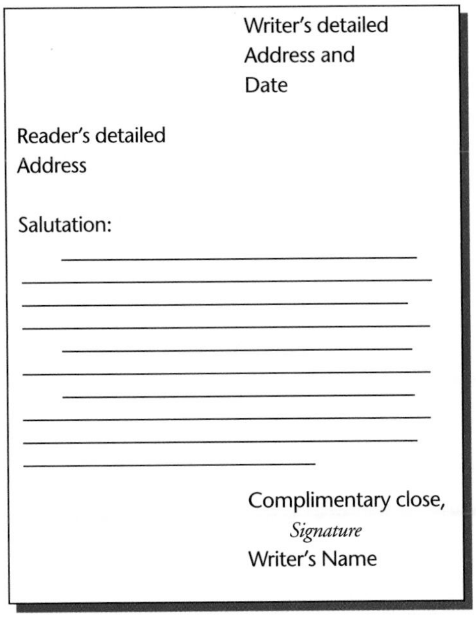

Writer's detailed
Address and
Date

Reader's detailed
Address

Salutation:

_____
_____
_____
_____
_____
_____
_____
_____

Complimentary close,
Signature
Writer's Name

Figure 22.1   A Semi-block Format

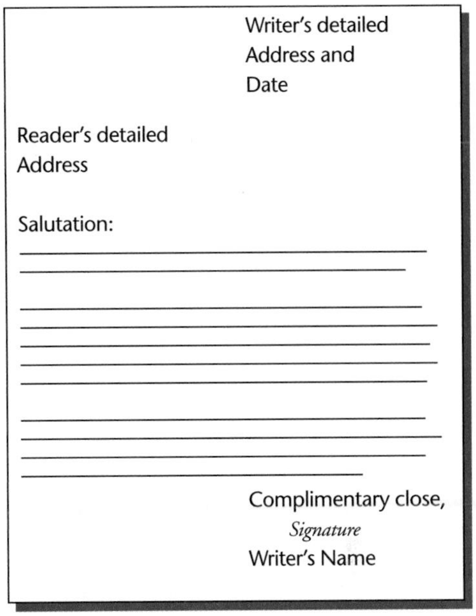

Writer's detailed
Address and
Date

Reader's detailed
Address

Salutation:

_____
_____
_____
_____
_____
_____
_____

Complimentary close,
Signature
Writer's Name

Figure 22.2   A Block Format

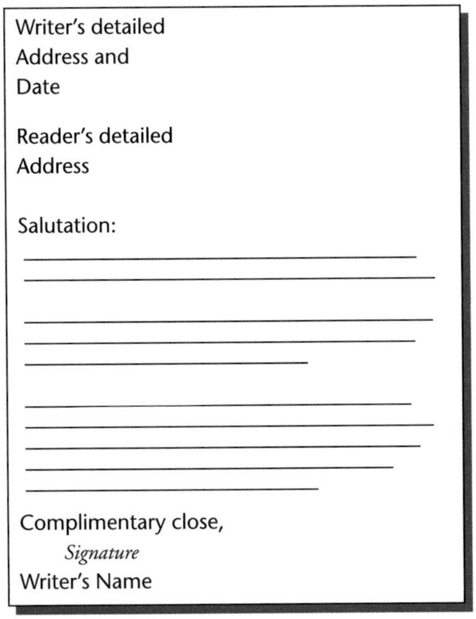

Writer's detailed
Address and
Date

Reader's detailed
Address

Salutation:

_____
_____
_____
_____
_____
_____
_____

Complimentary close,
Signature
Writer's Name

Figure 22.3   A Full-block Format

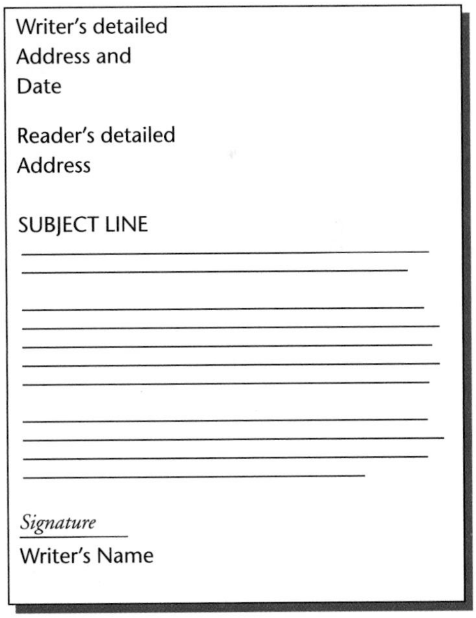

Writer's detailed
Address and
Date

Reader's detailed
Address

SUBJECT LINE

_____
_____
_____
_____
_____
_____
_____

Signature
Writer's Name

Figure 22.4   A Simplified Format

BASIC ELEMENTS OF LETTERS

Business letters have traditionally included five elements (in order from top to bottom): heading and date, inside address, salutation, letter text, and closing.

Heading and Date.   Your personal business letters start with your detailed address (but not your name) at the top of the page, followed by the date of the letter.

> 127 Marchbank Avenue
> Barrie ON   K9M 7H3
>
> June 23, 2003

If you're sending a letter on company stationery, place the date a line or two below the letterhead.

> **ROCKWOOD INDUSTRIES CO. LTD.**
> 1222 Terminal Road Rockwood ON   N0B 2K3 (413) 554-7863
>
> June 23, 2003

Canadian usage includes four ways of writing addresses:

1. 127 Marchbank Avenue
   Barrie ON K9M 7H3

2. 127 Marchbank Avenue
   Barrie, ON K9M 7H3

3. 127 Marchbank Avenue
   Barrie, Ontario
   K9M 7H3

4. 127 Marchbank Avenue
   Barrie, Ont.
   K9M 7H3

Acceptable form #1, above, has been instituted by Canada Post Corporation to help its computerized optical scanners operate. Notice that Canada Post's address standard eliminates internal punctuation. Actually, Canada Post's "optimum requirements" for the address on the envelope (the "outside address") look like this:

127 MARCHBANK AVENUE
BARRIE ON K9M 7H3

NOTE   *The majority of businesses, with the exception of some government offices, use the second acceptable form of writing addresses. Check with your employer. This book's sample letters use acceptable forms 1, 2, and 3.*

Inside Address.   Place the inside address two to four spaces below the date and abutting the left margin. Wherever possible, address your letter to a specifically named reader, and include your reader's job title. Using a form of address such as "Mr." or "Ms." before the name is optional.

Inside address format	Mr. Saul Kaufman, General Manager	*or,*	Bluenose Engineering Co.

Inside address format

Mr. Saul Kaufman, General Manager   *or,*
Bluenose Engineering Co.
1774 Robie Street
Halifax NS B3H 3G7

Bluenose Engineering Co.
1774 Robie Street
Halifax NS B3H 3G7

Attention: Mr. Saul Kaufman
                      General Manager

> **NOTE**   *Depending on the length of your letter, adjust the vertical placement of the heading and inside address to achieve a page that appears balanced.*

Salutation format

Salutation.   The salutation usually appears two line spaces below the inside address. A standard salutation begins with *Dear* and ends with a colon (*Dear Mr. Kaufman:*). If you don't know the person's name, an attention line (*Attention: General Manager*) is preferable to using the position title (*Dear General Manager:*).

No satisfactory guidelines exist for addressing several people within an organization. *Gentlemen* or *Dear Sirs* shows implied bias. *Ladies and Gentlemen* sounds too much like the beginning of a speech. *Dear Sir or Madam* is too old-fashioned. *To Whom It May Concern* is vague and impersonal. Your best bet is to use an attention line (*Attention: Personnel Department*) and eliminate the salutation (use a simplified format).

Letter Text.   Begin the text of your letter two spaces below the salutation or subject line. Workplace letters typically include: (1) a brief *introduction* paragraph (two or three sentences) that identifies your purpose and connects with the reader's interest; (2) one or more paragraphs that present the *details* of your message; and (3) a *closing* paragraph that sums up and encourages action. Some letters provide *background* information immediately after the introduction.

Keep your paragraphs short, usually fewer than eight lines. If a paragraph goes beyond eight lines, or if the paragraph contains detailed supporting facts or examples, consider using bulleted or numbered lists to make the paragraph readable. On average, letter paragraphs should not exceed 60 words per paragraph.

Closing.   The closing, placed one line space below the last line of text, includes three components in traditional formats: complimentary close, signature, and writer's name and position.

Complimentary close

Signature

Writer's identity

Yours truly,

Maris McGovern (signature)

Maris McGovern
Sales Manager

Yours truly and *Sincerely* are the most commonly used complimentary closes. Others, in order of decreasing formality, include:

Respectfully,
Cordially,

Best wishes,
Warmest regards,
Regards,
Best,

Align the three-part closing with the letter's heading.

If you are representing a company or group that bears legal responsibility for the correspondence, key the company's name in full caps two spaces below your complimentary closing; place your keyed name and title four spaces below the company name and sign in the triple space between.

Yours truly,

ROCKWOOD INDUSTRIES

MBaxter

Mary Baxter
Research Coordinator

SPECIALIZED PARTS OF LETTERS

Some letters require one or more of the following specialized parts. *Examples appear in sample letters in this chapter and in Chapter 23.*

Other format elements

Attention Line.   Use an attention line when you direct a letter to a specific department or position within an organization but don't know the reader's name. (Or, use it in the simplified letter format.)

Rockwood Industries Inc.
335 – 11th Avenue S.W.
Calgary AB T2R 1L9

Attention: Director of Research and Development
or,
ATTENTION: Director of Research and Development

Subject Line. Because it announces the topic of your letter, the subject line is a good device for attracting a busy person's attention.

Subject: Improvements in Client Service
or,
SUBJECT: Improvements in Client Service

Place the subject line below the salutation with one line space before and after it.

Dear Mr. Patrese:

SUBJECT: Improvements in Client Service

If you are using a simplified style letter, place the subject line below the inside address with two line spaces before and after it.

Initials.   If someone keys your letter, your initials (in caps) and his or hers (in lower case letters) should appear below the writer's keyed name, flush with the left margin. This practice is disappearing because the overwhelming majority of writers key their own letters these days.

> J. Mansonneau
> Manager
>
> JM/to

Enclosure Notation.   When other documents accompany your letter, add an enclosure notation one line below the initials (or writer's name and position), flush with the left margin. State the number of enclosures.

> Enclosure
> Enclosures 2
> Encl. 3

Distribution Notation.   If you distribute copies of your letter to other readers, insert the notation *Copy,* or *c,* or *pc,* or *Distribution,* one line below the previous line (such as an enclosure line).

> Copy: B. Grammel
>
> c: M. Henderson
>
> Distribution: Hamilton Better Business Bureau
>   Hamilton Chamber of Commerce
>   Ontario Ombudsman

Most copies are distributed on an FYI (for your information) basis, but writers sometimes use the distribution notation to maintain a paper trail or to signal that the letter is being shared with others (e.g., superiors or legal authorities).

 The notation bc *means "blind copy." It appears on copies other than the original and indicates that the recipient of the letter is not aware that a copy is being sent to others.*

DESIGN FACTORS

Letter design factors

Design starts with a choice of formats. Currently, the format most favoured by Canadian business writers is the full-block layout, but the simplified format is gaining in popularity. Both these forms look businesslike and eliminate the need to tab and centre.

Additional design factors enable workplace letters to appear inviting, accessible, and professional:

Quality Stationery.   Use high-quality 20-pound or 24-pound, 21.5 cm x 28 cm (8$^1/_2$" x 11") white paper with a minimum fibre content of 25 percent.

Hundreds of varieties of coloured, specially textured papers are available, but you should exercise caution in using anything other than white stationery.

Uniform Margins and Spacing. When using stationery without a letterhead, frame your letter with 2.5–3.75 cm (1"–1$\frac{1}{2}$") margins, depending on the amount of space required by the letter's text. Strive for a balanced look. Use single spacing within paragraphs and double spacing between paragraphs.

Page Continuation Format. If your letter continues beyond a first page, begin each additional page with a notation identifying the addressee, date, and page number.

Saul Kaufman   June 25, 2004   Page 2

If there's sufficient space, you can stack the notation at the left margin.

Saul Kaufman
June 25, 2004
Page 2

Begin the text two spaces below the page continuation notation. Never use an additional page solely for the closing section. Instead, reformat the letter so that the closing appears on the first page or so that at least two lines of text appear above the closing on the subsequent page.

Envelope Preparation. Your 24 cm x 10.5 cm (9$\frac{1}{2}$" x 4$\frac{1}{4}$") envelope (also called a #10 envelope) should be of the same quality as your stationery. Place your reader's name and address at a fairly central point on the envelope (13–15 line spaces from the top of the envelope and 1.25 cm ($\frac{1}{2}$") left of the horizontal centre of the envelope. Place your own name and address in the upper left corner. Single-space these elements. Most likely, your word processor will have an envelope printing function that will automatically place these elements in the correct location on the envelope. (See your printer's operating manual for instructions.)

MEMO USAGE AND FORMAT

*Memos have
many uses*

Memoranda (usually called "memos") are used within organizations for a wide variety of messages. Originally, memos were intended for relatively brief messages, but now they're also used for longer messages, including informative and analytical reports of up to four pages, as Chapter 21 demonstrates. Memos are also used for proposals and other persuasive messages.

A major form of communication in most organizations, memos leave a paper trail of directives, enquiries, instructions, requests, recommendations, and daily reports needed to run an organization.

Organizations rely heavily on memos to trace decisions and responsibilities, track progress, and recheck data. Therefore, any memo you write can have far-reaching ethical and legal implications. Be sure your memo includes the date and your initials or signature. Also, make sure your information is specific, unambiguous, and accurate. Finally, remember that your memos must provide the information and analysis that the reader needs, but *no more* than the reader needs.

FORMAT

Basically, there are two varieties of memos: intra-office (within an office) and inter-office (between offices of the same firm). Both use the same type of format, with minor variations.

Memohead.   Intra-office memos usually feature the designation, **Memorandum** (or **Memo**) at the top of the page, sometimes in conjunction with the company's name and logo, sometimes by itself. (See Figure 21.6, page 505 and Figure 21.7, page 509.) Inter-office memos usually name the company at the top of the page, followed by the term "Inter-office Memo." (See Figure 3.4, page 50.)

Heading Guides.   Four heading guides are mandatory: date line, receiver's name and title, sender's name and title, and subject line.

> **DATE:**   June 25, 2004
>
> **TO:**   John Tarnowski, Purchasing Coordinator
>
> **FROM:**   May Krienke, Explorations Chief
>
> **RE:**   **Cost Estimates for the Churchill River Project**

Provide the titles held by you and your reader, even if you are both well aware of each other's position. Why? First, it's a formal courtesy used within most organizations. Second, the title designations provide a record that may prove useful in the future.

Additional heading information might include file locators (FILE, or OUR FILE, or YOUR FILE), the sender's phone number or e-mail address, or a distribution line. Distribution lines more often appear at the end of a memo, but if you prefer to include your distribution list in the heading, incorporate that list in the TO section or place a DISTRIBUTION line between the TO and FROM sections. If you want to direct copies of a memorandum to certain personnel, include a COPIES or COPY TO line after the TO section.

Margins and Spacing.   Use block format for memos: Do not indent paragraphs. Instead, single-space within paragraphs and double-space between paragraphs. Leave 2.5 cm (1") left and right margins. Do not right justify paragraphs; leave a ragged right edge. Do not worry about balancing memos on the page; start the heading guides one or two line spaces below the memohead and continue until the memo is complete.

Normally, memo writers leave a line space between each pair of heading guides. Also, it's common practice to tab the information following the heading guides so that the information is aligned. (See the above example of heading guides usage.)

Signature versus Initials.   Many memo writers simply sign their initials beside their name in the FROM section (see Figure 21.11 on page 521). Others choose the more formal (and, they say, the more businesslike) practice of signing their name or placing their initials at the end of the memo.

J. Mansonneau (signature)

J. Mansonneau

This practice has the added advantage of identifying the writer on each page of a two-page memo.

Attachments.   An effective way of including detailed useful or corroborative data is to attach it to the memo. If such material were included in the body of the data, it might interfere with the smooth development and easy reading of that memo, so it's better placed where the readers can read it if they want to do so. For example, an internal proposal concerning a proposed expansion of the firm's advertising program might attach a demographic analysis of the firm's target market and some sample newspaper ads.

Graphic Highlighting.   Memo readers are usually busy people who want easily understood documents. One way to provide readability is to use bulleted or numbered lists of information and ideas. Headings also make documents easier to read. Tables or columns of information can present material concisely and clearly. So, remember to use combinations of these techniques, even for relatively short memos.

Page Continuation Format.   For memos, use the page continuation format that you would use for letters. (See page 536.)

FACSIMILE TRANSMISSION

A fax (short for facsimile) machine electronically scans a page, converts the scanned image to digital code, and sends the code through a telephone cable to another fax machine, which decodes the digital message so that it can print an exact copy (facsimile) of the scanned page. Fax transmissions can also be generated by your word processor and sent via fax modem. For example, Microsoft Word allows you to send a Word document via fax; simply go the **File** menu, choose *Send To...* and then, *Fax Recipient*, and follow the instructions. Other computer systems have similar capabilities.

Faxed messages offer the advantage of speedy delivery for both internal memos and external letters. The technology also provides a confirmation that the receiving fax machine has indeed received the message. However, faxes also have some disadvantages. They don't transmit colour, and often the lines and even letters in faxed messages are broken or blurred. In addition, certain kinds of fax paper break down and thus do not provide a permanent record unless photocopied by the recipient. So, if time is of the essence, send a fax message, but if you want your reader to have a crisp, professional-looking copy on file, send the original copy as a follow-up, keeping a copy of the original for your files.

NOTE   *Recipients with fax modems in their computers can print faxes on their laser printer, avoiding the "paper breakdown" problem.*

FORMAT

Documents transmitted by fax usually are accompanied by a cover sheet or transmittal sheet. The actual format for such sheets varies widely, but they should include:

Hints for fax formatting

1. The name and fax number of the receiver
2. The name and fax number of the sender
3. The number of pages sent, including the cover sheet
4. The name and phone number of the person to contact if the fax isn't satisfactorily sent

Figure 22.5 shows a sample cover sheet.

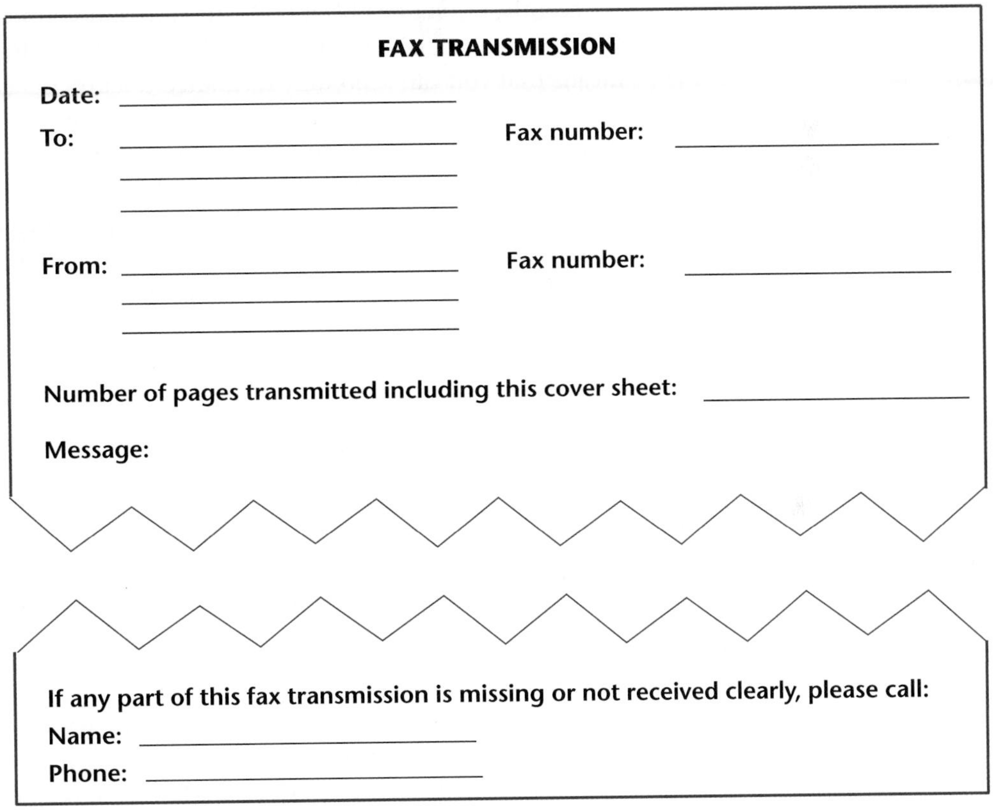

FAX TRANSMISSION

Date: _____

To: _____   Fax number: _____
    _____
    _____

From: _____   Fax number: _____
      _____
      _____

Number of pages transmitted including this cover sheet: _____

Message:

If any part of this fax transmission is missing or not received clearly, please call:

Name: _____

Phone: _____

Figure 22.5   A Fax Transmission Cover Sheet

ELECTRONIC MAIL

Perhaps the most widely used application on the Internet is electronic mail, known simply as e-mail. As early as 1997, the Electronic Messaging Association estimated that Internet users sent more than 2 trillion messages that year. Who knows what that figure might be this year?

E-mail carries routine, day-to-day communication and connects users to discussion forums on listservs and usenets. E-mail transmits an electronic document via networked computer terminals to recipients in the same building or across the globe. Specific codes direct the message to any electronic mailbox designated by the sender, or to all the mailboxes on a mailing list. Alerted by an audio signal or on-screen message, the recipient opens the on-screen mailbox, reads the message, and then either responds, files the message, prints it out, forwards it, or deletes it. E-mail messages can be exchanged instantly or at the convenience of the communicating parties.

E-MAIL FORMAT

In effect, e-mail is a new hybrid of written correspondence and one-on-one conversation. Most e-mail messages are relatively casual in tone and format, reflecting the democratic, "free" nature of the medium. However, as uses for e-mail increase and as software becomes more sophisticated, e-mail messages are starting to look more formal and writers are taking more care with their phrasing, especially in business settings. Until recently, e-mail has been best suited to simple messages. Now, however, new software with enriched formatting and the ability to attach more complex documents means that you can send fully formatted documents via e-mail.

In late 2001, Mykon Communications surveyed the e-mail practices of 45 Canadian organizations (engineering firms; forest products companies; financial firms; educational institutions; service industries; manufacturers; retailers; high-tech firms; environmental consultants; federal, provincial, and municipal government agencies; biotech firms; and others). The survey revealed a growing trend toward formal e-mails. Indeed, e-mail format seems to have evolved into three main levels of formality:

◆ *Personal, brief notes,* similar to the unstructured, casual messages sent in the early stages of e-mail development. These notes, often carelessly written and containing spelling errors, are fine for chatting but are not acceptable for business and professional messages.

◆ *Memo style.* Internal e-mail messages tend to look like standard, informal memos, with carefully constructed block-format paragraphs. Shorter internal e-mails tend to look less formal than the ones sent externally, especially if the message is only one or two paragraphs. However, longer internal e-mails sometimes take on characteristics of formal memos and letters: bulleted or numbered lists, bolding and italic text, spell-checked text, and even bolded headings.

Some internal e-mails even use salutations ("Hello," "Good afternoon," "Greetings") and complimentary closes ("Regards," "Best regards," "Cheers," "Sincerely"). Still, 84 percent of respondents to the Mykon survey said that they and their fellow workers tend to send less formal-looking e-mails for internal messages.

◆ *Letter style.* Formal business e-mails sent to clients, service providers, or associates often have the characteristics of formal business letters—bulleted or numbered lists, headings, bolding, and italics help clarify meaning and impart a professional look.

Respondents to the Mykon survey indicate they use salutations: 40 percent use formal greetings such as "Dear Mr...," while 60 percent said that they use less formal salutations at least part of the time. The formal address is used most often in making an initial business contact. Letter-style e-mails also use complimentary closes—70 percent of the survey respondents regularly use such closes as "Regards," "Best regards," and "Sincerely."

Alternatively, external e-mails are used to introduce the reader to attached formal documents (in Word, WordPerfect, or PDF formats). PDF is particularly favoured by engineering firms and others who want to prevent readers from altering an attached document's findings and recommendations.

GUIDELINES

for E-mail Format

1. *Use a clear subject line to identify your topic* ("Subject: Request for Beta Test Data for Project # 16"). This line helps recipients decide whether to read the message immediately, and helps the recipient file and retrieve the message for later reference.

2. *Refer clearly to the message to which you are responding* ("Here are the Project 16 Beta test data you requested on October 10").

3. *Choose the degree of formality that reflects your reader and your purpose;* the more important the message, the more formal the format. Short, abrupt messages such as the following do not require salutations or closings: "The attached file summarizes the tender process you requested. If you'd like to discuss the process, please call."

4. *Use block format.* Don't indent paragraphs. Keep paragraphs and sentences short.

5. *Where appropriate in formal e-mails, use graphic highlighting* (headings, bullets, numbered lists, bolding, and italics) to improve readability and impart professionalism.

6. *Where appropriate, use formal salutations and closings* ("Dear Sir," "Sincerely"), but use less formal greetings and closings for most e-mails ("Hello," "Regards").

7. *Close with a signature section that names you, your company, or department;* your telephone and fax number; and any other information the recipient might consider relevant. If you have the technology, include your electronic signature.

8. *Do not write in FULL CAPS,* unless you want to SCREAM at the recipient!

Figure 22.6 illustrates a note-style e-mail, and Figure 22.7 shows a letter-style e-mail. Figure 22.8, on page 550, illustrates an e-mail inquiry.

E-MAIL BENEFITS

Compared with phone, fax, or conventional mail (or even face-to-face conversation, in some cases), e-mail offers benefits:

E-mail facilitates communication and collaboration

♦ *E-mail is fast, convenient, efficient, and relatively unintrusive.* Unlike conventional mail, which can take days to travel, e-mail travels instantly. Although a fax network can transmit printed copy rapidly, e-mail eliminates paper shuffling, dialing, and a host of other steps. Moreover, e-mail makes for efficiency by eliminating "telephone tag." It is less intrusive than the telephone, leaving the choice of when to read and respond to a message entirely up to the recipient.

    Thus, it is hardly surprising that in the Mykon e-mail survey, 78 percent of respondents said that e-mail messages have replaced at least half of their organization's letters and memos. Of those same respondents, 60 percent said that e-mails have replaced at least 50 percent of their phone calls.

Reply Reply All Forward 🖨 🗐 ✖ ⇧ ⇩ Follow Up A ▾

This message was sent with high priority.

From: **Marie Szarbynski**

To: Markus Anstenson
Sent: February 11, 2004   11:13 AM
Subject: **Updates for PURE Anti-virus Software**

Please see the forwarded enquiry from customer Bart Nickel. I can't answer his questions until I know which version he's purchased. Will you please check the registration records to see which version he has and when he registered it? If he has Version 4.0, can you tell me why he would have received an updated message?

If possible, I'd like to respond to his enquiry before noon, so your prompt reply would be appreciated.

Marie

Figure 22.6 A Note-style E-mail

- *E-mail is democratic.* With few exceptions, e-mail messages appear as plain print on a screen, with no special typestyles, fancy letterheads, paper design, or paper texture—enabling readers to focus on the message instead of the medium. E-mail also allows for transmission of messages by anyone at any level in an organization to anyone at any other level. For instance, the mail clerk conceivably could e-mail the company president directly, whereas a conventional memo or phone call would be routed through the chain of management or screened by administrative assistants (Goodman 33–35). In addition, people who are ordinarily shy in face-to-face encounters may be more willing to express their views in an e-mail conversation.
- *E-mail can foster creative thinking.* E-mail dialogues involve give-and-take, much like a conversation. Writers feel encouraged to express their thoughts spontaneously, thinking as they write, without worrying about page design, paragraph structure, perfect phrasing, or the like. The focus is on conveying your meaning to the recipients, who in turn will respond with thoughts of their own. This relatively free exchange of views can lead to all sorts of new insights or ideas (Bruhn 43).
- *E-mail is excellent for collaborative work and research.* Collaborative teams keep in touch via e-mail, and researchers contact people who have the answers they need. Especially useful for collaborative work is the e-mail function that enables documents or electronic files of any length to be attached and sent for downloading by the receiver.

E-MAIL PRIVACY ISSUES

Gossip, personal messages, or complaints about the boss or a colleague—all might be read by unintended receivers. Employers often claim legal right to monitor *any* of their company's information, and some of these claims can be legitimate:

🗨 Reply   🗨 Reply All   📄 Forward   🖨 📑 ✖   ⇧ ⇩   📨 Follow Up   A ▾

⇧ This message was sent with high priority.

From: Marie Szarbynski [custserv@pure.com]

To: Bart Nickel [bartn@peg.net]
Sent: February 11, 2004 11:48 AM
Subject: Your Enquiry about PURE Anti-virus Software Updates

Dear Mr. Nickel,

Thank you for purchasing our PURE anti-virus software. We've checked our records, from which we've learned that you purchased Version 3.0; since its release in March 2001, we've released Version 4.0, on October 1, 2003.
Future Computers should have sold you Version 4.0, which features automatic updates sent through the Internet; we asked all our retailers to discontinue Version 3.0 as of October 1, 2003.

Version 3.0 V.S. Version 4.0
I can answer your question, as follows:
• Version 3.0 does not contain automatic Internet updates. It does, however, contain date-specific prompts to maintain coverage against new viruses; to do so, you e-mail PURE (as the prompt directs) and you will be shown how to upgrade your virus protection. Unfortunately, Version 3.0 is no longer capable of handling some of the new types of viruses, and is not compatible with Version 4.0's software platform.
• Version 4.0 automatically updates your computer's anti-virus system through your computer's Web connection, whenever new viruses are detected. That service is free for your first year of use and $10 for each succeeding year.
• Version 4.0, when updated in your computer, detects and eliminates all new viruses (and all previously detected viruses) as they arise.
• Our software engineers have devised master detection software that continuously searches for new viruses on the major networks (Netscape, AOL, etc.).

Your Options
We have advised Future Computers to accept your Version 3.0 for full credit toward Version 4.0, if you would like to do that. The suggested retail price for Version 4.0 is $59.95, plus taxes. We recommend you upgrade to Version 4.0; Version 3.0 can detect all of the viruses that were around before mid-2003. However, Version 3.0 is incapable of handling some especially new nasty viruses that seem to have been designed to bypass the previous standard for anti-virus software platforms. Also, Version 4.0 provides up-to-the-minute protection against new viruses.

In future, if we must change Version 4.0s platform, you'll be able to get full credit for your old version when you choose the new one. Please contact me if you have any other questions.

Sincerely,
Marie Szarbynski

Customer Service Representative
PURE Software Inc.
1125 B Gordon Road

Figure 22.7   A Letter-style E-mail

Monitoring of
e-mail by an
employer is illegal

In some instances it may be proper for an employer to monitor e-mail, if it [the employer] has evidence of safety violations, illegal activity, racial discrimination, or sexual improprieties, for instance. Companies may also need access to business infor-mation, whether it is kept in an employee's drawer, file cabinet, or computer e-mail. (Bjerklie "E-Mail," 15)

E-mail privacy can be compromised in other ways as well:

E-mail offers
no privacy

◆ Everyone on a group mailing list—intended reader or not—automatically receives a copy of the message.
◆ Even when deleted from the system, messages often live on for years, saved in a backup file.
◆ Anyone who gains access to your network and your private password can read your document, alter it, use parts of it out of context, pretend to be its author, forward it to whomever, plagiarize your ideas, or even author a document or conduct illegal

activity in your name. (One partial safeguard is encryption software, which scrambles the message, enabling only those who possess the special code to unscramble it.)

E-MAIL QUALITY ISSUES

Free exchange via computer screen adds to the *quantity* of information exchanged, but not always to its *quality*. As the following examples show, information quality can be compromised by e-mail communication:

E-mail does not always promote quality in communication

◆ The ease of sending and exchanging messages can generate overload and junk mail—a party announcement sent to 300 employees on a group mailing list, or the indiscriminate mailing of a political statement to dozens of newsgroups (*spamming*).

GUIDELINES

for
**Using E-mail
Effectively**

Recipients who consider an e-mail message poorly written, irrelevant, offensive, or inappropriate will only end up resenting the sender. These guidelines offer suggestions for effective e-mail use.

1. *Check and answer your e-mail daily.* Like an unreturned phone call, unanswered mail annoys people. If you're really busy, at least acknowledge receipt and respond later.

2. *At work, consider e-mail to be correspondence, not conversation.*

3. *Use e-mail to reach people quickly.* Most e-mails contain relatively brief, informal messages, but you can also use e-mail to convey serious messages in some depth.

4. *Don't use e-mail to send confidential information.* Avoid complaining, evaluating, criticizing, or saying anything private.

5. *Don't use the company e-mail network for personal correspondence or for anything not work related.* Increasingly, employers monitor their e-mail networks.

6. *Check your distribution list before each mailing* to be sure the message reaches all intended primary and secondary readers but no unintended ones.

7. *Assume your e-mail correspondence is permanent and could be read by anyone anytime.* **Electronic mail is not a private medium.** Also, ask yourself whether you've written anything you couldn't say to that person face to face. Avoid spamming (sending junk mail) and flaming (making rude remarks).

8. *Before you forward an incoming message to other recipients, obtain permission from the sender.* Assume that everything you receive is the sender's private property.

9. *Limit your message to a single topic.* Remain focused and concise. (Yours may be just one of many messages confronting the recipient.) Don't ramble.

10. *Limit your topic to a single screen, if possible.* Don't force recipients to scroll needlessly.

11. *Carefully check the spelling, grammar, and tone.* Even short messages reflect your image.

- ◆ Some electronic messages may be poorly edited and long-winded.
- ◆ Off-the-cuff messages or responses might offend certain recipients. E-mail users often seem less restrained about making rude remarks (*flaming*) than they would be in a face-to-face encounter.
- ◆ Recipients might misinterpret the tone. *Emoticons* or *Smileys*, punctuation cues that signify pleasure :-), displeasure :-(, sarcasm ;-), anger :-< and other emotional states, offer some assistance but are not always an adequate or appropriate substitute for the subtle cues in spoken conversation. Also, common e-mail abbreviations (FYI, BTW, HAND—which mean "for your information," "by the way," and "have a nice day") might strike some readers as too informal.

GUIDELINES

for Choosing E-mail versus Paper or the Telephone

E-mail is excellent for reaching a lot of people quickly with a relatively brief, informal message. But there are often good reasons to put something on paper, or to speak directly with the recipient. Consider the following guidelines.

1. *Don't use e-mail when a more personal medium is preferable.* Sometimes an issue is best resolved by a phone call, or even by voice mail. In those circumstances, e-mail might imply that the sender can't be bothered to speak directly with the recipient.

2. *Don't use e-mail for a detailed discussion.* In contrast to a rapid-fire e-mail message, preparing a paper document is more deliberate, giving you a chance to choose words carefully and to revise. Also, a well-crafted letter or memo is likely to be read more attentively. For in-house recipients, and others who can easily receive and print your form of word-processed documents, attach your paper document to an introductory e-mail. By doing so, you include the best features of both media.

INTERPERSONAL ELEMENTS OF WORKPLACE CORRESPONDENCE

In addition to presenting the reader with an accessible and inviting design, effective correspondence enhances the relationship between writer and reader. Various interpersonal elements forge a *human* connection, as the following advice explains.

Focus on Your Reader's Interests: The "You" Perspective.   In speaking face to face, you unconsciously modify your statements and expression as you read the listener's signals: a smile, a frown, a raised eyebrow, a nod. In a telephone conversation, a voice provides cues that signal approval, dismay, anger, or confusion. Writing a letter, memo, fax, or e-mail, however, carries a major disadvantage; you can easily forget that a flesh-and-blood person will be reacting to what you are saying—or seem to be saying.

Workplace correspondence displaying a "you" perspective subordinates the writer's interests to those of the reader. In addition to focusing on what is important to the reader, the "you" perspective conveys respect for the reader's feelings and attitudes.

In Brief SPAMMING

Spam is defined as "unsolicited e-mail, sent en masse and often promoting an electronic sex site, pyramid-type scheme, or niche product."[3] By clicking onto e-mail lists, spammers can send thousands, even millions of e-mails simultaneously. Spam can be very annoying; worse, the large potential volume of simultaneous e-mails can crash Internet service provider (ISP) systems.

Legislation may eventually make it possible to convict illegal purveyors of spam, but other measures are needed to protect against receiving such messages. Simon Tuck offers the following advice for individual Internet users (D1, D5):

- *Lie low.* Industry officials say the best way to avoid spam is to stay off the spammers' lists by keeping your e-mail address to friends and business contacts. Spammers use automated programs to reap addresses from chat rooms and electronic bulletin boards.
- *Filters.* Use filters that recognize and eliminate spam before it reaches your e-mail account. You can filter out messages with certain words such as "sex" or "rich" or go even further and set up an account that accepts mail only from a predetermined set of addresses.
- *Code names.* Use a different on-line identity from your e-mail address. That way you can protect your privacy from stalkers and fool the spammers' programs that dig for addresses.
- *No directories.* Don't list yourself in the member directory of your ISP, which can be used like a telephone book by spammers.
- *Gripe.* Complain about spam and spammers to your ISP and make sure it does something about your complaint.
- *Avoid bombs.* Industry officials say "mail bombings"—sending a huge number of e-mails to crash a computer system—only provokes spammers into taking worse actions.
- *Don't respond.* Some spams ask you to send a return e-mail to get yourself off their list. Some mass e-mailers are making a genuine offer—but others use the response to confirm that they've found a live address.

3. Source: Simon Tuck "Canning Spam," page D1 and page D5, Globe and Mail, Thursday, October 15, 1998.

To achieve a "you" perspective, put yourself in the place of the person who will read your correspondence; ask yourself how the reader will react to what you have written. Even a single word, carelessly chosen, can offend. In writing to correct a billing error, for example, you might feel tempted to say this:

A needlessly offensive tone

> Our record keeping is very efficient and so this obviously is your error.

Such an accusatory tone might be appropriate after numerous failed attempts to achieve satisfaction on your part, but in your initial correspondence it will alienate the reader. The following is a more considerate version:

A tone that conveys the "you" perspective

> If my paperwork is wrong, please let me know and I will send you a corrected version immediately.

Instead of indicting the reader, this second version conveys respect for the reader's viewpoint.

Use Plain English. Workplace correspondence too often suffers from *letterese,* those tired, stuffy, and overblown phrases some writers think they need to make their communications seem important. Here is a typically overwritten closing sentence:

Letterese

> Humbly thanking you in anticipation of your kind cooperation, I remain
> Faithfully yours,

Although no-one *speaks* this way, some writers lean on such heavy prose instead of simply writing this:

Clear phrasing

> I will appreciate your cooperation.

Here are a few of the many old standards that are popular because they are easy to use but that make correspondence unimaginative and boring:

Letterese	Plain English
As per your request	As you requested
Contingent upon receipt of	As soon as we receive
I am desirous of	I want, I would like
Please be advised that I	I
This writer	I
In the immediate future	Soon
In accordance with your request	As you requested
Due to the fact that	Because
I wish to express my gratitude	Thank you

Be natural. Write as you would speak in a classroom or office.

Anticipate Your Reader's Reaction. Like any effective writing, good correspondence does not just happen. It is the product of a deliberate process. As you plan, write, and revise, answer these questions:

1. *What do I want the reader to do, think, or feel after reading this correspondence?* (Do you want him or her to offer you a job, give advice or information, answer an enquiry, follow instructions, grant a favour, enjoy good news, accept bad news?)
2. *What facts will my reader need?* (Do you need to give measurements, dates, costs, model numbers, enclosures, other details?)
3. *To whom am I writing?* (Do you know the reader's name? When possible, write to a person, not a title.)
4. *What is my relationship to my reader?* (Is the reader a potential employer, an employee, a person doing a favour, a person whose products are disappointing, an acquaintance, an associate, a stranger?)

Answer those four questions *before* drafting correspondence. After you have a draft, answer the following three questions, which pertain to the *effect* of your correspondence. Will readers be inclined to respond favourably?

5. *How will my reader react to what I've written?* (With anger, hostility, pleasure, confusion, fear, guilt, resistance, satisfaction?)
6. *What impression of me will my reader get from this correspondence?* (That you are intelligent, courteous, friendly, articulate, pretentious, illiterate, confident?)
7. *Am I ready to sign my correspondence with confidence?* (Think about it.)

Send correspondence only when you have answered each question to your satisfaction.

Decide on a Direct or Indirect Plan. The reaction you anticipate from your reader should determine the organizational plan of your correspondence: either *direct* or *indirect*.

- ◆ Will the reader feel pleased or neutral about the message?
- ◆ Will the message cause resistance, resentment, or disappointment?

Each reaction calls for a different organizational plan. The direct plan puts the main point right in the first paragraph, followed by the explanation. Use the direct plan when you expect the reader to react with approval or when you want the reader to know immediately the point of your letter (e.g., in good-news, enquiry, or application letters—or other routine correspondence).

If you expect the reader to resist or to need persuading, consider an indirect plan. Give the explanation *before* the main point (as in refusing a request, admitting a mistake, or requesting a pay raise). An indirect plan might make readers more tolerant of bad news or more receptive to your argument.

Whenever you consider using an indirect plan, think carefully about its ethical implications. Never try to deceive the reader—and never create an impression that you have something to hide.

STRUCTURES FOR WORKPLACE CORRESPONDENCE

Trial and error, along with perceptive analysis of readers and purposes for letters, memos, and e-mails, will allow you to find the best ways to organize your workplace correspondence. The best structure for each message will depend on the circumstances. You can start by using the audience/purpose analysis form to choose the content of the message. Then you can adapt the action structure described in Chapter 21 for virtually any letter, memo, or e-mail you write.

Let's review that structure:

1. *Action opening:* Connect with the reader's interest and (where appropriate) summarize the full message in one or two sentences.
2. *Background:* Provide any information the reader needs to understand the main message that follows. Many letters and memos don't need this section; in others, you'll be able to fit a one-sentence background overview into the introductory paragraph.

3. *Details:* Present the main message. (Answer the reader's questions; describe your idea; present your information and analysis; argue your case; explain your point of view.)
4. *Action Closing:* Request the desired action from your reader and/or describe the action you intend to take. Provide information that the reader needs to perform the requested action.

Now let's see some examples of how that structure can work for common workplace messages. The scope of this book does not allow coverage of every kind of letter or memo you might eventually write in your working life, but a careful study of the following examples will help you learn how to adapt the action structure to many other kinds of messages.

Notice that some messages require a *direct approach:* the letter or memo's main point or bottom line is provided in the opening paragraph because you've realized that's what the reader wants, and you've concluded that your own purpose will not be jeopardized by doing so. Other messages require an *indirect plan,* where the opening paragraph previews the document's structure and the main point comes at the end of the document.

The enquiry described in Table 22.1 (and later illustrated by Figure 22.8) employs a direct pattern; the reader will want to immediately know the message's main purpose. Then, seeing the potential for business, the reader will be motivated to read carefully.

Table 22.1 Enquiry (Letter or E-mail) Structure

Section	Situation: Enquiry about updates for PURE anti-virus software Reader: PURE Inc. Client Services Department
Action Opening	State purpose for writing/general nature of enquiry (paragraph 1)
Background	Say how long you've had the PURE software installed; say whether you purchased a CD-ROM version or downloaded it from the Internet (paragraph 2)
Details	Ask detailed questions about the updates: ◆ range of virus types detected ◆ methods of receiving updates/frequency of updates ◆ company's main strategy for discovering viruses ◆ costs of updates/types of payment accepted by PURE (paragraph 3 uses numbered or bulleted list)
Action Closing	Provide e-mail address and phone number; ask PURE rep. to respond; give deadline for response, if that's an issue (paragraph 4)

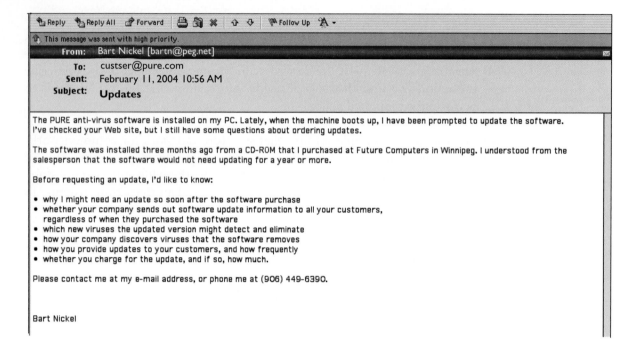

Figure 22.8   A Sample E-mail Enquiry

Figure 22.8 illustrates the phrasing that might be used for the message that Table 22.1 outlines. The situation outlined in the following table *also* uses a direct pattern: stating the requested action and its contribution to the *reader's* goals will get the reader's attention and help hold that attention throughout the memo.

Table 22.2   Request Memo Structure

Section	Situation: Request for funds to design a company website Reader: General Manager Writer: Sales Representative
Action Opening	The main action that needs to be taken and the primary reason for that action, connect with reader's interests relative to this action <div align="right">(paragraph 1)</div>
Background	Circumstances leading to this situation or events affecting the situation <div align="right">(paragraph 2)</div>
Details	Detailed description of what is being requested or needs to be done: ◆ staffing required ◆ equipment and software required and whether the company currently has these requirements in inventory ◆ time lines and schedule ◆ costs (one or more paragraphs)
Action Closing	Request the specific desired action or reader's approval for the writer's proposed action (final paragraph)

Figure 22.9, on page 551, illustrates the phrasing that might be used for the message outlined above.

MEMO   **Magnum Mine Machinery**

DATE: June 23, 2004

TO: Kordell Dobson, General Sales Manager

FROM: Marc Bessier, Sales Representative   *M B*

RE: **Company website**

In Monday's monthly sales meeting, you mentioned the possibility of using e-commerce as a way of building our market base. Recent developments in my sales territory reinforce the need to establish a company website as a first step toward e-commerce.

Background

My territory primarily consists of mining operations in northern Ontario and northern Manitoba. Because many of the mines are quite remote, they use the Internet to keep in touch with the world and to shop for equipment and supplies. I've been using the mail, phone calls, and fax messages to send product information and answer their enquiries, but I'm on the road 7 days out of 10, so it's difficult for me to respond promptly.

Several purchasing officers have told me that they'd prefer to go to the Internet for product information, prices, and availability. And they'd like to order on the Net, too.

Action Required

Our chief competitor, Allan-Price, has set up a home page. There's not much on it right now, and it's not very well organized, but Allan-Price has a presence on the Web, as Marco Corrazini of Canway's Musquean Mine pointedly told me on the phone yesterday. I think we have to keep up with Allan-Price.

What will establishing a website require? I called a friend at Merced Industrial Machines to learn how his company set up its top-quality site. He told me that some of the Merced head office people had computer experience, so they tried to do the work themselves. In the end, though, they had to call in a consultant, who charged about $5,000 to design and build the site. In addition, Merced purchased about $1,500 worth of software. And it continues to pay a part-time Webmaster $400 per month to update and troubleshoot the site.

I contacted that same consultant, June Paschke, yesterday. She has three years' experience as a Web-page designer. She said that building a site for us would take about the same amount of time (10 working days) and cost about the same as the Merced contract. Of course, she would have to meet with us before presenting a detailed proposal of her work plan and fees.

Authorization

May I arrange a meeting next Monday for June Paschke to discuss our needs and her solutions with you, me, and our other three regional sales representatives? We'll all be in town for the AGM. Also, may I meet with you to discuss my possible involvement in the project? I have a special interest in a website project because of its potential for building business. I'll be in Cochrane and Kapuskasing for the next two days, but I'll check frequently for messages on my pager, 689-4352.

Figure 22.9   A Request

Next, the following structure for a claim letter (also know as an adjustment letter or a complaint letter) also uses a direct pattern. The writer immediately gets to the point.

Table 22.3 Claim Letter (or E-mail) Structure

Section	Situation: Claim for damages Reader: Manager, Shipping Department, Genesis Computer Systems Co. (an Edmonton wholesaler) Writer: Home-based computer consultant, Fort McMurray, Alberta
Action Opening	State the problem or the action requested (in this case, claim for a system damaged in shipment because of poor packing) (paragraph 1)
Background	Name the original order no., the packing slip no., the invoice no., and the method and date of shipment (paragraph 2)
Details	Details of the claim: ◆ Describe the nature and extent of the equipment damage and estimate the damage. ◆ Refer to how the damaged equipment is being returned to the supplier. (paragraph 3) ◆ Describe the loss of business (the customer went to another dealer). ◆ Refer to enclosed documents.
Action Closing	Close politely, but firmly—say what you expect to be done, and confirm that you desire continued business relations with the reader (paragraph 4)

As the preceding tables show, the direct, four-part action structure can be adapted for a variety of situations. Some of those situations, such as a refusal of a request, will require an indirect pattern. Table 22.4 presents a suggested structure for refusing a request.

When Kordell Dobson writes to refuse Marc Bessier's request (Figure 22.10 on page 554), note the phrasing Kordell uses.

LENGTH OF WORKPLACE CORRESPONDENCE

All letters and memos should be as short as you can make them. Restrict them to one page, if you can. Although one-page messages are not always possible—see several samples in Chapter 21, for example—you can restrict some messages to one page by placing the main message in a cover letter and the remaining details in enclosures. For instance, a complex order for supplies might be organized in order sheets enclosed with a cover letter which:

- authorizes the purchase, designates the delivery method, and names the source of your information about the supplies
- quickly summarizes the nature of the order and refers to the enclosed order "forms" (which provide details of items, quantities, order numbers, prices, taxes, shipping, and overall costs)
- closes by saying how you will pay and when you expect delivery

Table 22.4   Refusal Message Structure

Section	Situation: Respond to request for funds to design a company website Reader: Sales Representative Writer: General Manager
Action Opening	Express appreciation for the sales rep's initiative; acknowledge that there has recently been discussion of creating a company website (paragraph 1)
Background	Without actually saying "no" just yet, explain the circumstances that preclude creating a website at this time (perhaps the firm is in the process of contracting a Web consultant or joining forces with another firm, or perhaps the company is analyzing its entire marketing strategy) (paragraph 2)
Details	Soften the bad news by: ◆ placing it in the middle of the paragraph ◆ using the passive voice (not, "We cannot grant your request at this time," but, "Funds are therefore not available for this project at this time.") ◆ focusing on the reasons for the refusal, not on the refusal itself (part of paragraph 2, or perhaps a new paragraph)
Action Closing	Thank the reader for his or her commitment and ideas; encourage the reader to pursue an interest in website design; assure the reader that he or she will be consulted when the project is next discussed (final paragraph)

A FINAL WORD

When you write workplace correspondence, write efficiently. Very few one-page messages should take more than 45 minutes to plan, compose, and polish. You will, however, take longer if you have to rewrite major sections of letters or memos. That kind of rewriting can be avoided by clear thinking in the early stages of the process.

So, remember:

◆ Consider your audience's needs, interests, and priorities and the purpose of the document.
◆ Use your audience/purpose analysis to choose the content and arrange that content.
◆ Work from an outline that identifies the content and/or purpose of each paragraph.
◆ Revise the content and structure before writing sentences and paragraphs.

You'll be amazed at how efficiently you'll write.

Also remember that how you use workplace correspondence will affect your credibility within your organization. Your colleagues' and supervisors' perceptions of your worth will be affected by the degree of care and skill you put into your letters, memos, and e-mails.

MEMO   Magnum Mine Machinery

DATE: June 23, 2004

TO: Marc Bessier, Sales Representative

FROM: Kordell Dobson, General Sales Manager

RE: **Your Ideas for a Company Website**

Thank you for your initiative in promoting the creation of our own website. Over the past few months there have indeed been some recent discussions about establishing a presence on the Web.

I'm going to be in Montreal until next week, so I'm faxing this memo to your hotel in Timmins. Next week my preparations for the AGM and my participation in the AGM's follow-up meetings will occupy my time.

We are in negotiations to merge with a Western Canadian firm, in order to give both firms a national presence. (I'm sure you've heard the rumours.) If indeed that merger happens, we will combine our strategies and resources with that firm on many fronts. In the meantime, all new projects are on hold, including the creation of new marketing vehicles and e-commerce. We haven't discussed the possibility of establishing a website with our potential merger partner; too many other issues need to be resolved first.

I appreciate your commitment to the firm's goals. Please continue your interest in Web-page design, so that you can contribute to future discussion and actions in this exciting method of communicating with our present and future customers. Perhaps we can grab a few minutes to discuss your ideas, after the CEO's address on Monday.

Kordell

Kordell Dobson

Figure 22.10   A Refusal of a Request

EXERCISES

1. Design a letterhead and a memohead for a business that you would like to start.

2. Collect samples of workplace writing—letters, memos, and e-mail messages. In those samples, look for instances of letterese, clichés, and wordy and unclear phrasing. Improve the phrasing where necessary.

3. Check the tone of memos and letters that you examine. Do you see phrasing that might antagonize the reader? Do you see phrasing that places the writer's needs and interests ahead of the reader's? If so, re-phrase.

4. Examine workplace letters and memos to see if they use a variation of the four-part action structure. (See pages 548 to 552 and Chapter 21.) If the action structure has not been used for a given message, would the message be more effective if it were restructured?

5. Monitor your next composition of a one-page letter, memo, or e-mail. Record how long you take to complete each stage of the writing process. (See Chapter 3.)

COLLABORATIVE PROJECT

In conjunction with other members of your class, survey correspondence practices of businesses in your part of Canada.

Ask about:

◆ which page format (block, full-block, simplified) the business prefers its employees to use, and whether those employees are encouraged to use attention lines and/or subject lines

◆ which addressing format is favoured by the business

◆ whether the business allows its employees to use abbreviations in external messages

Also, ask for a sample letter and a sample internal memo.

Members of your group could conduct the survey in person. If so, send a copy of the survey questions with an accompanying letter first, and then telephone for an appointment. Emphasize that information in the sample letters and memos will be held confidential. Also offer to make a copy of the survey's statistical results available to each participating business.

Alternatively, the entire survey could be conducted by mail.

Note: All letters sent during the survey should be approved by an editing team appointed by the class.

CHAPTER 23

Job-search Communications

Technology has added new tools for job seekers, but today's job searches require the same basic approach and communication skills that have been used for some time. This chapter discusses seven steps in a successful job search:

1. Assessing what you have to offer
2. Identifying which kinds of jobs to pursue
3. Researching the job market
4. Producing résumés and other job-search documents
5. Devising a campaign strategy of contacting employers
6. Performing well in interviews
7. Responding to job offers

Being informed about careers and specific positions in your field, preparing to meet professional requirements, and then following a systematic job-search strategy will increase your future *job satisfaction.*

SELF-ASSESSMENT INVENTORY

An objective, organized self-appraisal

Assessing what you have to offer employers is not easy to do, especially if you have not received detailed, accurate job performance appraisals from previous employers. Self-assessment is even more difficult for those of you who have had little job experience to use as a basis for evaluating your skills and aptitudes. Assessing your package of employable characteristics requires objectivity and an organized approach. Fortunately, tools such as the self-inventory shown in Figure 23.1 help provide that structured objectivity.

Apparent to Audience	**Subtle or Not Obvious**
1. Strengths	2. Strengths
3. Limitations	4. Limitations

Figure 23.1   A Self-inventory

COMPLETING A SELF-INVENTORY

The self-inventory chart's four sections (quadrants) can be completed in any sequence. The chart can be very useful if you fill each quadrant with lists of your transferable skills (see next section), work habits, attitudes, and/or knowledge, which make up the full package you bring to an employer. You can use this chart to choose content for your résumés, or to help write an application letter, or to prepare for a job interview.

Readily apparent to an employer

Quadrant 1.   Quadrant 1, "apparent strengths," features those qualities that an employer would discover by reading your résumé, interviewing your references, or by asking you certain questions in an interview. For example, you may have had several years of work experience in which you developed skills required by the position you're applying for. Or, your college major may have developed skills required by the position. By highlighting such strengths in your résumé and cover letters, you can create the image you want to project. Reviewing quadrant 1 will also help you prepare your strategy for job interviews.

Not necessarily apparent

Quadrant 2.   The "subtle strengths" in quadrant 2 will be more difficult to project to employers. These strengths may include such qualities as your loyalty, capacity for long hours and hard work, or desire to succeed in your career. Such qualities are difficult to build into traditional (reverse chronological) résumés, but you could use some of the following techniques to highlight your subtle strengths:

- Start your résumé with a profile section that summarizes the main positive features you bring to a particular position.
- Use a functional résumé or hybrid résumé that includes sections of relevant skills and attitudes.
- Attach reference letters that mention your less obvious strengths.
- Refer to your "quiet strengths" in your application letter.

Readily apparent— will concern an employer

Quadrant 3.   Quadrant 3 includes those characteristics that you and the employer will readily identify as potential limitations on your ability to do the job. The most common limitation for applicants trying to enter a given field is lack of experience in that area. You can be certain that these limitations will be considered when the employer chooses which applicants to interview, and that the interviewer will either ask about these limitations or will wait to see what you have to say about them.

So, what can you do? Trying to avoid discussing your obvious limitations does not work. You need to directly address the issue by showing that you have indeed developed the skills and understanding needed for the job, through college work placements, realistic class projects, volunteer work, and/or leisure activities. (If you have not in fact developed the requirements for the position, you should not apply!)

Not likely apparent

Quadrant 4.   Quadrant 4 houses those limitations that the employer will not discover unless you blurt them out. You may have an aversion to writing the daily reports required by your job, for example. Or you may be a "night person" who has difficulty getting to work on time in the morning. Or you may not like the customer service component of a position that otherwise appeals very much to you. Or perhaps your perfectionist tendencies bring you into conflict with others.

There are two main issues here, an ethical issue and a practical one. Ethically, you should not present yourself for a position where you are not prepared to work wholeheartedly. You should also not apply for a position if you have attitudes that you are not prepared to change and that you know will hamper your performance in the job. On the other hand, if you are aware of a potential limitation and you are in the process of overcoming the problem, there may be no need to reveal the problem when the interviewer asks you to comment about your limitations.

YOUR TRANSFERABLE SKILLS

Employers look for these skills

Another very useful tool in determining your employable characteristics is a list of the transferable skills that you have previously developed. As Figure 23.2 demonstrates, such a list includes proof ("validating experience") as well as the skills themselves.

Skill	Above Average	Average	Below Average	Validating Experience
Programming in Java	✔			• Programming project in CoSci 356—mark: 92% • Co-op work experience at Fundy Industries—good evaluation
Supervising/coaching/ leading	✔			• Coached Jr. lacrosse 3 yrs • Bell captain at Lakeshore Lodge for two summers
Report writing		✔		• Completed 5 major reports at college—avg grade: 70% • Collaborated on documentation report at Fundy Industries—involved in all phases of 95-page report
Keying	✔			• Timed at 64 words per minute
Problem solving	✔			• Won software troubleshooting competition 2 yrs at college

Figure 23.2   A Segment of a Transferable Skills Inventory

Many of the transferable skills valued by employers are communication skills: writing, speaking, listening, reading, and interpersonal. Such skills are featured in the "Critical Skills Required of the Canadian Workforce" identified by the Corporate Council on Education, a program sponsored by The Conference Board of Canada. (See the In Brief box on page 561 for more details.)

PERSONAL JOB ASSESSMENT

The long-term goal for your job search should be job satisfaction, so after you have thoroughly examined what makes you employable, you should choose what kinds of positions to pursue. Before drawing up a list of positions for which to apply, you will benefit from thinking about your life goals and reasons for working. After all, most of us spend a large portion of our lives at work.

YOUR CAREER ORIENTATION

Why do you work?

What motivates you to work? What benefits and satisfactions most attract you to certain types of positions? Knowing the answer to these questions may help you choose your career path and help you in your job-search activities.

Here is a set of categories that reflect what workers hope to "get" from their jobs:

1. *Getting ahead.* This worker looks for advancement within an organization and starts looking elsewhere if the organization doesn't have such opportunities.
2. *Getting rich.* This worker is primarily interested in a position's financial benefits. Even though he or she may find the work stimulating, this person will take less interesting positions if the pay is better.
3. *Getting secure.* Some workers will sacrifice organizational status and high salaries for long-term security.
4. *Getting control.* This employee wants to control his or her work day to reduce stress and to be more productive. Sometimes this person equates "control" with controlling others.
5. *Getting high on work.* This person loves the work he or she does and would likely do it for no pay if a paycheque were not necessary.
6. *Getting balanced.* Many employees stay in jobs that no longer challenge them because they see their work as just one part of their lives; their families and their leisure activities or community activities are as important or more important than their work.

In Brief EMPLOYABILITY SKILLS PROFILE: THE CRITICAL SKILLS REQUIRED OF THE CANADIAN WORKPLACE

ACADEMIC SKILLS

Those skills that provide the basic foundation to get, keep, and progress on a job and to achieve the best results

Canadian employers need a person who can:

Communicate

- Understand and speak the languages in which business is conducted
- Listen to understand and learn
- Read, comprehend, and use written materials, including graphs, charts, and displays
- Write effectively in the language in which business is conducted

Think

- Think critically and act logically to evaluate situations, solve problems, and make decisions
- Understand and solve problems involving mathematics, and use the results
- Use technology, instruments, tools, and information systems effectively
- Access and apply specialized knowledge from various fields (e.g., skilled trades, technology, physical sciences, arts, and social sciences)

Learn

- Continue to learn for life

PERSONAL MANAGEMENT SKILLS

The combination of skills, attitudes, and behaviours required to get, keep, and progress on a job and to achieve the best results

Canadian employers need a person who can demonstrate:

Positive Attitudes and Behaviours

- Self-esteem and confidence
- Honesty, integrity, and personal ethics
- A positive attitude toward learning, growth, and personal health

- Initiative, energy, and persistence to get the job done

Responsibility

- The ability to set goals and priorities in work and personal life
- The ability to plan and manage time, money, and other resources to achieve goals
- Accountability for actions taken

Adaptability

- A positive attitude toward change
- Recognition of and respect for people's diversity and individual differences
- The ability to identify and suggest new ideas to get the job done creatively

TEAMWORK SKILLS

Those skills needed to work with others on a job and to achieve the best results

Canadian employers need a person who can:

Work with Others

- Understand and contribute to the organization's goals
- Understand and work within the culture of the group
- Plan and make decisions with others and support the outcomes
- Respect the thoughts and opinions of others in the group
- Exercise "give and take" to achieve group results
- Seek a team approach as appropriate
- Lead when appropriate, mobilizing the group for high performance

---

Source: This document was developed by the Corporate Council on Education, a program of the National Business and Education Center, The Conference Board of Canada.

In Brief   TODAY'S NEW WORKER

Barbara Moses, a career development specialist, has put a new spin on what she calls "today's new worker." She describes six profiles of "new workers":

1. **Independent thinkers:** Want to own or build their own work; impatient with corporate norms; little allegiance to the company; detest endless meetings.

2. **Lifestylers:** Determined to balance outside interests and responsibilities with career. Motto: "I work to live, not live to work."

3. **Personal developers:** Evaluate their work in terms of whether they're being challenged; will take career risks if they thus develop new skills; identify with their work, not with their employer.

4. **Careerists:** Ambitious; aspire to management roles; motivated by prestige and status.

5. **Authenticity seekers:** Motto: "I gotta be me." Won't sacrifice their own personality in order to play a corporate role; can be creative, but difficult to manage if employer demands conformity to corporate norms.

6. **Collegiality seekers:** Associate strongly with their team or work group and derive much of their identity from belonging to it; what's important is working with people they enjoy; not happy working by themselves.

Source: Barbara Moses, "The Challenge: How to satisfy the new worker's agenda," *Globe and Mail*, p.B15, Nov. 10, 1998.

CHOOSING POSITIONS TO PURSUE

Choosing what types of work to do involves a simple four-step process:

What type of work suits you best?

- ◆ *Step 1.* List the positions you would like to have, the jobs you would like to do. Don't worry about your qualifications for these positions just yet.
- ◆ *Step 2.* List the positions that you could currently handle, based on your self-inventory of employable characteristics and transferable skills. If you're close to completing an educational or training program, you could also list those positions you'll soon be able to handle successfully.
- ◆ *Step 3.* Check which positions are on both lists and organize them on the basis of two criteria: (1) which positions appeal to you the most, and (2) which positions best suit your qualifications. It's important to be realistic at this point, but it's also important to try to identify what you consider ideal employment. You might also list positions for which you'll be qualified after you gain more work experience.
- ◆ *Step 4.* Confirm that you have included the full range of positions open to people with your interests and qualifications. For example, a person with a degree in technical writing might consider working as a Web-page designer and Webmaster, or a B.Sc. graduate with a major in botany might look at developing new varieties of plants for a commercial greenhouse, or a civil engineering technologist might explore the opportunities for a consultant contracting for small towns in a rural region.

The Internet has many sites that help people choose career paths. For example, you could check

- ◆ The Keirsey Temperament Sorter, **at www.Keirsey.com/cgi-bin/Keirsey/ newkts.cgi**
- ◆ The Princeton Review Career Quiz, at **www.review.com/career/careerquiz home.cfm?careers=6**

So, now you have a list of what types of positions to pursue. The next task in your job search is to learn what's available.

JOB MARKET RESEARCH

Again, you need an organized approach:

- ◆ Identify information needs.
- ◆ Identify information sources.
- ◆ Use a variety of research tools to ensure that you find all available positions, including those in the "hidden job market."

IDENTIFY INFORMATION NEEDS

Unfortunately, wanting to hold a particular position does not guarantee that such a position is available. You'll need to research the job market to learn several key things:

Know the job market

- ◆ which of your desired positions are available now, and which might be available in the near future
- ◆ which companies are expanding and thus requiring additional workers
- ◆ which companies have internship programs or will accept "volunteer employees"
- ◆ what qualifications are required for specific positions
- ◆ when and how companies recruit for seasonal employees and/or permanent employees
- ◆ which firms contract some or all of their work, and how you can monitor these firms' ongoing contract requests
- ◆ which business trends and expansion of government or commercial operations can be anticipated for the foreseeable future

Your first research step, then, should be to decide the scope of your information search. In your current job search, do you need to know only which companies are hiring, or will soon be hiring, people like you? Or, are you interested in the full range of information listed in the previous paragraph? Your information requirements will determine which of the following job market information sources you use.

IDENTIFY INFORMATION SOURCES

Know how to research the market

Literally hundreds of information banks can help you find employers and jobs. Some of these will list available positions; others will help you direct your spoken or written enquiries.

The following information sources should provide valuable leads:

Useful information sources

1. **Campus Placement Office.** Most universities and colleges have placement offices. Those institutions that offer co-op work programs provide information about co-op employers; also, evaluation reports written by former co-op students are on file.
2. **Canada Employment Centres** and **Job Futures** publications.
3. **Career Planning Annual** published by the University and College Placement Association provides information on employer-members who recruit at the college and university level.

4. **College Placement Annual** published by the College Placement Council contains a list of companies in both Canada and the U.S. who are seeking college and university graduates.

5. **Canadian Trade Index** provides information about Canadian manufacturers.

6. **Dun & Bradstreet Canadian Key Business Directory** has information about businesses in Canada.

7. **Standard & Poor's Register of Corporations, Directors, and Executives** contains an alphabetical listing of more than 35,000 corporations in Canada and the U.S., detailing products and services, officers, and telephone numbers.

8. **Financial Post's 100 Best Companies** groups companies according to their service or product and includes information on work environment and benefits.

9. **Canadian Miner's Handbook** includes information about the Canadian mining industry.

10. **Lumberman's Green Book** provides information on the forest industry.

11. **Newspaper feature articles** focus on companies and executives, expansion plans, new products, and key appointments.

12. **Trade or professional journals and associations** list opportunities, trade shows, seminars, and meetings that provide opportunities for networking.

13. **Job postings in government publications**.

14. Dozens of **Internet sites** provide information about career choices, labour markets, companies, job postings, and job search skills. Some sites offer services for a fee (and some even provide free services, such as posting résumés). Your school's website will likely link to many employment sites. If not, you might start with some of the following sites:

> **www.monster.com and www.monster.ca**
> **www.kenevacorp.mb.ca**
> **www.headhunter.net**
> **www.jobtrack.com**
> **www.workinfonet.ca**
> **www.careerexchange.com**
> **www.careeredge.org**
> **www.rileyguide.com**
> **www.cacee.com**

In addition to information gleaned from websites such as the ones listed above, you can learn a great deal about some companies by visiting their home pages. Besides telling about their operations, many companies list job openings on their Web pages.

ENQUIRIES

Enquiries can elicit valuable information

Statistics Canada's 1997 survey of job hunting techniques revealed that 69 percent of Canadian job seekers contacted employers directly in the early stages of a job search. Many of those contacts were enquiries about potential openings.

Why Enquire?   You should enquire about potential employment if you don't know if a company will have a position for someone like you **and** if you're not sure whether the company's available position would suit your interests and qualifications. Also, a well-crafted enquiry will tell you which additional skills you will have to develop.

The following advice applies particularly to letter enquiries, but much of the advice could also be applied to telephone and in-person enquiries.

Your main challenge in sending unsolicited enquiries is getting the reader to respond. To meet that challenge, you must do two main things:

How to improve your chances of getting a reply

1. You must show your reader that you have something to offer the company. The best way to do this is to quickly review your qualifications and refer your reader to an enclosed résumé for more details. Remember, though, that you're not applying for a named position in this letter, so do not write a sales pitch. Save that for any application letter that you later send to this same reader.

2. You must make it easy for the reader to respond. Even the kindest, best-intentioned reader will place your enquiry at the bottom of a priority list if a response to your letter requires a lot of time and effort.

Tips.   Here are some tips for reducing your reader's effort and thus increasing the chances of getting a response to your enquiry:

- Use a structure that the reader can easily follow. (See chapter 22, page 548, for details.)
- Place your questions in a numbered or bulleted word list.
- Make sure those questions follow a logical order. (The first logical question for most job enquiries will be whether the employer has, or anticipates having, openings in your field.)
- Make each question absolutely clear. (Show your letter to others before sending it.)
- Ask a reasonable number of questions; five seems to be the limit.
- Do not ask questions that you could easily answer by reading the company's annual report, visiting its website, or phoning the company's personnel office.
- Use appropriate tone and phrasing: positive, assertive, polite, concise, business-like, fresh (no clichés), and energetic (active verbs).
- Make it easy for the reader to respond: Where feasible, offer to phone or visit your reader to get the information.
- Display the "you" attitude by focusing on the nature of the work involved, rather than on how you might benefit.

The enquiry illustrated in Figure 23.3 on page 566 was sent by a civil engineering technology student who was responding to an invitation he had received while on co-op job placement.

Chris Hendsbee's accompanying résumé is shown in Figure 23.9, on pages 578 and 579.

NETWORKING

Surprisingly, in 1997 only 22 percent of Canadian job seekers contacted friends, relatives, and/or former colleagues, students, or teachers as part of a job search. That statistic is slightly higher for better-educated job seekers: In 1997, about 25 percent of university graduates asked friends and relatives to pass on information about job opportunities.

453B Gordon Drive
Kelowna BC   V1W 2T1
January 27, 1999

Mr. Miro Betts
International Survey Systems
1199 – 11th S.W.
Calgary AB   T2R 1K7

Dear Mr. Betts:

In late October, at Greenwood Consultants in Nelson, your demonstration of real-time kinematics further kindled an interest I've developed in GPS surveying techniques. Your invitation to contact ISS this winter has led to this enquiry.

As you may remember, I'm studying civil engineering technology at Okanagan University College. Also, my interest in GPS has prompted me to read widely on the subject. That's partly why I was so enthusiastic about the accuracy of the equipment you demonstrated in Nelson. Now, I'm eager to get a chance to work with a company such as yours.

Consequently, I'm wondering:

1. whether ISS will have an opening for someone like me, this May to September,
2. what the work would entail and what equipment might be used,
3. what qualifications ISS would require, and
4. whether I should upgrade my qualifications to be ready to work for ISS this summer. (Please see the attached résumé for my current qualifications.)

Would it be possible for me to meet with you or another representative to discuss what I might contribute to ISS? I am willing to travel wherever you set up a meeting in Alberta or B.C. You can contact me at my e-mail address, chris@silk.net, or you can leave a phone message at (250) 876-3422, and I'll return your call as soon as I return from class.

Sincerely,

Chris Hendsbee

Chris Hendsbee

Enclosure: Résumé

Figure 23.3   A Sample Job Enquiry

The other three-quarters of job seekers missed a very powerful two-way communication tool. Let's examine the power of networking.

Jenny Roy, age 25, recently graduated from the Environmental Sciences program at the University of Waterloo. After graduation, she landed a four-month contract with Cambrian Consultants, a private firm investigating the link between water quality and fish stocks in Georgian Bay. But that contract ran its course and Jenny had had no luck finding other employment through newspaper ads, Internet sites, environmental journals, or her enquiries to a variety of private and government operations.

In desperation, she turned to several people who had been influential in her life:

- her mother, Adrienne Lavois, who worked as an administrative secretary in the Department of Indian Affairs in Ottawa
- her uncle, Pierre Riley, a retired geologist in Montreal
- her mentor, Dr. Howard Cash, who taught freshwater biology at the University of Waterloo
- a former classmate, Sandy Travers, who worked as a lab assistant at the University of Guelph
- her high school volleyball coach, Nancy St. Jean, who ran a fishing lodge on Lake Temagami

Initially, these five people agreed to watch for possible positions for Jenny. They offered to distribute résumés for her and to talk to their friends, relatives, business acquaintances, and colleagues. At that point, Jenny had five people "searching" for job leads. Figure 23.4 shows how she might have diagrammed her network:

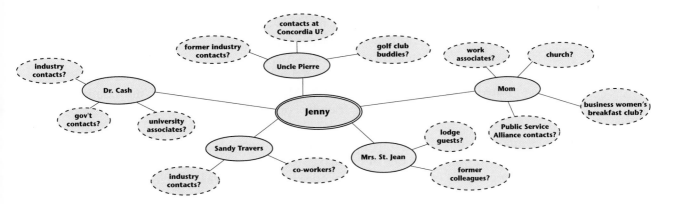

Figure 23.4   A Basic Network

Jenny originally thought only in terms of her five front-line searchers. She didn't anticipate how far the network could develop. Indeed, she didn't know that dozens of strangers would be contributing to her job search. Let's see how just one of the contact lines developed (Figure 23.5):

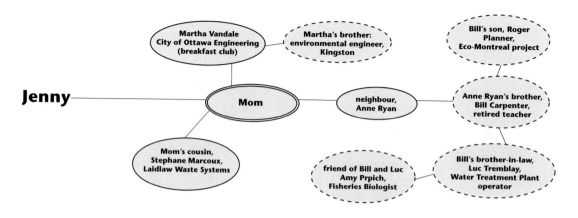

Figure 23.5 An Expanded Network

In addition to Jenny's mother, eight other people became aware of a bright, eager, young person looking for work. For various reasons, all of them were on alert for suitable employment opportunities and were prepared to pass their information along. The amazing thing is that only two of them, Jenny's mother's neighbour and Jenny's mother's cousin, have ever met Jenny! ▲

If you think that the basic assumption behind the networking concept is naive, and if you say that there's no motivation for people to actively seek employment for someone whom they've never met, remember two things:

Why networking works

1. The person farther down the network (Bill Carpenter's brother-in-law, for example) is not really doing a favour for Jenny; he's doing it for his friend (or brother-in-law, in this case).
2. Each person in the network only has to pass on information, which is not a time-consuming or onerous task. Most often, this type of information is passed along during casual, friendly conversation.

Most jobs are not advertised

Jenny's situation, though fictional, is common enough. She can get a tip about an opportunity from any one of the many "researchers" in her network. Thus, she has a strong chance of tapping into the huge hidden job market—the vast majority of available positions are not formally advertised.

To further confirm the power of networking, let's look at a rather dramatic real-life example. The central figure's name has been changed, but the details of his story are accurate.

In 1990, Jim Landon lost his sawmill job when Simpson Timber closed its operation in Hudson Bay, Saskatchewan. Jim had worked at the mill since leaving high school in 1970. In a job-search program sponsored by his employer, Jim learned how to use networking for his job search. It appealed to Jim: He'd used the concept successfully in raising funds for minor hockey in his community.

Basic contact information

Jim Landon

Box 309 Hudson Bay SK
S0D 3T4
Phone: (306) 334-7429

Names positions, but doesn't restrict his chances

<u>Preferred Position</u>
Production line worker, loader, or yard worker in sawmill or plywood mill

Quick summary of relevant experience. Shows potential for adapting to a variety of positions

<u>Experience</u>
20 years at Simpson Timber, Hudson Bay (Mill to close this summer.) Have performed all sawmill production jobs.
Expert in sorting and grading lumber.
Skilled with wide variety of power tools and equipment.

<u>Personal Characteristics</u>
Hard working and conscientious. Loyal.
Eager to earn living for family of four.

Suggests high level of motivation

Will relocate. Willing to work any hours.

Figure 23.6   Snapshot Résumé

Jim tied the networking idea to a rather novel approach: He produced snapshot résumés on 8 cm × 9 cm (3" × 3½") filing cards, as shown in Figure 23.6.

Jim worked part-time pumping gas and diesel fuel at the Co-op station on the Yellowhead Highway, which runs through Hudson Bay. He gave copies of his mini-résumé to truckers on their way from Quebec and Ontario to British Columbia. He asked the truckers to leave the résumés at truck stops in Alberta and British Columbia.

A month after his campaign began, Jim received a call from a mill near Sparwood, British Columbia. The mill manager's brother had found a copy of Jim's card résumé on a bulletin board in Fernie, 29 kilometres down the road from Sparwood. After going to Sparwood for an interview, Jim accepted a job offer. ▲

PRODUCING JOB-SEARCH MATERIALS

Jim Landon's card is just one kind of sales material produced by job searchers. The most common sales tool is called a *résumé*. It provides an objective, organized record of key facts about the applicant and summarizes the applicant's skills.

USES FOR RÉSUMÉS

Résumés accompany letters of introduction, application letters, and enquiries. You can send them by fax or e-mail; you can post e-résumés on the Internet; you can send them as follow-ups to self-introduction telephone calls; and you can leave them at each employer in a campaign of visiting business offices and other places of employment. You can also use a résumé to accompany a business proposal.

PREPARING YOUR RÉSUMÉ

Create more than
one résumé

You may need two or more résumés, depending on the range of positions you decide to pursue. Indeed, it's a good idea to modify your basic résumé to match each position for which you apply. Different positions emphasize certain requirements more than others, so you might place your education first in one résumé and your work experience first in another. A third résumé might expand the amount of detail regarding your volunteer experiences, if you've developed relevant skills through that experience.

Basic Considerations. Employers generally spend 15 to 45 seconds initially scanning a résumé. They look for an obvious and persuasive answer to the question, "What can you do for us?"

Employers are impressed by a résumé that:

1. looks good (conservative, tasteful, uncluttered, on quality paper)
2. reads easily (headings, typeface, spacing, and punctuation that provide clear orientation)
3. provides information the employer needs to decide whether to interview the applicant

Employers generally discard résumés that are mechanically flawed, cluttered, sketchy, or difficult to follow. Don't leave readers guessing or annoyed; make the résumé perfect.

Most résumés organize the information within these categories:

- name and contact information
- job and career objectives
- educational background
- work experience
- personal data
- interests, leisure pursuits, volunteer activities, awards, and skills
- references

Select and organize material to emphasize what you can offer. Don't just list *everything*; be selective. (We're talking about *communicating*, not just delivering information.) Don't abbreviate, because some readers might not know the referent and because other readers may infer that you're lazy and take shortcuts. Use punctuation to clarify and emphasize, not to be "artsy."

Begin your résumé well before you plan to use the document as part of your job search. You'll need time to do a first-class job. Your final version can be duplicated for various similar targets, but each new type of job requires a new résumé that is tailored to fit the advertised demands of the job. Most job seekers find it useful to create two or three different résumés that have different emphases and different lengths, so that these can be quickly modified to suit positions that suddenly appear. **Caution:** Never invent or misrepresent credentials. Your résumé should make you look as good as the facts allow. Distorting the facts is unethical and counter-productive. Companies routinely investigate claims made in résumés; people who have lied have been fired.

Name and Address.   In a heading, include your full name, mailing address, phone number, and e-mail address. Make your name stand out by using a larger type size and by bolding it. If you anticipate an address change after a certain date, include both your current and future addresses and the date of the change. Make it easy for an employer to contact you!

Job and Career Objectives.   Your job market research should give you a good idea of the specific jobs for which you *realistically* qualify. Resist the impulse to be all things to all people. Be prepared to have different statements of objectives to meet the requirements of different job descriptions.

The key to a successful résumé is the image of you it projects. That image should *accurately* reflect your package of qualities and qualifications, but it should also match the position's requirements. For example, the following Career Objective appeared in a résumé submitted to a municipal engineering department by a civil engineering technologist with 11 years' work experience and several courses in business administration.

Career Objective

Project planning and administration for a municipal engineering office, with responsibility for cost-effective solutions to urban traffic and transportation problems.

To save space, you can omit your statement of career objectives and include it in the accompanying letter.

Educational Background.   If your education is more substantial than your work experience, place it first for emphasis. Begin with your most recent school and work backward, listing degrees, diplomas, and schools attended beyond high school (unless the high school's prestige, its program, or your achievements warrant its inclusion). List the courses that have directly prepared you for the job you seek. Where applicable, name co-op work placements or special projects that have helped you develop relevant skills and knowledge.

Work Experience.   If you have solid experience that relates to the job you seek, list it before your education. Begin by listing your most recent experience and then work backward, listing each significant job you've had. Describe your exact duties in each job, indicating promotions. Where applicable, describe skills developed in certain positions.

The standard heading, "Work Experience," might not adequately describe your situation. If that's the case, here are some alternative ways of highlighting your valuable experience:

Use effective, appropriate headings

- ◆ Provide two categories, "Related Experience" and "Other Work Experience." In the first category, include related volunteer experience, as long as it is substantial enough to merit inclusion. In the second category, list those positions not related to the position for which you're currently applying. Include details of non-related work experience to show that you've had significant experience and to highlight such qualities as loyalty, leadership ability, or ability to learn new skills quickly.

◆ Feature a special category called "Volunteer Experience." A person's experience in a service club or other organization will develop skills needed for a particular position, sometimes at a very high level.
◆ Create a separate category called either "Practicum Experience" or "Co-op Experience" if you've had significant work experience through a college work placement.

Personal Data.   An employer cannot legally discriminate on the basis of gender, age, religion, race, national origin, disability, or marital status. Therefore, you aren't required to provide this information. But if you believe that any of this information could advance your prospects, by all means include it.

Personal Interests, Activities, Awards, and Skills.   List leisure activities, sports, and other pastimes. Employers use this information to learn whether you will easily fit into an existing work team. Also list memberships and offices held in teams and organizations. Employers value team skills and leadership potential. If your volunteer experience is not directly related to the position for which you're applying, include that volunteer work in this personal activity section; employers know that people with well-rounded lifestyles are likely to take an active interest in their jobs. But be selective—list only those items that reflect qualities employers seek.

References.   List three to five people *who have agreed to provide* strong, positive assessments of your qualifications and personal qualities. Some of their support will come in the form of requested letters or e-mail responses. Other references will be asked to complete reference forms. Or, you could request reference letters from previous employers, teachers, or others who have closely observed your work habits and skills. Still other reference requests will come in the form of telephone calls from employers who are doing reference checks.

Some employers use references to narrow the list of applicants to a shortlist. Others will contact references after interviews have been conducted. Usually, the motivation for these follow-up calls is to confirm interviewers' perceptions or to investigate aspects that troubled the interviewers during the interview.

Often, then, your references' comments can make or break your application. So, choose those references carefully:

How to choose references

◆ Select references who can speak with authority about your ability and character.
◆ Choose among previous employers, professors, and respected community figures who know you well enough to *comment concretely* on your behalf. (Opinions without detailed, concrete observations will not usually convince alert employers.)
◆ Do not choose members of your family or friends not in your field.
◆ When asking someone to be a reference, don't simply ask, "Could you please act as one of my references?" This question leaves the person little chance to refuse, but this person might not know you well or might be unimpressed by your work and therefore might provide a watery reference that does you more harm than good. Instead, ask: "Do you know me and my work well enough to provide a strong reference?" This second approach gives your respondent the option of declining gracefully or it elicits a firm commitment to a positive recommendation.

Your listing of references should include each person's name, position title or occupation, full address, phone number, e-mail address, or other method of contacting the person. You might also indicate the nature of the person's reference (work, academic, character, supervisory) if that's not obvious from your listing of the person's occupation.

NOTE

There is some controversy about whether to include references in a résumé. Some job-search counsellors suggest a line saying, "References available upon request." These advisors argue that the employer will then call you to get those references and you'll have an opportunity to make a positive impression. The problem is that if other qualified applicants have listed references in their résumés, the employer will save time and effort by interviewing them instead of you.

Counsellors at colleges that have job placement offices will sometimes advise graduates to list the address of the placement office where employers can get your list of references. The problem with that arrangement is that you lose control of which references are provided for certain applications.

ORGANIZING YOUR RÉSUMÉ

Three main organizational patterns

Organize your résumé in the order that conveys the strongest impression of your qualifications, skills, and experience. Depending on your background, you can arrange your material in reverse chronological order, functional order, or a combination of both.

Reverse Chronological Organization.   In a reverse chronological résumé you list your most recent experience first, moving backward through your earlier experiences. Use this arrangement to show a pattern of job experiences or progress along a specific career path. Reverse chronology suits applicants who have a well-established record of relevant experience and education. Many employers like this traditional organization because they're used to it and they find it easy to read. Marcel Dionne's résumé in Figure 23.7 on page 576 shows an example of the traditional reverse chronological order.

Functional Organization.   In a functional résumé, you emphasize skills, abilities, and achievements that relate specifically to the job for which you are applying. Use this arrangement if you have limited job experience, gaps in your employment record, or are changing careers. See Carol Hampton's résumé in Figure 23.8 on page 577 for an example.

Combined Organization.   Most employers prefer chronologically ordered résumés because they are easier to scan. However, electronic scanning of résumés calls for a more functional pattern. One alternative is a modified-functional résumé, which preserves the logical progression that employers prefer but which also highlights your abilities and job skills (as in Christopher Hendsbee's résumé shown in Figure 23.9 on pages 578 and 579).

IMAGE PROJECTION

Project an appropriate image

When you plan, write, edit, and periodically adjust your résumés, evaluate them according to how well they project your desired image. If, for example, you want to be seen as energetic, active, and enthusiastic, be sure to include information about your volunteer activities and outdoor leisure pursuits. Further project the desired image by choosing active verbs such as "led," "organized," "built," maintained," and "completed."

Also, have your résumé project an image consistent with the position's requirements. For example, if you're applying for a position where you'd spend half your time in the field collecting samples and the other half assessing those samples and reporting the results, use language and provide facts that reveal your writing skills and the meticulous side of your nature, as well as your physical fitness.

ELECTRONIC RÉSUMÉS

Increasingly, employers are favouring electronic résumés over paper documents; Forrester Research predicts that, "by 2002, all large companies, 60 percent of medium-sized, and 20 percent of small companies will recruit on-line" (Kerr C10). This explosion of on-line recruiting has been abetted by electronic headhunters such as Recruitsoft and E-cruiter, but many companies now post job positions on their corporate websites.

Companies see many advantages:

◆ They can save up to 75 percent of the average cost of $6,000 (U.S.) of recruiting and hiring employees, according to Recruitsoft President, Louis Tetu (Ray E7).
◆ The hiring cycle is dramatically shortened; hundreds of applicants can be screened, candidates can be interviewed, and an employee hired, all in two or three days, instead of in two to six weeks.
◆ Sophisticated software screening helps companies find the best available talent, even among employed workers not actively seeking a new job.

In future, on-line résumés will appear as hypertext documents with links to the applicant's placement dossier, publications, and support material. Certain links (such as references) can be password-protected so that only employers can access these links. Multi-media résumés will incorporate sound, video, and animation.

Electronic Presentation of Résumés.   Hard copy résumés still work for many situations, but to tap into the total job market, you'll need to adapt your résumé to be scanned by an electronic scanner or to be placed on-line. Electronic storage of scanned or on-line résumés offers employers an efficient way to screen countless applicants, to compile a database of applicants for later openings, and to evaluate all applicants fairly.

How Scanning Works.   The computer captures and searches the printed or on-line image for keywords (nouns instead of traditional "action verbs"). Those résumés containing the most keywords ("hits") make the final cut (Pender).

How to Prepare an E-résumé.   Using nouns as keywords, list all your skills, credentials, and job titles. (Help-wanted ads are a useful source for keywords.)

- *List specialized skills:* Marketing, C++ programming, database management, user documentation, Internet collaboration, software development, graphic design, hydraulics, fluid mechanics, editing, surveying, soil testing, water-quality monitoring, job-site management.
- *List general skills:* Teamwork coordination, conflict management, oral communication, report and proposal writing, troubleshooting, bilingual in French and English.
- *List credentials:* Student member, Institute of Electrical and Electronic Engineers, certified PADI Open Water diver, B.S. Electrical Engineering, top 5 percent of class.
- *List job titles:* Manager, director, supervisor, intern, coordinator, project leader, technician, trainer.

If you lack skills or experience, emphasize your personal qualifications: analytical skills, energy, efficiency, flexibility, imagination, motivation.

In preparing a scannable résumé, use a plain typeface, 10 to 14 point type, and white paper. Use boldface or full caps for emphasis. Avoid fancy fonts, typefaces, underlining, bullets, slashes, dashes, parentheses, or ruled lines (Pender). Left-justify all information within the résumé: many searchable résumé systems do not understand indenting or columns. Avoid a two-column format, which often is jumbled by scanners that read across the page. Figure 23.10 on page 580 illustrates a computer-scannable résumé.

Detailed practical advice about how to prepare and post e-résumés on the Internet may be found at Susan Ireland's website, **susanireland.com.**

Remember, an electronic résumé does not have to be as attractive as your "normal" résumé. Personnel departments that use computer databases are looking for content, not appearance.

GUIDELINES

for Sending
E-résumés

- *Always ask before faxing a scannable résumé format.* Many personnel departments are not automated and would prefer to see your formatted résumé.
- ***Be cautious about sending résumés via e-mail.* The receiving software may not be able to "read" the document you send. Do not send your résumé as part of the e-mail itself; attach the résumé in a common format such as Word or WordPerfect. (But be sure that your receiver's software is compatible with the document you're attaching.)**

Marcel Dionne

144 Avenue Champlain
Quebec PQ   G7E 1R6
Cell Ph. 451-3565   E-mail: mdionne@avoir.ca

EDUCATION
1995–1999

Software Engineering (Bachelor degree)

University of Waterloo

Learned to apply a full range of software engineering principles through scenarios based on concurrent object-oriented software system designs.

Projects: 2nd year—helped Prof. Warsaw develop tracking software for Kitchener Trucking's "smart tire" program
3rd year—developed software reliability model and automatic detection of software failures for Kitchener Board of Education
4th year—re-designed maintenance scheduling software for the University of Waterloo's boiler heating system

EXPERIENCE
Nov. 1999
to present

Software Consultant (contract position)

City of Quebec

Work with City engineers to re-design Quebec City's transit management software. Co-responsible for designs and responsible for implementing the program, including training 12 city employees to work with the software. The project will be complete June 10, 2000.

Summer Relief Worker

Summers
1995–1998,
and May to
Oct., 1999

Canada Courier, Montreal and Hamilton

Worked in all aspects of the parcel delivery business: truck driver, dispatch worker, delivery tracking troubleshooter, order desk clerk, packing department clerk. A highlight of this experience was the opportunity to troubleshoot problems with Canada Courier's parcel tracking software. Re-wrote the software to make it more reliable.

PROFILE

Interests include distance running, computers, and computer-generated music Single. 24 years old. Capable of working long, productive hours.

REFERENCES

Professor John Marks
Computer Engineering Dep't.
University of Waterloo
Ph. (519) 888-4532, ex.5430
E-mail: marks@coulomb.uwaterloo.ca

Bill Spender
Operations Manager
Canada Courier
1780 Boulevard St. Laurent
Montreal PQ   G9T 2W2
Ph. (514) 986-2987

Professor Viktor Warsaw
Computer Engineering Dep't.
University of Waterloo
Ph. (519) 888-4532, ex.5444
E-mail: warsaw@coulomb.uwaterloo.ca

Lise Tremblay
Engineering Contract Manager
City of Quebec
Place Centrale
Quebec PQ   G7E 4R4
Ph. (418) 655-2517

Figure 23.7   A Reverse Chronological Format Résumé (compressed to fit on one page)

Carol Hampton

196-4068 Minster Avenue
Toronto ON   M6W 2R5
Phone: (416) 445-5333
E-mail: comphamp@collect.ca

OBJECTIVE

To apply computer installation, configuration, and troubleshooting skills to a networked environment

SKILLS AND ABILITIES

Analytical

- Have shown an aptitude for determining client network requirements
- Proficient at solving math-based problems (top marks in Mathematics in first two semesters of College program)
- Proficient at discovering programming errors

Design

- Won city-wide competition for high school Web-page design, May 2000
- Proficient in desktop publishing operations—currently design advertising for George Brown College's student newspaper

Research

- Researched comparative study of firewall software packages for Professional Communications II class
- For a 30-minute class oral presentation, researched the San Francisco Project's development of Java applications
- Industry experience with Farnham Industries, Mississauga—found ready-made solutions for problems with a networked configuration. Saved the firm $6,000, the cost of a custom-designed solution. (Summer experience, 2001)

Interpersonal

- Worked closely with the staff at Farnham Industries to implement network solutions. Gained the trust and cooperation of staff who were at first reluctant to try new methods, especially those introduced by a "student."
- Engaged in peer counselling in high school and at George Brown College

EDUCATION

Currently completing the last semester of the two-year **Computer System Technician** diploma program at George Brown College, Toronto. Qualified to function as a user support specialist, network administrator, or business network trainer.

REFERENCES

Terry MacArthur
Systems Analyst Consultant
Toronto
(416) 442-3454
tersystem@toronto.ca

Jens Husted
Computer Networking Instructor
Information Technology Department
George Brown College
(416) 415-2010
husted@gbc/scitech.ca

Marjorie Prystai-Alvarez
Personnel Coordinator
Farnham Industries
Mississauga
(905) 657-7439
staff@farnham.on.ca

Figure 23.8   **A Functional Format Résumé**

Christopher Hendsbee

453B Gordon Drive
Kelowna BC   V1W 2T1
Phone: (250) 876-3422
E-mail: chris@silk.net

Objective	GPS surveying position, with a chance to use and further develop current skills
Education	**Civil Engineering Technology**

Okanagan University College, Kelowna BC   1997–present
The program stresses practical applications of civil engineering theory.

Experience   **Surveyor**

Deep Woods Consultants, Nelson BC   May–Nov, 1998
Surveyed forestry roads, property lines, bridge sites, and topographics.
Became familiar with many types of total stations and data collectors.
Led and supervised a surveying crew. Responsible for all the notes.

Equipment Operator

Lucky Logging, Keremeos BC   July–Aug, 1997
Operated a 966 Cat loader and stacked processed logs. Operated the excavator
with a Styer processing head, and worked on the grapple skidder.
Built landings and fire guards.
Greased and repaired equipment.

Release Driver

Sterile Insect Release Program, Osoyoos BC   Aug–Sept, 1996
Followed a marked route on an ATV. Every morning the manager would drop the
sterile moths and I would release them over my route. Recorded field notes and
addressed questions and problems raised by farmers.

Labourer

Worked as a ranch hand, fencing company assistant, and assistant orchardist.

Special Skills   Proficient with Windows 98 and NT 4.0. Also proficient with software programs:
WordPerfect 2000; Microsoft Works (spreadsheets, word processing, databases,
charts/graphs, communications); Microsoft Office 2000 (Word, Excel, Access,
PowerPoint); Harvard Graphics presentation package; Microsoft Publisher;
AutoCAD version 14, AutoCAD Light; and MiniCAD
Proficient with compasses
Very familiar with GPS
Proficient at operating heavy equipment
Proficient at operating chainsaws and a large variety of hand and power tools

. . . 2

Figure 23.9   A Combination Format Résumé: Chronological and Functional   *(continued)*

Christopher Hendsbee/Page 2

Interests and Activities	Archery (bronze medal in the BC Summer Games), hunting, hiking, swimming, weight lifting, trap shooting
Accreditations Licences, and Memberships	S100 Wild Fire Suppression Certificate Hunters licence Avalanche course Driver's licence Level one first aid Firearms acquisition certificate Student member, Applied Science Technologists and Technicians
References	**Dave Muster** Owner, Lucky Logging, Nelson BC RR#1 Newhouse Road Keremeos BC V0X 1D0 Ph. (250) 499-5609 **Jean McElroy** Survey Manager Deep Woods Consultants Nelson BC V1L 5T9 Ph. (250) 825-4565 **Bill Waterburn** SIR Supervisor RR#1 Barcello Road Cawston BC V8X 5Y2 Ph. (250) 499-2366 **Bob Bradley** Department Chair Civil Engineering Technology Okanagan University College 1000 K.L.O. Road Kelowna BC V1Y 4X8 Ph. (250) 762-5445 (ext. 4590)

Figure 23.9 A Combination Format Résumé: Chronological and Functional

Carol Hampton
196-4068 Minster Avenue
Toronto ON   M6W 2R5
Phone: (416) 445-5333
E-mail: comphamp@collect.ca

OBJECTIVE
To apply computer installation, configuration, and troubleshooting skills in collaboration with others.

QUALIFICATIONS
Programming in Java, HTML, Logo, Pascal, C++. Network design and troubleshooting. Determining client network requirements. Advanced computer mathematics. Web-page design. Newspaper advertising design. Internet research. Usability testing. Desktop publishing of installation and maintenance manual. Writing, designing, and testing hardware upgrade manuals. Research on comparative study of firewall software packages.

EXPERIENCE
FARNHAM INDUSTRIES, 7800 CONCEPT COURT, MISSISSAUGA. Solutions for problems with a networked configuration. Network solution implementation. Liaison with staff. Positive relations. Development of trust and cooperation.

EDUCATION
GEORGE BROWN COLLEGE, TORONTO. **Computer System Technician** diploma.
Qualified as a user support specialist, network administrator, or business network trainer.

REFERENCES
Terry MacArthur
Systems Analyst Consultant
Toronto
(416) 442-3454
tersystem@toronto.ca

Marjorie Prystai-Alvarez
Personnel Coordinator
Farnham Industries
Mississauga
(905) 657-7439
staff@farnham.on.ca

Jens Husted
Computer Networking Instructor
Information Technology Department
George Brown College
(416) 415-2010
husted@gbc/scitech.ca

Figure 23.10   A Computer-scannable Résumé

CONTACTING EMPLOYERS

Use your resources wisely and effectively

It's important to have a plan for contacting employers. Much time and money can be wasted on pursuing the wrong positions or pursuing them at the wrong time. You need to plan whom you contact, how you contact them, in what order, and when. You also need to follow up initial contacts. Pick suitable times to contact employers by phone or in person; for example, most managers don't want to get job enquiries at the end of their work week. Don't drop off a résumé at a restaurant during its lunch or dinner rush. Try to pick a time when you know that the recipient will have time to speak with you. If possible, call a receptionist to learn the best time to call or appear.

CAMPAIGN STRATEGIES

The job-search campaign you wage will vary with your circumstances. If you search for a new job while employed, your search will be more selective and discreet than if you are unemployed or at college.

Searching for a satisfying job involves a carefully planned campaign of contacting employers. The following excerpt from a campaign log (Figure 23.11) illustrates the required planning and detailed record keeping. Such a log becomes especially important if your campaign involves many contacts over a prolonged period.

Employer	Contact	Response	Interview	Follow-up	Status	Comments
Trendmark Electronics, Kanata	Enquiry letter/ Mar. 17 (see file)	Mar. 28; phone call Jim Lund, R&D encouraged me to apply	Not yet	Mar. 29—sent thank-you letter; asked to tour Ottawa plant	Hopeful— J.D. said my qualifications "looked fine"	Phone by April 10 if no further news
Ames Research, Hull	Applied—letter Mar. 21 for Electronics Researcher	None so far	Not yet	Not yet	Highly desirable but doubtful; many experienced people will apply	Will send positive follow-up letter if no response by April 4— prepare letter!

Figure 23.11   An Excerpt from a Campaign Log

Many employment advisors recommend tiered campaigns. For example, Tom Smith, a 25-year old graduate of the Industrial Electronics program at Saskatoon's Kelsey Institute, used the following strategy to win a position that would give him job satisfaction after his May graduation.

Tier 1:

Because of his military supervisory service, Tom wanted to start his electronics career as a quality control supervisor or production shift supervisor. He found four firms that had such positions; three in Saskatoon and one in Calgary. Using the library and the Internet, he set out to learn all he could about the four companies, their business prospects, and their range of positions. He used his contact network, which included friends (and friends of friends) who worked at three of the companies, to try to learn about possible openings.

He devised a detailed campaign of introduction letters, follow-ups, and requests for informational interviews, to exhaust every possibility of being employed by one of the four companies. He placed a time limit of six weeks on Tier 1 activities before moving on to Tier 2.

Tier 2:

Tom's secondary interest was in electronics research, so he identified several firms in Halifax, Moncton, Ottawa, Saskatoon, Calgary, Montreal, and Vancouver engaged in such research. He did some preliminary research on these firms, but delayed further action until Tier 1 activities were completed.

Tier 3:

In case the first three months of job searching at his Tier 1 and Tier 2 levels proved fruitless, Tom was prepared to go to his backup plan, Tier 3. Here, Tom would look for a position as a glider flight instructor (developed in the military and as a personal interest) or as an electronics research assistant.

Result:

Tom did not have to go to Tier 2 or Tier 3 of his plan. In week 2 of his campaign, one of the Saskatoon firms advertised a quality control position; also, early in week 3, Tom learned that the other three firms were sufficiently interested in Tom's qualifications to invite him for an exploratory meeting. The Calgary firm brought him back for a second meeting and then for a formal interview for a production shift manager position.

He was also formally interviewed for the quality control position in Saskatoon. One week later, he was offered both positions for which he had been formally interviewed. He accepted the production manager job in Calgary, though the pay was lower, because that company would give Tom more opportunities to pursue a management career. ▲

Perhaps *your* next campaign will not need to be as elaborate as Tom's. But his approach had one major advantage—he gave himself every chance to land the kind of position that would give him *maximum job satisfaction* at that stage of his career.

CONTACT METHODS

An enquiry letter (or in-person enquiry) offers a good method of getting noticed by an employer. Indeed, it's not uncommon for the authors of well-written enquiries to be invited for exploratory interviews; sometimes, an employer will offer a position to an impressive interviewee, even if the employer hadn't intended on hiring just yet.

Still, the enquiry is primarily a research tool, not a method of selling your qualifications. (See the advice about enquiries earlier in this chapter.) So, if you already know that you would like to work for a particular employer, but it doesn't seem that this employer currently has openings, you may want to introduce yourself.

A letter of introduction is quite straightforward. You introduce your qualifications and your interest in the company. You name the position(s) that would appeal to you and refer the reader to the pertinent experience in your enclosed résumé. The action closing should indicate when you're available and how the reader can easily contact you. Perhaps you'll also request a meeting to discuss future employment opportunities.

An introduction letter does not apply for a specific available position, so you can't present a full sales pitch showing your suitability for a given position. But you can say enough about yourself and your career goals to engage the reader's interest. The introduction letter should be short, though; you have to know when to stop and let your résumé speak for you.

An application letter is more formal and persuasive in tone. You will write an application when you know that a position is available and when you're certain that you have what it takes to fulfill the position's requirements. Each application must be tailored to the specific position for which you're applying, so you have to think clearly about the letter's content and phrasing.

First, you need to remember that *unsolicited applications* require a different approach from *solicited applications*.

Unsolicited Applications.   If your contact network has informed you of an opening, and if that position has not been formally advertised, your main challenge is to deal with a reader who is not anticipating your application. It's possible that the reader doesn't even *want* applications; the reader may be using his or her own contacts to find candidates. If that's the case, you must get and hold the reader's interest by immediately appealing to the reader's needs and preferences.

Finding the right appeal is not easy. For starters, you'll have to research the position, to learn what type of work is done, which qualifications are essential, and, if possible, what characteristics the employer would like the new employee to have. If the employer has not yet drawn up a set of position requirements or employee characteristics, that employer may be receptive to a wider range of candidates than if the position had been advertised. The following is an unsolicited application scenario.

Unsolicited Application

Appeal to the reader of an unsolicited application letter

Royale Ltee., a Montreal-based courier company, is expanding from metropolitan Montreal to all of southern Quebec and southern Ontario. It has recently been purchased by a trucking consortium, Bouchard Transport, which has injected cash into the operation.

The writer of the following letter, Marcel Dionne, is a 24-year old graduate of the University of Waterloo. His cousin, Michelle Legaree, who works as an accountant for Bouchard Transport, gives Marcel the background details and says that she's heard that Royale Ltee. will need someone to help develop the logistics of expanding the courier operation. Marcel worked four summers for Canada Courier when he was at university, so he has a good idea of what Michelle means. His recent degree in software engineering has led to an eight-month contract to help redesign Quebec City's transit management software, but that contract will finish in three weeks. Acting quickly, he learns what he can about Royale Ltee. Among other things, he learns that Jean-Guy Ryan is the general manager, at age 29, and that Ryan and the other Royale Ltee. managers have been successful running an informal, democratic style of operation that encourages employees to suggest ways of improving service and controlling operating costs. ▲

Figure 23.12 on page 584 shows how Marcel's unsolicited application letter tries to appeal to Jean-Guy Ryan's needs and priorities. Marcel's accompanying résumé appears on p. 576.

144 Avenue Champlain
Quebec PQ   G7E 1R6
May 18, 2000

M. Jean-Guy Ryan, General Manager
Royale Ltee.
2408 Boulevard Laurentian
Montreal PQ   G6H 5T5

Dear Monsieur Ryan:

Michelle Legaree, an accountant with Royale's parent company, Bouchard Transport, has told me about your expansion plans. Please consider my offer to help your courier firm develop software and logistics to allow a rapid, smooth transition from an urban courier to a regional operation.

My background in software engineering, my experience with Canada Courier, and my current work for the City of Quebec place me in a unique position to contribute to your firm. The enclosed résumé provides details of my software engineering degree; you'll notice that most of my class projects focused on developing solutions for logistical problems. That focus resulted from my experience at Canada Courier in Montreal and Hamilton; as a summer relief worker, I became familiar with all aspects of the delivery business. In my last stint with Canada Courier, I was asked to troubleshoot problems with its parcel tracking software. That experience has helped me manage several very difficult logistical issues in Quebec City's transit management software.

A growing company needs energetic, innovative people who wish to grow with the organization. The enclosed reference letters show that I match that description. Two of the references listed in my résumé, Professor Marks and Canada Courier's Bill Spender, will be able to comment on my problem-solving abilities and work ethic.

I'll be available for employment in three weeks, when my project contract in Quebec City finishes. However, my supervisor has given his approval for me to go to Montreal to discuss my application for a position with your firm, at any time you set. I really hope you do invite me for an interview because I believe that my skills and interests match Royale's upcoming needs. You can contact me at my e-mail address (on the résumé) or at my cell phone, 451-3565.

Sincerely,

Marcel Dionne

Marcel Dionne

Enclosure: Résumé

Figure 23.12   An Unsolicited Application

Solicited Applications.   Solicited applications, such as the one illustrated in Figure 23.13, require a similar structure to the unsolicited letter. However, the *opening paragraph* will be more straightforward—you simply apply for the position by name, you refer to the advertisement, and you preview your sales pitch.

Show that you understand the position's requirements

The application letter's *sales pitch* depends on your accurate analysis of the advertised position's requirements. The employer has carefully chosen those requirements and will expect your application to show how you can fulfill them. Therefore, if your reading of the position description leaves you with questions, write or phone the employer to get answers. Your main challenge is to prove that you have what the employer needs, so you *must* understand those needs.

When you construct your sales argument, look at your application from the employer's point of view. As in all business letters, answer the reader's question: *What's in it for me?* Two hints for answering that question are:

Show that you can meet the position's requirements

1. Sell your qualifications and personal qualities as a *package*. The whole is greater than the sum of its parts, but only if you show how one set of skills complements and strengthens another set. For example, a former McDonald's manager who has earned a computer science degree will be better able to apply that computer knowledge because of previous business experience.
2. Highlight that package with a phrase or fact that identifies you and you only. Marketers refer to this technique as "positioning your product." You "position" yourself in a special place in the reader's mind. That positioning idea should reflect the strongest connection between your package and the employer's set of requirements.

Three letters in this chapter illustrate successful positioning:

1. The enquiry letter on page 566 positions the writer's enthusiasm for GPS surveying.
2. The unsolicited application in Figure 23.12 on page 584 presents three faces of the writer's special collection of relevant skills. (The opening sentence in paragraph 2 even uses the term "unique position.")
3. The solicited application in Figure 23.13 on page 587 positions the applicant's unusual and special combination of electronics training, leadership skills, and communications skills.

CAUTION   *Do not use Tom's positioning statement unless you can prove that you know "exactly" what the position requires.*

In phrasing your sales pitch, project a *you* attitude by:

◆ discussing the position's requirements *before* matching your qualifications to those requirements
◆ using objective, third-person phrasing wherever possible ("the enclosed résumé lists that related experience") or "you" phrasing ("you can reach me at . . .," rather than "I can be reached at . . .")

Close the sale!

Application letters require a *positive, business-like closing*. Say when you're available for work or for an interview, and indicate the easiest way for the reader to reach you. Request an interview, or express your desire for an interview, because gaining an interview invitation is the reason you wrote this letter! Finally, you might close by stating your desire for the position or by reiterating your key argument for considering your application.

Figure 23.13 on page 587 shows Tom Smith's response to an opening for a quality control supervisor. This is the advertisement to which Tom replied:

Quality Control Supervisor

The company: Altitude Electronics specializes in electronic instrumentation and controls for the aerospace industry—satellites, launch rockets, and high-altitude weather balloons. Its award-winning products result from superior design and meticulous attention to detail in the various stages of the manufacturing process.

The position: You will be one of two quality control supervisors. You will test products, report deficiencies, supervise production-worker methods, and work in conjunction with the Production Manager to train production line assemblers.

The successful candidate will have: A two-year Electronics diploma from an accredited Canadian college, or equivalent. Demonstrated leadership skills and above-average communications skills (interpersonal, oral, written, listening). Familiarity and some skill in fabrication, component assembly, soldering, and product finishing. Demonstrated skills in electronics troubleshooting and CAD/CAM.
Apply to   Dirk Benefeld, General Manager, Altitude Electronics
          2310 Hanselman Avenue
          Saskatoon SK   S7L 6A4

Tom's first response was to think, "It's as if they wrote this ad for me!" Then he analyzed the advertisement as follows:

- The position requires a combination of technical skills and communication skills. That's exactly what I have to offer, so I should be able to make a good case.
- The ad expresses pride in the company's accomplishments ("award-winning," "superior design"), so I should emphasize my own drive to succeed.
- Mr. Benefeld is on the Industrial Electronics industry advisory board, so I shouldn't have to sell the program to him, just remind him of how practical and comprehensive it is.
- Probably my competitive advantage is my background of leadership positions and communications experience, so that should be stressed in the résumé as well as in the letter. And I'll choose my references and reference letters to emphasize the leadership communications.
- The ad mentions "meticulous attention to detail," so I'd better show that side of my character.

108-3118D 33rd Street West
Saskatoon SK   S7L 6K3
May 7, 2001

Mr. Dirk Benefeld, General Manager
Altitude Electronics
2310 Hanselman Avenue
Saskatoon SK   S7L 6A4

Dear Mr. Benefeld:

Re: Quality Control Supervisor

Please consider my application for the quality control supervisor position which you advertised on page D18 of yesterday's Star-Phoenix. This position requires exactly what I have to offer.

As the enclosed résumé details, I'm about to graduate from Kelsey's Industrial Electronics program, which provides a comprehensive and practical overview of current electronics applications. The course also teaches us how to think and how to keep abreast of the rapidly changing electronics field. My marks have consistently been in the top three of the class in each of the six trimesters of the program. This reflects my work ethic and my attention to detail. It also reflects the solid theoretical grounding I received from two years of engineering studies at the University of Saskatchewan.

Instructor Al Schlatter, who's listed as a reference in my résumé, has agreed to tell you about my fabrication, assembly, troubleshooting, and CAD skills.

From my enquiries, I've learned that your quality control position requires strong leadership communications skills. The attached reference letters comment on my performance as President of the Kelsey Students' Association this year. As President, I've been able to hone skills which I've developed as a cadet leader and glider pilot instructor for the Department of National Defence.

I'd appreciate the opportunity to discuss the quality control supervisor position with you. If you agree that my qualifications match the position's requirements, you can contact me, Monday through Saturday, 7.30 a.m. to 11.00 p.m., by calling the Students' Association office, 652-0980. I can be available for an interview at any time you arrange. I'll be available for work after May 29, the final day of exams.

Yours sincerely,

Tom Smith

Tom Smith

Enclosures: Résumé
                Industrial Electronics Course Overview and Grades
                Reference letters (3)

Figure 23.13   A Solicited Application

EMPLOYMENT INTERVIEWS

Prepare thoroughly

When your application is successful, you will be invited to an interview, sometimes to a series of interviews. (Weyerhaueser Canada, for example, often subjects applicants to two days of interviews and tests.) Sometimes, just one person will interview you. More often, you can expect to be quizzed by a panel. At a minimum, expect to be interviewed by your potential immediate supervisor and by someone from the company's personnel division. Smaller firms might be represented by the firm's manager, a project supervisor, and a member of the work team that has the open position.

INTERVIEW PREPARATION

How to prepare

Careful preparation is the key to a productive interview. Prepare by learning about the company in trade journals and industrial indexes. (Learn about the company's products or services, history, prospects, branch locations.) Request company literature, including its annual report if the company is publicly traded. Speak with people who know about the company.

If you're applying to a government or municipal agency, try to learn what will affect the growth or downsizing of that agency, and try to learn if there's room for personal growth or career advancement within the agency. Also, learn about the agency's range of services. If you can't find information of this sort before the interview, ask questions during the interview.

Prepare also by thinking about your reasons for wanting the job and what you have to offer the employer. Your self-inventory and list of transferable skills will help you prepare. Also, look at your résumé from the interviewer's point of view—think of the questions you would ask if you were the interviewer. If you have difficulty thinking of such questions, have friends go through your résumé to look for issues that will elicit questions.

As you prepare, you will find the following "game plan" helpful.

Interview Preparation

1. Strengths to emphasize, in order of priority

2. Subtle strengths I need to bring out

3. Apparent limitations I should be ready to counteract

4. Subtle limitations I need not introduce

5. My overall strategy for making my case (for projecting my image)

6. Things I want to learn about the job and the company

Questions

It's possible to anticipate many of the questions you'll be asked at an interview. You can prepare for obvious questions like:

- Why do you want to work here?
- What do you know about our company?
- What do you see as your biggest weakness? biggest strength?
- What type of supervisor do you prefer?
- Do you prefer to work under close supervision in a structured environment, or more independently within broadly stated guidelines?
- What are your short-term and long-time career goals?

You can anticipate three categories of questions: information questions, high-risk questions, and opportunity questions.

Categories of interview questions

Information Questions.   These easy questions come early in the interview and enable you to provide factual background information about your education and experience. The interviewer will assess how you present yourself, what you know about the firm, and how you "fit" the culture of the organization. (*Why did you apply for this job? Tell me about your program at the college.*)

High-risk Questions.   These questions can destroy your chances for a job if handled poorly. These questions probe what the interviewer believes to be your potential weaknesses related to the job requirements. If you can successfully deflect these questions in the middle of the interview, then you will have an opportunity to sell yourself in the next stage. (*Why did you leave your previous job? What difficulties have you faced in previous jobs? What are your limitations?*)

Opportunity Questions.   These questions usually are asked at the end of the interview and provide the opportunity for you to convince the interviewer that you can make a contribution to the organization. Your answers should be bold (not arrogant) and confident. (*What are your strengths? How can you contribute to our product or service? How do you see yourself fitting into our organization?*)

Also, you will have a chance to ask your own questions. Have a list of written questions with you, in an unobtrusive notebook. Refer to this notebook when you're asked if you have questions. Focus on the work you would be doing and the conditions of employment—ratio of office work and field work, travel involved, level of responsibility, opportunities for further training and skill development. If your questions have already been answered in the interview's give-and-take, mention them anyway; show that your primary interest is in the work.

If at all possible, do not ask questions about salary and benefits. After you receive an offer of employment, you can discuss employment benefits. Or, if you must know, call the company's personnel office. Usually, however, these issues will not pose a problem because a company representative will review salary and benefits in a closing stage of the interview.

ANSWERS

Good answers present appropriate levels of detail

Answer questions directly and fully. Some closed-ended questions such as, *Are you familiar with Microsoft Office Pro?* probably only merit a short response of one to three sentences. However, the majority of interview questions are open-ended, as in, *What are your main strengths?* Such questions need to be answered fully enough to satisfy the interviewer's interest, though not so fully that you totally exhaust that interest. If you don't provide enough detail to support your responses, the interviewer will doubt your assertions. So, give examples of your strengths, skills, and achievements; back up your statements with reasons and specific details; recount incidents that establish that you've used certain techniques or completed certain tasks in the past.

Interviews can place you under pressure, so two kinds of preparation are essential:

Job interview hints

1. Anticipate questions and rehearse ways to answer them.
2. Practise using impromptu speaking techniques. Chapter 24 describes impromptu organizing methods that will allow you to control nervousness and to perform under pressure.

Here are some other hints for successful job interviews:

- If possible, research the range of candidates who might apply for the same position as you. Know your competition, and how you compare to them.
- Dress appropriately for the interview—as if this were your most important day at work.
- Arrive early for the interview. Occasionally, the interview team will give you a task to perform or a case study before you're called into the interview room.
- Remember that you'll be the "guest" and the lead interviewer will be the "host," so don't be too socially aggressive—wait for the host to initiate a handshake; wait to be asked to take a chair; allow the interviewers to guide the conversation.
- Maintain eye contact much of the time, but don't stare.
- Maintain a relaxed but alert posture. Don't slouch. Don't fidget. Show your interest by leaning forward. Smile when appropriate. Be yourself, your best self.
- Answer questions truthfully; skilled interviewers will ask the same controversial question or probe the same issue in different ways to see if your answers are consistent.
- When you don't know the answer to a question, say so. In some cases, you could explain how you would determine the answer to the question.
- Don't blurt things you don't need to mention. (See Quadrant 4 of the self-inventory.)
- Don't be afraid to allow silence. An interviewer may stop talking, to observe your reaction and to induce you into revealing your inner thoughts.
- Remember to smile, and show other non-verbal responsiveness to your interviewers. You'll improve your chances of getting the job—no-one likes to work with a grouch!
- When the interviewer says the interview is ending, or hints that it's drawing to a close (closes the question folder or puts away his or her pen), don't wear out your welcome. Restate your interest, ask when you might expect further word, thank the interviewer, and leave.
- In general, treat the interview as potentially one of the most important conversations in your life!

FOLLOW-UPS TO INTERVIEWS

Within a few days after the interview, refresh the employer's memory with a letter restating your interests. Here's an example:

> Thank you for the opportunity to discuss your technologist position on Wednesday.
>
> Learning about your planned plant expansion has strengthened my desire to work for Kraftsteel Industries. Also, our conversation has confirmed my belief that I could contribute to your company's continuing growth, through my CAD skills and design capabilities.
>
> If you require further information, please call my pager number, 767-9901.

RESPONDING TO JOB OFFERS

If all goes well, your strategy will result in an offer for a position in your first tier of choices. If so, you'll likely have no difficulty accepting with enthusiasm. If you receive an offer by phone, ask when you'll receive a written offer. Respond to that letter with a formal *letter of acceptance*. Your response may serve as part of your contract, so spell out the terms you are accepting. Here's an acceptance letter, written by the same person as the above follow-up letter.

> I am delighted to accept your offer of the mechanical engineering technologist position at your Hampstead plant, with a starting salary of $32,400. I understand that there will be a six-month probationary period and that I will be eligible for raises after one year.
>
> As you requested, I will phone Pat Larsen in your Human Resources office for instructions on reporting date, physical exam, and employee orientation.
>
> I look forward to a satisfying career with Kraftsteel Industries.

You may have to refuse offers, perhaps because you learned things during the job interview that gave you reason to believe that the position was not for you, or perhaps because you have accepted another position that better meets your current job objectives. So, even if you refuse by telephone, write a prompt, cordial letter of refusal, explaining your reasons and allowing for future possibilities. The writer of the above acceptance letter phrased a refusal this way:

> Although I was impressed with your new hydro-forming technology and the efficiency of your auto parts plant, I am unable to accept your offer of a position as an assistant shift foreman at your plant.
>
> I have accepted a position with Kraftsteel Industries because Kraftsteel has offered me the chance to join its design team. CAD design, as I mentioned in our recent interview, is one of my main interests.
>
> In the future, if opportunities arise with your hydro-forming design group, I would appreciate the chance to be considered for a position with that group.
>
> Thank you for your interest in me and for your courtesy.

In Brief EVALUATING A JOB OFFER

Fortunately, most organizations will not expect you to accept or reject an offer on the spot. You probably will be given at least a week to make up your mind. Although there is no way to remove all risks from this career decision, you will increase your chances of making the right choice by thoroughly evaluating each offer—weighing all the advantages against all the disadvantages of taking the job.

The Organization

Background information on the organization—be it a company, government agency, or non-profit concern—can help you decide whether it is a good place for you to work.

Is the organization's business or activity in keeping with your own interests and beliefs? It will be easier to apply yourself to the work if you are enthusiastic about what the organization does.

How will the size of the organization affect you? Large firms generally offer a greater variety of training programs and career paths, more managerial levels for advancement, and better employee benefits than do small firms. Large employers also have more advanced technologies in their laboratories, offices, and factories. However, jobs in large firms tend to be highly specialized—workers are assigned relatively narrow responsibilities. On the other hand, jobs in small firms may offer broader authority and responsibility, a closer working relationship with top management, and a chance to clearly see your contribution to the success of the organization.

Should you work for a fledgling organization or one that is well established? New businesses have a high failure rate, but for many people, the excitement of helping create a company and the potential for sharing in its success more than offset the risk of job loss. It may be almost as exciting and rewarding, however, to work for a young firm that already has a foothold on success.

Does it matter to you whether the company is private or public? A private company may be controlled by an individual or a family, which can mean that key jobs are reserved for relatives and friends. A public company is controlled by a board of directors responsible to the stockholders. Key jobs are open to anyone with talent.

Is the organization in an industry with favourable long-term prospects? The most successful firms tend to be in industries that are growing rapidly.

Where is the job located? If it is in another city, you need to consider the cost of living, the availability of housing and transportation, and the quality of educational and recreational facilities in the new location. Even if the place of work is in your area, consider the time and expense of commuting and whether it can be done by public transportation.

Where are the firm's headquarters and branches located? Although a move may not be required now, future opportunities could depend on your willingness to move to these places. It frequently is easy to get background information on an organization simply by accessing its website or telephoning its public relations office. A public company's annual report to the stockholders tells about its corporate philosophy, history, products or services, goals, and financial status. Most government agencies can furnish reports that describe their programs and missions. Press releases, company newsletters or magazines, and recruitment brochures also can be useful. Ask the organization for any other items that might interest a prospective employee.

Background information on the organization also may be available at your public or school library. If you cannot get an annual report, check the library for reference directories that provide basic facts about the company, such as earnings, products and services, and number of employees.

Stories about an organization in magazines and newspapers can tell a great deal about its successes, failures, and plans for the future. You can identify articles on a company by looking under its name in periodical or computerized indexes—such as the *Business Periodicals Index*. It probably will not be useful to look back more than two or three years.

The library also may have government publications that present projections of growth for the industry in which the organization is classified. Long-term projections of employment and output for industries, covering the entire economy, are developed by the Department of Labour and Statistics Canada. Trade magazines also have frequent articles on the trends for specific industries.

Career centres at colleges and universities often have information on employers that is not available in libraries. Ask the career centre librarian how to find out about a particular organization. The career centre may have an entire file of information on the company.

The Nature of the Work

Even if everything else about the job is good, you will be unhappy if you dislike the day-to-day work. Determining in advance whether you will like the work may be difficult. However, the more you find out about it before accepting or rejecting the job offer, the more likely you are to make the right choice. Ask yourself questions like the following.

Does the work match your interests and make good use of your skills? The duties and responsibilities of the job should be explained in enough detail to answer this question.

How important is the job in this company? An explanation of where you fit in the organization and how you are supposed to contribute to its overall objectives should give an idea of the job's importance.

Are you comfortable with the supervisor?

Do employees seem friendly and cooperative?

Does the work require travel?

Does the job call for irregular hours?

How long do most people who enter this job stay with the company? High turnover can mean dissatisfaction with the nature of the work or something else about the job.

The Opportunities

A good job offers you opportunities to grow and move up. It gives you chances to learn new skills, increase your earnings, and rise to positions of greater authority, responsibility, and prestige.

The company should have a training plan for you. You know what your abilities are now. What valuable new skills does the company plan to teach you?

The employer should give you some idea of promotion possibilities within the organization. What is the next step on the career ladder? If you have to wait for a job to become vacant before you can be promoted, how long does this usually take? Employers differ on their policies regarding promotion from within the organization. When opportunities for advancement do arise, will you compete with applicants from outside the company? Can you apply for jobs for which you qualify elsewhere within the organization or is mobility within the firm limited?

The Salary and Benefits

Wait for the employer to introduce these subjects. Most companies will not talk about pay until they have decided to hire you. In order to know if their offer is reasonable, you need a rough estimate of what the job should pay. You may have to go to several sources for this information. Talk to friends who recently were hired in similar jobs. Ask your teachers and the staff in the college placement office about starting pay for graduates with your qualifications. Scan the help-wanted ads in newspapers and on the Internet. Check the Internet, the library, or your school's career centre for salary surveys. If you are considering the salary and benefits for a job in another geographic area, make allowances for differences in the cost of living, which may be significantly higher in a large metropolitan area than in a smaller city, town, or rural area. Use the research to come up with a base salary range for yourself, the top being the best you can hope to get and the bottom being the least you will take. An employer cannot be specific about the amount of pay if it includes commissions and bonuses. The way the plan works, however, should be explained. The employer also should be able to tell you what most people in the job earn.

Also take into account that the starting salary is just that, the start. Your salary should be reviewed regularly—many organizations do it every 12 months. If the employer is pleased with your performance, how much can you expect to earn after one, two, or three or more years?

Don't think of your salary as the only compensation you will receive—consider benefits. Benefits can add a lot to your base pay. Health insurance and pension plans are among the most important benefits. Other common benefits include life insurance, paid vacations and holidays, and sick leave. Benefits vary widely among smaller and larger firms, among full-time and part-time workers, and between the public and private sectors. Find out exactly what the benefit package includes and how much of the costs you must bear.

Asking yourself these kinds of questions won't guarantee that you will make the best career decision—only hindsight could do that—but you probably will make a better choice than if you act on impulse.

Source: Adapted excerpts from U.S. Department of Labor. *Tomorrow's Jobs*. Washington, D.C.: GPO, 1995.

Checklist FOR EDITING AND REVISING JOB-SEARCH CORRESPONDENCE

Use this checklist to refine the content, arrangement, and style of your letters and résumés.

Content

- Is the letter addressed to a specifically named person?
- Does the letter contain all of the standard parts?
- Does the letter have all needed specialized parts?
- Have you given the reader all necessary information?
- Have you identified the name and position of your reader?

Arrangement

- Does the introduction immediately engage the reader and lead naturally to the body?
- Are transitions between letter parts clear and logical?
- Does the conclusion encourage the reader to act?
- Is the format correct?
- Is the design acceptable?

Style

- Is the letter in conversational language (free of letterese)?
- Does the letter reflect a "you" perspective throughout?
- Does the tone reflect your relationship with your reader?
- Is the reader likely to react favourably to this letter?
- Is the style throughout clear, concise, and fluent?
- Is the letter grammatical? (see Appendix A)
- Does the letterís appearance enhance your image?

Résumés

- Have you adapted the content, structure, and format of your résumé to suit the specific position you're pursuing?
- Have you made sure that your résumé is absolutely free of errors in content, spelling, and grammar?
- Have you used section headings that accurately identify your package of employable characteristics?
- Have you made it easy for your reader to find key pieces of information?
- Does your résumé's appearance enhance your image?

EXERCISES

1. Complete a self-inventory and a list of transferable skills. Use these self-assessment tools to:

   ◆ identify the employable characteristics that form the core package you'll "sell" for any position you may pursue

   ◆ identify which skills, qualifications, and personal qualities you will emphasize for particular positions

   ◆ prepare for questions job interviewers will ask you

2. Develop a job-search strategy for your next job search.

   ◆ List the types of positions you should pursue.

   ◆ Determine which positions are currently available and which may become available soon.

   ◆ Decide which position(s) to pursue in the first tier of your campaign and the methods you'll use to contact employers.

3. Design and write two or more résumés that you can adapt for the range of positions you expect will interest you.

4. Analyze an advertisement that describes a position that interests you. Which qualifications does the employer's advertisement stress? How? Do you have the qualifications and personal attributes required by the position? If you have relevant skills that have not been developed (and demonstrated) through directly relevant work experience, how will you show that you can indeed perform the tasks required in the position?

5. Write a letter applying for an advertised position. Show how your combined qualifications, skills, and personal qualities match the employer's stated needs.

6. Write a letter applying for a non-advertised position. Show that you understand the position's requirements and that you can meet them. Place your primary appeal in the opening paragraph and elaborate in subsequent paragraphs.

7. Prepare a strategy sheet for a real or imagined job interview. As you do so, plan answers for questions likely to be asked.

COLLABORATIVE PROJECT

In conjunction with other members of your class, prepare an inventory of positions that may be available to graduates of your program of study. Alternatively, prepare an inventory of summer part-time jobs.

Among other information, provide:

◆ descriptions of the positions, their related duties, and required qualifications

◆ expected dates of availability and whether the positions will be advertised externally

◆ background information about the companies offering the positions

◆ salary ranges and benefit packages

◆ names of contact persons, their telephone or fax numbers, and mailing or e-mail addresses

Oral Presentations

Think about the kinds of situations where your speaking skills will lead to success at work. Include informal situations (your firm's general manager stops you in the hallway to ask how a project is progressing) as well as formal situations (you make a half-hour sales presentation to an important client).

Think about opportunities you'll have to

- persuade people to consider, perhaps even accept, your ideas
- provide listeners with the information they need
- foster the image you want and need to project (knowledgeable? skillful? dynamic? competent? supportive?)
- help listeners solve problems, find answers, or resolve issues
- maintain or improve working relationships with your colleagues
- show your clients that you value their business
- explain how things work
- instruct others how to perform tasks
- maintain or build your status within the organization

Now, take your reflections a step further. Think about your reactions to a speaker you've recently heard in a classroom or meeting room. What judgments did you make about that speaker's competence based on the way the speaker spoke? To what degree did the speaker's presentation style affect your willingness to believe the speaker's information and agree with the speaker's ideas? Did you "buy" what the speaker had to say?

As a result of these sets of reflections, are you now more aware of how effective presentation skills could advance or hinder your career?

SPEAKING SITUATIONS FOR TECHNICAL PEOPLE

Many engineers, technologists, technicians, scientists, and other "technical workers" prefer to design, test, build, or improve things, rather than to talk or write about the details of their work. But often these same people can succeed only by communicating their information, instructions, explanations, and analyses. Also, they must communicate effectively to get the facilities, staff, or equipment they need.

Here is a random selection of situations where a technical person's speaking skills could be very important:

- answering key questions in a job interview
- presenting an informal proposal at a department meeting
- introducing a sewer line plan to residents who don't want the line extended into their rural subdivision
- announcing a layoff to the employees affected
- selling a formal proposal to a client
- presenting a technical paper or describing a successful project at a convention

Your degree of success in such situations depends upon your speaking skills and the preparation you put into each presentation.

FACTORS IN SUCCESS

Ask almost anyone about the number one factor in speaking successfully and that person will reply, "confidence." Or, possibly, the response might reflect a negative view—"you have to overcome fear and anxiety to speak well." Certainly, many people fear public speaking—Peter Urs Bender, Canada's guru of business presentations, points to a survey that found people fear public speaking more than death.[1] It seems that cartoonist Scott Adams has heard of the same survey:

Source: DILBERT reprinted by permission of United Feature Syndicate, Inc.

Fear, anxiety, nervousness—these feelings are understandable. After all, others judge us by the way we speak and our work effectiveness frequently depends on our ability to explain and persuade. However, there's no point in dwelling on non-productive feelings. Instead, we need to focus on the factors that make us successful.

You already know the secret to effective speaking—confidence! However, if you currently don't have much confidence, you might find that statement rather hollow. If that's the case, the following hints will probably interest you.

How to build confidence

1. *Have something to say*. Usually, you'll be asked to speak if your experience or role means that you ought to have information and insights that the audience will be glad to hear. But even if you're not an expert on a subject, you can research that subject so that you have plenty to offer. The net result is that your confidence level should rise because you know that you have a worthwhile message.

2. *Know your purpose*. Being clear about the purpose of your presentation really helps you focus on the key ideas and information to stress. When you have that focus, when you know exactly what you're trying to accomplish, and when you know how to achieve your purpose, you'll feel a surge of confidence.

---

1. Peter Urs Bender, *The Secrets of Power Presentations*. 5th edition. Toronto: The Achievement Group, 1991.

Here are some common purposes for talks: inform, explain, persuade, entertain, teach, involve emotions, inspire. Usually, you will try to combine two or more purposes.

3. *Know your audience.* Knowing your listeners' needs and interests should help to further boost your confidence because you can shape your talk to meet your audience's expectations and priorities. Some presenters also prefer to speak to people whom they know because they feel supported by a familiar group of faces. However, the main advantage of a known audience is that you can anticipate audience reactions and you can plan accordingly.

4. *Know the speaking environment and situation.* You can reduce pre-talk jitters and the chance of being flustered during a presentation if you know what to expect. Will the audience have heard other speakers before you get up? Will you be introduced as the highlight speaker? If you're making a sales presentation or oral proposal, how many people will be at the meeting and who will be the prime decision makers? Also, will your competitors have already presented their sales pitches or will they present later?

   You'll also find it useful to *scout the room in advance.* Learn the location of the AC outlets, the lectern, the viewing screen, and the light switches. Plan your placement of VCRs, data projection units, or overhead projectors. Check whether sight lines are obscured by pillars or by your audio visual (AV) placement. Determine where you can move around. Check that the seating arrangements suit your planned presentation. Above all, get a "feel" for the room. You'll feel more confident.

5. *Know how to perform.* This last factor is the most difficult to manage because it takes practice to develop one's performing style. However, here's a good starting point: **BE YOURSELF!** Whatever style you develop, it should grow out of your normal conversational patterns. Depending on your personality and the style required for the situation, you will eventually be able to move to one or more of the following levels.

   a. **Conversational level.** At its best, the conversational style is intimate, relaxed, and natural, but not particularly forceful. It may include bursts of liveliness, but generally stays rather low-key, with just enough volume and projection to allow the audience to hear.

   b. **Heightened conversational level.** Speaking at an augmented conversational level may be the best level for beginning speakers. At this level, the speaker pumps up the volume. He or she increases the energy level and emphasizes key words more forcefully. The speaker is somewhat animated and involved with the audience.

   c. **Performance level.** Now, the speaker is more dramatic. He or she uses vocal emphasis, pacing, pauses, varying volume, and varying vocal tones to really engage listener interest and to signal the speaker's full meaning. Both speaker and listener are aware that this is a performance, but the speaker retains his or her natural spontaneity.

   d. **Oratorical level.** This is suitable for large audiences only. In every way possible, the speaker uses dramatic techniques and theatrical gestures to enflame audience interest and to amplify the speaker's message. This is definitely not a normal way of speaking, but the oratorical style works for political rallies, big religious meetings, and calls to war.

So, where do you fit on that scale? Perhaps you can operate on any one of the levels, depending on the situation. To assess your speaking strengths and limitations, complete Exercise 5 on page 619.

FOUR PRESENTATION STYLES

There are four main ways to deliver oral presentations:

1. You can read a prepared script.
2. You can memorize the whole talk.
3. You can speak impromptu, with little or no prior preparation.
4. You can speak extemporaneously, with a rehearsed set of note cards.

Only highly trained professionals are proficient at *reading* presentations. Untrained readers lose eye contact with their audience, so the audience feels excluded. Also, most people do not read well—their delivery sounds artificial and the audience quickly becomes bored. Further, a scripted presentation allows little chance for revision in mid-delivery.

For several reasons, then, you should not read a speech, not even a complex technical presentation. Your audience for such a presentation would prefer a handout along with your comments about how to read and interpret that document. Still, your job might require you to read a prepared statement. If so, use a double-spaced, large-type script with extra space between paragraphs. Rehearse until you are able to glance up from the script periodically without losing your place.

Like scripted deliveries, most *memorized* speeches don't sound natural. They sound, well, "memorized." Also, you cannot change the content or tone of the speech, even if audience reactions make it obvious that you need to take a different approach. And, of course, the pressure of giving a speech easily plays havoc with your memory. You only have to forget one phrase for whole sections of the memorized talk to temporarily disappear from your memory.

The *impromptu* style brings what the previous two styles lack—spontaneity and real contact with the audience. However, "speaking off the cuff" has its own dangers, most notably the lack of a clear line of development. Impromptu speakers can ramble and thus make very little sense to their listeners. Also, the impromptu approach can put so much pressure on you that you can't retrieve from memory what you know and believe. Further, you run the risk of phrasing things rather awkwardly and ineffectively.

The *extemporaneous* style makes the most sense for nearly all presentations because it provides the benefits of careful preparation (research, structured outline, strong introduction and conclusion, planned transitions, and rehearsed AV aids), as well as the direct contact and spontaneity that this method allows.

DELIVERING EXTEMPORANEOUS TALKS

Briefly, here's how the extemporaneous technique works. The speaker

The extemporaneous process

1. chooses appropriate material and organizes that material in an outline
2. carefully prepares an introduction and conclusion
3. transfers the whole talk to a series of brief notes on note cards
4. rehearses with the notes and with the planned AV materials
5. refers to notes while delivering the talk

The rehearsals are critical to success. In the rehearsals, the speaker can

How to rehearse

◆ find the most effective ways of expressing points
◆ develop clear bridges between various parts of the talk
◆ learn the best way to integrate AV aids and supporting examples or stories
◆ determine how long the planned talk takes to deliver and modify the content accordingly
◆ modify the note cards where necessary

It's important to rehearse thoroughly, but not to the point where you memorize the talk. At that point, you can get trapped into saying things one way only, and thus lose your ability to respond to audience feedback. One of the major advantages of the properly rehearsed talk is that you know what you are about to say and have the freedom to use the best way to phrase ideas to reach a given audience at a given moment. You talk directly to the audience with occasional glances at your note cards.

In the end, speaking directly to the audience is the best way of overcoming stage fright. If you concentrate on getting your message to your audience, and if you heed your audience's reactions, you soon forget about yourself and your natural stage fright. You focus on communicating!

IMPROMPTU SPEAKING

Although impromptu speaking is not recommended for any situation where you have adequate time to prepare, certain situations occur where impromptu skills can help you achieve your goals. Job interviews come immediately to mind, for example. Also, you'll frequently be asked questions after your presentations or you'll be asked for your opinion during meetings.

THERE'S NOTHING TO FEAR BUT FEAR ITSELF

Quite simply, many people don't perform well as impromptu speakers because of panic. Suddenly, the pressure is on and your brain freezes. Later, perhaps just a few seconds later, you know exactly what you should have said, but it's too late. You have spouted some gibberish, and feel like a fool.

HERE'S WHAT TO DO

Flustered speakers need something that will allay the panic long enough to get their brain moving again. At the same time, they need a device to comb through their

How to overcome
panic

brain to retrieve what they know about the subject. And finally, that magic "something" must be used to make sense of what the speakers have found in their brain.

Sound impossible? Actually, the solution is really quite simple and effective—the speaker needs an *instant organizing device* such as a *statement-elaboration pattern*. If you're asked in a job interview why you chose to study at a certain university, you might pause for a second and say:

> "I chose this university mainly because the graduates from my program are qualified for a wide range of positions, and most of those graduates get employed in their field."

The above statement buys you a little time to think, but more importantly you have an idea and you have an organizing method! The statement-elaboration pattern allows you to expand upon each of the main parts of the statement:

- when and how you chose your university and which other colleges you investigated
- the "wide range of positions"
- some of the "employed graduates" you know (or statistics which you may happen to know)

Knowing that you have a "method" allows the panic to subside long enough for you to retrieve and organize what you really do have to say. You're not stuck! Now, this method requires practice before you can get really proficient with it, but you'll be amazed at how well you can teach yourself to use this pattern and the following impromptu organizing patterns.

USE ADDITIONAL IMPROMPTU PATTERNS

1. *Chronological.* Time patterns are built into our language and thinking. For example, we readily understand the *past/present/future* pattern. Or, we can organize our thoughts according to a series of dates that lead up to the present.

   > This technique could work for a 28-year-old female who has completed an engineering program. In a job interview she is asked why she chose engineering. She might reply that starting when she was 10 years old, she used to help her father repair equipment at the family-run amusement arcade. Then, when she was 14, she won a science contest with a design for a gravity pump for the family home's fish pond. When she was 19 . . .

   And, so it goes. Chronological patterns are easy to use because we organize our brains with certain milestones and accompanying memories.

2. *Narrative.* Similarly, we can use story patterns to retrieve key bits of information from our brains and then use the same story (or "narrative") to organize our thoughts on a subject. For example:

   > If the female engineering grad in the job interview mentioned above were asked to explain whether she could effectively supervise a work crew on a forest road deactivation project, she could tell a story about a parallel experience she had during a work placement last year. At various points she could interrupt her narrative to point out the skills she demonstrated during that earlier experience.

3. *Topical arrangement.* In a job interview for a technologist position, a 37-year-old graduate of an engineering technology program might be asked about his strengths relative to the position being discussed. He might say:

> "I can answer that by discussing three main areas: my previous factory experience, my talent for computer drafting and design, and the valuable experience I gained during my four-month work placement at your firm last year." Then, he would discuss each topic in turn.

4. *Cause-effect.* This two-part structure is described by its name. It can start with causative factors and work toward the results of those factors, or it can start with a description of a situation and work back to the causes of that situation. For example:

> The head of a highway maintenance team was recently asked why a certain highway had so many potholes during the previous winter. He began by briefly describing the physical features of some of the more common types of potholes and then recounted the special factors that result in those types of potholes, factors that were particularly prevalent that winter.

Now, before we get lost in a consideration of potholes, let's remember the main benefit of that highway maintenance spokesperson's use of the cause-effect pattern—long after listeners forgot the details of his explanation, they remembered that he seemed sensible and trustworthy!

PREPARING YOUR PRESENTATION

Plan the presentation systematically, to stay in control and build confidence. For our limited purposes here, we will assume your presentation is extemporaneous.

RESEARCH YOUR TOPIC

Do your homework. Be prepared to support each assertion, opinion, conclusion, and recommendation with evidence and reason. Check your facts for accuracy. Your listeners expect to hear a knowledgeable speaker; don't disappoint them.

Begin gathering material well ahead of time. Use summarizing techniques from Chapter 9 to identify and organize major points. If your presentation is a spoken version of a written report, most of your research has been done, but you may need to introduce material not found in the written version in order to appeal to listeners rather than readers.

AIM FOR SIMPLICITY AND CONCISENESS

Keep your presentation short and simple. Boil the material down to a few main points. Listeners' normal attention span is about 20 minutes. After that, they begin tuning out. Time yourself in practice sessions and trim as needed. If the material requires a lengthy presentation, plan a short break, with refreshments if possible, about halfway.

ANTICIPATE AUDIENCE QUESTIONS

Consider those parts of your presentation that listeners might question or challenge. You might need to clarify or justify information that is new, controversial, disappointing, or surprising.

OUTLINE YOUR PRESENTATION

Review Chapter 10 for organizing and outlining strategies. When you're preparing a talk, especially a lengthy, complex presentation, you can use the same kinds of organizing techniques as in a written document. However, the patterns in a written report might not suit an oral presentation. For example, see the report outlines on pages 454–56, Chapter 19. That outline suits a written technical report, but not an oral version of the report. If Art Basran, the report's author, is asked to present his report orally to Jackson Mining's Board of Directors, he might revise his working outline as follows:

INTRODUCTION

Connect with audience

- Express appreciation for opportunity to address Board.
- Review Jackson Mining's requirements and the four transport alternatives and analysis method.

Provides background

- Show **Overhead** of 4 alternatives.
- Show **Overhead** of guidelines for analysis.
- Discuss **handout:** background info re: 4 haul alternatives.

CONCLUSION

Provides bottom line early in talk

- The diesel and electric train options give much better results than diesel trucks and electric train fleets.
- Which is better, diesel train option or electric train option?
- Show **Overheads** of Table 4: Diesel Trains as Best Alternative and Table 5: Electric Trains as Best Alternative.
- Figures are actually closer than Tables 4 and 5 might indicate at first.
- *MMT Consulting cannot recommend which specific option to choose because of the factors beyond the scope of this report, which you must decide.*

Provides transition to the detailed analysis

- **My purpose here today:**
  Help you understand the analysis behind our findings so you can make the best decision for your operation.

COST ANALYSIS

Annual Costs

Discusses (and shows) highlights of the analysis, not full details

- *Fuel consumption*: Why important, and how calculated (Refer to Appendix A of report). Show **Overhead** of Table 1: Tractive and Grade Resistances for Diesel Trucks.
- *Labour*: Method of calculating. Refer to Appendix B.
- *Maintenance*: Quick reminder only.

Capital Costs

- *Transportation units*: Quick overview of calculation method.
- *Construction costs*: Show 3 main factors—**Overhead.**
  Show calculation method—**Overhead.**
  Comment re: maintenance buildings.

Communication System

◆ Name components (brief).

Cost Comparisons

◆ Show **Overhead** of Table 3.
◆ Explain Table 3 and lead into Equivalent Uniform Annual Costs.

Equivalent Uniform Annual Costs

◆ Why we chose EUAC comparisons.
◆ Show **Overhead** of Figure 1: Show how to read it.
  Refer to Appendix D (report **handouts**).
  Explain the sharp decline in EUAC.

> Stop for questions at this point

POTENTIAL FOR EXPANSION

◆ Refer to conversation with Jessica Proctor re: flexibility (potential for expansion).
◆ Use **Overheads** to show how:
  – an immediate increase in production of 23.3% could be achieved for diesel trucks
  – diesel and electric trains could handle up to a 53.5% increase in production
  – electric fleets have room for 130% increase in production

LEAD-OUT

◆ Wish Board success with this venture.
◆ Indicate MMT's willingness to oversee road construction.
◆ Invite further questions.

Maintains business connection with audience

CAREFULLY PREPARE YOUR INTRODUCTION AND CONCLUSION

Intriguing introductions

Speech *introductions* are very important. You must gain audience attention, establish your credibility, and lead in to the content of your talk. In the process, you will establish the mood of your presentation. Three groups of related techniques for immediately engaging your listeners are listed in Table 24.1.

Table 24.1   Effective Methods for Engaging Listeners

Establish the Topic:	Grab Audience Attention:	Connect with Your Audience
◆ Refer directly to subject.	◆ Refer to recent events.	◆ Establish common ground.
◆ Provide a pre-summary.	◆ State a startling fact or idea.	◆ Promise a reward.
◆ Delimit the topic.	◆ Give illustration or story.	◆ Present your credentials.
◆ Provide background info.	◆ Ask a rhetorical question.	◆ Justify the topic.
◆ Define terms.		

The *conclusion* to your talk provides your last chance to confirm your main point. Sometimes you may also want to finish on an emotional high or persuade your listeners to act in a certain way. And, no matter what the purpose of your talk has been, your closing phrasing should signal a strong sense of conclusion; you should never have to tell your audience, "That's the end of my talk."

Here are some techniques for closing strongly:

Strong conclusions

- Summarize your main points.
- Request a specific action from your audience.
- Challenge the audience to reach certain goals.
- Declare *your* intended action.
- Quote a memorable phrase or piece of poetry.
- Finish with a powerful example or story.
- Repeat a key phrase from the introduction.
- Show how you've fulfilled a promise made in the introduction.

Some combination of two or more of the above techniques will suit any presentation.

BUILD BRIDGES FOR YOUR LISTENERS

Most audience members don't listen attentively all the time; they tune in and out. When they *do* tune back in, you need to help them understand what direction the talk has taken while they were "away." Not all audience members wander, but even the best listeners find it difficult to follow speakers who do not provide bridges from one section of a talk to the next.

The following bridging techniques can help the best and worst of listeners stay on your idea track:

Keep listeners on track

1. *Build bridges at the beginning.* Preview the presentation's structure so that your listeners know where you're headed and can anticipate the relationships among the topics you'll be discussing. If your presentation is lengthy and complex, an agenda in poster form or a handout would help the audience.

2. *Use the organization structure that best suits the topic.* A presentation about overcoming speaker nervousness could use a problem-solution pattern. A presentation about the development process for a residential subdivision could employ a chronological pattern, perhaps with key deadline dates assigned to each stage of the process.

3. *Build word bridges during the presentation.* Transitional phrases can take several forms:

   - Confirm the direction of the presentation. (Now let's consider some other methods of reducing waste.)
   - Change the direction of the presentation. (We've identified the problem; now it's time to find a solution.)
   - Lead into an example or a detailed list of evidence. (If you want assurance that this program actually works, here's a case for you to consider . . .)
   - Develop an idea further. (In addition to their contribution to the local ecology, these marsh lands help local tourism . . .)

4. *Use repetition to confirm key ideas and to re-orient wayward listeners.* You could repeat a key phrase (such as Martin Luther King's "I have a dream"). Or, you could repeat an idea without making the repetition obvious (*Nothing to fear but fear itself; You're your own worst critic; Believe in your abilities as much as we do*). You could summarize the points you've just made in the preceding section—these *stage summaries* help those who've not listened as attentively as they might have.

PLAN YOUR VISUALS

Visuals increase listeners' interest, focus, understanding, and memory. Select visuals that will clarify and enhance your talk—without making you fade into the background.

Decide Where Visuals Will Work Best.   Use visuals to emphasize a point and enhance your presentation whenever *showing* would be more effective than merely *telling.*

Decide Which Visuals Will Work Best.   Will you need numerical or prose tables, graphs, charts, graphic illustrations, computer graphics? How fancy should these visuals be? Should they impress or merely inform? Use the visual planning sheet in Chapter 12 to guide your decisions.

List Your Visuals in Your Notes.   You can list your visuals as in Art Basran's outline on pages 453 and 456. Or, you can create two columns, with the content notes in the left-hand column and the corresponding visuals in the right-hand column.

Decide How Many Visuals Are Appropriate.   Use an array of lean, simple visuals that present material in digestible amounts instead of one or two over-stuffed visuals that people end up staring at endlessly.

Fit each visual to the situation

Decide Which Visuals Are Achievable.   Fit each visual to the situation. The visuals you select will depend on the room, the equipment, and the production resources available.

How large is the room and how is it arranged? Some visuals work well in small rooms but not large ones, and vice versa. How well can the room be darkened? Which lights can be left on? Can the lighting be adjusted selectively? What size should visuals be to be seen clearly by the whole room? (A smaller, intimate room usually is better than a room that is too big and cavernous.)

What hardware is available (slide projector, opaque projector, overhead projector, film projector, videotape player, data projection unit, projection screen)? How far in advance does this equipment have to be requested? What graphics programs are available? Which is best for your purpose and listeners?

What resources are available for producing the visuals? Can drawings, charts, graphs, or maps be created as needed? Can transparencies (for overhead projection) be made or slides collected? Can handouts be keyed and reproduced? Can multi-media displays be created?

Select Your Media.   Fit the medium to the situation. Which medium or combination is best for the topic, setting, and listeners? How fancy do listeners expect this to be? Which media are appropriate for this occasion? Examples appear on pages 609–610.

Fit the medium to the situation

- For a weekly meeting with colleagues in your department, scribbling on a blank transparency, chalkboard, or dry-erase markerboard might suffice.
- For interacting with listeners, you might use a whiteboard or chalkboard to record audience responses to your questions.
- For immediate orientation, you might begin with a poster or flip chart sheet listing key visuals/ideas/themes to which you will refer repeatedly.
- For helping listeners take notes, absorb technical data, or remember complex material, you might distribute a presentation outline as a preview or provide handouts.
- For a presentation to investors, clients, or upper management, you might require polished and professionally prepared visuals, including computer graphics, multimedia, and state-of-the-art technology.

Figure 24.1 presents the various common media in approximate order of availability and ease of preparation.

PREPARE YOUR VISUALS

As you prepare visuals, focus on economy, clarity, and simplicity.

Be Selective.   Use a visual only when it truly serves a purpose. Use restraint in choosing what to highlight with visuals. Try not to begin or end the presentation with a visual. At those times, listeners' attention should be focused on the presenter instead of the visual.

Make Visuals Easy to Read and Understand.   Think of each visual as an image that flashes before your listeners. They will not have the luxury of studying the visual at leisure. Listeners need to know at a glance what they are looking at and what it means. The following are guidelines for achieving readability.

GUIDELINES

for Readable Visuals

- *Make visuals large enough to be read anywhere in the room.*
- *Don't cram too many words, ideas, designs, or type styles onto a visual.*
- *Keep wording and images simple.*
- *Boil your message down to the fewest words.*
- *Break things into small sections.*
- *Summarize with key words, phrases, or short sentences.*
- *Use 18–24 point type size and sans serif typeface (White, Great Pages 80).*

In addition to being able to *read* the visual, listeners need to *understand* it. Following are guidelines for achieving clarity.

Whiteboard/Chalkboard

Uses • simple, on-the-spot visuals
 • recording audience responses
 • very informal settings
 • small, well-lighted rooms

Tips • copy long material in advance
 • write legibly
 • make it visible to everyone
 • use washable markers on whiteboard
 • speak to the listeners—not to the board

Poster

Uses • overviews, previews, emphasis
 • recurring themes
 • formal or informal settings
 • small, well-lighted rooms

Tips • use 51 cm x 76.5 cm (20" x 30")
 posterboard or larger
 • use intense, washable colours
 • keep each poster simple and uncrowded
 • arrange/display posters in advance
 • make each visible to the whole room
 • point to what you are discussing

Flip Chart

Uses • a sequence of visuals
 • back-and-forth movement
 • formal or informal settings
 • small, well-lighted rooms

Tips • use an easel pad and easel
 • use intense, washable colours
 • work from a storyboard
 • check your sequence beforehand
 • point to what you are discussing

Handouts

Uses • present complex material
 • help listeners follow along
 • help listeners take notes
 • help listeners remember

Tips • staple or bind the packet
 • number the pages
 • try saving for the end
 • if you must distribute up-front, ask
 listeners to await instructions
 before reading the material

Figure 24.1   Selecting Media for Visual Presentations   *(continued)*

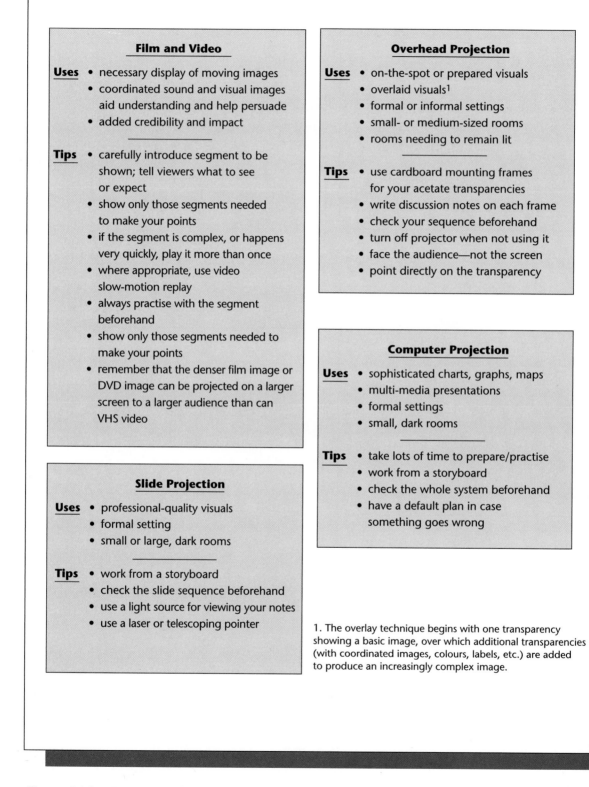

Film and Video

Uses
- necessary display of moving images
- coordinated sound and visual images aid understanding and help persuade
- added credibility and impact

Tips
- carefully introduce segment to be shown; tell viewers what to see or expect
- show only those segments needed to make your points
- if the segment is complex, or happens very quickly, play it more than once
- where appropriate, use video slow-motion replay
- always practise with the segment beforehand
- show only those segments needed to make your points
- remember that the denser film image or DVD image can be projected on a larger screen to a larger audience than can VHS video

Slide Projection

Uses
- professional-quality visuals
- formal setting
- small or large, dark rooms

Tips
- work from a storyboard
- check the slide sequence beforehand
- use a light source for viewing your notes
- use a laser or telescoping pointer

Overhead Projection

Uses
- on-the-spot or prepared visuals
- overlaid visuals[1]
- formal or informal settings
- small- or medium-sized rooms
- rooms needing to remain lit

Tips
- use cardboard mounting frames for your acetate transparencies
- write discussion notes on each frame
- check your sequence beforehand
- turn off projector when not using it
- face the audience—not the screen
- point directly on the transparency

Computer Projection

Uses
- sophisticated charts, graphs, maps
- multi-media presentations
- formal settings
- small, dark rooms

Tips
- take lots of time to prepare/practise
- work from a storyboard
- check the whole system beforehand
- have a default plan in case something goes wrong

1. The overlay technique begins with one transparency showing a basic image, over which additional transparencies (with coordinated images, colours, labels, etc.) are added to produce an increasingly complex image.

Figure 24.1   Selecting Media for Visual Presentations

GUIDELINES

**for
Understandable
Visuals**

- *Display only one point per visual—unless previewing or reviewing (White 79).*
- *Give each visual a title that announces the topic.*
- *Use colour, sparingly, to highlight key words, facts, or the bottom line.*
- *Use the brightest colour for what is most important (White, Great Pages 78–79).*
- *Label each part of a diagram or illustration.*
- *Proofread each visual carefully.*

When your material is extremely detailed or complex, distribute handouts to each listener.

Look for Alternatives to Word-filled Visuals.   Instead of presenting mere overhead versions of printed pages, explore the full *visual* possibilities of your media. For example, anyone who tries to write a verbal equivalent of the visual message in Figure 24.2 soon will appreciate the power of images in relation to words alone. Whenever possible, use drawings, graphs, charts, photographs, and other visual representation discussed in Chapter 12.

**Figure 24.2
Images More
Powerful Than
Words**

Source:
"Distribution of
the World's
Water," as
appeared in *WWF
Atlas of the
Environment* by
Geoffrey Lean and
Don Hinrichsen.
Copyright © 1994
Banson Marketing
Ltd. Reprinted
with permission.

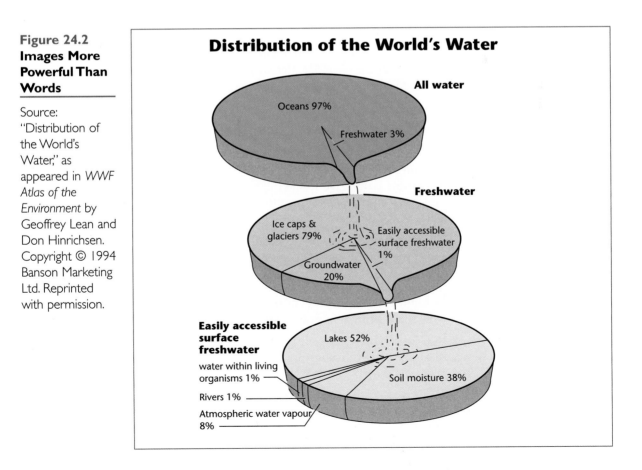

Use all Available Technology.   Using desktop publishing systems and presentation software such as PowerPoint™ or Inspiration™, you can create professional-quality visuals and display technical concepts. Using hypertext and multi-media systems, you can create dynamic presentations that appeal to the listener's multiple senses. Using an automatic, remote-controlled transparency feeder and a laser pointer (pencil-sized), you can deliver a smooth and elegant presentation. These are just a few of the possibilities inherent in the technology.

Check the Room and Setting Beforehand.   Make sure that you have enough space, electrical outlets, and tables for your equipment. If you will be addressing a large audience by microphone and plan to point to features on your visuals, be sure the microphone is movable. Pay careful attention to lighting, especially for whiteboards, chalkboards, flip charts, and posters. Don't forget a pointer if you need one.

PREPARE NOTE CARDS

Note cards build your confidence—you know that you won't be stuck for something to say. Think of note cards as insurance against going blank in front of your audience. More importantly, note cards help keep you from wandering. When you work from a structured set of notes, there's less chance of going off on tangents.

GUIDELINES

for Creating
Note Cards

1. Use card stock or heavy paper (32 lb. or heavier), so the notes don't shake or droop.

2. Choose a size that suits you. Probably the smallest effective size is 7.5 cm x 12.5 cm (3" x 5"). Remember that the larger the card, the more you can place on it.

3. Use no more than 10 cards. Four to six cards will suffice for most talks.

4. Write on one side of each card. Number the cards sequentially in the upper right-hand corner.

5. Use point form to remind yourself of the things you want to say. Write large, legible points. Do not print your speech out in full; you'll be tempted to read the material, and if you succumb to the temptation, you'll lose your audience.

6. Write sentences for the few critically important statements that you want to get exactly right (no more than three to five per talk).

7. Include:

   ◆ headings or names of sections of the talk
   ◆ your main points, in point form
   ◆ key statistics, facts, or quotes that you may not remember
   ◆ reminders to yourself—Breathe! Don't pace! Pause here. Slow down! Look at them!
   ◆ AV usage cues—overhead: Table 1; draw on whiteboard; dim lights
   ◆ time limits for certain parts of the presentation

However, you do need to discipline yourself. Notes will help only if you regularly check them to make sure that you're still on course and that you haven't omitted key points. And you do need to rehearse with the cards so that you're sure that you can read them and that you're comfortable with them. Rehearsal builds confidence.

Figure 24.3 illustrates note cards that Art Basran might use for his presentation to the Jackson Mining Board:

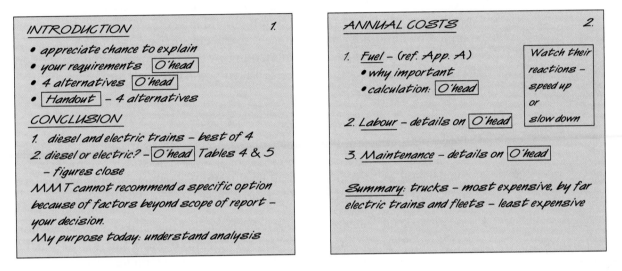

Figure 24.3 **Sample Note Cards**

REHEARSE YOUR TALK

Reread the advice on page 601 to refresh your memory of how and what to rehearse. Pay particular attention to the introduction and conclusion, both of which must be presented smoothly and confidently. If possible, videotape one of your rehearsed deliveries to determine what might be improved. Then, use the evaluation sheet on page 621 to guide your improvements.

 NOTE   *Most speakers find that their rehearsal time is about 80 percent of actual performance time, unless nervousness causes them to forget material or to speak too quickly.*

DELIVERING YOUR PRESENTATION

You have planned and prepared carefully. Now consider the following simple steps to make your actual presentation enjoyable instead of terrifying.

1. *Connect With Your Audience*
   A successful presentation involves relationship building between presenter and audience.

   ◆ *Get to Know Your Audience.* Try to meet some audience members before your presentation. We all feel more comfortable with people we know. Don't be afraid to smile.

◆ *Display Enthusiasm and Confidence.* Audience members prefer lively, engaging speakers. So, overcome your shyness; research indicates that shy people are seen as less credible, trustworthy, likeable, attractive, and knowledgeable. Also, use downward inflections at the end of sentences, to project self-confidence.

◆ *Be Reasonable.* Don't make your point at someone else's expense. If your topic is controversial (layoffs, policy changes, downsizing), decide how to speak candidly and persuasively with the least chance of offending anyone. Avoid personal attacks. For example, a groundwater consultant discussing his or her evaluation of the town's water supply should assess the system's strengths and limitations, and not criticize those who manage that system.

◆ *Don't Preach.* Speak like a person talking—not like someone giving a sermon. Use we, you, your, and our to establish commonality with the audience. Avoid jokes or wisecracks.

2. **Adjust Your Presentation Style**
   Pages 599 and 600 of this chapter discuss four levels of speaking; try to operate at least at a heightened conversational level style and move to a performance level or even an oratorical level when you're comfortable at that level and when the occasion demands a more dramatic approach.

◆ *Use Natural Movements and Reasonable Postures.* Move and gesture as you normally would in conversation (except for more dramatic moments), and maintain reasonable postures. Avoid foot shuffling, pencil tapping, swaying, slumping, fidgeting, or nervous pacing.

◆ *Adjust Volume and Rate.* Adjust your volume to the size of the audience and to the situation—for example, you'll need to speak louder when a projector fan is running. When you use an amplified microphone for a large audience, speak more intimately, as if to one person or to a small group.

   Nervousness causes some speakers to gallop along and mispronounce things. Slow down and enunciate clearly. Vary your rate to reflect the changing mood or content of the talk.

◆ *Maintain Eye Contact.* Look directly into listeners' eyes. With a small audience, eye contact is one of your best connectors. As you speak, establish eye contact with as many listeners as possible. With a large group, maintain eye contact with those in the first rows. Establish eye contact immediately—before you even begin to speak—by looking around.

◆ *Avoid Vocal Fillers.* We use fillers such as "uhh," "eh," "like," and "um" in normal conversation out of habit, often to non-verbally signal that we're not finished speaking and that we don't want to be interrupted. However, frequent use of such fillers can really annoy audiences.

◆ *Vary Your Tone.* Good speakers skillfully use the right tones to signal the emotional content of their messages. A variety of vocal tones helps listeners stay awake!

3. **Manage the Communications Flow**
   Do everything you can to keep things running smoothly.

◆ *Be Responsive to Listener Feedback.* Assess listener feedback continually and adjust the talk as needed. If you are labouring through a long list of facts or figures and people begin to doze or fidget, you might summarize. Likewise, if frowns, raised eyebrows, or questioning looks indicate confusion, skepticism, or indignation, you can backtrack with a specific example or explanation. You might also prepare more than one example to illustrate a given point, so that you can select the best example to match the audience's current mood.

◆ *Stick to Your Plan.* Say what you came to say, then summarize and close—politely and on time. Don't punctuate your speech with digressions that pop into your head. Unless a specific anecdote was part of your original plan to clarify a point or increase interest, avoid excursions. We often tend to be more interested in what we have to say than our listeners are!

◆ *Leave Listeners with Something to Remember.* Before ending, take a moment to summarize the major points and re-emphasize anything of special importance. Are listeners supposed to remember something, have a different attitude, take a specific action? Let them know!

◆ *Allow Time for Questions and Answers.* If questions suit the format of the event, at the very beginning, tell your listeners that a question-and-answer period will follow. Observe the following guidelines for managing listener questions diplomatically and efficiently.

SPEAKING AT MEETINGS

A recent Internet search for sites discussing "effective meetings" led to 1.6 **million** results! Clearly, there's a market for advice on how to conduct effective meetings. Here, we focus on preparing for meetings and on the skills required to speak successfully at meetings.

THE MEETING MANAGER'S ROLE

In most cases, a leader is responsible for calling meetings, setting the goals for a meeting, and distributing agendas. That leader might run meetings as part of his or her job duties, or a task group might elect one of its members to organize and chair its meetings.

Setting the Agenda. Group members often contribute agenda items, but the chairperson must organize those items into a smooth-flowing agenda. That agenda should do more than simply list topics; it should help participants prepare for their roles in the meeting. Figure 24.4 shows such an agenda, which involves employees of Mountain Environmental Consultants. Present at the meeting will be the firm's managing partner (Rene Aubois), two members of the surveying team (Orvald Tomsen and Jas Dhaliwal), and three members of an environmental assessment team (Joe Silveira, engineer, and Rita Cherneski and Chris Dohrmann, technologists).

Chairing the Meeting. Usually, the designated meeting manager chairs the meeting. This person introduces the meeting's goals, provides necessary background, and establishes the maximum meeting length. The chair also keeps the discussion on track and gives everyone a chance to speak. The chair should summarize consensus decisions and guide debates, from the time that motions are made to the final decision. In situations where participants regulate themselves, the chair guides the discussion; in other situations, the chair must wield his or her authority to control undisciplined participants.

Notice that the agenda in Figure 24.4 reflects an inclusive, democratic leadership style that encourages participation. Some leaders, however, favour a more autocratic style in which meeting participants have less influence than the leader.

Meeting purpose: Establish a procedure for an environmental assessment of the proposed Chetwekoo Beach reclamation project

Place and time: Mountain Environmental board room. 7:30 a.m., 27 March 2002

Item:	Responsible:	Supporting information:	Discussion time:
1. Announcements and project overview	Rene	Attached Appendix A	3 to 5 mins.
2. Geological overview	Joe	Provided by Joe at meeting	3 to 5 mins.
3. Potential surveying times and problems	Orvald and Jas	Provided by Orvald & Jas, if necessary	3 to 5 mins.
4. History of two similar assessments we've done	Joe	n.a.	3 to 5 mins.
5. **Discussion:** procedure for surveying and for assessing the impact of reclaiming the beach	All—review all required resources and M.O.E. regulations	All: bring pertinent documents	15 to 25 mins.
6. **Action plan:** for the environmental assessment project	All		15 to 25 mins.
Follow-up: write business proposal for submission to the Chetwekoo Regional District	Rene, with input from all	Material generated during meeting	**Deadline for proposal:** April 7

Figure 24.4   Sample Meeting Agenda

THE MEETING PARTICIPANT'S ROLE

Many meeting participants simply show up at meetings, with no prior preparation. The participants in the Mountain Environmental meeting (Figure 24.4) apparently will not have such an option. Still, participants should not need a totally structured agenda to spur them to prepare.

You can get ready to contribute to meetings by

◆ making notes about points you'd like to make about certain agenda items
◆ gathering relevant information and bringing pertinent documents
◆ preparing visuals to help support our points

In other words, you should prepare to do more than simply occupy a chair at meetings. When you do speak, the following guidelines will help you contribute to a productive meeting.

GUIDELINES

for Speaking Effectively at Meetings

1. *Plan and rehearse your comments* when you're responsible for introducing the discussion or for reporting to the group.

2. *Prepare visuals to illustrate and support your points.* Effective visuals can reduce meeting time and improve the chances of reaching consensus. Designated presenters should particularly consider using visuals, but other participants might bring their own visuals.

3. *Speak only if you can contribute meaningfully to the discussion.* Do not talk just to hear your voice, or to maintain your status within the group.

4. *Don't waste time during discussions.* Do not repeat what someone else has already said. Don't repeat what you said earlier. Don't ramble while you discover what you have to say—know the basic structure of your point before you speak.

5. *Listen carefully to others' comments.* Also, if you're not certain of another's intended meaning, ask for clarification before responding.

6. *Express your views tactfully.* Focus on the facts and how the facts relate to the item being discussed. Don't attack others, even though it may be necessary to disagree with others' conclusions or opinions.

7. *In sales meetings, focus on providing value to your listeners.* Speak to your listeners' needs, interests, and priorities.

GUIDELINES

for Managing Listener Questions

◆ *Announce a specific time limit* for the question period to avoid prolonged debates.
◆ *Listen carefully to each question.*
◆ *If you can't understand a question,* ask that it be rephrased.
◆ *Repeat every question,* to ensure that everyone hears it.
◆ *Be brief in your answers.*
◆ *If you need extra time to answer a question,* arrange for it after the presentation.
◆ *If anyone attempts lengthy debate,* offer to continue after the presentation.
◆ *If you can't answer a question,* say so and move on.
◆ *End the session with,* "We have time for one more question," or some such signal.

In Brief ORAL PRESENTATIONS FOR CROSS-CULTURAL AUDIENCES

Imagine you've been assigned to represent your company at an international conference or before international clients (e.g., of passenger aircraft or computers). As you plan and prepare your presentation, remain sensitive to various cultural expectations.

For example, some cultures might be offended by a presentation that gets right to the point without first observing formalities of politeness, well wishes, and the like.

Certain communication styles are welcomed in some cultures but considered offensive in others. In southern Europe and the Middle East, people expect direct and prolonged eye contact as a way of showing honesty and respect. In Southeast Asia, this may be taken as a sign of aggression or disrespect (Gesteland 24). A sampling of the questions to consider:

◆ Should I smile a lot or look serious? (Hulbert, *Overcoming* 42)

◆ Should I rely on expressive gestures and facial expressions?

◆ How loudly or softly, rapidly or slowly should I speak?

◆ Should I come out from behind the lectern and approach the audience or keep my distance?

◆ Should I get right to the point or take plenty of time to lead into and discuss the matter thoroughly?

◆ Should I focus on only the key facts or on all the details and various interpretations?

◆ Should I be assertive in offering interpretations and conclusions, or should I allow listeners to reach conclusions on their own?

◆ Which types of visuals and which media might work or not work?

◆ Should I invite questions from this audience, or would this be offensive?

To account for language differences, prepare a handout of your entire script for distribution after the presentation, along with a copy of your visuals. This way, your audience will be able to study your material at leisure.

EXERCISES

1. In a memo to your instructor, identify and discuss the kinds of oral reporting duties you expect to encounter in your career.

2. Design an oral presentation for your class. (Base it on a written report.) Make a sentence outline, and a storyboard that includes at least three visuals. Practise with a tape recorder or a friend. Use the peer evaluation sheet (page 623) to evaluate your delivery.

3. Observe a lecture or speech, and evaluate it according to the peer evaluation sheet. Write a memo to your instructor (without naming the speaker), identifying strong and weak areas and suggesting improvements.

4. In an oral presentation to the class, present your findings, conclusions, and recommendations from the analytical report assignment in Chapter 19.

5. Prepare a speaker self-assessment. You'll find the self-assessment form on pages 621 and 622 very useful if the assessment is done objectively and thoroughly. To improve objectivity, either enlist the aid of one or more committed, objective observers, or videotape a presentation and analyze it with the help of the self-assessment form. Alternatively, you can view and analyze the tape together with some supportive critics.

    You may find the definitions, comments, and advice on page 622 helpful.

6. Use the following criteria to evaluate the Power-Point slides in Figure 24.5 (page 620):

    ◆ Are the slides visually appealing?
    ◆ Are all visuals simple? Does each slide spotlight only one major point?
    ◆ Have the same type size and typeface been used for all major headings?
    ◆ Is the page design consistent from one slide to the next?
    ◆ Do foreground colours stand out from background colours?
    ◆ Does each slide present only a few points? (One often-quoted rule stipulates that each slide should have no more than seven lines, with no more than seven words per line.)

COLLABORATIVE PROJECT

Exercises 2, 3, 4, and 6 may be done as collaborative projects.

When time counts and lives depend on your tools, you want the best tool for the job

ASSESSMENT CRITERIA

♦ Cost comparison

♦ Material specifications and Design

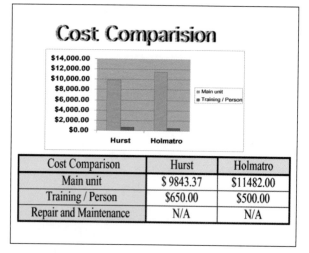

Cost Comparision

Cost Comparison	Hurst	Holmatro
Main unit	$ 9843.37	$11482.00
Training / Person	$650.00	$500.00
Repair and Maintenance	N/A	N/A

Material Specifications And Design

HURST

- Lightweight, compact design
- Field replaceable tips
- Dual pilot check valve
- "Deadman" control

Material Specifications And Design

HOLMATRO

- Light weight/easy to handle
- Twist handle switch
- Built-in check valves
- Protective rubber cap

CONCLUSION

Cost comparison   **HURST**

Material specifications and design   **HURST**

Figure 24.5 Sample PowerPoint Slides Courtesy of Dean Ashby, Sandy Hagel, Matt Nakazawa, and Brandin Slonski.

SPEAKER SELF-ASSESSMENT

Item	Description	Audience Reactions?	Improvements?
Appearance Physical presence Style of dress/grooming			
Voice Timbre Volume/projection Resonance			
Vocal Delivery Clear sounds Tonal range/skill Tonal variety Inflections Emphasis Pacing: rate & variety Vocal fillers			
Other Non-verbals Eye contact Facial expressions Body movements Hand/arm gestures Nervous habits			
Language Use Slang Clear phrasing Appropriate vocabulary level			
Performance Level Conversational Heightened Performance Oratorical			

Appearance

Physical presence . . . . Sheer size gives an advantage, but smaller people can compensate by being physically active.

Dress/grooming . . . . Your appearance is particularly important in a talk's opening moments.

Voice

Timbre . . . . The nature of your vocal cords: bass? reedy? shrill? birdlike? throaty?

Volume/projection . . . . "Volume" ranges from a whisper to a shout, but a whisper can be projected nearly as far as a shout.

Resonance . . . . To achieve more resonance, vibrate your vocal cords more vigorously.

Vocal Delivery

Clear sounds? . . . . To avoid slurring sounds, take care to say **each sound** distinctly.

Tonal range/skill . . . . Can you convey a full range of emotions by changing vocal tones?

Tonal variety . . . . It's essential to vary your tones and avoid the dreaded monotone.

Inflections . . . . Downward inflections at the end of statements sound confident and authoritative, but upward inflections make you sound unsure.

Emphasis . . . . Emphasizing keywords is a distinguishing feature of "public" speaking.

Delivery rate . . . . Speaking quickly is okay **if** you emphasize keywords and speak clearly.

Pacing: variety . . . . Effective speakers vary their pace and use pauses to emphasize points.

Vocal fillers . . . . "Like," "uh," "um," "okay," and "eh" are the most common fillers used by Canadians; these fillers can distract listeners.

Other Non-verbals

Eye contact . . . . If you want to keep audience interest, you **must** look at your listeners.

Facial expressions . . . . Lively facial expressions show interest in your listeners.

Body movements . . . . These can vary with the situation, but they should be controlled in business presentations.

Hand/arm gestures . . . . These should be used naturally, not in a contrived or wooden manner.

Nervous habits . . . . Our bodies reveal our real feelings, so we can't eliminate nervous signs completely; however, we should stifle those which distract audiences.

Language Use

Slang? . . . . Slang is inappropriate in all formal business settings.

Clear phrasing? . . . . Public talks require well-conceived, clear descriptions and explanations.

Level of vocabulary . . . . Carefully match your level of vocabulary to your audience, without talking down to them, or overusing jargon.

Peer Evaluation Sheet for Oral Presentations

Presentation Evaluation for (name/topic) _____

Comments

Content

☐ Began with a clear purpose. _____

☐ Showed command of the material. _____

☐ Supported assertions with evidence. _____

☐ Used adequate and appropriate visuals. _____

☐ Used material suited to this audience's
needs, knowledge, concerns, and interests. _____

☐ Acknowledged opposing views. _____

☐ Gave the right amount of information. _____

Organization

☐ Presented a clear line of reasoning. _____

☐ Used transitions effectively. _____

☐ Avoided needless digressions. _____

☐ Summarized before concluding. _____

☐ Was clear about what the listeners
should think or do. _____

Style

☐ Seemed confident, relaxed, and likable. _____

☐ Seemed in control of the speaking situation. _____

☐ Showed appropriate enthusiasm. _____

☐ Pronounced, enunciated, and spoke well. _____

☐ Used appropriate gestures, tone,
volume, and delivery rate. _____

☐ Had good posture and eye contact. _____

☐ Answered questions concisely and convincingly. _____

Overall professionalism: Superior_____ **Acceptable** _____ **Needs work** _____

Evaluator's signature:_____

Websites and On-line Documentation

Except for manuals and reference books, paper documents (letters, memos, reports, proposals) usually are structured to be read in linear sequence: front-to-back. Information builds on the information that precedes it. E-mail documents, often printed out at their destination, usually display this linear structure.

But some types of electronic documents enable users to design their own information sequence as they search for specific chunks of information or merely browse through parts of a topic in no particular order, as one might read a newspaper or encyclopedia (Grice and Ridgway 35–43). Because readers design their own information sequence, these documents are written in small, discrete modules that can stand alone in meaning.

This chapter introduces types of electronic documents essential to workplace communication.

ON-LINE DOCUMENTATION

People who use computers in their jobs need instructions and training for operating the systems and understanding their equipment's many features. On-line documentation is designed to support specific tasks and provide answers to specific questions.

Although computers come with printed manuals, the computer itself is becoming the preferred training medium. In computer-based training, documentation on the screen itself explains how the system works and how to use it. Examples of on-line help include:

Types of on-line documentation

- ◆ error messages and troubleshooting advice
- ◆ reference guides to additional information or instructions
- ◆ tutorial lessons that include interactive exercises with immediate feedback
- ◆ help and review options to accommodate different learning styles

Instead of leafing through a printed manual, users find what they need by keying a simple command, clicking a mouse, using a help menu, or following an electronic prompt.

The documentation itself might appear (1) in dialogue boxes that ask the user to input a response or click on an option, or (2) pop-up or balloon help that appears when the user clicks on an icon or points to an item on the screen for more information. (For examples, explore the on-line help resources on your own computer.)

HYPERTEXT

One extension of on-line documentation increasingly used in tutorial and training software is *hypertext*, information in electronic form designed for non-linear reading. Unlike printed text, designed to be read front-to-back, a hypertext document offers various informational paths through the material, the particular path (or paths) determined by what and how much a reader wants to know. *Hypermedia* expand the applications of hypertext by adding graphics, sound, video, and animation.

In addition to on-line documentation, hypertext is used as a research and reference tool for navigating complex cross-references and retrieving information electronically. In a hypertext system, a topic can be explored from any angle, at any level of detail. Assume, for instance, that you are using a hypertext database to research the AIDS epidemic. The database contains chunks of related topics organized in a network (or web) of files linked electronically (Horton 22), as illustrated in Figure 25.1.

Hypertext accommodates enquiry in various directions

After accessing the initial file ("The AIDS Epidemic"), you navigate the network in any direction, choosing which file to open and where to go next. The files themselves might be printed words, graphics, sound, video, or animation. Freed from the fixed-page sequence of a printed document, you customize the direction of your search.

Because hypertext offers multiple layers of information, from general to specific (Figure 25.2), you also customize the depth of your search.

Communication specialist William Horton explains the instructional power achieved through hypertext layering:

Hypertext accommodates enquiry at various depths

> *Paper . . . lacks depth. All information must be on the same level or layer. With hypertext, however, the screen can have deeper reserves of information. These deeper layers do not clutter the screen, but are available if needed. ("Is Hypertext" 25)*

Persons navigating a hypertext document invent their own "text" and so can discover combinations, relationships, and chains of knowledge impossible to express in the sequential pages of a printed document (Bernstein 42).

**Figure 25.1
Topics (or Files) in a Hypertext Network are Linked Electronically**

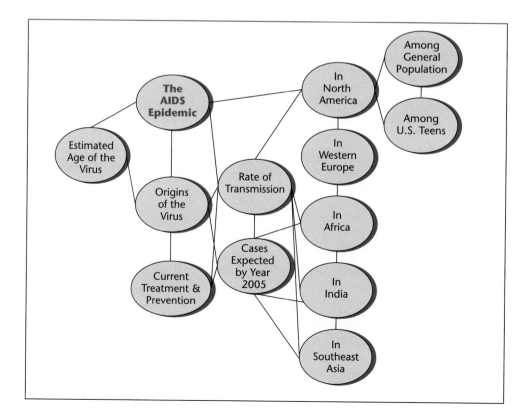

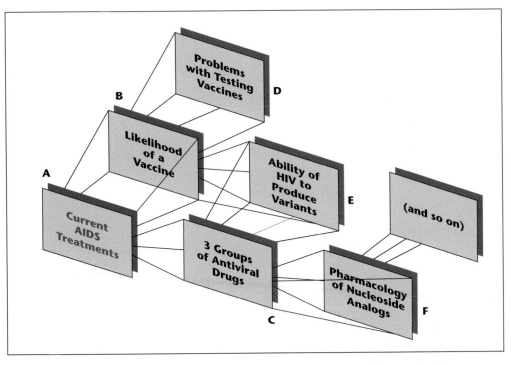

Figure 25.2 Hypertext Topics Can Be Layered to Serve Different Audiences
Source: Visual adapted from Horton, William. "Is Hypertext the Best Way to Document Your Product?" *Technical Communications* 39.1 (1991):25. Used with permission from the Society for Technical Communication, Arlington, VA.

Following is an application of hypertext instruction in the auto industry:

A hypertext
application

A mechanic can zoom from a picture of a car engine to a video of a specific malfunctioning part, and then move directly to the text that tells how to fix it. (Morse 7)

For all their potential as research and instructional tools, hypertext documents carry limitations:

Limitations of
hypertext
documents

- ◆ Confronting countless possible paths through the material, readers can get lost in "hyperspace." (*Where do I go next? How do I get out? How do I organize?*) Ease of navigation depends on how effectively individual chunks (or "nodes") of information are segmented and linked.
- ◆ Readers may resist a document that leaves them responsible for organizing their own learning. They generally rely on the writer to help organize their thinking (Horton, "Is Hypertext" 26).
- ◆ Instead of always enhancing learning, hypertext documents can in fact interfere with understanding and impede performance (Barfield, Haselkorm and Weatbrook 22, 27). Users in one study took longer to read a hypertext document than the equivalent paper document (Rubens 36).
- ◆ Some hypertexts can be too highly structured, thus leaving users with *fewer* navigational choices than they would have with a printed document (Selber).

DECISIONS IN CREATING HYPERTEXT DOCUMENTS

- How much information should be included on one screen?
- What level of detail should be presented?
- How can each chunk be written so it can be read in any order and still make sense?
- How can the material best be linked for easiest navigation of various possible paths?
- Which combination of media should be employed (printed words, speech, sound effects, animation, music)?

These problems can be addressed by communicators making the right decisions (Horton "Is Hypertext" 27; Nickels-Shirk 191).

Creating hypertext documents usually is not a one-person job; it often requires writers, graphic artists, computer specialists, animators, and the like. This is one example of how communication technology makes collaboration both possible and necessary.

The most rapidly expanding application of hypertext occurs via the Internet, on the World Wide Web. (Chapter 6 covers Internet research.)

THE WEB

Like a CD-ROM or electronic database, the Web offers a collection of electronic documents and multi-media. But hypertext enables the Web to link information in non-linear patterns, providing countless routes to be explored—worldwide—according to the user's special needs (Hunt 377). Some unique characteristics of the Web (December 371–72) are:

How the Web differs from other media

- *The Web is interactive.* Users construct their own hypertextual path through the material and can respond/add to the message.
- *The Web allows reciprocal use.* Besides obtaining information, users can also provide it.
- *The Web is porous.* A work can be entered at various points because a website usually offers multiple files that are linked.
- *The Web is ever changing.* A Web page or site is continually "a work in progress"—not only in its content but also in the technology itself (software, hardware, modems, servers). Unlike paper, software, or CD-ROM, the Web has no "final state."

These features enable Web users to discover and create their own connections among an endless array of ideas.

NOTE   *Keep in mind that Web pages, like all on-line screens, take at least 25 percent longer to read than paper documents. One possible solution: high-resolution screens that are as readable as paper copy. These should be widely available and affordable within a decade (Neilsen "Be Succinct" 1).*

In Brief HOW WEBSITES ENHANCE WORKPLACE TRANSACTIONS

The Web is a tool for advertising, learning about new products or companies, updating product information, or ordering products (Teague 236, 238). Each organization advertises its services and products via its own *home page*, a type of electronic billboard that introduces the organization and provides links to additional pages users can explore as they wish.

Specific Benefits

◆ *Visibility.* A site attracts business by establishing a presence in markets worldwide.
◆ *Access.* A site is accessible 24 hours a day (Dulude 47).
◆ *Customer relations.* Through enhanced customer service and support and rapid response, a site increases customer satisfaction and enhances a company's caring persona (Hoger, Cappel, and Myerscough 41).
◆ *Efficiency.* Two-way, real-time communication enables sudden problems, errors, or areas of danger to be broadcast rapidly. The audience can control the viewing of messages and respond immediately. On-screen instructions (for example, for assembling a modem) can be enhanced with high-resolution, 3-D graphics; parts can be colour-coded for assembly; and material can be updated instantly (Dulude 49–60).
◆ *Economy.* The cost of an Internet/Web bank transaction drops from over a $1 to roughly

1 cent; the cost of processing an airline ticket drops from $8 to $1 (IBM 13). A site enables mass publishing. Radically reduced printing, mailing, and distribution costs facilitate mass publishing. Also, an advertiser can embed deeper and deeper levels of product details, without consuming extra page space. Ultimately, as the cost of business transactions drops, so does the number of required employees.
◆ *Data gathering.* Tracking software provides customer data by recording who uses the website, how often they use it, and exactly where they go. Employees access reference materials from journal and trade magazines, access addresses of researchers, and remain current about legal issues and government regulations (Ritzenthaler and Ostroff 17–18).
◆ *Information sharing.* Intranets and extranets increase the flow of ideas up and down, and down and up, from outside to inside the company and vice versa. Knowledge audits identify who knows what; this information is then listed in the company intranet directory ("yellow pages").
◆ *Collaboration.* Company sites help reduce the length/need for face-to-face meetings. And people who do meet are better prepared because they have shared information beforehand.

ELEMENTS OF A USABLE WEBSITE

Although more diverse than typical users of paper documents, Web users share common expectations. The following are basic usability indicators for a website.

ACCESSIBILITY

Users expect a site that is easy to enter, exit, and navigate. Instead of reading word for word, they tend to skim, looking for key material without having to scroll through pages of text. They look for chances to interact and they want to download material at a reasonable speed.

WORTHWHILE CONTENT

Users expect the site to contain all the explanations they need (help screens, links, etc.). They want material that is accurate and constantly up-to-date. They expect

In Brief WEB APPLICATION IN MAJOR COMPANIES

◆ For training employees in rapidly changing job skills, companies increasingly rely on distance-learning programs offered by colleges and universities. Such programs include e-mail correspondence with faculty, on-line discussion groups, assignments and examples downloaded from the school website, and searches of virtual libraries (for special training or MBA work, etc.).

◆ General Electric expects to save $500 million over three years by purchasing via the Internet (McWilliams and Stepanik 170+).

◆ NASA posts RFPs (requests for proposals) on an engineering website, to which bidders and contractors can respond via e-mail—thereby speeding the whole process, by eliminating fax, phone, and copying time (Machlis 45).

◆ The Volvo Corporation is connecting all its branches, warehouses, and truck dealerships in Sweden and the U.S. via a global network to keep track of parts, specifications, and product updates. This allows authorized employees worldwide access to company databases so that "all data will be available anywhere." (Hamblen 51+)

clear error messages that spell out appropriate corrective action. They look for links to other, high-quality sites as indicators of credibility. They look for an e-mail address and other contact information being prominently displayed.

SENSIBLE ARRANGEMENT

Users always want to know where they are and where they are going. They expect a recognizable design and layout, with links easily navigated forward or backward, back links to the home page, and no dead ends. They look for navigation bars and hot buttons explicitly labelled ("Company Information," "Ordering," "Job Openings," and so on).

Instead of a traditional introduction, discussion, and conclusion, users expect the punch line right up front. Because they hate to scroll, users often get only what is on the first screen.

GOOD WRITING AND PAGE DESIGN

Users expect a writing style that is easy to read and error-free. They look for concise pages that are quick to skim, with short sentences and paragraphs, headings, and bulleted lists. Instead of having to wade through overstatement and exaggeration to "get at the fact," readers expect restrained, impartial language (Neilsen "Be Succinct" 2). Table 25.1 on page 631 illustrates the impact of good writing on usability.

GOOD GRAPHICS AND SPECIAL EFFECTS

Some users look for images or multi-media special effects—as long as these are not excessive or gratuitous. Since other users often disable their browser's visual capability (to save memory and downloading time), they look for a prose equivalent of each visual (*visual/prose redundancy*). They expect to recognize each icon and screen element—hot buttons, links, help options, and so on.

Table 25.1   The Impact of Good Writing on Usability

Source: Neilsen, Jakob. "Reading on the Web." Oct. 1997. Alertbox. http://www.useit.com/alertbox/9710a.html (8 Aug. 1997).

Site Version	Sample Paragraph	Usability Improvement (relative to control condition)
Promotional writing (control condition) using the "marketese" found on many commercial websites	Nebraska is filled with internationally recognized attractions that draw large crowds of people every year, without fail. In 1996, some of the most popular places were Fort Robinson State Park (355,000 visitors), Scotts Bluff National Monument (132,166), Arbor Lodge State Historical Park & Museum 100,000), Carhenge (86,598), Stuhr Museum of the Prairie Pioneer (60,002), and Buffalo Bill Ranch State Historical Park (28,446).	0% (by definition)
Concise text with about half the word count as the control condition	In 1996, six of the best-attended attractions in Nebraska were Fort Robinson State Park, Scotts Bluff National Monument, Arbor Lodge State Historical Park & Museum, Carhenge, Stuhr Museum of the Prairie Pioneer, and Buffalo Bill Ranch State Historical Park.	58%
Scannable layout using the same text as the control condition in a layout that facilitated scanning	Nebraska is filled with internationally recognized attractions that draw large crowds of people every year, without fail. In 1996, some of the most popular places were: ◆ Fort Robinson State Park (355,000 visitors) ◆ Scotts Bluff National Monument (132,166) ◆ Arbor Lodge State Historical Park & Museum (100,000) ◆ Carhenge (86,598) ◆ Stuhr Museum of the Prairie Pioneer (60,002) ◆ Buffalo Bill Ranch State Historical Park (28,446)	47%
Objective language using neutral rather than subjective, boastful, or exaggerated language (otherwise the same as the control condition)	Nebraska has several attractions. In 1996, some of the most-visited places were Fort Robinson State Park (355,000 visitors), Scotts Bluff National Monument (132,166), Arbor Lodge State Historical Park & Museum (100,000), Carhenge (86,598), Stuhr Museum of the Prairie Pioneer (60,002), and Buffalo Bill Ranch State Historical Park (28,446).	27%
Combined version using all three improvements in writing style together: concise, scannable, and objective	In 1996, six of the most-visited places in Nebraska were: ◆ Fort Robinson State Park ◆ Scotts Bluff National Monument ◆ Arbor Lodge State Historical Park & Museum ◆ Carhenge ◆ Stuhr Museum of the Prairie Pioneer ◆ Buffalo Bill Ranch State Historical Park	124%

In Brief HOW SITE NEEDS AND EXPECTATIONS DIFFER ACROSS CULTURES

Despite its United States origin, the Internet rapidly has become international and cross-cultural. And yet, countries vary greatly in their level of "Internet maturity," with much of the world several years behind North America (Neilsen, "Global" 2). A useful international site therefore reflects careful regard for cost, clarity, and cultural sensitivity.

Cost

High telephone rates in many countries hike Internet costs. In Japan, for example—whose Internet use ranks second to that in North America—monthly cost for a one-hour daily on-line is more than double the North American cost for unlimited access (Neilsen "Global"). A usable site therefore omits graphics that are slow to load.

Clarity

To facilitate access and avoid misunderstandings, international communication via the Internet incorporates measures like these:

◆ Sites often provide home-page versions in various languages (or links to a translation package).

◆ Time zones, currencies, and other units of measurement differ (10 a.m. in Vancouver is 6 p.m. in London, 7 p.m. in Stockholm, or 3 a.m. in Tokyo). In arranging real-time interactions (e.g., an on-line conference), the host specifies the recipient's time as well as the home time (Neilsen, "International" 1).

◆ Because the value of a "dollar" in countries such as Australia, U.S., Singapore, or Zimbabwe differs from the value of the Canadian dollar, businesses specify "$12.50 Canadian," and so on. Also, offering payment options in the culture's own currency helps avoid currency-exchange ambiguities (Hodges 18–19).

◆ A date listed as "6/10/08" might be confusing in other cultures. Preferable would be "10 June 2008" or "June 10, 2008."

◆ Temperature measurements are specified as "Fahrenheit" or "Celsius."

Cultural Sensitivity

A site truly "international" in ambiance—and not merely "North American"—enables anyone, anywhere to feel at home (Neilsen, "Global" 2). For example, it avoids sarcasm or irreverence (which some cultures consider highly offensive), and exclusive references or colloquialisms such as "bear markets," "the wild west," and "Stanley Cup."

SCRIPTING A WEB DOCUMENT WITH HTML

HTML (hypertext markup language) is a computer scripting language that can be understood by all Web browsers and that specifies the positioning of each element on a Web page: text, art, headings, lists, hot-buttons, etc. For any type of computer or operating system, HTML provides a common "interface" enabling all users on a network to create, access, exchange, or edit information.

An HTML document is coded by use of *tags*: the command, enclosed in angle brackets, appears on both ends of the content to be acted upon. The tagged command plus the related text are known as an *HTML element*.

A typical HTML element

<TITLE>English 266 Technical Writing</TITLE>

Figure 25.3 shows basic HTML commands. Figure 25.4 shows a Pearson Education Web page; Figure 25.5 shows that page's partial HTML script (obtained from "View Source" on the desktop menu).

NOTE   *To visit a website offering HTML advice, instruction, and useful links, go to **http://trace.wisc.edu/world/web/**. Also, keep in mind that WYSIWYG ("what you see is what you get") editing programs, such as Adobe PageMill™, Microsoft FrontPage™, or Symantec Visual Page™, largely eliminate the need for HTML coding by hand. These authoring tools provide step-by-step instructions and templates for creating Web pages, positioning graphics and other page elements, and adding clip art, among other things.*

Java enhances
HTML documents

HTML produces only static pages, with limited possibilities for data display. However, a programming language called *Java* can be embedded in an HTML script to provide "dynamic" content (graphics, motion, and sound). Java-enhanced Web pages allow more sophisticated, interactive applications such as simulations and computer-based training.

Symantec Visual Page™ largely eliminates the need for HTML coding by hand. This authoring tool provides step-by-step instructions and templates for creating Web pages, positioning graphics and other page elements, adding clip art, etc.

NOTE   *Organizational websites generally are developed by a Web team: content developers, graphic designers, programmers, and managers. Whether or not you are an actual team member, expect to play a collaborative role in your organization's site development and maintenance.*

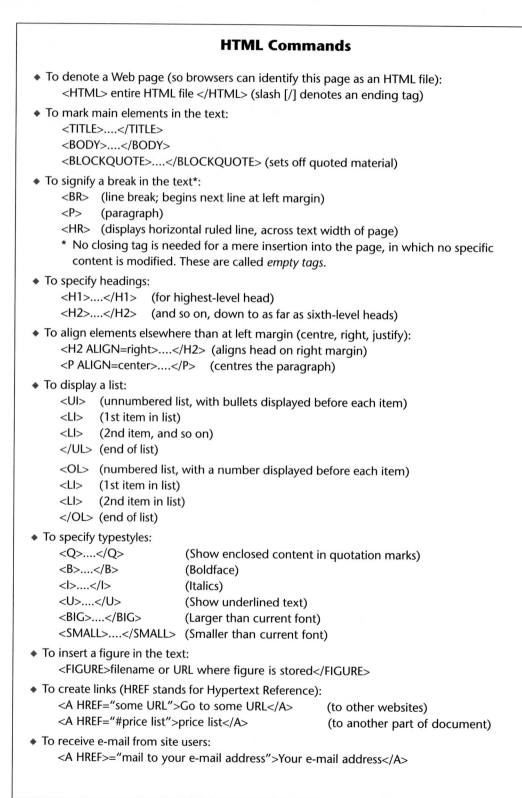

Figure 25.3   A Sample of HTML Commands

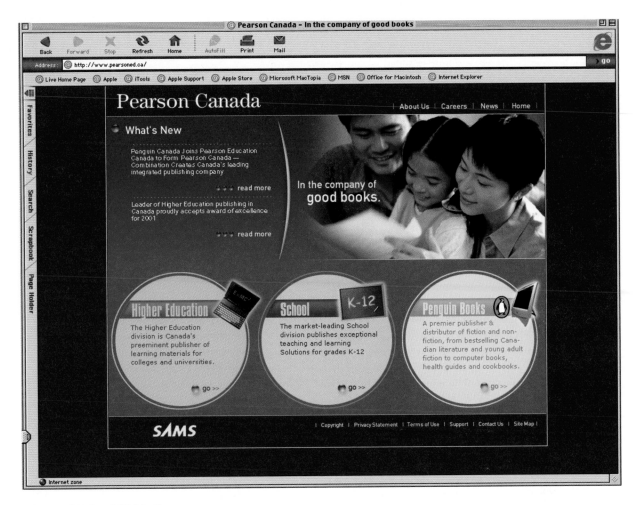

Figure 25.4   A Web Page

```
© Pearson Canada - In the company of good books

HTML: Pearson Canada - In the company of good books

<!DOCTYPE HTML PUBLIC "-//W3C//DTD HTML 4.01 Transitional//EN">
<html>
<head>
<title>Pearson Canada - In the company of good books</title>
<meta http-equiv="Content-Type" content="text/html; charset=iso-8859-1">
<meta name="description" content="Pearson Canada is a premier trade publisher with numerous bestselling fiction, nonfiction and co
<meta name="keywords" content="Pearson Canada, book, boook, textbook, text, Canadian book, trade, college, university, school, pub
<meta http-equiv="pragma" content="no-cache">

<script language="JavaScript">
<!--
function MM_swapImgRestore() { //v3.0
  var i,x,a=document.MM_sr; for(i=0;a&&i<a.length&&(x=a[i])&&x.oSrc;i++) x.src=x.oSrc;
}

function MM_preloadImages() { //v3.0
  var d=document; if(d.images){ if(!d.MM_p) d.MM_p=new Array();
    var i,j=d.MM_p.length,a=MM_preloadImages.arguments; for(i=0; i<a.length; i++)
    if (a[i].indexOf("#")!=0){ d.MM_p[j]=new Image; d.MM_p[j++].src=a[i];}}
}

function MM_findObj(n, d) { //v4.0
  var p,i,x;  if(!d) d=document;  if((p=n.indexOf("?"))>0&&parent.frames.length) {
    d=parent.frames[n.substring(p+1)].document; n=n.substring(0,p);}
  if(!(x=d[n])&&d.all) x=d.all[n]; for (i=0;!x&&i<d.forms.length;i++) x=d.forms[i][n];
  for(i=0;!x&&d.layers&&i<d.layers.length;i++) x=MM_findObj(n,d.layers[i].document);
  if(!x && document.getElementById) x=document.getElementById(n); return x;
}

function MM_swapImage() { //v3.0
  var i,j=0,x,a=MM_swapImage.arguments; document.MM_sr=new Array; for(i=0;i<(a.length-2);i+=3)
   if ((x=MM_findObj(a[i]))!=null){document.MM_sr[j++]=x; if(!x.oSrc) x.oSrc=x.src; x.src=a[i+2];}
}
//-->
</script>
<style>
a:visited {  font-family: Verdana, Arial, Helvetica, sans-serif; font-size: 11px; color: #FFFF99; text-decoration: none}

a:hover {  font-family: Verdana, Arial, Helvetica, sans-serif; font-size: 11px; color: #FFFF99; text-decoration: underline}

a:link {  font-family: Verdana, Arial, Helvetica, sans-serif; font-size: 11px; color: #FFFFFF; text-decoration: none}

a:active {  font-family: Verdana, Arial, Helvetica, sans-serif; font-size: 11px; color: #FFFFFF; text-decoration: none}

a {  font-family: Verdana, Arial, Helvetica, sans-serif; font-size: 11px; color: #FFFF99; text-decoration: none}

td {  font-family: Verdana, Arial, Helvetica, sans-serif; font-size: 11px}

font {  font-family: Verdana, Arial, Helvetica, sans-serif; font-size: 11px}

.links_sponsor {  font-family: Verdana, Arial, Helvetica, sans-serif; font-size: 10px; font-style: normal; font-weight: bold; colo
</style>
</head>

<body bgcolor="#003366" text="#000000" leftmargin="0" topmargin="0" marginwidth="0" marginheight="0" onLoad="MM_preloadImages('ima
<div align="center">
  <table width="764" border="0" cellspacing="0" cellpadding="0">
    <tr>
      <td bgcolor="#FFFFFF"><img src="images/corporate/white_pixel.gif" width="1" height="1"></td>
      <td width="762" valign="top">
        <table width="762" border="0" cellspacing="0" cellpadding="0">
    <tr>
      <td><img src="images/corporate/top_logo.gif" width="495" height="53"></td>
      <td width="73"><a href="about.html" onMouseOut="MM_swapImgRestore()" onMouseOver="MM_swapImage('about','','images/corporate,
      <td width="67"><a href="careers.html" onMouseOut="MM_swapImgRestore()" onMouseOver="MM_swapImage('career','','images/corpora
      <td width="54"><a href="news.html" onMouseOut="MM_swapImgRestore()" onMouseOver="MM_swapImage('news','','images/corporate/ne
      <td width="73"><a href="index.html" onMouseOut="MM_swapImgRestore()" onMouseOver="MM_swapImage('home','','images/corporate/h
    </tr>
  </table>
  <table width="762" border="0" cellspacing="0" cellpadding="0">
    <tr>
      <td width="1" background="images/corporate/blue_pixel.gif"><img src="images/corporate/blue_pixel.gif" width="1" height="1">
      <td background="images/corporate/home_page_bg_mid.gif" bgcolor="#003399" valign="top">
        <table width="320" border="0" cellspacing="0" cellpadding="0">
          <tr>
            <td valign="top">
              <table width="292" border="0" cellspacing="0" cellpadding="4">
                <tr>
                  <td height="34"><img src="images/corporate/whats_new.gif" width="132" height="31" vspace="0"><br>
                  </td>
                </tr>
                <tr>
                  <td valign="top">
                    <table width="285" border="0" cellspacing="0" cellpadding="0">
                      <tr>
```

Figure 25.5   The Partial HTML Script for Figure 25.4

GUIDELINES

for Creating a
Website

Planning Your Site

1. *Identify the site's intended audience.* Are they potential customers seeking information; people purchasing a product or service; or customers seeking product support, updates, or troubleshooting advice (Wilkinson 33)? Will different audiences be seeking different material?

2. *Identify the site's purpose.* Is the purpose to publish information, sell a product, promote an idea, solicit customer feedback, advertise talents, or create goodwill? Should the site convey the image of a "cool" cutting-edge company (or individual), displaying skill with the latest Web technologies (animation, interaction, fancy design)? For specific ideas, look for and examine other sites that display the features you seek.

3. *Identify what the site will contain.* Will it display only print documents or graphics, audio, and video as well? Will links be provided and, if so, how many and to where? Will user feedback be solicited and, in what form: survey questions, e-mail comments, or the like?

4. *Identify the level of user interaction.* Will this be a document-only site, offering no interaction beyond downloading and printing? Will it offer dynamic marketing (Dulude 69): on-line questions and answers, technical support, downloadable software, on-line catalogues? Will users be able to download documents, software, or documentation? Will an e-mail button be included?

Laying Out Your Pages

1. *Design your pages to guide the user.* Highlight important material with headings, lists, typestyles, colour, and white space. Remember that too much white space causes excessive scrolling. Use sans serif typefaces. Limit page size to 30 k, to speed downloading.

2. *Use graphics that download quickly.* Avoid excessive complexity and colour, especially in screen background. Keep maximum image size below 30 k. Create an individual file for each graphic and use thumbnail sketches on the home page, with links to the larger images, each in its own file.

3. *Include text-only versions of all visual information.* Roughly 20 percent of users turn off the graphics function on their browsers (Gannon 22–23).

4. *Provide orientation.* Structure the content to reflect its relative importance and how often it's used. Place the material most important to your readers right up front and create links to more detailed information. Date each page to announce the exact time of each update—or include a "What's New" head, so readers can stay abreast of changes.

5. *Provide navigational aids.* Keep links logical and always link back to the home page. Don't overwhelm the user with excessive choices. Label each link explicitly (for example, "Product Updates" instead of merely "Click here").

6. *Define and shape the content.* Use hypertext to chunk information into subtopics, each in a digestible node and link it—but remember that hypertext takes longer to download and print (Neilsen "Be Succinct" 1–2). Structure each hypertext node as an "inverse pyramid," in which you begin with the conclusion (Neilsen, "How Users Read" 1). The inverted pyramid works like a newspaper article, in which the major news/conclusion appears first (e.g., "The jury deliberated only two hours before returning a guilty verdict"), followed by the details (Neilsen, "Inverted Pyramids"). Last but not least, use restraint: Give users the opportunity to receive less information (Outing 2). Think hard about what users need and give them only that.

7. *Sharpen the style.* Make the on-line text at least 50 percent shorter than its hard copy equivalent. Try to summarize (see Chapter 9). Use short sentences and paragraphs. Avoid "marketese" or promotional language that exaggerates ("breakthrough," "revolutionary," "cutting edge").

8. *Check your site.* Double check the accuracy of all numbers, dates, data, etc., and check for broken links and correct spelling, grammar, and so on.

9. *Attend to legal considerations.* Have your legal department approve all material before you post it (Wilkinson 33). Obtain written permission before linking to other Web pages or borrowing any graphic element from another site. Display a privacy notice that explains how each transaction is being recorded, collected, and used. To help protect your own intellectual property, display a copyright notice on every page of the site (Evans, *Whose Website* 48, 50).

10. *Test your site for usability.* Test for usability with unfamiliar users (beta testing), and keep track of their problems and questions. What do users like or dislike? Can they navigate effectively to get to what they need? Are the icons recognizable? Is the site free of needless complexity or interactivity? Test your document with various browsers to be sure it can be downloaded.

11. *Maintain your site.* Review the site regularly, update often, and redesign as needed. If the site accommodates e-mail queries, respond within one business day (Dulude 117).

In Brief PRIVACY ISSUES ON THE WEB

Information sharing between computers is what makes the Internet and World Wide Web possible. For instance, when a user visits a site, the host computer needs to know what browser is being used. Also, for improved client service, a host site can track the links visitors follow, the files they open or download, and the pages they visit most often (Reichard 106).

Too often, however, more information gets "shared" than the user intended. For instance, the host computer can record the user's domain name, place of origin, and usage patterns (James-Catalano 32).

Some servers and sites display privacy notices explaining how usage patterns or transactions are being recorded, collected, and used. But this offers only limited protection. Any Internet transaction is routed through various browsers and servers and can be intercepted anywhere along the way.

In the workplace, monitoring increasingly compromises employee privacy. As early as 1993—with monitoring technology in its infancy—an estimated 20 million American workers were subject to computer monitoring (Karaim 72). Some types of workplace monitoring, of course, have legitimate purposes.

Legitimate Purposes for Monitoring a Workplace Website

◆ *Troubleshooting.* Monitoring software (Alert-Page™, Net.Medic™) can scan a company site for broken lines, and identify server glitches, software bugs, modem problems, or faulty hardware connections (Reichard 106).

◆ *Productivity.* Companies track intranet use for the number of queries per employee, types of questions asked, by whom, and the length of time required for employees to find what they need. These data help decide whether the search mechanism (user interface) can be improved or whether on-line documents can be organized or written more clearly (Cronin 103).

◆ *Security.* Software can track employees' visits to other websites, as well as files opened for recreational or personal use, e-mail sent and received, and even provide snapshots of an employee's computer screen (Karaim 72). They can deny access to unauthorized Web sites and also inform the employer about the employee's attempt. Such monitoring can be a justifiable precaution against employee theft, drug abuse, security violations (such as publishing trade secrets on the Internet)—or wasted time. For example, businesses lose millions of worker hours yearly to computer game-playing by employees (Hutheesing 369). Beyond its legitimate uses, monitoring also carries potential for the abuse of personal privacy.

Privacy Abuses in Workplace Monitoring

◆ Employers have more freedom to violate employee privacy than the police (Karaim 72). Andre Bacard, author of *The Computer Privacy Handbook,* points out that supervisors can "tap an employee's phones, monitor her e-mail, watch her on closed-circuit TV, and search her computer files, without giving her notice" (qtd. in Karaim 72).

◆ Some companies notify their employees that their electronic transactions are subject to monitoring, but many do not.

◆ Even face-to-face transactions are subject to monitoring: An electronic "Active Badge" tracks employees as they move about their work site, recording how much time they spend in the bathroom or at the water cooler and who they talk to during the work day (Karaim 72).

Checklist FOR WEBSITE USABILITY

Use this checklist to assess website usability.

Accessibility

◆ Is the site easy to enter, navigate, and exit?
◆ Is required scrolling kept to a minimum?
◆ Is downloading speed reasonable?
◆ If interaction is offered, is it useful—not superfluous?
◆ Does the site avoid overwhelming the user with excessive choices?

Content

◆ Are all needed explanations, error messages, and help screens provided?
◆ Is the time of each update clearly indicated?
◆ Is everything accurate and up-to-date?
◆ Are links connected to high-quality sites?
◆ Does everything belong (nothing excessive or superfluous or needlessly complex)?
◆ Is an e-mail button or other contact method prominently displayed?
◆ Does the content accommodate international users?

Arrangement

◆ Is the key part of the message on the first page?
◆ Are navigation bars, hot buttons, and help options clearly displayed and explicitly labelled?

◆ Are links easily navigated—backward and forward—with back links to the home page?

Writing and Page Design

◆ Is the text easy to scan, with short sentences and paragraphs, and do headings, lists, type styles, and colour highlight important material?
◆ Is the overall word count roughly one-half of the hard copy equivalent?
◆ Is the tone reasonable and restrained—free of overstatement and "marketese"?

Graphics and Special Effects

◆ Is each graphic easy to download?
◆ Is each graphic backed up by a text-only version?
◆ Is each graphic or special effect necessary?

Legal Considerations

◆ Does the site display a privacy notice that explains how the transaction is being recorded, collected, and used?
◆ Does each page of the site display a copyright notice?
◆ Has written permission been obtained for each link to other sites and for each graphic element borrowed from another site?
◆ Has all posted material received legal approval?

EXERCISES

1. Consult pages 639–640 and evaluate a website for usability. You might select a site at your school or place of employment. You might begin by deciding on specific information you seek (such as "internship opportunities," "special programs," "campus crime statistics," or "average SAT scores of admitted students") and use this as a basis for assessing the site's accessibility, content, arrangement, and so on.

   Complete your evaluation and report any problems or suggested improvements in a memo to a designated decision maker. (Your instructor might ask different class groups to evaluate the same site and to compare their findings in class.)

2. Download and print pages from a website. Edit these pages to improve layout and writing style. Submit copies to your instructor.

3. Examine websites from three or four competing companies (e.g., computer makers IBM™, Apple™, Gateway™, Dell™, and Compaq™, or auto makers). Which site do you think is the most effective, the least effective, and why? Report your findings in a memo to your colleagues.

COLLABORATIVE PROJECTS

Any of the exercises may be used as collaborative projects.

APPENDIX

A

Review of Grammar, Usage, and Mechanics

COMMON SENTENCE ERRORS

EFFECTIVE PUNCTUATION

TRANSITIONS AND OTHER CONNECTORS

EFFECTIVE MECHANICS

No matter how vital and informative a message may be, its credibility is damaged by basic errors. Any of these errors—an illogical, fragmented, or run-on sentence; faulty punctuation; or a poorly chosen word—stands out and mars otherwise good writing. Not only do such errors annoy the reader, but they also speak badly for the writer's attention to detail. Your career will make the same demands for good writing that your English classes do. The difference is that evaluation (grades) in professional situations usually affects promotions, reputation, and salary.

The Correction Symbol Table, on the last page of this book, lists correction symbols.

COMMON SENTENCE ERRORS

Any piece of writing is only as good as each of its sentences. Here are common sentence errors, with suggestions for easy repairs.

SENTENCE FRAGMENT

frag

A sentence expresses a logically complete idea. Any complete idea must have a subject and a verb and must not depend on another complete idea to make sense. Your sentence *might* contain several complete ideas, but it *must* contain at least one!

> Although Mary was nervous, she grabbed the line, and she saved the sailboat.
>
> *(incomplete idea)*   *(complete idea)*   *(complete idea)*

If the idea is not complete—if your reader is left wondering what you mean—you probably have omitted some essential element (the subject, the verb, or another complete idea). Such a piece of a sentence is a *fragment*.

> Grabbed the line. *(a fragment because it lacks a subject)*
>
> Although Mary was nervous. *(a fragment because, although it has a subject and a verb, it needs to be joined with a complete idea)*

The only exception to the sentence rule applies when we give a command (Run!), in which the subject (you) is understood. Logically complete, this statement is properly called a sentence. So is this one:

> Sam is an electronics technician.

Readers cannot miss your meaning: Somewhere is a person; the person's name is Sam; the person is an electronics technician. Suppose instead we write:

> Sam an electronics technician.

This statement is not logically complete, therefore not a sentence. The reader is left asking, "What about Sam the electronics technician?" The verb—the word that makes things happen—is missing. By adding a verb we can easily change this fragment to a complete sentence.

Simple verb	Sam **is** an electronics technician.
Verb and adverb	Sam, an electronics technician, **works hard.**
Dependent clause, verb, and subjective complement	**Although he is well paid,** Sam, an electronics technician, **is not happy.**

Do not, however, mistake the following statement—which seems to have a verb—for a complete sentence:

Sam being an electronics technician.

Such "ing" forms do not function as verbs unless accompanied by such other verbs as **is, was,** and **will be.** Again, readers are confused unless you complete your idea with an independent clause.

Sam, being an electronics technician, checked all **circuitry.**

Likewise, remember that the "to + verb" form does not function as a verb.

To become an electronics technician.

The meaning is unclear unless you complete the thought.

To become an electronics technician, **Sam had to pass an exam.**

Sometimes we inadvertently create fragments by adding certain words (**because, since, if, although, while, unless, until, when, where,** and others) to an already complete sentence, transforming our independent clause (complete sentence) into a dependent clause.

Although Sam is an electronics technician.

Such words subordinate the words that follow them so that an additional idea is needed to make the first statement complete. That is, they make the statement dependent on an additional idea, which must itself have a subject and a verb and be a complete sentence. (See "Faulty Subordination.") Now we have to round off the statement with a complete idea (independent clause).

Although Sam is an electronics technician, **he hopes to be an engineer.**

NOTE *Be careful not to use a semicolon or a period, instead of a comma, to separate elements in the preceding sentence. Because the dependent clause depends on the independent clause for its meaning, you need only a pause (symbolized by a comma), not a break (symbolized by a semicolon), between these ideas. In fact, many fragments are created when too strong a mark of punctuation (period or semicolon) severs the connection between a dependent and an independent clause. (See the later discussion of punctuation.)*

Here are some fragments from technical documents. Each is repaired in more than one way. Can you think of other ways of making these statements complete?

Fragment	She spent her first week on the job as a researcher. **Compiling information from digests and journals.**
Correct	She spent her first week on the job as a researcher, compiling information from digests and journals. She spent her first week on the job as a researcher. She compiled information from digests and journals.
Fragment	**Because the employee was careless.** The new computer was damaged.
Correct	Because the employee was careless, the new computer was damaged. The employee's carelessness resulted in damage to the new computer.

COMMA SPLICE

cs

In a comma splice, two complete ideas (independent clauses), which should be *separated* by a period or a semicolon, are incorrectly *joined* by a comma:

Comma splice	Sarah did a great job, she was promoted.

You can choose among several possibilities for correcting this error:

1. Substitute a period followed by a capital letter:

> Sarah did a great job. She was promoted.

2. Substitute a semicolon to signal the relationship:

> Sarah did a great job; she was promoted.

3. Use a semicolon with a connecting adverb (a transitional word):

> Sarah did a great job; **consequently,** she was promoted.

4. Use a subordinating word to make the less important sentence incomplete, thereby dependent on the other:

> **Because** Sarah did a great job, she was promoted.

5. Add a connecting word after the comma:

> Sarah did a great job, **and** she was promoted.

Your choice of construction will depend, of course, on the exact meaning or tone you want to convey. The following comma splice can be repaired in the ways described above.

Comma splice	This is a new technique, therefore, some people mistrust it.
Correct	This is a new technique. Some people mistrust it.
	This is a new technique; therefore, some people mistrust it.
	Because this is a new technique, some people mistrust it.
	This is a new technique, **so** some people mistrust it.

RUN-ON SENTENCE

ro

The run-on sentence, a cousin to the comma splice, crams too many ideas without needed breaks or pauses.

| Run-on | The hourglass is more accurate than the water clock because water in a water clock must always be at the same temperature to flow at the same speed since water evaporates and must be replenished at regular intervals, thus being less effective than the hourglass for measuring time. |

Like a runaway train, this statement is out of control. Here is a corrected version:

| Revised | The hourglass is more accurate than the water clock because water in a water clock must always be at the same temperature to flow at the same speed. Also, water evaporates and must be replenished at regular intervals. These temperatures and volume problems make the water clock less effective than the hourglass for measuring time. |

FAULTY COORDINATION

coord

Give equal emphasis to ideas of equal importance by joining them, within simple or compound sentences, with coordinating conjunctions: **and, but, or, nor, for, so,** and **yet.**

This course is difficult **but** worthwhile.
My horse is old **and** grey.
We must decide to support **or** reject the manager's proposal.

But do not confound your meaning by coordinating excessively.

| Excessive coordination | The climax in jogging comes after a few miles **and** I can no longer feel stride after stride **and** it seems as if I am floating **and** jogging becomes almost a reflex **and** my arms **and** legs continue to move **and** my mind no longer has to control their actions. |
| Revised | The climax in jogging comes after a few miles when I can no longer feel stride after stride. By then I am jogging almost by reflex, nearly floating, my arms and legs still moving, my mind no longer having to control their actions. |

Notice how the meaning becomes clear when the less important ideas (**nearly floating, arms and legs still moving, my mind no longer having**) are shown as dependent on, rather than equal to, the most important idea (**jogging almost by reflex**)—the idea that contains the lesser ones.

Avoid coordinating ideas that cannot be sensibly connected:

Faulty	Josh had a drinking problem **and** he dropped out of school.
Revised	Josh's drinking problem depressed him so much that he couldn't study, so he quit school.

Faulty	I was late for work **and** wrecked my car.
Revised	Late for work, I backed out of the driveway too quickly, hit a truck, and wrecked my car.

Instead of *try and,* use *try to.*

Faulty	I will try and help you.
Revised	I will try to help you.

FAULTY SUBORDINATION

sub

Proper subordination shows that a less important idea is dependent on a more important idea. By using subordination you can combine simple sentences into complex sentences and emphasize the most important idea. Consider these complete ideas:

Joe studies hard. He has severe math anxiety.

Because these ideas are expressed as simple sentences, they appear to be coordinate (equal in importance). But if you wanted to indicate your opinion of Joe's chances of succeeding in math, you would need a third sentence: **His disability probably will prevent him from succeeding,** or **His willpower will help him succeed.** To communicate the intended meaning concisely, combine ideas and subordinate the one that deserves less emphasis:

Despite his severe math anxiety *(subordinate idea),* Joe studies hard *(independent idea).*

This first version suggests that Joe will succeed. Below, subordination is used to suggest the opposite meaning:

Despite his diligent studying *(subordinate idea),* Joe has severe math anxiety *(independent idea).*

A dependent (or subordinate) clause in a sentence is signalled by a subordinating conjunction: **because, so, if, unless, after, until, since, while, as,** and **although,** among others. Be sure to place the idea you want emphasized in the independent clause; do not write

Although Moira is receiving excellent medical treatment, she is seriously ill.

if you mean to suggest that Moira has a good chance of recovering.

Do not coordinate when you should subordinate:

> **Weak** Television viewers can relate to an athlete they idolize and they feel obliged to buy the product endorsed by their hero.

Of the two ideas in the sentence above, one is the cause, the other the effect. Emphasize this relationship through subordination:

> **Revised** Because television viewers can relate to an athlete they idolize, they feel obliged to buy the product endorsed by their hero.

When combining several ideas within a sentence, decide which is most important, and subordinate the other ideas to it—do not merely coordinate:

> **Faulty** This employee is often late for work, and he writes illogical reports, and he is a poor manager, and he should be fired.
>
> **Revised** Because this employee is often late for work, writes illogical reports, and has poor management skills, **he should be fired.** *(The last clause is independent.)*

Do not overstuff sentences by subordinating excessively:

> **Overstuffed** This job, which I took when I graduated from college, while I waited for a better one to come along, which is boring, where I've gained no useful experience, makes me anxious to quit.
>
> **Revised** Upon college graduation, I took this job while waiting for a better one to come along. Because I find it boring and have gained no useful experience, I am eager to quit.

FAULTY AGREEMENT—SUBJECT AND VERB

s/v agr

The subject should agree in number with the verb. We are not likely to use faulty agreement in short sentences, where subject and verb are not far apart. Thus, we are not likely to say, "Jack eat too much" instead of "Jack eats too much," but in more complicated sentences—those in which the subject is separated from its verb by other words—we sometimes lose track of the subject-verb relationship.

> **Faulty** The lion's **share** of diesels **are** sold in Europe.

Although **diesels** is closest to the verb, the subject is **share,** a singular subject that must agree with a singular verb.

> **Correct** The lion's **share** of diesels **is** sold in Europe.

Agreement errors are easy to correct when subject and verb are identified.

> **Faulty** A **system** of lines **extend** horizontally to form a grid.
>
> **Correct** A **system** of lines **extends** horizontally to form a grid.

A second problem with subject-verb agreement occurs when we use indefinite pronouns such as **each, everyone, anybody,** and **somebody.** They function as subjects and usually take a singular verb.

Faulty	**Each** of the crew members **were** injured.
Correct	**Each** of the crew members **was** injured.
Faulty	**Everyone** in the group **have** practised long hours.
Correct	**Everyone** in the group **has** practised long hours.

Agreement problems can be caused by collective nouns such as **herd, family, union, group, army, team, committee,** and **board.** They can call for a singular or plural verb—depending on your intended meaning. When denoting the group as a whole, use a singular verb.

Correct	The **committee meets** weekly to discuss new business.
	The editorial **board** of this magazine **has** high standards.

To denote individual members of the group, however, use a plural verb.

Correct	Not all members of the editorial **board are** published authors.

Yet another problem occurs when two subjects are joined by **either . . . or** or **neither . . . nor.** Here, the verb is singular if both subjects are singular, and plural if both subjects are plural. If one subject is plural and one is singular, the verb agrees with the subject closest to the verb.

Correct	Neither **Al** nor **Bill works** regularly.
	Either apples or **oranges are** good vitamin sources.
	Either Felix or his **friends are** crazy.
	Neither the boys nor their **father likes** the home team.

If, on the other hand, two subjects (singular, plural, or mixed) are joined by **both . . . and,** the verb will be plural. Whereas **or** suggests "one or the other," **and** announces a combination of the two subjects, thereby requiring a plural verb.

Correct	**Both** Hal and Will **are** resigning.
	The **book and** the **briefcase appear** expensive.

A single **and** between singular subjects makes for a plural subject.

FAULTY AGREEMENT—PRONOUN AND REFERENT

pro agr

A pronoun can make sense only if it refers to a specific noun (its referent or antecedent), with which it must agree in gender and number.

Correct	**Tao** lost **his** blueprints.
	The **workers** complained that **they** were treated unfairly.

Some instances, however, are not so obvious. When an indefinite pronoun such as **each, everyone, anybody, someone,** and **none** serves as the pronoun referent, the pronoun itself is singular.

Correct	**Anyone** can get **his or her** degree from that college.
	Each candidate described **her** plans in detail.

FAULTY OR VAGUE PRONOUN REFERENCE

pro ref

Whenever a pronoun is used, it must refer to one clearly identified referent; otherwise, your message will be confusing.

Ambiguous	**Sally** told **Sarah** that **she** was obsessed with her job.
Correct	Sally told Sarah, "I'm obsessed with my job."
	Sally told Sarah, "I'm obsessed with your job."
	Sally told Sarah, "You're obsessed with [your, my] job."
	Sally told Sarah, "She's obsessed with [her, my, your] job."

Avoid using **this, that,** or **it**—especially to begin a sentence—unless the pronoun refers to a specific antecedent (referent).

Vague	He drove away from his menial **job,** boring **lifestyle,** and damp **apartment,** happy to be leaving **it** behind.
Correct	He drove away, happy to be leaving behind his menial job, boring lifestyle, and damp apartment.
Vague	The problem with our **defective machinery** is compounded by the **operator's incompetence. That** annoys me!
Correct	I am annoyed by the problem with our defective machinery as well as by the new operator's incompetence.

FAULTY PRONOUN CASE

ca

A pronoun's case (nominative, objective, or possessive) is determined by its role in a sentence: as subject, object, or indicator of possession.

If the pronoun serves as the subject of a sentence (**I, we, you, she, he, it, they, who**), its case is *nominative.*

She completed her graduate program in record time.

Who broke the chair?

When a pronoun follows a version of **to be** (a linking verb), it explains (complements) the subject, and so its case is nominative.

It was **she.**

The chemist who perfected our new distillation process is **he.**

If the pronoun serves as the object of a verb or a preposition (**me, us, you, her, him, it, them, whom**), its case is *objective.*

Object of the verb	The employees gave **her** a parting gift.
Object of the preposition	Several colleagues left with **him.**
	To **whom** do you wish to complain?

If a pronoun indicates possession (**my, mine, ours, your, yours, his, her, hers, its, their, theirs, whose**), its case is *possessive*.

The brown briefcase is **mine.**

Her offer was accepted.

Whose opinion do you value most?

Here are some frequent errors in pronoun case:

Faulty	**Whom** is responsible to **who**? *(The subject should be nominative and the object should be objective.)*
Correct	**Who** is responsible to **whom?**
Faulty	The debate was between Marsha and **I.** *(As object of the preposition, the pronoun should be objective.)*
Correct	The debate was between Marsha and **me.**
Faulty	**Us** board members are accountable for our decisions. *(The pronoun accompanies the subject, "board members," and thus should be nominative.)*
Correct	**We** board members are accountable for our decisions.
Faulty	A group of **we** managers will fly to the convention. *(The pronoun accompanies the object of the preposition, "managers," and thus should be objective.)*
Correct	A group of **us** managers will fly to the convention.

Hint: By deleting the accompanying noun from the two latter examples, we can easily identify the correct pronoun case ("We . . . are accountable"; "A group of us . . . will fly").

NOTE *Frequently, pronouns ending in "self" are incorrectly used as subjects or objects, as in the example below.*

Faulty	Myself and three others attended the meeting.
Correct	"Three others and I," or "Four of us."
Faulty	Send the results to myself.
Correct	Instead, say "to me."

Hint: "Self" ending pronouns should be used only to refer to previously mentioned nouns or pronouns:

I completed the report myself.

The board members have only **themselves** to blame for the contract impasse.

Send the test results to **myself.** ("to me")

FAULTY MODIFICATION

dm/mm

A sentence's word order (syntax) helps determine its effectiveness and meaning. Words or groups of words are modified by adjectives, adverbs, phrases, or clauses. Modifiers explain, define, or add detail to other words or ideas. Prepositional phrases, for example, usually define or limit adjacent words:

> the foundation **with the cracked wall**
>
> the repair job **on the old Ford**
>
> the journey **to the moon**

As do phrases with "-ing" verb forms:

> the student **painting the portrait**
>
> **Opening the door,** we entered quietly.

Phrases with "to + verb" form limit:

> **To succeed,** one must work hard.

Some clauses limit:

> the person **who came to dinner**

Problems with word order occur when a modifying phrase begins a sentence and has no word to modify.

> **Dangling modifier**   **Dialling the telephone,** the cat ran out the open door.

The cat obviously did not dial the telephone, but because the modifier **Dialling the telephone** has no word to modify, the word order suggests that the noun beginning the main clause *(cat)* names the one who dialled the phone. Without any word to join itself to, the modifier *dangles*. By inserting a subject, we can repair this absurd message.

> **Correct**   As Moe dialled the telephone, the cat ran out the open door.

A dangling modifier also can obscure your meaning.

> **Dangling modifier**   After completing the student financial aid application form, the Financial Aid Office will forward it to the appropriate agency.

Who completes the form—the student or the financial aid office?

Here are some other dangling modifiers that make the message confusing, inaccurate, or downright absurd:

> **Dangling modifier**   **While walking,** a cold chill ran through my body.
>
> **Correct**   While **I** walked, a cold chill ran through my body.

Dangling modifier	Impurities have entered our bodies **by eating chemically processed foods.**
Correct	Impurities have entered our bodies by **our** eating chemically processed foods.
Dangling modifier	**By planting different varieties of crops,** the pests were unable to adapt.
Correct	By planting different varieties of crops, **farmers** prevented the pests from adapting.

The order of adjectives and adverbs in a sentence is as important as the order of modifying phrases and clauses. Notice how changing word order affects the meaning of these sentences:

I **often** remind myself of the need to balance my chequebook.

I remind myself of the need to balance my chequebook **often.**

Be sure that modifiers and the words they modify follow an order that reflects your meaning.

Misplaced modifier	Cal keyed another memo on our computer **that was useless.** *(Was the computer or the memo useless?)*
Correct	Cal keyed another useless memo on our computer. *or* Cal keyed another memo on our useless computer.
Misplaced modifier	He read a report on the use of non-chemical pesticides **in our conference room.** *(Are the pesticides to be used in the conference room?)*
Correct	In our conference room, he read a report on the use of non-chemical pesticides.
Misplaced modifier	She volunteered **immediately** to deliver the radioactive shipment. *(Volunteering immediately, or delivering immediately?)*
Correct	She immediately volunteered to deliver . . . *or* She volunteered to deliver immediately . . .

FAULTY PARALLELISM

par

To reflect relationships among items of equal importance, express them in identical grammatical form:

Correct	. . . We here highly resolve . . . that government **of the people, by the people, for the people** shall not perish from the earth.

The statement above describes the government with three modifiers of equal importance. Because the first modifier is a prepositional phrase, the others must be also. Otherwise, the message would be garbled, like this:

Faulty	We here highly resolve . . . that government **of the people, which the people created and maintain, serving the people** shall not perish from the earth.

If you begin the series with a noun, use nouns throughout the series; likewise for adjectives, adverbs, and specific types of clauses and phrases.

Faulty	The new apprentice is **enthusiastic, skilled,** and **you can depend on her.**
Correct	The new apprentice is **enthusiastic, skilled,** and **dependable.** *(all subjective complements)*
Faulty	In his new job, he felt **lonely** and **without a friend.**
Correct	In his new job, he felt **lonely** and **friendless.** *(both adjectives)*
Faulty	She plans **to study** all this month and **on scoring well** in her licensing examination.
Correct	She plans **to study** all this month and **to score well** in her licensing examination. *(both infinitive phrases)*
Faulty	She **sleeps** well and **jogs** daily, **as well as eating** high-protein foods.
Correct	She **sleeps** well, **jogs** daily, and **eats** high-protein foods. *(all verbs)*

To improve coherence in long sentences, repeat words that introduce parallel expressions:

Faulty	Before buying this property, you should decide whether you will settle down and raise a family, travel for a few years, or pursue a graduate degree.
Correct	Before buying this property, you should decide whether **to settle** down and raise a family, **to travel** for a few years, or **to pursue** a graduate degree.

SENTENCE SHIFTS

shift

Shifts in point of view damage coherence. If you begin a sentence or paragraph with one subject or person, do not shift to another.

Shift in person	When **you** finish the job, **one** will have a sense of pride.
Correct	When **you** finish the job, **you** will have a sense of pride.
Shift in number	**One** should sift the flour before **they** make the pie.
Correct	**One** should sift the flour before **one** makes the pie. (Or better: Sift the flour before making the pie.)

Do not begin a sentence in the active voice and then shift to passive.

| Shift in voice | **He delivered** the plans for the apartment complex, and the building site **was also inspected by him.** |
| Correct | **He delivered** the plans for the apartment complex and also **inspected** the building site. |

Do not shift tenses without good reason.

| Shift in tense | She **delivered** the blueprints, **inspected** the foundation, **wrote** her report, and **takes** the afternoon off. |
| Correct | She **delivered** the blueprints, **inspected** the foundation, **wrote** her report, and **took** the afternoon off. |

Do not shift from one mood to another (as from imperative to indicative mood in a set of instructions).

| Shift in mood | **Unscrew** the valve and then steel wool **should be used** to clean the fitting. |
| Correct | **Unscrew** the valve and then **use** steel wool to clean the fitting. |

Do not shift from indirect to direct discourse within a sentence.

| Shift in discourse | Jim wonders **if he will get the job** and **will he like it**? |
| Correct | Jim wonders **if he will get the job** and **if he will like it**. Will Jim get the job, and will he like it? |

EFFECTIVE PUNCTUATION

p.

Punctuation marks are like road signs and traffic signals. They govern reading speed and provide clues for navigation through a network of ideas; they mark intersections, detours, and road repairs; they draw attention to points of interest along the route; and they mark geographic boundaries. In short, punctuation marks give us a simple way of making ourselves understood.

END PUNCTUATION

The three marks of end punctuation—period, question mark, and exclamation point—work like a red traffic light by signalling a complete stop.

./

Period.   A period ends a sentence. Periods end some abbreviations.

| Ms. | Assn. | Dr. |
| M.D. | Inc. | B.A. |

Periods serve as decimal points for figures.

$15.95
2.14%

?/

Question Mark.   A question mark follows a direct question.

> Where is the balance sheet?

Do not use a question mark to end an indirect question.

> **Faulty**   He asked if all students had failed the test?
>
> **Correct**   He asked if all students had failed the test.
> *or*
> He asked, "Did all students fail the test?"

!/

Exclamation Point.   Because exclamation points symbolize strong feeling, don't overuse them. Otherwise you might seem hysterical or insincere.

> **Correct**   Oh, no!
> Pay up!

Use an exclamation point only when expression of strong feeling is appropriate.

SEMICOLON

;/

A semicolon usually works like a blinking red traffic light at an intersection by signalling a brief but definite stop.

Semicolon Separating Independent Clauses.   Semicolons separate independent clauses (logically complete ideas) whose contents are closely related and are not connected by a coordinating conjunction.

> The project was finally completed; we had done a good week's work.

The semicolon can replace the conjunction–comma combination that joins two independent ideas.

> The project was finally completed, and we were elated.
> The project was finally completed; we were elated.

The second version emphasizes the sense of elation.

Semicolons Used with Adverbs as Conjunctions and Other Transitional Expressions.   Semicolons accompany conjunctive adverbs and other expressions that connect related independent ideas (**besides, otherwise, still, however, furthermore, consequently, therefore, in contrast, in fact,** or the like).

> The job is filled; however, we will keep your résumé on file.
> Your background is impressive; in fact, it is the best among our applicants.

Semicolons Separating Items in a Series.   When items in a series contain internal commas, semicolons provide clear separation between items.

> We are opening branch offices in the following cities: Halifax, Nova Scotia; Moncton, New Brunswick; Montreal, Quebec; and Windsor, Ontario.
>
> Members of the survey crew were Laura Joe, a geologist; Hector Lightweight, a draftsperson; and Mary Shelley, a graduate student.

COLON

Like a flare in the road, a colon signals you to stop and then proceed, paying attention to the situation ahead, the details of which will be revealed as you move along. Usually, a colon follows an introductory statement that requires a follow-up explanation.

> We need the following equipment immediately: a voltmeter, a portable generator, and three pairs of insulated gloves.
>
> She is an ideal colleague: honest, reliable, and competent.
>
> Two candidates clearly are superior: Don and Marsha.

With the exception of **Dear Sir:** and other salutations in formal correspondence, colons follow independent (logically and grammatically complete) statements. Because colons, like end punctuation and semicolons, signal a full stop, they never are used to fragment a complete statement.

> **Faulty**   My plans include: finishing college, travelling for two years, and settling down in Edmonton.

No punctuation should follow "include."

Colons can introduce quotations.

> The supervisor's message was clear enough: "You're fired."

A colon normally replaces a semicolon in separating two related, complete statements when the second statement explains or amplifies the first.

> His reason for accepting the lowest-paying job offer was simple: He had always wanted to live in the Northwest.

The statement following the colon explains the "reason" mentioned in the statement preceding the colon.

COMMA

The comma is the most frequently used—and abused—punctuation mark. Unlike the period, semicolon, and colon, which signal a full stop, the comma signals a *brief pause*. The comma works like a blinking yellow traffic light, for which you slow down without stopping. Never use a comma to signal a *break* between independent ideas; it is not strong enough.

Comma as a Pause Between Complete Ideas.   In a compound sentence where a coordinating conjunction (**and, or, nor, for, but**) connects equal (independent) statements, a comma usually precedes the conjunction.

> This is a high-paying job, but the stress is high.
>
> This vacant shop is just large enough for our juice bar, and the location is excellent for walk-in customer traffic.

Without the conjunction, these statements would suffer from a comma splice, unless the comma were replaced by a semicolon or period.

Comma as a Pause Between an Incomplete and a Complete Idea.
A comma usually appears between a complete and an incomplete statement in a complex sentence to show that the incomplete statement depends for its meaning on the complete statement. (The incomplete statement cannot stand alone, separated by a break such as a semicolon, colon, or period.)

> **Because he is a plump person,** Mel diets often.
>
> **When he eats too much,** Mel gains weight.

Above, the first idea is made incomplete by a subordinating conjunction (**since, when, because, although, where, while, if, until**), which here connects a dependent with an independent statement. The first (incomplete) idea depends on the second (complete) for wholeness. When the order is reversed (complete idea followed by incomplete), the comma usually is omitted.

> Mel diets often **because he is a plump person.**
>
> Mel gains weight **when he eats too much.**

Because commas take the place of speech signals, reading a sentence aloud should tell you whether or not to pause (and use a comma).

Commas Separating Items (Words, Phrases, or Clauses) in a Series.
Use a comma to separate items in a series.

> **Ann, Ned, Shelley,** and **Philip** are joining us on the hydroelectric project.
>
> The office was **beige, blue,** and **burgundy**.
>
> She works hard **at home, on the job,** and even **during her vacation**.
>
> The employee claimed **that the hours were long, that the pay was low, that the work was boring,** and **that the supervisor was paranoid.**

Use no commas when *or* or *and* appears between all items in a series.

> She is willing to work in Winnipeg or Regina or even in Toronto.

Add a comma when *or* or *and* is used only before the final item in the series.

> Our luncheon special for Thursday will be rolls, steak, beans, and ice cream.

Without the comma, the sentence might cause readers to conclude that beans and ice cream is an exotic new dessert.

Comma Setting off Introductory Phrases.   Infinitive, prepositional, or verbal phrases introducing a sentence usually are set off by commas.

Infinitive phrase	**To be or not to be,** that is the question.
Prepositional phrase	**In Rome,** do as the Romans do.
Participial phrase	**Moving quickly,** the army surrounded the enemy.

When an interjection introduces a sentence, it is set off by a comma.

Oh, is that the final verdict?

When a noun in direct address introduces a sentence, it is set off by a comma.

Mary, you've done a great job.

Commas Setting off Non-restrictive Elements.   A restrictive phrase or clause limits or defines the subject in such a way that deleting the modifier would change the meaning of the sentence.

All candidates **who have work experience** will receive preference.

The clause, **who have work experience,** defines **candidates** and is essential to the meaning of the sentence. Without this clause, the meaning would be entirely different.

All candidates will receive preference.

This next sentence also contains a restriction:

All candidates **with work experience** will receive preference.

The phrase, **with work experience,** defines **candidates** and thus specifies the meaning of the sentence. Because this phrase *restricts* the subject by limiting the category, **candidates,** it is essential to the sentence's meaning and so is not separated from the sentence by commas.

A non-restrictive phrase or clause does not limit or define the subject; a nonrestrictive element could be deleted without changing the basic meaning of the sentence.

Our draftsperson, **who has little experience,** is highly competent.
This house, **riddled with carpenter ants,** is falling apart.

In each of those sentences, the modifying phrase or clause does not restrict the subject; each could be deleted:

Our new draftsperson is highly competent.
This house is falling apart.

A non-restrictive clause or phrase is set off from the sentence by commas.

Commas Setting off Parenthetical Elements. Items that interrupt sentence flow are called parenthetical and are enclosed by commas. Expressions such as **of course, as a result, as I recall,** and **however** are parenthetical and may denote emphasis, afterthought, clarification, or transition.

Emphasis	This deluxe model, **of course,** is more expensive.
Afterthought	Your report format, **by the way,** was impeccable.
Clarification	The loss of my job was, **in a way,** a blessing.
Transition	Our warranty, **however,** does not cover tire damage.

Direct address is parenthetical.

Listen, **my children,** and you shall hear . . .

A parenthetical expression at the beginning or the end of a sentence is set off by a comma.

Naturally, we will expect a full guarantee.

My friends, I think we have a problem.

You've done a good job, **Jim.**

Yes, you may use my name in your advertisement.

Commas Setting off Quoted Material. Quoted items included within a sentence are often set off by commas.

The customer said, **"I'll take it,"** as soon as he laid eyes on our new model.

Commas Setting off Appositives. An appositive, a word or words explaining a noun and placed immediately after it, is set off by commas.

Lindsay O'Shea, **our new president,** is overhauling all personnel policies.

Alpha waves, **the most prominent of the brain waves,** are typically recorded in a waking subject whose eyes are closed.

Please make all cheques payable to Sam Sawbuck, **company treasurer.**

Commas Used in Common Practice. Commas set off the day of the month from the year, in a date.

May 10, 2006

They set off numbers in three-digit intervals.

11,215

6,463,657

They set off street, city, and province in an address.

> The bill was sent to Jaspal Singh, 184 Sea Street, Victoria, BC   V9W 4D6.

When the address is written vertically, however, the omitted commas are those which would otherwise occur at the end of each address line.

> Jaspal Singh
> 184 Sea Street
> Victoria, BC V9W 4D6

Commas set off an address or date in a sentence.

> Room 3C, Margate Complex, is the site of our newest office.
> December 15, 2007, is my retirement date.

They set off degrees and titles from proper names.

> Roger P. Cayer, M.D.
> Gordon Browne, Jr.
> Sandra Mello, Ph.D.

Commas Used Erroneously.   Avoid needless or inappropriate commas. You are probably safer using too few commas than too many. Reading sentences aloud is one way to identify inappropriate pauses.

> **Faulty**   As I opened the door, he told me, that I was late. *(separates the indirect from the direct object)*
>
> The universal symptom of the suicide impulse, is depression. *(separates the subject from its verb)*
>
> This has been a long, difficult, project. *(separates the final adjective from its noun)*
>
> Poon, Henri, and Parvis, are joining us on the design phase of this project. *(separates the final subject from its verb)*
>
> An employee, who expects rapid promotion, must quickly prove her or his worth. *(separates a modifier that should be restrictive)*
>
> I spoke in a conference call with Jaswinder, and Gemma. *(separates two words linked by a coordinating conjunction)*
>
> The room was, 18 feet long. *(separates the linking verb from the subjective complement)*
>
> We painted the room, red. *(separates the object from its complement)*

APOSTROPHE

ap/

Apostrophes serve three purposes: to indicate the possessive, a contraction, and the plural of numbers, letters, and figures.

Apostrophe Indicating the Possessive.   At the end of a singular word, or of a plural word that does not end in *s,* add an apostrophe plus *s* to indicate the possessive.

> The people's candidate won.
> The chain saw was Lyle's.
> The men's locker room burned.
> I borrowed Doris's book.
> Have you heard Olga Charles's speech?

Do not use an apostrophe to indicate the possessive form of either singular or plural pronouns:

> The book was hers.
> Ours is the best sales record.
> The fault was theirs.

At the end of a plural word that ends in *s,* add an apostrophe only.

> the cows' water supply
> the Jacksons' wine cellar

At the end of a compound noun, add an apostrophe plus *s.*

> my father-in-law's false teeth

At the end of the last word in nouns of joint possession, add an apostrophe plus *s* if both own one item.

> Elke and Tran's lakefront cottage

Add an apostrophe plus *s* to both nouns if each owns specific items.

> Elke's and Tran's passports

Apostrophe Indicating a Contraction.   An apostrophe shows that you have omitted one or more letters in a phrase that is usually a combination of a pronoun and a verb.

> I'm (I am)   they're (they are)
> he's (he is)   we're (we are)
> you're (you are)   who's (who is)

Don't confuse *they're* with *their* or *there.*

> **Faulty**   there books
>   Their now leaving.
>   living their

Correct   their books
They're now leaving.
living there

Remember the distinction in this way:

Their boss knows they're there.

Don't confuse **it's** and **its**. **It's** means "it is." **Its** is the possessive.

It's watching its reflection in the pond.

Don't confuse **who's** and **whose**. **Who's** means "who is," whereas **whose** indicates the possessive.

Who's interrupting whose work?

Other contractions are formed from the verb and the negative.

isn't (is not)   can't (cannot)

don't (do not)   haven't (have not)

won't (will not)   wasn't (was not)

Apostrophe Indicating the Plural of Numbers, Letters, and Figures.
Use an apostrophe only when its absence would create ambiguity.

The 6's look like smudged G's, the 9's are illegible, and the %'s are unclear.

QUOTATION MARKS

Quotation marks set off exact words borrowed from another speaker or writer. At the end of a quotation the period or comma is placed within quotation marks.

"Hurry up," he whispered.
She told me, "I'm depressed."

The colon or semicolon is always placed outside quotation marks:

Our contract clearly defines "middle-management personnel"; however, it does not state salary range.
You know what to expect when Honest Al offers you a "bargain": a piece of junk.

Sometimes a question mark is used within a quotation that is part of a larger sentence. (Do not follow a question mark with a comma.)

"Can we stop the flooding?" enquired the supervisor.

When the question mark or exclamation point is part of the quotation, it belongs within quotation marks, replacing the comma or period.

"Help!" he screamed.

She asked Dal, "Can't we agree about anything?"

When the question mark or exclamation point denotes the attitude of the person quoting instead of the person quoted, it belongs outside the quotation mark.

Why did he wink and tell me, "It's a big secret"?

He actually accused me of being an "elitist"!

When quoting a passage of 50 words or longer, indent the entire passage five spaces and single-space between its lines to set it off from the text. Do not enclose the indented passage in quotation marks.

Use quotation marks around titles of articles, book chapters, poems, and unpublished reports.

The enclosed article, "The Job Market for College Graduates," should provide some helpful insights.

The title of a published work—book, journal, newspaper, brochure, or pamphlet—should be underlined or italicized.

Finally, use quotation marks (with restraint) to indicate your ironic use of a word.

He is some "friend"!

ELLIPSES

The dots in a row (. . .) indicate that you have omitted some material from a quotation. If the omitted words come at the end of the original sentence, a fourth dot indicates the period. Use several dots centred in a line to indicate that a paragraph or more has been left out. Ellipses help you save time and zero in on the important material within a quotation.

"Three dots . . . indicate that you have omitted some material . . . . A fourth dot indicates the period. . . . Several dots centred in a line . . . indicate a paragraph or more. . . . Ellipses help you . . . zero in . . . .

ITALICS

In longhand writing, indicate italics by *underlining*. On a word processor, use italic print for titles of books, periodicals, films, newspapers, and plays; for the names of ships; for foreign words or scientific names; for emphasizing a word (use sparingly); for indicating the special use of a word.

The Gage Canadian Dictionary is a handy reference tool.

The *Lusitania* sank rapidly.

She reads the *Globe and Mail* often.

My only advice is *caveat emptor.*

Bacillus anthracis is a highly virulent organism.

Do *not* inhale these spores, under any circumstances!

Our contract defines a *full-time employee* as one who works a minimum of 35 hours weekly.

PARENTHESES

Use commas normally to set off parenthetical elements, dashes to give some emphasis to the material that is set off, and parentheses to enclose material that defines or explains the statement that precedes it.

> This organism requires an anaerobic (oxygenless) environment.
>
> The cost of manufacturing our Beta II fuel cells has increased by 10 percent in one year. (See Appendix A for full cost breakdown.)
>
> This new three-colour model (made by Ilco Corporation) is selling well.

Notice that material within parentheses, like all other parenthetical material discussed earlier, can be deleted without harming the logical and grammatical structure of the sentence.

Also, use parentheses to enclose numbers or letters that segment items of information in a series.

> The three basic steps in this procedure are (1) . . . , (2) . . . , and (3) . . . .

BRACKETS

Use brackets within a quotation to add material that was not in the original quotation but is needed for clarification. Sometimes a bracketed word will provide an antecedent (or referent) for a pronoun.

> "She [Melanchuk] was the outstanding candidate for the job."

Brackets can enclose information from some other location within the context of the quotation.

> "It was in early spring [April 2, to be exact] that the tornado hit."

Use brackets to correct a quotation.

> "His report was [full] of mistakes."

Use *sic* ("thus" or "so") when quoting a mistake in spelling, usage, or logic.

> Her assistant's comment was clear: "She don't [sic] want any of these."

DASHES

Dashes can be effective indicators of meaning—as long as they are not overused. Make dashes by selecting the en dash (–) or em dash (—) from the special characters feature of your word processor. Alternatively, make an en dash using a space, a hyphen, and another space; and make an em dash or by placing two hyphens side by side. Parentheses de-emphasize enclosed material; dashes emphasize it.

Used selectively, dashes can provide emphasis, but they are not a substitute for all other punctuation. When in doubt, do not use a dash.

Dashes can denote an afterthought.

> Have a good vacation—but don't get sunstroke.

They can enclose an interruption in the middle of a sentence.

> This building's designer—I think it was Wright—was, above all, an artist.
> Our new team—Wong, Tse, and Lau—already is compiling outstanding statistics.

Although they can often be used interchangeably with commas, dashes dramatize a parenthetical statement more than commas do.

> Julie, a true friend, spent hours helping me rehearse for my interview.
> Julie—a true friend—spent hours helping me rehearse for my interview.

Notice the added emphasis in the second version.

HYPHEN

Use a hyphen to join compound modifiers (two or more words preceding the noun as an adjective), but not compound nouns.

> the rough-hewn wood
> the well-written report
> the all-too-human error
> a three-part report

Do not hyphenate these same words if they *follow* the noun

> The wood was rough hewn.
> The report is well written.
> The error was all too human.

Hyphenate an adverb–participle compound preceding a noun.

> the high-flying glider

Do not hyphenate compound modifiers if the adverb ends in **-ly**.

> the finely tuned engine

Hyphenate most words that begin with the prefix **-self**. (Check your dictionary.)

> self-reliance
> self-discipline
> self-actualizing

Hyphenate to avoid ambiguity.

> re-creation (a new creation)
> recreation (leisure activity)

Hyphenate words that begin with **ex-** only if **ex-** means "past."

> ex-employee
> expectant

Hyphenate all fractions, along with ratios that are used as adjectives and that precede the noun.

> a two-thirds majority
> In a four-to-one ratio they defeated the proposal.

Do not hyphenate ratios if they do not immediately precede the noun.

> The proposal was voted down four to one.

Hyphenate compound numbers from twenty-one through ninety-nine.

> Thirty-eight windows were broken.

Hyphenate a series of compound adjectives preceding a noun.

> The subjects for the motivation experiment were fourteen-, fifteen-, and sixteen-year-old students.

Use a hyphen to divide a word at the right-hand margin. Consult your dictionary for the correct syllable breakdown:

> com-puter
> comput-er

Actually, it is best to avoid altogether this practice of dividing words at the ends of lines.

TRANSITIONS AND OTHER CONNECTORS

trans

Transitions help make your meaning clear, signalling to readers that you are in a specific time or place, that you are giving an example, showing a contrast, shifting gears, or concluding your discussion. Here are some common transitions and the relations they indicate:

Addition	I am majoring in naval architecture; **furthermore,** I spent three years crewing on a racing yawl.

moreover	and
in addition	again
also	as well as

Place	Here is the switch that turns on the stage lights. **To the right** is the switch that dims them.

beyond	to the left
over	nearby

	under	adjacent to
	opposite to	next to
	beneath	where

Time   The crew will mow the ball field this morning; **immediately afterward,** we will clean the dugouts.

first	the next day
next	in the meantime
second	in turn
then	subsequently
meanwhile	while
at length	since
later	before
now	after

Comparison   Our reservoir is drying up because of the drought; **similarly,** water supplies in neighbouring towns are low.

likewise
in the same way
in comparison

Contrast   Felix worked hard; **however,** he received poor grades.

however	but
nevertheless	on the other hand
yet	to the contrary
still	notwithstanding
in contrast	conversely

Results   Jack fooled around; **consequently,** he was fired.

thus	thereupon
hence	as a result
therefore	so
accordingly	as a consequence

Example   Competition for part-time jobs is fierce; **for example,** 80 students applied for the sales associate's job at Sears.

for instance	namely
to illustrate	specifically

Explanation   She had a terrible semester; **that is,** she flunked four courses.

in other words	in fact
simply stated	put another way

Summary or conclusion   Our credit is destroyed, our bank account is overdrawn, and our debts are piling up; **in short,** we are bankrupt.

in closing	to sum up
to conclude	all in all
to summarize	on the whole
in brief	in retrospect
in summary	in conclusion

Pronouns serve as connectors, because a pronoun refers back to a noun in a preceding clause or sentence.

> As the **crew** neared the end of the project, **they** were all willing to work overtime to get the job done.
>
> **Low employee morale** is damaging our productivity. **This** problem needs immediate attention.

Repetition of key words or phrases is another good connecting device—as long as it is not overdone. Here is another example of effective repetition:

> Overuse and drought have depleted our water supply critically. Because of our **depleted supply,** we need to enforce strict **water**-conservation measures.

Here, the repetition also emphasizes a critical problem.

Because this next paragraph lacks transitions, sentences seem choppy and awkward:

> **Choppy**   Technical writing is a difficult but important skill to master. It requires long hours of work and concentration. This time and effort are well spent. Writing is indispensable for success. Good writers derive pride and satisfaction from their effort. A highly disciplined writing course should be part of every student's curriculum.

Here is the same paragraph rewritten to improve coherence:

> **Revised**   Technical writing is a difficult but important skill to master. It requires long hours of work and concentration. This time and effort, **however,** are well spent **because** writing is an indispensable tool for success. **Moreover,** good writers derive pride and satisfaction from their effort. A highly disciplined writing course, **therefore,** should be part of every student's curriculum.

Besides increasing coherence *within* a paragraph, transitions and connectors emphasize relationships *between* paragraphs by linking related groups of ideas. Here are two transitional sentences that could serve as conclusions for some paragraphs or as topic sentences for paragraphs that would follow; or they could stand alone for emphasis as single-sentence paragraphs:

> Because the A-12 filter has decreased overall engine wear by 15 percent, it should be included as a standard item in all our new models.
>
> With the camera activated and the watertight cover sealed, the diving bell is ready to be submerged.

Topic headings, like those in this book, are another connecting device. A topic heading is both a link and a separation between related yet distinct groups of ideas.

Sometimes a whole paragraph can serve as a connector between major sections of your report. Assume that you have just completed a section in a report on the advantages of a new oil filter and are now moving to a section on selling the idea to the buying public. Here is a paragraph you might write to link the two sections:

> Because the A-12 filter has decreased overall engine wear by 15 percent, it should be included as a standard item in all our new models. However, tooling and installation adjustments will add roughly $100 to the list price of each model. Therefore, we have to explain to customers the filter's long-range advantages. Let's look at ways of explaining these advantages.

Notice that this transitional paragraph *contains* transitional expressions as well.

EFFECTIVE MECHANICS

Correctness in abbreviation, capitalization, use of numbers, and spelling is an important sign of your attention to detail.

ABBREVIATIONS

ab

Whenever you abbreviate, consider your audience; never use an abbreviation that might confuse your reader. Abbreviations are often inappropriate in formal writing. When in doubt, write the word out.

Abbreviate some words and titles when they precede or immediately follow a proper name.

> **Correct**   Ms. Seligman   Mr. Trautwein
> Dr. Jekyll   Raymond Dumont, Jr.
> St. Simeon   Loretta Della Savo, Ph.D.

Do not, however, write abbreviations such as these:

> **Faulty**   Tess is a Dr.
> Pray hard and you might become a St.

In general, do not abbreviate military, religious, and political titles.

> **Correct**   Reverend Ormsby
> Captain Hook
> Prime Minister Jean Chrétien

Abbreviate time designations only when they are used with actual times.

> **Correct**   A.D. 576
> 400 B.C.
> 5:15 a.m.

Do not abbreviate these designations when they are used alone.

> **Faulty**   Plato lived sometime in the B.C. period.
> She arrived in the a.m.

In formal writing, do not abbreviate days of the week, months, words such as **street** and **road,** or names of disciplines such as **English**. Avoid abbreviating provinces, such as **Que.** for **Quebec**; countries, such as **Can.** for **Canada**; and book parts such as **Chap.** for **Chapter**, **pp.** for **pages**, and **fig.** for **figure**.

Use **no.** for **number** only when the actual number is given.

> **Correct**   Check switch no. 3.

Abbreviate a unit of measurement only when it appears often in your report and is written out in full on first use. Use only abbreviations you are sure readers will understand. Abbreviate items in a visual aid only if you need to save space.

COMMON ABBREVIATIONS FOR UNITS OF MEASUREMENT

AC	alternating current	kn	knot
amp; A	ampere	kw	kilowatt
Å	angstrom	kwh	kilowatt hour
az	azimuth	l; L	litre
bbl	barrel	lat	latitude
BTU	British Thermal Unit	lb	pound
C	Celsius, coulomb	lin	linear
cal	calorie	long	longitude
cc	cubic centimetre	log	logarithm
cd	candela	m	metre
circ	circumference	m^2	square metre
cm	centimetre	m^3	cubic metre
cm^2	square centimetre	M	nautical mile
cm^3	cubic centimetre	max	maximum
CPS	cycles per second	mg	milligram
cu ft	cubic foot	mi	mile
cu m	cubic metre	min	minute
dB	decibel	ml; mL	millilitre
DC	direct current	mm	millimetre
dm	decimetre	mo	month
doz	dozen	mol	mole
DP	dewpoint	mph	miles per hour
eV	electronvolt	N	newton
F	Fahrenheit; farad	oct	octane
fbm	foot board measure	oz	ounce
fl oz	fluid ounce	Pa	pascal
FM	frequency modulation	psf	pounds per square foot
freq	frequency	psi	pounds per square inch
ft; '	foot	qt	quart
ft lb	foot pound	r	roentgen
gal	gallon	rpm	revolutions per minute
GPM	gallons per minute	s	second
gr; g	gram	sp gr	specific gravity
ha	hectare	sq	square
hp	horsepower	t	ton; tonne
hr; h	hour	temp	temperature
Hz	hertz	tol	tolerance
in; "	inch	ts	tensile strength
IU	international unit	V	volt
J	joule	VA	volt ampere

K	kelvin	W	watt
ke	kinetic energy	wk	week
kg	kilogram	WL	wavelength
km	kilometre	yd	yard
km^2	square kilometre	yr	year

COMMON ABBREVIATIONS FOR REFERENCE IN MANUSCRIPTS

anon.	anonymous	fig.	figure
app.	appendix	i.e.	that is
b.	born	illus.	illustrated
©	copyright	jour.	journal
c., ca.	about (c. 1999)	l., ll.	line(s)
cf.	compare	ms., mss.	manuscript(s)
ch.	chapter	no.	number
col.	column	p., pp.	page(s)
d.	died	pt., pts.	part(s)
ed.	editor	rev.	revised or review
e.g.	for example	sec.	section
esp.	especially	sic	thus, so (to cite an error in the quotation)
et al.	and others		
etc.	and so on	trans.	translation
ex.	example	vol.	volume
f. or ff.	the following page or pages		

For abbreviations of other words, consult your dictionary. Most dictionaries list abbreviations at the front or back or alphabetically with the word entry.

CAPITALIZATION

cap

Capitalize these items: proper nouns, titles of people, books and chapters, languages, days of the week, the months, holidays, names of organizations or groups, races and nationalities, historical events, important documents, and names of structures or vehicles. In titles of books, films, and so on, capitalize first and following words, except articles or short prepositions.

A Tale of Two Cities	the Chevrolet Corvette
Protestant	Russian
Wednesday	Labour Day

the *Queen Elizabeth II*	DuPont Chemical Company
the CN Tower	Senator John Pasteur
April	France
Charlottetown	the War of 1812
the Charter of Rights	the Motor Vehicle Act

Do not capitalize the seasons, names of college classes (**senior, junior**), or general groups (**the younger generation**, or **the leisure class**).

Capitalize adjectives that are derived from proper nouns.

Chaucerian English
Roman numerals

Capitalize titles preceding a proper noun but not those following:

Provincial Premier Penny Nguyen
Penny Nguyen, provincial premier

Capitalize words such as **street**, **road**, **corporation**, **college** only when they accompany a proper noun.

George Brown College
High Street
the Rand Corporation

Capitalize **north, south, east,** and **west** when they denote specific locations, not when they are simple directions.

the South
the Northwest
Turn east at the next set of lights.

Begin all sentences with capitals.

NUMBERS

#

If numbers can be expressed in one or two words, you can write them out or you can use the numerals.

fourteen	14
eighty-one	81
ninety-nine	99

For larger numbers, use numerals.

4,364 or 4 364   2,800,357 or 2 800 357
543   200 million
$3^1/_4$

Use numerals to express decimals, precise technical figures, or any other exact measurements. Numerals are more easily read and better remembered than numbers that are spelled out.

50 kilowatts   15 pounds of pressure
14.3 milligrams   4,000 rpm or 4 000 rpm

Express these in numerals: dates, census figures, addresses, page numbers, exact units of measurement, percentages, ages, times with a.m. or p.m. designations, and monetary and mileage figures.

page 14   1:15 p.m.
18.4 kilos   9 metres
115 kilometres   12 litres
the 9-year-old tractors $15
15.1 percent or 15% (financial documents and statistics)

Do not begin a sentence with a numeral.

Six hundred students applied for the 102 available jobs.

Do not use numerals to express approximate figures, time not designated as a.m. or p.m., or streets named by numbers less than 100.

about seven hundred fifty
four fifteen
108 East Fifth Street

If one number immediately precedes another, spell out the first and use a numeral for the second:

Please deliver twelve 5-metre rafters.

Works Cited

Adams, Gerald R., and Jay D. Schvaneveldt. *Understanding Research Methods.* New York: Longman, 1985.

The Aldus Guide to Basic Design. Seattle, WA: Aldus Corporation, 1988.

American Psychological Association. *Publication Manual of the American Psychological Association.* 4th ed. Washington: APA, 1994.

"Are We in the Middle of a Cancer Epidemic?" *University of California at Berkeley Wellness Letter* 10.9 (1994): 4–5.

Bailey, Edward P. *Writing Clearly: A Contemporary Approach.* Columbus, OH: Merrill, 1984.

Baker, Russ. "Surfer's Paradise." *Inc.* Nov. 1997: 57+.

Barbour, Ian. *Ethics in an Age of Technology.* New York: Harper, 1993.

Barfield, Woodrow, Mark Haselkorn, and Catherine Weatbrook. "Information Retrieval with a Printed User's Manual and with Online Hypercard Help." *Technical Communication* 37.1 (1990): 22–27.

Barker, Larry J. et al. *Groups in Process.* 3rd ed. Englewood Cliffs, NJ: Prentice Hall, 1987.

Barnes, Shaleen. "Evaluating Sources Checklist." 10 June 1997. Information Literacy Project. On-line posting. (http://www.2lib/umassd.edu/library2/INFOLIT/prop.html) 23 June 1998.

Barnett, Arnold. "How Numbers Can Trick You." *Technology Review* Oct. 1994: 38–45.

Barnum, Carol, and Robert Fisher, "Engineering Technologists as Writers: Results of a Survey." *Technical Communication* 31.2 (1984): 9–11.

Bashein, Barbara, and Lynne Markus. "A Credibility Equation for IT Specialists." *Sloan Management Review* 38.4 (Summer 1997): 35–44.

Baumann, K. E. et al. "Three Mass Media Campaigns to Prevent Adolescent Cigarette Smoking." *Preventive Medicine* 17 (1988): 510–30.

Beamer, Linda. "Learning Intercultural Communication Competence." *Journal of Business Communication* 29.3 (1992): 285–303.

Bedford, Marilyn S., and F. Cole Stearns. "The Technical Writer's Responsibility for Safety." *IEEE Transactions on Professional Communication* 30.3 (1987): 127–32.

Begley, Sharon. "Is Science Censored?" *Newsweek* 14 Sept. 1992.

Benson, Phillipa J. "Visual Design Considerations in Technical Publications." *Technical Communication* 32.4 (1985): 35–39.

Bernstein, Mark. "Deeply Intertwingled Hypertext: The Navigation Problem Reconsidered." *Technical Communication* 38.1 (1991): 41–47.

Bjerklie, David. "E-Mail: The Boss Is Watching." *Technology Review* 14 Apr. 1993: 14–15.

Blackwell, Tom. "'You Are Not God,' Walkerton Health Watchdog Told." *Ottawa Citizen,* final ed. 11 Jan. 2001, A3.

Blatchford, Christie. "Koebel Knew It Killed, But Drank It Anyway." *National Post,* natl. ed. 21 Dec. 2000, A1.

Blum, Deborah. "Investigative Science Journalism." *Field Guide for Science Writers*. Eds. Deborah Blum and Mary Knudson. New York: Oxford, 1997: 86–93.

Bogert, Judith, and David Butt. "Opportunities Lost, Challenges Met: Understanding and Applying Group Dynamics in Writing Projects." *Bulletin of the Association for Business Communication* 53.2 (1990): 51–53.

Boiarsky, Carolyn. "Using Usability Testing to Teach Reader Response." *Technical Communication* 39.1 (1992): 100–02.

Branscum, Deborah. "bigbrother@the.office.com." *Newsweek* 27 Apr. 1998: 78.

Brody, Herb. "Great Expectations: Why Technology Predictions Sometimes Go Awry." *Technology Review* July 1991: 38–44.

Brownell, Judi, and Michael Fitzgerald. "Teaching Ethics in Business Communication: The Effective/Ethical Balancing Scale." *Bulletin of the Association for Business Communication* 55.3 (1992): 15–18.

Bruhn, Mark J. "E-Mail's Conversational Value." *Business Communication Quarterly* 58.3 (1995): 43–44.

Bryan, John. "Down the Slippery Slope: Ethics and the Technical Writer as Marketer." *Technical Communication Quarterly* 1.1 (1992): 73–88.

Burghardt, M. David. *Introduction to the Engineering Profession*. New York: Harper, 1991.

Burnett, Rebecca E. "Substantive Conflict in a Cooperative Context: A Way to Improve the Collaborative Planning of Workplace Documents." *Technical Communication* 38.4 (1991): 532–39.

Buseil, Christopher, and Tom Maegelin. *Researching Online*. New York: Addison, 1998.

Callahan, Sean. "Eye Tech." *Forbes ASAP* 7 June 1993.

Canadian Press. "Doctor Faces Rough Ride at Walkerton Inquiry." *Ottawa Citizen*, 8 Jan. 2001, A5.

————. "Notifying Government about Tainted Walkerton Water Standard Practice for Lab," 20 Oct. 2000. On-line database: Canadian MAS FullTEXT Elite. Accessed June 22, 2001.

————. "Ontario Ministry of Environment Wins Inaugural Code of Silence Award," May 26, 2001. On-line database: Canadian MAS FullTEXT Elite. Accessed June 22, 2001.

Carliner, Saul. "Demonstrating Effectiveness and Value: A Process for Evaluating Technical Communication Products and Services." *Technical Communication* 44.3 (1997): 252–65

Caswell-Coward, Nancy. "Cross-Cultural Communication: Is It Greek to You?" *Technical Communication* 39.2 (1992): 264–66.

Chauncey, Caroline. "The Art of Typography in the Information Age." *Technology Review* Feb./Mar. (1986): 26+.

Christians, Clifford G. et al. *Media Ethics: Cases and Moral Reasoning*. 2nd ed. White Plains, NY: Longman, 1978.

Clark, Gregory. "Ethics in Technical Communication: A Rhetorical Perspective." *IEEE Transactions on Technical Communication* 30.3 (1987): 190–95.

Clement, David E. "Human Factors, Instructions, and Warnings, and Product Liability." *IEEE Transactions on Professional Communication* 30.3 (1987): 149–56.

Cochran, Jeffrey K. et al. "Guidelines for Evaluating Graphical Designs." *Technical Communication* 36.1 (1989): 25–32.

Cole-Gomolskli. "Users Loathe to Share Their Know-How." *Computerworld* 17 Nov. 1997: 6.

Coletta, W. John. "The Ideologically Biased Use of Language in Scientific and Technical Writing." *Technical Communication Quarterly* 1.1 (1992): 59–70.

Communication Concepts, Inc. "Electronic Media Poses New Copyright Issues." *Writing Concepts.* Reprinted in *INTERCOM* [Newsletter of the Society for Technical Communication] Nov. 1995: 13+.

Consumer Product Safety Commission. *Fact Sheet No. 65.* Washington: GPO, 1989.

Cooper, Lyn. "Listening Competency in the Workplace: A Model for Training." *Business Communication Quarterly* 60.4 (Dec. 1997): 75–84.

Cotton, Robert, ed. *The New Guide to Graphic Design.* Secaucus, NJ: Chartwell, 1990.

Council of Biology Editors. *Scientific Style and Format: The CBE Manual for Authors, Editors, and Publishers.* 6th ed. Chicago: Cambridge UP, 1994.

Cronin, Mary J. "Knowing How Employees Use the Intranet Is Good Business." *Fortune* 21 July 1997: 103.

Cross, Mary. "Aristotle and Business Writing: Why We Need to Teach Persuasion." *Bulletin of the Association for Business Communication* 54.1 (1991): 3–6.

Crossen, Cynthia. *Tainted Truth: The Manipulation of Fact in America.* New York: Simon, 1994.

Dana-Farber Cancer Institute. *Facts and Figures about Cancer.* Boston: Dana-Farber 1995.

Daugherty, Shannon. "The Usability Evaluation: A Discount Approach to Usability Testing." *INTERCOM* Dec. 1997: 16–20.

Davenport, Thomas H. *Information Ecology.* New York: Oxford, 1997.

Debs, Mary Beth, "Collaborative Writing in Industry." *Technical Writing: Theory and Practice.* Eds. Bertie E. Fearing and W. Keats Sparrow. New York: Modern Language Assn., 1989: 33–42.

———. "Recent Research on Collaborative Writing in Industry." *Technical Communication* 38.4 (1991): 476–85.

December, John. "An Information Development Methodology for the World Wide Web." *Technical Communication* 46.3 (1996): 369–75.

Dombrowski, Paul M. "Challenger and the Social Contingency of Meaning: Two Lessons for the Technical Communication Classroom." *Technical Communication Quarterly* 1.3 (1992): 73–86.

Dowd, Charles. "Conducting an Effective Journalistic Interview." *INTERCOM* May 1996: 12–14.

Dragga, Sam, and Gwendolyn Gong. *Editing: The Design of Rhetoric.* Amityville, NY: Baywood, 1989.

Dulude, Jennifer. "The Web Marketing Handbook." Thesis. University of Massachusetts Dartmouth, 1997.

Dumont, R. A. "Writing, Research, and Computing." *Critical Thinking: An SMU Dialogue,* 1988.

Dumont, R. A., and J. M. Lannon. *Business Communications*. 3rd ed. Glenview, IL: Scott, 1990.

Dyson, Esther. "Intellectual Value." *Wired* July 1995: 134+.

Dyson, Esther, and Nicholas Negroponte. "How Smart Agents Will Change Selling." *Forbes ASAP* 28 Aug. 1995: 95.

"Earthquake Hazard Analysis for Nuclear Power Plants." *Energy and Technology Review* June 1984: 8.

Elbow, Peter. *Writing without Teachers*. New York: Oxford, 1973.

"Electronic Mentors." *The Futurist* May 1992: 56.

Elliot, Joel. "Evaluating Web Sites: Questions to Ask." 18 Feb. 1997. On-line posting. List for Multimedia and New Technologies in Humanities Teaching. (http:// www.learnnc.org/ documents/webeval.html) 9 Mar. 1997.

Evans, James. "Legal Briefs." *Internet World* Feb. 1998: 22.

———. "Whose Web Site Is It Anyway?" *Internet World* Sept. 1997: 46+.

Evenson, Brad. "Bacterial Strain One of World's Most Dangerous." *National Post*, natl. ed., 26 May 2000, A9.

Facklemann, Kathleen. "Science Safari in Cyberspace." *Science News* 152.50 (1997): 397–98.

Facts and Figures about Cancer. Boston: Dana-Farber Cancer Institute, 1990.

Felker, Daniel B., et al. *Guidelines for Document Designers*. Washington: American Institutes for Research, 1981.

Figgins, Ross. "The Future of Business Communication Technology: Where Are We Headed, Captain?" *Communication and Technology: Today and Tomorrow*. Ed. Al Williams. Denton, TX: Assn. for Business Communication, 1994. 123–42.

Fineman, Howard, "The Power of Talk." *Newsweek* 8 Feb. 1993: 24–28.

Finkelstein, Leo, Jr. "The Social Implications of Computer Technology for the Technical Writer." *Technical Communication* 38.4 (1991): 466–73.

Fisher, Anne. "My Team Leader Is a Plagiarist." *Fortune* 27 Oct. 1997: 291–92.

Florman, Samuel, quoting Donald D. Rikard. "Toward Liberal Learning for Engineers." *Technology Review* Mar. 1986: 18–25.

Franke, Earnest A. "The Value of the Retrievable Technical Memorandum System to the Engineering Company." *IEEE Transactions on Professional Communication* 32.1 (Mar. 1989): 12–16.

Fugate, Alice E. "Mastering Search Tools for the Internet." *INTERCOM* Jan. 1998: 40–41.

Gannon, Joseph. "From GUI Guru to Web Weaver: Making the Transition." *INTERCOM* Dec. 1997: 21–25.

Garfield, Eugene, "What Scientific Journals Can Tell Us about Scientific Journals." *IEEE Transactions on Professional Communication* 16.4 (1973): 200–02.

Gartiganis, Arthur. "Lasers." *Occupational Outlook Quarterly* Winter 1984: 22–26.

Gesteland, Richard R. "Cross-Cultural Compromises." *Sky* May 1993: 20+.

Gibaldi, Joseph. *MLA Handbook for Writers of Research Papers*. 4th ed. New York: Modern Language Association., 1995.

Gibaldi, Joseph, and Walter S. Achtert. *MLA Handbook for Writers of Research Papers*. 3rd ed. New York: Modern Language Assn., 1988.

Gilbert, Nick, "1-800-ETHIC." *Financial World* 16 Aug. 1994: 20+.

Gilsdorf, Jeanette W. "Executives' and Academics' Perception of the Need for Instruction in Written Persuasion." *Journal of Business Communication* 23.4 (1986): 55–68.

———. "Write Me Your Best Case for . . ." *Bulletin of the Association for Business Communication* 54.1 (1991): 7–12.

Girill, T. R. "Technical Communication and Art. *Technical Communication* 31.2 (1984): 35.

———. "Technical Communication and Ethics." *Technical Communication* 34.3 (1987): 178–79.

———. "Technical Communication and Law." *Technical Communication* 32.3 (1985): 37.

Glidden, H. K. *Reports, Technical Writing, and Specifications.* New York: McGraw, 1964.

Golen, Steven et al. "How to Teach Ethics in a Basic Business Communications Class." *Journal of Business Communication* 22.1 (1985): 75–84.

Goodall, H. Lloyd, Jr., and Christopher L. Waagen. *The Persuasive Presentation.* New York: Harper, 1986.

Goodman, Danny. *Living at Light Speed.* New York: Random, 1994.

Goubil-Gambrell, Patricia. "Designing Effective Internet Assignments in Introductory Technical Communication Courses." *IEEE Transactions on Professional Communication* 39.4 (1996): 224–31.

Gouran, Dennis S. et al. "A Critical Analysis of Factors Related to Decisional Processes Involved in the Challenger Disaster." *Central States Speech Journal* 37.3 (1986): 119–35.

Grassian, Esther. "Thinking Critically about World Wide Web Resources." 20 Aug. 1997. UCLA College Library. (http://library.ucla.edu/ librairies/college/instruct/critical.htm) 25 Oct. 1997.

Gribbons, William M. "Organization by Design: Some Implications for Structuring Information." *Journal of Technical Writing and Communication* 22.1 (1992): 57–74.

Grice, Roger A. "Document Development in Industry." *Technical Writing: Theory and Practice.* Ed. Bertie E. Fearing and W. Keats Sparrow. New York: Modern Language Assn., 1989. 27–32.

———. "Focus on Usability: Shazam!" *Technical Communication* 42.1 (1995): 131–33.

Grice, Roger A., and Lenore S. Ridgway. "Presenting Technical Information in Hypermedia Format: Benefits and Pitfalls." *Technical Communication Quarterly* 4.1 (1995): 35–46.

Grossman, Wendy M. "Downloading as a Crime." *Scientific American* Mar. 1998: 37.

Gruman, Galen. "The Paper Chase." *MACWORLD* Apr. 1995: 126–31.

Guffey, Mary Ellen et al. *Business Communication.* 2nd Canadian ed. Toronto: Nelson, 1999.

Hafner, Kate, "Have Your Agent Call My Agent." *Newsweek* 27 Feb. 1995: 76–77.

Hall, Judith G. "Medicine on the Web." *Technology Review* Mar./Apr. 1998: 60–61.

Halpern, Jean W. "An Electronic Odyssey." *Writing in Nonacademic Settings.* Eds. Dixie Goswami and Lee Odell. New York: Guilford, 1985: 157–201.

Hamblen, Matt. "Volvo Taps AT&T for Global Net." *Computerworld* 1 Dec. 1997: 51+.

Hammett, Paula. "Evaluating Web Resources." 29 Mar. 1997. Ruben Salazar Library, Sonoma State University. On-line posting. (http://libweb.sonoma.edu,resources/eval.html) 26 Oct. 1997.

Harcourt, Jules. "Teaching the Legal Aspects of Business Communication." *Bulletin of the Association for Business Communication* 53.3 (1990): 63–64.

Harris, Robert. "Evaluating Internet Research Sources." 17 Nov. 1997. Online posting. (http://www.sccu.edu/faculty/R_Harris/eval8it.htm) 23 June 1998.

Hart, Geoff. "Accentuate the Negative: Obtaining Effective Reviews through Focused Questions." *Technical Communication* 44.1 (1997): 52–57.

Hartley, James. *Designing Instructional Text.* 2nd ed. London: Kogan Page, 1985.

Haskin, David. "Meetings without Walls." *Internet World* Oct. 1997: 53–60.

———. "The Extranet Team Play." *Internet World* Aug. 1997: 57–60.

Hauser, Gerald. *Introduction to Rhetorical Theory.* New York: Harper, 1986.

Hayakawa, S.I. *Language in Thought and Action.* 3rd ed. New York: Harcourt, 1972.

Haynes, Kathleen J. M., and Linda K. Robertson. "An Application of Usability Criteria in the Classroom." *Technical Writing Teacher* 18.3 (1991): 236–42.

Hays, Robert. "Political Realities in Reader/Situation Analysis." *Technical Communication* 31.1 (1984): 16–20.

Hein, Robert G. "Culture and Communication." *Technical Communication* 38.1 (1991): 125–26.

Hill-Duin, Ann. "Terms and Tools: A Theory and Research-Based Approach to Collaborative Writing." *Bulletin of the Association for Business Communication* 53.2 (1990): 45–50.

Hodges, Mark. "Is Web Business Good Business?" *Technology Review* Aug./Sept. 1997: 22+.

Hoger, Elizabeth, James J. Cappel, and Mark A. Myerscough. "Navigating the Web with a Typology of Corporate Uses." *Business Communication Quarterly* 61.2 (1998): 39–47.

Holler, Paul F. "The Challenge of Writing for Multimedia." *INTERCOM* [Newsletter of the Society for Technical Communication] July/Aug. 1995: 25.

Hopkins-Tanne, Janice. "Writing Science for Magazines." *A Field Guide for Science Writers.* Eds. Deborah Blum and Mary Knudson. New York: Oxford, 1997: 17–26.

Horton, William. "Is Hypertext the Best Way to Document Your Product?" *Technical Communication* 38.1 (1991): 20–30.

———. "Mix Media, Not Metaphors." *Technical Communication* 41.4 (1994): 781–83.

Howard, Tharon. "Property Issue in E-Mail Research." *Bulletin of the Association for Business Communication* 56.2 (1993): 40–41.

Huff, Darrell. *How to Lie with Statistics.* New York: Norton, 1954.

Hulbert, Jack E. "Developing Collaborative Insights and Skills." *Bulletin of the Association for Business Communication* 57.2 (1994): 53–56.

———. "Overcoming Intercultural Communication Barriers." *Bulletin of the Association for Business Communication* 57.2 (1994): 41–44.

Humphreys, Donald S. "Making Your Hypertext Interface Usable." *Technical Communication* 40.4 (1993): 754–61.

Hunt, Kevin. "Establishing a Presence on the World Wide Web: A Rhetorical Approach." *Technical Communication* 46.4 (1996): 376–87.

Hutheesing, Nikhil. "Who Needs the Middleman?" *Forbes* 28 Aug. 1995.

IBM Corporation. "IBM Solutions" [Advertisement] 1997.

James-Catalano, "The Virtual Library." *Internet World* June 1995: 26+.

Jameson, Daphne A. "Using a Simulation to Teach Intercultural Communication in Business Communication Courses." *Bulletin of the Association for Business Communication* 56.1 (1993): 3–11.

Janis, Irving L. *Victims of Groupthink: A Psychological Study of Foreign Policy Decisions and Fiascos.* Boston: Houghton, 1972.

Johannesen, Richard L. *Ethics in Human Communication.* 2nd ed. Prospect Heights, IL: Waveland, 1983.

Journet, Debra. Unpublished review of *Technical Writing.* 3rd ed.

Karaim, Reed. "The Invasion of Privacy." *Civilization* Oct./Nov. 1996: 70–77.

Kawasaki, Guy. "Get Your Facts Here." *Forbes* 23 Mar. 1998: 156.

————. "The Rules of E-Mail." *MACWORLD* Oct. 1995: 286.

Kelley-Reardon, Kathleen. *They Don't Get It Do They?: Communication in the Workplace—Closing the Gap between Women and Men.* Boston: Little, 1995.

Kelman, Herbert C. "Compliance, Identification, and Internalization: Three Processes of Attitude Change." *Journal of Conflict Resolution* 2 (1958): 51–60.

Kerr, Ann. "Sophisticated Software Does the Job for On-line Recruiters," *Globe and Mail,* 30 June 2000, C10.

Keyes, Elizabeth. "Typography, Color, and Information Structure." *Technical Communication* 40.4 (1993): 638–54.

Kiely, Thomas. "The Idea Makers." *Technology Review* Jan. 1993: 31–40.

Kipnis, David, and Stuart Schmidt. "The Language of Persuasion." *Psychology Today* Apr. 1985: 40–46. Reprinted in Raymond S. Ross, *Understanding Persuasion.* 3rd ed. Englewood Cliffs, N J: Prentice, 1990.

Kirsh, Lawrence. "Take It from the Top." *MACWORLD* Apr. 1986: 112–15.

Kohl, John R., et al. "The Impact of Language and Culture on Technical Communication in Japan." *Technical Communication* 40.1 (1993): 62–72.

Kotulak, Ronald. "Reporting on Biology of Behavior." *A Field Guide for Science Writers.* Eds. Deborah Blum and Mary Knudson. New York: Oxford, 1997: 142–51.

Kraft, Stephanie. "Whistleblower Bill's Holiday Adventures." *The Valley Advocate* [Northhampton, MA] 6 Jan. 1994: 5–6.

Kremers, Marshall. "Teaching Ethical Thinking in a Technical Writing Course." *IEEE Transactions on Professional Communication* 32.2 (1989): 58–61.

Lambert, Steve. *Presentation Graphics on the Apple® Macintosh.* Bellevue, WA: Microsoft Corporation, 1984.

Lang, Thomas A., and Michelle Secic. *How to Report Statistics in Medicine.* Philadelphia: American College of Physicians, 1997.

LaPlante, Alice. "Brainstorming." *Forbes ASAP* 25 Oct. 1993: 45–61.

————. "Imaging Your Sea of Data." *Forbes ASAP* 29 Aug. 1994: 37–41.

Larson, Charles U. *Persuasion: Perception and Responsibility*. 7th ed. Belmont, CA: Wadsworth: 1995.

Lavin, Michael R. *Business Information: How to Find It, How to Use It*. 2nd ed. Phoenix, AZ: Oryx, 1992.

Lederman, Douglas. "Colleges Report Rise in Violent Crime," *Chronicle of Higher Education* 3 Feb. 1995, sec. A: 31–42.

Leki, Ilona. "The Technical Editor and the Non-native Speaker of English." *Technical Communication* 37.2 (1990): 148–52.

Levi, Stephen. "Optimism about the Net." *MACWORLD* July 1994: 179–80.

Lewis, Philip L., and N. L. Reinsch. "The Ethics of Business Communication." Proceedings of the American Business Communication Conference. Champaign, IL., 1981. In *Technical Communication and Ethics*. Eds. John R. Brockman and Fern Rook. Washington: Society for Technical Communication, 1989: 29–44.

Littlejohn, Stephen W. *Theories of Human Communication*. 2nd ed. Belmont, CA: Wadsworth, 1983.

Littlejohn, Stephen W., and David M. Jabusch. *Persuasive Transactions*. Glenview, IL: Scott, 1987.

Machlis, Sharon. "Surfing into a New Career as Webmaster." *Computerworld* 1 Dec. 1997: 45+.

MacKenzie, Nancy. Unpublished review of *Technical Writing*. 5th ed.

Mackin, John. "Surmounting the Barrier between Japanese and English Technical Documents." *Technical Communication* 36.4 (1989): 346–51.

Maeglin, Thomas. Unpublished review of *Technical Writing*, 7th ed.

Martin, James A. "A Road Map to Graphics Web Sites." *MACWORLD* Oct. 1995: 135–36.

Martin, Jeanette S., and Lillian H. Chaney. "Determination of Content for a Collegiate Course in Intercultural Business Communication by Three Delphi Panels." *Journal of Business Communication* 29.3 (1992): 267–83.

Matson, Eric. "(Search) Engines." *Fast Company* Oct./Nov. 1997: 249–52.

Max, Robert R. "Wording It Correctly." *Training and Development Journal* Mar. 1985: 5–6.

McDonald, Kim A. "Some Physicists Criticize Research Purporting to Show Links between Low-Level Electromagnetic Fields and Cancer." *Chronicle of Higher Education* 3 May 1991, sec. A: 5+.

McGuire, Gene. "Shared Minds: A Model of Collaboration." *Technical Communication* 39.3 (1992): 467–68.

McWilliams, Gary, and Marcia Stepanik. "Taming the Info. Monster." *Business Week* 22 June 1998: 170+.

Meyer, Benjamin D. "The ABCs of New-Look Publications." *Technical Communication* 33.1 (1986): 13–20.

Meyerson, Moe. "Grand Illusions." *Inc. Tech* 2 (1997): 35–36.

Microsoft Word User's Guide: Word Processing Program for the Macintosh, Version 5.0. Redmond, WA: Microsoft Corporation, 1992.

Mirel, Barbara, Susan Feinberg, and Leif Allmendinger. "Designing Manuals for Active Learning Styles." *Technical Communication* 38.1 (1991): 75–87.

Mittelstaedt, Martin. "The Buck Stops Here." *Globe and Mail*, 30 June 2001, A1.

———. "Tory Didn't See Warning, Walkerton Probe Told." *Globe and Mail* 28 June 2001. *Globe and Mail* <u>on-line.</u> Retrieved 28 June 2001. http://www.globeandmail.com

———. "Tory 'Team' Made Cuts, Walkerton Probe Told." *Globe and Mail* 27 June 2001. *Globe and Mail* <u>on-line.</u> Retreived 28 June 2001. http://www.globeandmail.com

Mokhiber, Russell. "Crime in the Suites." *Greenpeace* May 1989: 14–16.

Monastersky, R. "Courting Reliable Science." *Science News* 153.16 (1998): 249–51.

———. "Do Clouds Provide a Greenhouse Thermostat?" *Science News* 142 (1992): 69.

Morgan, Meg. "Patterns of Composing: Connections between Classroom and Workplace Collaborations." *Technical Communication* 38.4 (1991): 540–42.

Morgenson, Gretchen. "Would Uncle Sam Lie to You?" *Worth* Nov. 1994: 53+.

Morse, June. "Hypertext—What Can We Expect?" *STC Intercom* [Newsletter of the Society for Technical Communication] Feb. 1992: 6–7.

Moses, Barbara. "The Challenge: How to Satisfy the New Worker's Agenda." *Globe and Mail* 10 Nov. 1998, B15.

Munger, David. Unpublished review of *Technical Writing*, 8th ed.

Nakache, Patricia. "Is It Time to Start Bragging about Yourself?" *Fortune* 27 Oct. 1997: 287–88.

Nantz, Karen S., and Cynthia L. Drexel. "Incorporating Electronic Mail with the Business Communication Course." *Business Communication Quarterly* 58.3 (1995) 45–51.

Neilsen, Jakob. "Be Succinct! (Writing for the Web)." 15 Mar. 1997. *Alertbox*. On-line posting. (http://www.useit.com/alertbox/9719a.html) 8 Aug. 1998.

———. "Global Web: Driving the International Network Economy." Apr. 1998. *Alertbox*. On-line posting. (http://www.useit.com/alertbox/9710a.html) 8 Aug. 1998.

———. "How Users Read on the Web." Oct. 1997. *Alertbox*. On-line posting. (http://www.useit.com/alertbox/9710a.html) 8 Aug. 1998.

———. "International Web Usability." Aug. 1996. Alertbox. On-line posting. (http://www.useit.com/alertbox/9710a.htm) 8 Aug. 1998.

———. "Inverted Pyramids in Cyberspace." June 1996. *Alertbox*. On-line posting. (http://www.useit.com/alertbox/9710a.html) 8 Aug. 1998.

Nelson, Sandra J., and Douglas C. Smith. "Maximizing Cohesion and Minimizing Conflict in Collaborative Writing Groups. *Bulletin of the Association for Business Communication* 53.2 (1990): 59–62.

Nickels-Shirk, Henrietta. "'Hyper' Rhetoric: Reflections on Teaching Hypertext." *Technical Writing Teacher* 18.3 (1991): 189–200.

"Notes." *Technology Review* July 1993: 72.

Nydell, Margaret K. *Understanding Arabs: A Guide for Westerners*. New York: Logan, 1987.

O'Connor, the Honourable Dennis R. *Report of the Walkerton Inquiry: The Events of May 2000 and Related Issues. (Part One: A Summary and The Full Report)*. Released 14 January 2002. Retrieved 14 January 2002 (http://www.walkertoninquiry.com).

"Olive Oil and Breast Cancer: How Strong a Connection?" *University of California at Berkeley Wellness Letter* 11.7 (1995): 1–2.

"On Line." *Chronicle of Higher Education* 21 Sept. 1992, sec. A: 1.

Ornatowski, Cezar M. "Between Efficiency and Politics: Rhetoric and Ethics in Technical Writing." *Technical Communication Quarterly* 1.1 (1992): 91–103.

Outing, Steve. "Does Your Site Contribute to Data Smog?" 28 May 1997. *Alertbox Editor and Publisher Interactive* On-line posting. (http://www.mediainfo.com/ephome/news/newsshtm/stop/st052897.htm) 8 Aug. 1997.

Pace, Roger C. "Technical Communication, Group Differentiation, and the Decision to Launch the Space Shuttle Challenger." *Journal of Technical Writing and Communication* 18.3 (1988): 207–20.

Pearce, Glenn, Iris Johnson, and Randolph Barker. "Enhancing the Student Listening Skills and Environment." *Business Communication Quarterly* 580.4 (Dec. 1995): 28–33.

Pender, Kathleen. "Dear Computer, I Need a Job." *Worth* Mar. 1995: 120–21.

Perelman, Lewis, J. "How Hypermation Leaps the Learning Curve." *Forbes ASAP* 25 Oct. 1993: 78+.

Perloff, Richard M. *The Dynamics of Persuasion.* Hillsdale, NJ: Erlbaum, 1993.

Peterson, Ivars. "Web Searches Fall Short." *Science News* 153.18 (1998): 286.

Peyser, Marc, and Steve Rhodes. "When E-Mail Is Oops-Mail." *Newsweek* 16 Oct. 1995: 82.

Pinelli, Thomas E. et al. "A Survey of Typography, Graphic Design, and Physical Media in Technical Reports." *Technical Communication* 33.2 (1986): 75–80.

Plumb, Carolyn, and Jan H. Spyridakis. "Survey Research in Technical Communication: Designing and Administering Questionnaires." *Technical Communication* 39.4 (1992): 625–38.

Porter, James E. "Truth in Technical Advertising: A Case Study." *IEEE Transactions on Professional Communication* 30.3 (1987): 182–89.

Powell, Corey. "Science in Court." *Scientific American* Oct. 1997: 32+.

Pugliano, Fiore. Unpublished review of *Technical Writing,* 5th ed.

Quimby, G. Edward. "Make Text and Graphics Work Together." *Intercom* Jan. 1996: 34.

Ray, Randy. "Web Expands Recruiting Role." *Globe and Mail* 10 Nov., 2000, E7.

Redish, Janice C., and David A. Schell. "Writing and Testing Instructions for Usability." *Technical Writing: Theory and Practice.* Eds. Bertie E. Fearing and W. Keats Sparrow. New York: Modern Language Assn., 1987: 61–71.

Redish, Janice C., et al. "Making Information Accessible to Readers." *Writing in Nonacademic Settings.* Eds. Lee Odell and Dixie Goswami. New York: Guilford, 1985.

Reichard, Kevin. "Web-Site Watchdogs." *Internet World* Dec. 1997: 106+.

Rensberger, Boyce. "Covering Science for Newspapers." *A Field Guide for Science Writers.* Eds. Deborah Blum and Mary Knudson. New York: Oxford, 1997: 7–16.

Richards, Thomas O. and Ralph A. Richards, "Technical Writing." Paper presented at the University of Michigan, 11 July 1941. Published by the Society for Technical Communication.

Riney, Larry A. *Technical Writing for Industry.* Englewood Cliffs: Prentice, 1989.

Ritzenthaler, Gary, and David H. Ostroff. "The Web and Corporate Communication: Potentials and Pitfalls." *IEEE Transactions on Professional Communication* 39.1 (1996): 16–20.

Rokeach, Milton. *The Nature of Human Values*. New York: Free, 1973.

Ross, Philip E. "Lies, Damned Lies, and Medical Statistics." *Forbes* 14 Aug. 1995: 130–35.

Ross, Raymond S. *Understanding Persuasion*. 3rd ed. Englewood Cliffs, NJ: Prentice, 1990.

Rothschild, Michael. "When You're Gagging on E-Mail." *Forbes ASAP* 6 June 1994: 25–26.

Rottenberg, Annette, T. *Elements of Argument*. 3rd ed. New York: St. Martin's, 1991.

Rowland, D. *Japanese Business Etiquette: A Practical Guide to Success with the Japanese*. New York: Warner, 1985.

Rowland Robert C. "The Relationship between the Public and the Technical Spheres of Argument: A Case Study of the Challenger Seven Disaster." *Central States Speech Journal* 37.3 (1986): 136–46.

Rubens, Philip M. "Reading and Employing Technical Information in Hypertext." *Technical Communication* 38.1 (1991): 36–40.

———. "Reinventing the Wheel?: Ethics for Technical Communicators." *Journal of Technical Writing and Communication* 11.4 (1981): 329–39.

Ruggiero, Vincent R. *The Art of Thinking*. 3rd ed. New York: Harper, 1991.

Ruhs, Michael A. "Usability Testing: A Definition Analyzed." *Boston Broadside* [Newsletter of the Society for Technical Communication] May/June 1992: 8+.

Samuelson, Robert. "Merchants of Mediocrity." *Newsweek* 1 Aug. 1994: 44.

Schafer, Arthur. "Harris's Take Doesn't Hold Water." *Globe and Mail* 4 July 2001, A13.

Schenk, Margaret T., and James K. Webster. *Engineering Information Resources*. New York: Decker, 1984.

Schwartz, Marilyn, et al. *Guidelines for Bias-Free Writing*. Bloomington, IN: Indiana UP, 1995.

Scott, James C., and Diana J. Green. "British Perspectives on Organizing Bad-News Letters: Organizational Patterns Used by Major U.K. Companies." *Bulletin of the Association for Business Communication* 55.1 (1992): 17–19.

Selber, Stuart. Unpublished review of *Technical Writing*, 7th ed.

Sherblom, John C., Claire F. Sullivan, and Elizabeth C. Sherblom, "The What, the Whom, and the Hows of Survey Research." *Bulletin of the Association for Business Communication* 56:12 (1993): 58–64.

Sherif, Muzapher et al. *Attitude and Attitude Change: The Social Judgment-Involvement Approach*. Philadelphia: Saunders, 1965.

Snyder, Joel. "Finding It on Your Own." *Internet World* June 1995: 89–90.

Spruell, Geraldine. "Teaching People Who Already Learned How to Write, to Write." *Training and Development Journal* Oct. 1986: 32–35.

Spyridakis, Jan H. "Conducting Research in Technical Communication: The Application of True Experimental Design." *Technical Communication* 39.4 (1992): 607–24.

Spyridakis, Jan H., and Michael J. Wenger. "Writing for Human Performance: Relating Reading Research to Document Design." *Technical Communication* 39.2 (1992): 202–15.

Stanton, Mike. "Fiber Optics." *Occupational Outlook Quarterly* Winter 1984: 27–30.

Staudennmaier, S. J. "Engineering with a Human Face." *Technology Review* July 1991: 66–67.

Stedman, Craig. "Extranet Brokers Access to Funds Data." *Computerworld* 20 Oct. 1997: 49+.

Steinberg, Stephen. "Travels on the Net." *Technology Review* July 1994: 20–31.

Stemmer, John. "Citing Internet Sources." 4 Mar. 1997. On-line posting. Political Science and Teaching List. (polpsrt@h-met.msu.edu) 22 April 1997.

Stevenson, Richard W. "Workers Who Turn in Bosses Use Law to Seek Big Rewards." *New York Times* 10 July 1989, sec A: 1.

Stone, Brad, and T. Trent Gegax. "Your Favorite Sites." *Newsweek* 20 Nov. 1995: 16.

Stone, Peter H. "Forecast Cloudy: The Limits of Global Warming Models." *Technology Review* Feb./Mar. 1992: 32–40.

Stonecipher, Harry. *Editorial and Persuasive Writing.* New York: Hastings, 1979.

Sturges, David L. "Internationalizing the Business Communication Curriculum." *Bulletin of the Association for Business Communication* 55.1 (1992): 30–39.

Teague, John H. "Marketing on the World Wide Web." *Technical Communication* 42.2 (1995): 236–42.

Templeton, Brad. "10 Big Myths about Copyright Explained." 29 Nov. 1994. On-line posting. Listserv law/copyright-FAQ/myths/part 1. BITNET. 6 May 1995.

Thrush, Emily A. "Bridging the Gap: Technical Communication in an Intercultural and Multicultural Society." *Technical Communication Quarterly* 2.3 (1993): 271–83.

Tuck, Simon. "Canning Spam." *Globe and Mail* 15 Oct. 1998, D1 and D5.

Unger, Stephen H. *Controlling Technology: Ethics and Responsible Engineer.* New York: Holt, 1982.

U.S. Air Force Academy. *Executive Writing Course.* Washington: GPO, 1981.

U.S. Department of Commerce. *Statistical Abstract of the United States.* Washington: GPO, 1994.

U.S. Department of Labor. *Tips for Finding the Right Job.* Washington: GPO, 1993.

———. *Tomorrow's Jobs.* Washington: GPO 1995.

U.S. General Services Administration. *Your Rights to Federal Records.* Washington: GPO, 1995.

"Using Icons as Communication." *Simply Stated* [Newsletter of the Document Design Center, American Institutes for Research] 75 (Sept./Oct. 1987): 1+.

van der Meij, Hans, and John M. Carroll. "Principles and Heuristics for Designing Minimalist Instruction." *Technical Communication* 42.2 (1995): 243–61.

Van Pelt, William. Unpublished review of *Technical Writing*, 3rd ed.

Varner, Iris I., and Carson H. Varner. "Legal Issues in Business Communications." *Journal of the American Association for Business Communication* 46.3 (1983): 31–40.

Vaughan, David K. "Abstracts and Summaries: Some Clarifying Distinctions." *Technical Writing Teacher* 18.2 (1991): 132–41.

Velotta, Christopher. "How to Design and Implement a Questionnaire." *Technical Communication* 38.3 (1991): 387–92.

Victor, David A. *International Business Communication.* New York: Harper, 1992.

Walker, Janice R. "MLA-Style Citations of Electronic Sources." April 1995. The Alliance for Computers and Writing On-line posting. (http://prarie-island.ttu.edu/ acw.html) 10 Jan. 1996.

Wallace, Bob. "Restaurant Franchiser Puts Intranet on Menu." *Computerworld* 10 Nov. 1997: 12.

Walter, Charles, and Thomas F. Marsteller. "Liability for the Dissemination of Defective Information." *IEEE Transactions on Professional Communication* 30.3 (1987): 164–67.

Warshaw, Michael. "Have You Been House-Trained?" *Fast Company* Oct. 1998: 46+.

Watkins, Beverly T. "Many Campuses Start Building Tomorrow's Electronic Library." *Chronicle of Higher Education* 2 Sept. 1992, sec. A: 1+.

Weinstein, Edith K. Unpublished review of *Technical Writing*, 5th ed.

Weiss, Edmond H. *How to Write a Usable User Manual.* Philadelphia: ISI, 1985.

Welz, Gary. "Job Seeking on the Net." *Internet World* May 1995: 52.

Weymouth, L. C. "Establishing Quality Standards and Trade Regulations for Technical Writing in World Trade." *Technical Communication* 37.2 (1990): 143–47.

White, Jan. *Color for the Electronic Age.* New York: Watson-Guptill, 1990.

———. *Editing by Design.* 2nd ed. New York: Bowker, 1982.

———. *Great Pages.* El Segundo, CA: Serif, 1990.

———. *Visual Design for the Electronic Age.* New York: Watson-Guptill, 1988.

Wicclair, Mark, R., and David K. Farkas. "Ethical Reasoning in Technical Communication: A Practical Framework." *Technical Communication* 31.2 (1984): 15–19.

Wickens, Christopher D. *Engineering Psychology and Human Performance.* 2nd ed. New York: Harper, 1992.

Wight, Eleanor, "How Creativity Turns Facts into Usable Information." *Technical Communication* 32.1 (1985): 9–12.

Wilkinson, Theresa A. "Defining Content for a Web Site." *INTERCOM* June 1998: 33–34.

Williams, Joseph. *Style.* 4th ed. New York: Harper, 1994.

Williams, Robert I. "Playing with Format, Style, and Reader Assumptions." *Technical Communication* 30.3 (1983): 11–13.

Winsor, D. A. "Communication Failures Contributing to the Challenger Accident: An Example for Technical Communicators." *IEEE Transactions on Professional Communication* 31.3 (1988): 101–07.

Wojahn, Patricia G. "Computer-Mediated Communication: The Great Equalizer between Men and Women?" *Technical Communication* 41.4 (1994): 747–51.

Wriston, Walter. *The Twilight of Sovereignty.* New York: Scribner's, 1992.

Yoos, George. "A Revision of the Concept of Ethical Appeal." *Philosophy and Rhetoric* 12.4 (1979): 41–58.

Young, Patrick. "Writing Articles for Science Journals." *A Field Guide for Science Writers.* Eds. Deborah Blum and Mary Knudson. New York: Oxford, 1997: 110–16.

Zinsser, William. *On Writing Well.* New York: Harper, 1980.

Index

Correction Symbols

Symbol	Meaning	Symbol	Meaning
ab	abbreviation	– —/	dashes
agr p	faulty pronoun/referent agreement	. . ./	ellipses
agr sv	faulty subject/verb agreement	!/	exclamation point
amb	ambiguity	-/	hyphen
appr	inappropriate diction	*ital*	italics
bias	biased tone	( )/	parentheses
ca	faulty pronoun case	./	period
cap	capitalization	?/	question mark
chop	choppy sentences	" / "	quotation marks
cl	clutter word	;/	semicolon
coh	paragraph coherence	*qual*	needless qualifier
cont	contraction	*red*	redundancy
coord	faulty coordination	*rep*	needless repetition
cs	comma splice	*ref*	faulty or vague pronoun reference
dgl	dangling modifier	*ro*	run-on sentence
euph	euphemism	*seq*	sequence of development in a paragraph
exact	inexact word	*sexist*	sexist usage
frag	sentence fragment	*shift*	sentence shift
gen	generalization	*st mod*	stacked modifiers
jarg	needless jargon	*str*	paragraph structure
len	paragraph length	*sub*	faulty subordination
lev	level of technicality	*th op*	"th" sentence openers
mng	meaning unclear	*trans*	transition
mod	misplaced modifier	*trite*	triteness
noun ad	noun addiction	*ts*	topic sentence
om	omitted word	*un*	paragraph unity
over	overstatement	*v*	voice
par	faulty parallelism	*var*	sentence variety
pct	punctuation	*w*	wordiness
ap/	apostrophe	*wo*	word order
[ ]/	brackets	*ww*	wrong word
:/	colon	#	numbers
,/	comma	¶	begin new paragraph